D1570438

Grundlehren der mathematischen Wissenschaften 237

A Series of Comprehensive Studies in Mathematics

Gottfried Köthe

Topological Vector Spaces II

Springer-Verlag
New York Heidelberg Berlin

Gottfried Köthe

Institut für Angewandte Mathematik
der Johann-Wolfgang-Goethe Universität
Frankfurt am Main
Federal Republic of Germany

AMS Subject Classification (1980): 46–02, 46 Axx, 46 Bxx, 46 Cxx, 46 Exx

Library of Congress Cataloging in Publication Data

Köthe, Gottfried, 1905–
Topological vector spaces.
(Grundlehren der mathematischen Wissenschaften 159, 237)
Translation of Topologische lineare Räume.
Bibliography: p.
1. Linear topological spaces. I. Title. II. Series.
QA322.K623 515′.73 78-84831

With 2 illustrations.

Printed in the United States of America.

9 8 7 6 5 4 3 2 1

ISBN 0-387-90440-9 Springer-Verlag New York Heidelberg Berlin
ISBN 3-540-90440-9 Springer-Verlag Berlin Heidelberg New York

Preface

In the preface to Volume One I promised a second volume which would contain the theory of linear mappings and special classes of spaces important in analysis.

It took me nearly twenty years to fulfill this promise, at least to some extent. To the six chapters of Volume One I added two new chapters, one on linear mappings and duality (Chapter Seven), the second on spaces of linear mappings (Chapter Eight). A glance at the Contents and the short introductions to the two new chapters will give a fair impression of the material included in this volume. I regret that I had to give up my intention to write a third chapter on nuclear spaces. It seemed impossible to include the recent deep results in this field without creating a great further delay.

A substantial part of this book grew out of lectures I held at the Mathematics Department of the University of Maryland during the academic years 1963–1964, 1967–1968, and 1971–1972. I would like to express my gratitude to my colleagues J. Brace, S. Goldberg, J. Horváth, and G. Maltese for many stimulating and helpful discussions during these years.

I am particularly indebted to H. Jarchow (Zürich) and D. Keim (Frankfurt) for many suggestions and corrections. Both have read the whole manuscript. N. Adasch (Frankfurt), V. Eberhardt (München), H. Meise (Düsseldorf), and R. Hollstein (Paderborn) helped with important observations.

Frankfurt, August 1979 G. Köthe

Contents of Vol. II

CHAPTER SEVEN
Linear Mappings and Duality

CHAPTER EIGHT
Spaces of Linear and Bilinear Mappings

Contents of Vol. I*

* Abbreviated.

CHAPTER FOUR
Locally Convex Spaces. Fundamentals

CHAPTER FIVE
Topological and Geometrical Properties of Locally Convex Spaces

CHAPTER SIX
Some Special Classes of Locally Convex Spaces

CHAPTER SEVEN

Linear Mappings and Duality

Continuous linear mappings between locally convex spaces are the subject of § 32. The most important result is the homomorphism theorem in § 32, 4. For (B)- and (F)-spaces much more can be said. § 33 contains a detailed investigation of these cases culminating in the homomorphism theorems for (B)- resp. (F)-spaces in § 33, 4. A lifting property for separable locally convex spaces leads to the theorem of SOBCZYK.

The following two paragraphs contain an exposition of some of the results on open-mapping and closed-graph theorems. § 34 starts with PTÁK's ideas and ends up with KŌMURA's closed-graph theorem and ADASCH's open-mapping theorem for barrelled spaces. Many other results are included, given by KALTON, KELLEY, and the ROBERTSONS. § 35 gives an account of DE WILDE's theory of webbed spaces and his closed-graph theorems, which are especially useful in applications. An optimal closed-graph theorem for ultrabornological spaces is obtained in § 35, 9.

Arbitrary linear mappings are studied in § 36. The introduction of the notion of the singularity and the regular contraction of a mapping reduces the investigation to the case of closable (regular) or even continuous mappings. A duality theory and an extension theory are presented. A second method of investigating arbitrary linear mappings uses the graph topology (§ 37). Both methods are applied to nearly open and open mappings. The open-mapping theorems obtained in this way are more general than the previous theorems. The cases of (B)- and (F)-spaces are treated in § 37, 7.

§ 38 contains applications to systems of linear equations, the existence and continuity of left, right, and two-sided inverse mappings, and an introduction to the problems of extending and lifting linear mappings.

§ 32. Homomorphisms of locally convex spaces

1. Weak continuity. Volume I contains very little information on linear mappings. We will now enter into a more systematic investigation of this topic and begin by recalling some of the basic facts.

Let E, F be topological vector spaces; then $\mathfrak{L}(E, F)$ is the vector space of all continuous linear mappings of E into F. Any $A \in \mathfrak{L}(E, F)$ has by § 15, 4. the natural decomposition

$$A = J\check{A}K, \tag{1}$$

where K is the canonical homomorphism of E onto $E/N[A]$, $\check{A}$ is a continuous one-one linear mapping of $E/N[A]$ onto $A(E)$, and J is the embedding of $A(E)$ into F. The product $J\check{A}$ will be denoted by $\hat{A}$.

A is a topological homomorphism of E into F if every open subset M of E has an image $A(M)$ which is open in $A(E)$. If A is also one-one, then A is a topological monomorphism. If further $A(E) = F$, then A is a topological isomorphism.

We will frequently omit "topological" when there is no danger of misunderstanding.

(2) *$A \in \mathfrak{L}(E, F)$ is a homomorphism if and only if one of the following conditions is satisfied:* (a) *A maps every neighbourhood of $\circ$ in E onto a neighbourhood of $\circ$ in $A(E)$;* (b) *$\check{A}$ is an isomorphism;* (c) *$\hat{A}$ is a monomorphism.*

The simple proof is left to the reader.

We consider now locally convex spaces E and F. Let E' and F' be their duals. We replace the initial topology on E resp. F by the weak topology $\mathfrak{T}_s(E')$ resp. $\mathfrak{T}_s(F')$. We proved in § 20, 4. the basic result

(3) *A linear mapping A of E into F is weakly continuous if and only if the adjoint A' maps F' into E'.*

A is weakly continuous if and only if A' is weakly continuous.

The weak topology on F' resp. E' is the topology $\mathfrak{T}_s(F)$ resp. $\mathfrak{T}_s(E)$. An equivalent formulation for (3) is

(4) *A linear mapping A of E into F is weakly continuous if and only if for every closed hyperplane $H \ni \circ$ in F the inverse image $A^{(-1)}(H)$ is closed in E.*

Proof. A hyperplane H in F containing $\circ$ is given by an equation $v(y) = 0$, where v is a linear functional in F and H is closed if and only if $v \in F'$ (§ 15, 9.(1)). The inverse image $A^{(-1)}(H)$ is determined by $ux = v(Ax) = (A'v)x = 0$. It is therefore closed if and only if $u = A'v$ is an element of E'. Now (3) follows from (2).

For a weakly continuous mapping A of E into F and its adjoints we have

(5)
$$A(E)^{\perp} = A(E)^{\circ} = N[A'],$$
$$A'(F')^{\perp} = A'(F')^{\circ} = N[A],$$

where $N[A]$ denotes the kernel of A. This is an immediate consequence of the relation $v(Ax) = (A'v)x$, $v \in F'$, $x \in E$.

By taking polars on both sides of (5) we get

(6)
$$\overline{A(E)} = N[A']^{\circ} = N[A']^{\perp},$$
$$\overline{A'(F')} = N[A]^{\circ} = N[A]^{\perp}.$$

We make the following remarks: i) Every $A \in \mathfrak{L}(E, F)$ is weakly continuous (§ 20, 4.(5)) and therefore has a weakly continuous adjoint. ii) In particular, (5) and (6) are true for continuous A too. iii) In (6) the closure of $A(E)$ may be taken in the initial topology of F or in the weak topology; the closure of $A'(F')$ is always the $\mathfrak{T}_s(E)$-closure.

An immediate consequence of (6) is

(7) *The range $A(E)$ of A is (weakly) dense in F if and only if $N[A'] = \mathfrak{o}$. The range $A'(F')$ of A' is weakly dense in E' if and only if $N[A] = \mathfrak{o}$.*

The relations (5) are also contained as special cases in the following proposition.

(8) *Let A be a (weakly) continuous mapping of E into F. Then*
(9) $A(M)^\circ = A'^{(-1)}(M^\circ)$ *holds for any subset M of E, and*
(10) $A^{(-1)}(N^\circ) = A'(N)^\circ$ *for any subset N of F'.*

Relation (9) was proved in Volume I, § 22, 7.(1). Replacing A by A' in (9) and using weak duality gives (10).

We give an application of (4).

(11) *Let E be locally convex and suppose that every sequentially closed hyperplane containing $\mathfrak{o}$ in E is closed. If A is a linear and sequentially continuous mapping of E into a locally convex space F, then A is weakly continuous.*

Proof. Let $H \supset (\mathfrak{o})$ be a closed hyperplane in F. Suppose $x_n \in A^{(-1)}(H)$ and $x_n \to x_0 \in E$. Then $Ax_n \to Ax_0$ and $Ax_0 \in H$; thus $x_0 \in A^{(-1)}(H)$. Therefore $A^{(-1)}(H)$ is sequentially closed and by hypothesis closed. The statement now follows from (4).

The assumption in (11) is satisfied for the weak dual $E'[\mathfrak{T}_s(E)]$ of a separable (F)-space E because of the theorem of Banach–Dieudonné (§ 21, 10.(7)).

2. Continuity. If A is continuous, then it is weakly continuous. The converse is not true. It is not difficult to get an exact description of the situation.

If E is locally convex, E' its dual, then a general method introducing other locally convex topologies on E is the following (§ 21, 1.). Choose a total and saturated class $\mathfrak{M}$ of weakly bounded subsets of E', and let $\mathfrak{T}_{\mathfrak{M}}$ be the topology of uniform convergence on the sets of $\mathfrak{M}$.

(1) *A weakly continuous linear mapping of E into F is continuous in the sense of $\mathfrak{T}_{\mathfrak{M}_1}$ on E and $\mathfrak{T}_{\mathfrak{M}_2}$ on F if and only if $A'(\mathfrak{M}_2) \subset \mathfrak{M}_1$.*

Proof. The sets $V = M_2^\circ$, $M_2 \in \mathfrak{M}_2$, form a base of $\mathfrak{T}_{\mathfrak{M}_2}$-neighbourhoods of $\mathfrak{o}$ in F. By 1.(10) we have $A^{(-1)}(V) = A^{(-1)}(M_2^\circ) = A'(M_2)^\circ$. Therefore $A^{(-1)}(V)$ is an $\mathfrak{T}_{\mathfrak{M}_1}$-neighbourhood in E if and only if $A'(M_2) \in \mathfrak{M}_1$.

We have already treated the special case in which both topologies are the Mackey topologies, in § 21, 4.(6); for the sake of completeness we state it again:

(2) *Weakly continuous and $\mathfrak{T}_k$-continuous linear mappings of E into F coincide.*

Two other special cases of (1) are given in

(3) *Every weakly continuous linear mapping A of E into F is strongly continuous and $\mathfrak{T}_{b^*}$-continuous.*

The first statement follows immediately from (1) since every weakly bounded subset M of F' has an image $A'(M)$ in E' with the same property.

$\mathfrak{T}_{b^*}$ is the topology of uniform convergence on the strongly bounded subsets of the dual space. Let N be strongly bounded in F'. Then we have for any bounded set M in E

$$\sup_{x \in M, u \in N} |(A'u)x| = \sup |u(Ax)| < \infty$$

since $A(M)$ is bounded in F. But then $A'(N)$ is strongly bounded in E' and (1) gives the desired result.

Observe that a strongly continuous mapping need not be weakly continuous. Since the dual space to $E[\mathfrak{T}_b(E')]$ is in general larger than E', there even exist strongly continuous linear functionals which are not weakly continuous. The same argument is valid for the topology $\mathfrak{T}_{b^*}$.

(4) *If $A \in \mathfrak{L}(E, F)$, then A' is weakly continuous, strongly continuous, $\mathfrak{T}_k$-continuous, and $\mathfrak{T}_c$-continuous.*

Since A' is weakly continuous, all the statements with the exception of the last one follow immediately from (2) and (3). Now $\mathfrak{T}_c$ is the topology of precompact convergence (§ 21, 6.) and by § 15, 6.(7) the continuous linear image $A(M)$ of a precompact set M is again precompact; hence A' is $\mathfrak{T}_c$-continuous by (1).

By § 5, 4.(4) it is often sufficient to define a continuous linear mapping only on a dense subspace:

(5) *Let E and F be topological vector spaces where F is complete. Then every $A \in \mathfrak{L}(E, F)$ has a uniquely determined linear and continuous extension $\tilde{A}$ defined on the completion $\tilde{E}$ of E, $A \in \mathfrak{L}(\tilde{E}, F)$.*

In § 17, 6.(6) we proved for normed spaces that A'' is an extension of A which maps the bidual E'' into the bidual F'' and satisfies $\|A''\| = \|A\|$. For locally convex spaces E and F we have the following situation:

(6) a) *Let $A \in \mathfrak{L}(E, F)$ be given. Then the double adjoint A'' of A is an extension of A which maps the bidual E'' into the bidual F'' and A'' is continuous in the sense of the natural topologies on the biduals.*

b) *A'' is also continuous for the topologies $\mathfrak{T}_s(E')$ and $\mathfrak{T}_s(F')$ on E'' and F'', respectively, and A'' is the uniquely determined weakly continuous extension of A to E''.*

a) A' is strongly continuous by (4), so it is in $\mathfrak{L}(F', E')$, where E', F' denote the strong duals. The adjoint A'' is thus a mapping of $(E')' = E''$ into F'' which is an extension of A.

By § 23, 4. a neighbourhood base of $\circ$ for the natural topology $\mathfrak{T}_n(E')$ on E'' consists of the sets $U^{\circ\circ}$, U an absolutely convex neighbourhood of $\circ$ in $E[\mathfrak{T}]$, where the last polar is taken in E''.

Starting with $A(U) \subset V$, we have $V^\circ \subset A(U)^\circ = A'^{(-1)}(U^\circ)$, $A'(V^\circ) \subset U^\circ$, $U^{\circ\circ} \subset A'(V^\circ)^\circ = A''^{(-1)}(V^{\circ\circ})$, and this means $A''(U^{\circ\circ}) \subset V^{\circ\circ}$, so that A'' is continuous.

b) A is also weakly continuous and the weak continuity of A'' follows as in a) if we use weak neighbourhoods U, V. Since E is weakly dense in E'', A'' is the uniquely determined extension of A to E''.

3. Weak homomorphisms. We begin with a proposition on arbitrary homomorphisms,

(1) *If A is a homomorphism of the locally convex space $E[\mathfrak{T}_1]$ into the locally convex space $F[\mathfrak{T}_2]$, then $A'(F')$ is weakly closed in E'.*

Let u_0 be a point of the weak closure of $A'(F')$. For all $z \in N[A]$ we have $u_0 z = 0$ by 1.(6). If we define $l(Ax) = u_0 x$, then l is therefore a uniquely determined linear functional on $A(E)$. It is continuous: $U = \{x; |u_0 x| < \varepsilon\}$ is an open $\mathfrak{T}_1$-neighbourhood of $\circ$ in E; since A is open, $A(U) = V$ is a $\mathfrak{T}_2$-neighbourhood of $\circ$ in $A(E)$. Therefore $|l(Ax)| < \varepsilon$ for all $Ax \in V$.

By the Hahn–Banach theorem l has an extension $v_0 \in F'$ and $v_0(Ax) = u_0 x$ for all $x \in E$; hence $u_0 = A' v_0$ and $A'(F')$ is weakly closed.

If $\mathfrak{T}_1$ and $\mathfrak{T}_2$ are the weak topologies, the converse of (1) is also true:

(2) *$A \in \mathfrak{L}(E, F)$ is a weak homomorphism if and only if $A'(F')$ is weakly closed in E'.*

Proof. Since A is continuous it is weakly continuous. So we have only to prove that A is weakly open if $A'(F')$ is weakly closed.

Let U be an absolutely convex open weak neighbourhood of $\mathfrak{o}$ in E. Since $A(U + N[A]) = A(U)$, we may assume $U \supset N[A]$. Then $U^\circ \subset N[A]^\circ = \overline{A'(F')} = A'(F')$. Now U° is bounded and finite dimensional in $A'(F')$, so U° is contained in the absolutely convex cover of finitely many $u_i \in A'(F')$, $i = 1, \ldots, k$. If U_0 denotes the open weak neighbourhood of $\mathfrak{o}$ in E given by $|u_i x| < 1$, $i = 1, \ldots, k$, then $U_0 \subset U$. There exist $v_i \in F'$ such that $u_i = A'v_i$. Let V be the open weak neighbourhood of $\mathfrak{o}$ in F given by $|v_i y| < 1$, $i = 1, \ldots, k$. From $v_i(Ax) = (A'v_i)x = u_i x$ follows $x \in U_0$ for all $Ax \in V$; therefore $A(U) \supset A(U_0) \supset V \cap A(E)$; A is weakly open.

(3) *A homomorphism $A \in \mathfrak{L}(E, F)$ is always a weak homomorphism.*

This is an immediate consequence of (1) and (2). We list some special cases:

(4) a) *$A \in \mathfrak{L}(E, F)$ is a weak monomorphism if and only if $A'(F') = E'$.*

b) *A is a weak homomorphism with weakly closed range if and only if this is true for A'.*

c) *A is a weak monomorphism with closed range if and only if A' is a weak homomorphism of F' onto E'.*

d) *A is a weak isomorphism if and only if A' is a weak isomorphism or if and only if $A(E) = F$ and $A'(F') = E'$.*

e) *A is one-one and $A(E) = F$ if and only if A' is a weak monomorphism onto a weakly dense subspace of E'.*

f) *A is a weak homomorphism with dense range in F if and only if A' is one-one and $A'(F')$ is weakly closed.*

Proof. a) follows from (2) and 1.(7), b) by applying (2) to A and A', c) from a) and b), d) from c) and 1.(7), e) from a) and 1.(7), and f) from (2) and 1.(7).

We remark that the range of a homomorphism is in general not closed. For example, the injection J of a normed space E into its completion $\tilde{E} \neq E$ is a monomorphism with range dense in $\tilde{E}$ but different from $\tilde{E}$. Applying e) to J' we see that J' is one-one and $J'(E') = E'$; but J' is not a weak isomorphism, since it is the identity on $E'[\mathfrak{T}_s(\tilde{E})]$ and maps it on $E'[\mathfrak{T}_s(E)]$.

However, if A is a monomorphism of a complete E into F, then the range $A(E)$ is closed in F, since $A(E)$ is isomorphic to E and therefore complete. This generalizes to homomorphisms in the following way:

(5) *Let A be a homomorphism of $E[\mathfrak{T}_1]$ in $F[\mathfrak{T}_2]$. The range $A(E)$ is closed if $(E/N[A])[\hat{\mathfrak{T}}_1]$ is complete.*

In particular, $A(E)$ is closed if A is a monomorphism and E is complete.

Examples. 1) Let J be the injection of a subspace $H[\mathfrak{T}]$ of $E[\mathfrak{T}]$ into E. J is a monomorphism; therefore ((4) a)) $J'(E') = H'$. If we identify H' with $E'/H^\perp$ (§ 22, 1.), then J' is the canonical mapping of E' onto $E'/H^\perp$.

If H is closed, then J' is a weak homomorphism ((4) c)), where the weak topology on $E'/H^\perp$ is $\mathfrak{T}_s(H)$. If H is not closed, J' is not a weak homomorphism.

2) Let H be a closed subspace of E, K the canonical homomorphism of E onto E/H. Then K' is by (4) b) and f) a weak monomorphism of $(E/H)'$ into E'. If we identify $(E/H)'$ with $H^\perp \subset E'$, then K' is the injection of $H^\perp$ in E' (§ 22, 2.).

3) By 1.(1) a continuous linear mapping A of E into F has the natural decomposition $A = J\check{A}K$. The corresponding decomposition of A' is

$$A' = K'\check{A}'J'. \tag{6}$$

Is this the natural decomposition of A' as a weakly continuous mapping of F' into E'?

By 1) J' is the canonical mapping of F' onto $A(E)' = F'/A(E)^\perp = F'/N[A']$ and this is a weak homomorphism if and only if $A(E) = \overline{A(E)}$. This is a first necessary condition. K' is, by 2), the injection of $N[A]^\perp = \overline{A'(F')}$ in E', so we find that $A'(F') = \overline{A'(F')}$ is a second necessary condition. If both these conditions are satisfied, then $\check{A}'$ is a weak isomorphism of $F'/N[A']$ onto $A'(F') = (E/N[A])'$. Thus we obtain

(7) *(6) is the natural decomposition of A' if and only if A is a weak homomorphism with closed range.*

4. The homomorphism theorem. In 3.(2) we gave a dual characterization of weak homomorphisms. Our aim is now to prove a corresponding theorem for arbitrary homomorphisms.

(1) *Let A be a weak homomorphism of the locally convex space $E[\mathfrak{T}_1]$ into the locally convex space $F[\mathfrak{T}_2]$ and let $\mathfrak{M}_1$ resp. $\mathfrak{M}_2$ be the class of equicontinuous subsets of E' resp. F'. If A is $\mathfrak{T}_1$–$\mathfrak{T}_2$-open, then $A'(\mathfrak{M}_2) \supset \mathfrak{M}_1 \cap A'(F') = \{M_1 \in \mathfrak{M}_1;\ M_1 \subset A'(F')\}$.*

Proof. It is sufficient to prove this for an arbitrary absolutely convex and weakly closed set $M_1 \in \mathfrak{M}_1 \cap A'(F')$ since $A'(F')$ is weakly closed by 3.(1). The polar M_1° is a closed $\mathfrak{T}_1$-neighbourhood of $\circ$ in E. Since A is $\mathfrak{T}_1$–$\mathfrak{T}_2$-open, $A(M_1^\circ)$ contains a closed $\mathfrak{T}_2$-neighbourhood of $\circ$ in $A(E)$,

$M_2^\circ \cap A(E)$, where $M_2 \in \mathfrak{M}_2$ is absolutely convex and weakly closed in F'. Applying $A^{(-1)}$ to $A(M_1^\circ) \supset M_2^\circ \cap A(E)$ gives $M_1^\circ + N[A] \supset A^{(-1)}(M_2^\circ) = A'(M_2)^\circ$ (1.(10)). But $M_1 \subset A'(F')$; hence $M_1^\circ \supset N[A]$ by 1.(5) and therefore $M_1^\circ = M_1^\circ + N[A] \supset A'(M_2)^\circ$. Taking polars in E' we obtain $M_1 \subset \overline{A'(M_2)}$. Now M_2 is weakly compact by the theorem of ALAOGLU–BOURBAKI; therefore $A'(M_2)$ is closed, so finally $M_1 \subset A'(M_2)$.

(2) *Let A be a weak homomorphism of $E[\mathfrak{T}_1]$ into $E[\mathfrak{T}_2]$. If $A'(\mathfrak{M}_2) \supset \mathfrak{M}_1 \cap A'(F')$, then A is $\mathfrak{T}_1$–$\mathfrak{T}_2$-open.*

We have to show that for every $\mathfrak{T}_1$-neighbourhood U of $\mathfrak{o}$ in E there exists a $\mathfrak{T}_2$-neighbourhood V of $\mathfrak{o}$ in F such that $A(U) \supset V \cap A(E)$. It is sufficient to prove this for U absolutely convex and closed and $U \supset N[A]$ since $A(U + N[A]) = A(U)$. Now $U^\circ = M_1$ is in $\mathfrak{M}_1$ and $M_1 \subset N[A]^\circ = A'(F')$ (here we used 3.(2)); therefore by assumption there exists $M_2 \in \mathfrak{M}_2$ such that $M_1 \subset A'(M_2)$. By 1.(10) we have $M_1^\circ \supset A^{(-1)}(M_2^\circ)$ and applying A on both sides we get $A(M_1^\circ) = A(U) \supset M_2^\circ \cap A(E)$, showing that A is $\mathfrak{T}_1$–$\mathfrak{T}_2$-open.

Combining (1), (2), and 3.(2) we have (GROTHENDIECK [11])

(3) *Homomorphism theorem: Let E and F be locally convex and $\mathfrak{M}_1$ resp. $\mathfrak{M}_2$ be the classes of equicontinuous sets in E' resp. F'.*

A linear continuous mapping A of E in F is a homomorphism if and only if a) *$A'(F')$ is weakly closed in E'*, b) $A'(\mathfrak{M}_2) \supset \mathfrak{M}_1 \cap A'(F')$.

If we use (1), (2), and 2.(1) we get the slightly different version

(4) *A weak homomorphism A of E in F is a homomorphism if and only if $A'(\mathfrak{M}_2) = \mathfrak{M}_1 \cap A'(F')$.*

The case of the Mackey topologies is of special interest.

(5) *Let $E[\mathfrak{T}_1]$ and $F[\mathfrak{T}_2]$ be locally convex and let $\mathfrak{T}_2$ be the Mackey topology $\mathfrak{T}_k(F')$.*

A linear continuous mapping A of E onto F is a homomorphism if and only if A is a weak homomorphism or if and only if $A'(F')$ is weakly closed.

Every weak homomorphism of E onto F is a $\mathfrak{T}_k$-homomorphism onto F and conversely.

First proof. The condition in the first statement is clearly necessary. Conversely, if A is a weak homomorphism onto F, then by 3.(4) c) A' is a weak monomorphism of F' onto $A'(F')$ and condition b) of (3) is satisfied since every absolutely convex and weakly compact set M in $A'(F')$ is the image of the set $A'^{(-1)}(M)$ with the same properties in F'. Hence the first statement in (5) is a special case of (3).

If $\mathfrak{T}_1$ and $\mathfrak{T}_2$ are both Mackey topologies, then a weak homomorphism is a linear $\mathfrak{T}_k$-continuous mapping (§ 21, 4.(6)) and the second statement in (5) is included as a special case in the first.

Second proof. Let $A \in \mathfrak{L}(E, F)$ be a weak homomorphism onto F; then A is a weak isomorphism of $E/N[A]$ onto F. On E, $\mathfrak{T}_1$ is coarser than $\mathfrak{T}_k(E')$ and by § 22, 2.(3) the quotient topology $\hat{\mathfrak{T}}_k$ on $E/N[A]$ is identical with the Mackey topology $\mathfrak{T}_k(N[A]^\circ)$. Therefore $\hat{\mathfrak{T}}_1$ is coarser than $\mathfrak{T}_k[N[A]^\circ]$ on $E/N[A]$, so that the weak isomorphism $\check{A}$ maps $\hat{\mathfrak{T}}_1$ onto a topology $\mathfrak{T}_1'$ on F which is coarser than $\mathfrak{T}_k(F')$. Now the image $A(U)$ of a $\mathfrak{T}_1$-open neighbourhood of $\mathfrak{o}$ in E is an open $\mathfrak{T}_1'$-neighbourhood of $\mathfrak{o}$ in F and therefore also an open $\mathfrak{T}_k(F')$-neighbourhood of $\mathfrak{o}$. This means that A is open, and, since by assumption A is continuous, that A is a homomorphism.

If we combine 3.(4) c) with the second statement of (5), we find immediately

(6) *Let A be a weakly continuous linear mapping of E into F. Then A' is a $\mathfrak{T}_k$-homomorphism of F' onto E' if and only if A is a weak monomorphism with closed range.*

The first statement of (5) is no longer true for arbitrary weak homomorphisms of E into F. The following example of Bourbaki shows that even a weak monomorphism with a dense range need not be a $\mathfrak{T}_k$-homomorphism.

Let $E = \varphi$, $E' = \omega$, and let z be an element of $(E')^*$ which does not lie in φ. Let F be the subspace $\varphi \oplus [z]$ of $(E')^*$. On E and F we define the topology $\mathfrak{T}_s(\omega)$, so that $E' = F' = \omega$. Since ω is bornological, the linear functional z is not bounded on all bounded subsets of ω. Therefore the class of all absolutely convex and $\mathfrak{T}_s(F)$-compact subsets of ω is a strict subclass of the class of all $\mathfrak{T}_s(E)$-compact subsets of ω. Since the latter class coincides with the class of all bounded subsets of ω, the Mackey topology on φ is strictly finer than the Mackey topology on F restricted to φ. Now, let J be the injection of φ into F. Since $J'(F') = E' = \omega$, J is a weak monomorphism onto the dense subspace φ of F, but it is not a $\mathfrak{T}_k$-homomorphism.

We note an application of (5):

(7) *Let A be a monomorphism of the complete semi-reflexive space E in the semi-reflexive space F. Then A' is a strong homomorphism of F' onto E'.*

By 2.(4) A' is a continuous mapping of $F'[\mathfrak{T}_b(F)]$ onto $E'[\mathfrak{T}_b(E)]$. Since E and F are semi-reflexive, the strong topologies on E' and F' coincide with the Mackey topologies. Hence $A'' = A$ and, since E is complete, $A(E)$ is closed and therefore $\mathfrak{T}_s(F')$-closed in F. By applying (5) to A' we obtain (7).

5. Further results on homomorphisms. So far we have investigated homomorphisms for topologies on E and F which are compatible with the dual systems $\langle E', E\rangle$ and $\langle F', F\rangle$. The situation becomes more involved when we include the strong topology. We give first two examples.

1. l^1 is a subspace of l^2. We equip both spaces with the norm topology of l^2. Let J be the injection of l^1 into l^2. Since $(l^1)' = (l^2)' = l^2$, J' is the identity on l^2 and is weakly continuous but is not a weak isomorphism of $l^2[\mathfrak{T}_s(l^2)]$ onto $l^2[\mathfrak{T}_s(l^1)]$. However, J' is a strong isomorphism since $\mathfrak{T}_b(l^1)$ and $\mathfrak{T}_b(l^2)$ coincide on l^2.

This shows that there exist strong isomorphisms which are weakly continuous but are not weak isomorphisms.

2. Let E be locally convex and H be a closed subspace of E such that the topology $\hat{\mathfrak{T}}_b(E')$ is strictly finer on E/H than $\mathfrak{T}_b(H^\perp)$; we constructed such an example in § 31, 7. Let K be the canonical homomorphism of E onto E/H. It is a weak but not a strong homomorphism. The adjoint K' is the injection of $(E/H)' = H^\perp$ into E'. If $\mathfrak{M}_1$ resp. $\mathfrak{M}_2$ is the class of weakly bounded subsets of E' resp. $H^\perp$, then $K'(\mathfrak{M}_2) = \mathfrak{M}_1 \cap K'(H^\perp)$. This means that K satisfies the analogous assumptions in 4.(4) and yet is not a strong homomorphism.

The homomorphism theorem is therefore no longer true for topologies which are strictly finer than the Mackey topology.

But it is possible to give a dual characterization of monomorphisms in the general case.

(1) *Let E and F be locally convex and $\mathfrak{M}_1 \supset \mathfrak{F}$ resp. $\mathfrak{M}_2 \supset \mathfrak{F}$ saturated classes of weakly bounded subsets of E' resp. F'. A map $A \in \mathfrak{L}(E, F)$ is a monomorphism of $E[\mathfrak{T}_{\mathfrak{M}_1}]$ in $F[\mathfrak{T}_{\mathfrak{M}_2}]$ if and only if the following conditions are satisfied:*

i) $A'(\mathfrak{M}_2) \subset \mathfrak{M}_1$;

ii) *every $M_1 \in \mathfrak{M}_1$ is contained in some $\overline{A'(M_2)}$, $M_2 \in \mathfrak{M}_2$.*

$\mathfrak{F}$ is always the class of all bounded subsets of finite dimension, so $\mathfrak{T}_{\mathfrak{F}}$ is the weak topology.

Proof. a) Necessity. For i) this is a consequence of 2.(1). Let M_1 be an absolutely convex weakly closed subset in $\mathfrak{M}_1$. Since A is open, there exists an absolutely convex and weakly closed $M_2 \in \mathfrak{M}_2$ such that $A(M_1^\circ) \supset M_2^\circ \cap A(E)$, so applying $A^{(-1)}$ gives $M_1^\circ \supset A'(M_2)^\circ$. From this follows $A'(M_2)^{\circ\circ} = \overline{A'(M_2)} \supset M_1$.

b) Sufficiency. From i) and 2.(1) it follows that A is $\mathfrak{T}_{\mathfrak{M}_1}$–$\mathfrak{T}_{\mathfrak{M}_2}$-continuous. From ii) it follows that $\bigcup_{M_2 \in \mathfrak{M}_2} \overline{A'(M_2)} \supset \bigcup_{M_1 \in \mathfrak{M}_1} M_1 = E'$; thus $\overline{A'(F')} = E'$ and A is one-one. It remains to prove that A is $\mathfrak{T}_{\mathfrak{M}_1}$–$\mathfrak{T}_{\mathfrak{M}_2}$-open. But, given $M_1 \in \mathfrak{M}_1$, there exists $M_2 \in \mathfrak{M}_2$ by ii) such that $M_1 \subset \overline{A'(M_2)}$. Therefore $M_1^\circ \supset A'(M_2)^\circ = A^{(-1)}(M_2^\circ)$ and $A(M_1^\circ) \supset M_2^\circ \cap A(E)$.

We have the following special cases of (1):

(2) *Let A be a continuous linear map of E in F, E and F locally convex. A' is a strong resp. $\mathfrak{T}_c$-monomorphism of F' in E' if and only if every bounded resp. precompact subset M_2 of F is contained in the closed image $\overline{A(M_1)}$ of a bounded resp. precompact set M_1 in E.*

This follows from (1) since A' is weakly continuous from F' in E' and $A'' = A$ satisfies i), because bounded resp. precompact sets have images with the same properties (§ 15, 6.).

We close with some remarks on extensions of homomorphisms.

(3) *Let E, F be locally convex and $A \in \mathfrak{L}(E, F)$, and let H be a dense subspace of E. If the restriction of A to H is a homomorphism resp. monomorphism of H in F, then A has the same property.*

Proof. Note first that $H' = E'$, A and its restriction have the same adjoint A', and the equicontinuous sets in E' are the same for H and E. Thus, since $A'(F')$ is $\mathfrak{T}_s(H)$-closed in E', it is also $\mathfrak{T}_s(E)$-closed. The statement follows now from 4.(3).

(4) *Let A be a homomorphism of $E[\mathfrak{T}_1]$ in $F[\mathfrak{T}_2]$. Assume that the natural topologies $\mathfrak{T}_{1n}$ resp. $\mathfrak{T}_{2n}$ on E'' resp. F'' are coarser than the Mackey topologies on E'' resp. F''. Then A'' is a homomorphism of $E''[\mathfrak{T}_{1n}]$ in $F''[\mathfrak{T}_{2n}]$.*

By 2.(6) A'' is continuous from $E''[\mathfrak{T}_{1n}]$ in $F''[\mathfrak{T}_{2n}]$. By our assumption on the topologies $(E'')' = E'$, $(F'')' = F'$, and the adjoint to A'' is A'. Since A is a homomorphism, $A'(F')$ is $\mathfrak{T}_s(E)$-closed and therefore also $\mathfrak{T}_s(E'')$-closed. So condition a) of 4.(3) is satisfied and condition b) is satisfied by A; and by the definition of the natural topologies A'' therefore satisfies b) also.

§ 33. Linear continuous mappings of (B)- and (F)-spaces

1. First results in normed spaces. So far we have investigated linear maps in general locally convex spaces. Much more information is available in the metrizable case, as we will see in this paragraph. We begin with some remarks on normed spaces.

Let E and F be normed spaces and $A \in \mathfrak{L}(E, F)$. Then A has a norm $\|A\|$ which was defined in § 14, 1.

Let $A = J\check{A}K = \hat{A}K$ be the natural decomposition of A (§ 32, 1.(1)). We have $\|J\| = 1$ for the injection J of $A(E)$ into F and $\|K\| = 1$ for the canonical homomorphism of E onto $E/N[A]$. We have also

$$\text{(1)} \qquad \|A\| = \|\check{A}\| = \|\hat{A}\|.$$

Trivially, $\|\check{A}\| = \|\hat{A}\|$ and $\|A\| \leqq \|J\|\|\check{A}\|\|K\| \leqq \|\check{A}\|$. Since $\check{A}\hat{x} = Ax$ for every $x \in \hat{x} \in E/N[A]$, it follows $\|\check{A}\hat{x}\| \leqq \|A\| \inf_{x \in \hat{x}} \|x\| = \|A\|\|\hat{x}\|$ for all $\hat{x}$ and so $\|\check{A}\| \leqq \|A\|$, which proves (1).

If A is a monomorphism of the normed space E into the normed space F, then $A^{(-1)}$ is a continuous and therefore bounded map of $A(E)$ onto E. From this follows

(2) *$A \in \mathfrak{L}(E, F)$, E and F normed spaces, is a monomorphism if and only if A is bounded from below, i.e., there exists $m > 0$ such that $\|Ax\| \geqq m\|x\|$ for all $x \in E$.*

Obviously we may define m by $\|A^{(-1)}\| = 1/m$.

The fundamental BANACH–SCHAUDER theorem (§ 15, 12.(2)) applied to (B)-spaces E and F says that $A \in \mathfrak{L}(E, F)$ is a homomorphism if and only if the range $A(E)$ is closed in F.

We give two examples to show that this is no longer true if E or F is normed but not complete.

a) Let E be normed and not complete and let J be the injection of E into $\tilde{E}$. Then J is a monomorphism but $J(E)$ is not closed in $\tilde{E}$.

b) Let E be l^1 and F be l^1 but with the norm of l^2. Then the identity map of E onto F is continuous and onto but not a monomorphism. F is normed but not complete.

Note that the range of a homomorphism between Banach spaces or even (F)-spaces is always closed; this follows from § 32, 3.(5) and the completeness of the quotient spaces of (F)-spaces. The real difficulty lies in the converse statement and it will be our aim to find more general classes of spaces for which the analogue of the BANACH–SCHAUDER theorem is true.

We now apply the duality theory of § 32 to the case of normed spaces.

(3) *Let E and F be normed spaces and A a homomorphism of E in F. Then A' is a strong homomorphism of F' in E'.*

Proof. A' is weakly and strongly continuous (§ 32, 2.(4)). The equicontinuous sets in E' and F' are the subsets of the multiples of the unit ball. Since A is a homomorphism, it follows from condition b) of § 32, 4.(3) that if M is the closed unit ball in F', then $A'(M)$ contains a ball of $A'(F')$. A' is therefore open in the sense of the strong topology.

The converse of (3) is true in the following form:

(4) *Let E be a* (B)*-space, F a normed space, and $A \in \mathfrak{L}(E, F)$. If A' is a strong homomorphism, then A is a homomorphism with closed range.*

Proof. By assumption there exists a closed ball N in F such that $A'(N^\circ) \supset M^\circ \cap A'(F')$, where M is the closed unit ball in E. Applying $A'^{(-1)}$ we get $N^\circ + N[A'] \supset A'^{(-1)}(M^\circ) = A(M)^\circ$. Taking polars gives $N \cap \overline{A(E)} \subset \overline{A(M)}$. Hence $N \cap \overline{A(E)} = N \cap \overline{A(M)}$. Since $A(M)$ is absolutely convex and F normed, one has $N \cap \overline{A(M)} = \overline{N \cap A(M)}$. It follows from $\overline{N \cap A(E)} = \overline{N \cap A(M)}$ that the image $A(M)$ of the unit ball is dense in the ball $N \cap A(E)$ of $A(E)$.

Repeating the argument in the second part of the proof of the Banach–Schauder theorem and using the completeness of E, one finds that $A((1 + \varepsilon)M)$ covers the ball $N \cap A(E)$; therefore A is open. That A has closed range follows from the completeness of $E/N[A]$.

As an example, in § 37, 6. will show that (4) is false if we suppose E to be only normed.

2. Metrizable locally convex spaces. We begin with an elementary characterization of homomorphisms.

Let E and F be locally convex and $A \in \mathfrak{L}(E, F)$. We say that A is sequentially invertible if for every sequence $y_n \in A(E)$ converging to zero there exists a sequence $x_n \in E$ such that $Ax_n = y_n$ and $x_n \to o$.

(1) *Let E and F be metrizable locally convex spaces. $A \in \mathfrak{L}(E, F)$ is a homomorphism if and only if A is sequentially invertible.*

Proof. A is a homomorphism if and only if $\check{A}$ is an isomorphism of $E/N[A]$ onto $A(E)$. Both are metrizable. If A is a homomorphism and $y_n = \check{A}\hat{x}_n \to o$, then $\hat{x}_n \to \hat{o}$. But there always exist $x_n \in \hat{x}_n$ such that $x_n \to o$, as can be seen by a diagonal procedure outlined in the proof of § 22, 2.(7). Therefore A is sequentially invertible.

On the other hand, if $y_n = Ax_n \to o$ and $x_n \to o$, then $\check{A}\hat{x}_n \to \hat{o}$ and $\hat{x}_n \to \hat{o}$, so $\check{A}^{(-1)}$ is continuous and $\check{A}$ is an isomorphism.

Our next result will be an application of the homomorphism theorem of § 32, 4. The topology of a metrizable locally convex space E is always the Mackey topology $\mathfrak{T}_k(E)$ (§ 21, 5.(3)). If H is a linear subspace of E, then H is metrizable in the induced topology $\mathfrak{T}_k(E')$ and therefore $\mathfrak{T}_k(E')$ coincides with $\mathfrak{T}_k(H')$.

(2) *The topologies $\mathfrak{T}_k(E')$ and $\mathfrak{T}_k(H')$ coincide on each subspace H of a metrizable locally convex space E.*

Remark. By § 22, 2. the equality of $\mathfrak{T}_k(E')$ and $\mathfrak{T}_k(H')$ may be formulated in the following way: If G is a weakly closed subspace of E', where E is metrizable locally convex, then every absolutely convex weakly compact subset of E'/G is the canonical image of an absolutely convex weakly compact subset of E'.

(3) *Let E, F be locally convex, F metrizable, and $A \in \mathfrak{L}(E, F)$. Then A is a homomorphism if and only if A is a weak homomorphism. If E and F are both metrizable, then homomorphisms, weak homomorphisms, and $\mathfrak{T}_k$-homomorphisms coincide.*

It is sufficient to prove the first assertion. If $A \in \mathfrak{L}(E, F)$ is a weak homomorphism, then it is in $\mathfrak{L}(E, A(E))$ and is a weak homomorphism onto $A(E)$. By (2) the topology on $A(E)$ induced by the topology of F is the topology $\mathfrak{T}_k(A(E)')$. By § 32, 4.(5) A is then a homomorphism onto $A(E)$, hence a homomorphism in F.

By 1.(3) A' is a strong homomorphism if A is a homomorphism of normed spaces. This is no longer true even for (F)-spaces. A positive result is the following special case of § 32, 4.(7):

(4) *If A is a monomorphism of the reflexive* (F)*-space E into the reflexive* (F)*-space F, then A' is a strong homomorphism of F' onto E'.*

Counterexamples. 1) In § 31, 5. an (FM)-space $E = \lambda[\mathfrak{T}]$ was constructed with a nonreflexive quotient space $E/N[A]$. The canonical mapping K of E onto $E/N[A]$ is a homomorphism, but the injection K' of $(E/N[A])' = N[A]^\perp$ into E' is not a strong homomorphism, since the strong topology of $N[A]^\perp$ is strictly finer on $N[A]^\perp$ than the strong topology of E.

2) In § 31, 7. we constructed a separable nondistinguished (F)-space $H = \lambda[\mathfrak{T}]$ which is a closed subspace of an (F)-space E. The topology $\hat{\mathfrak{T}}_b(E)$ on $E'/H^\perp$ is strictly finer than $\mathfrak{T}_b(H)$. The injection J of H into E is a monomorphism, but the canonical mapping J' of E' onto $E'/H^\perp$ is not a strong homomorphism.

The natural question whether A' is a $\mathfrak{T}_k$-homomorphism whenever A is a homomorphism has a negative answer even for (B)-spaces, as the following example shows.

By § 22, 4. there exists a closed subspace H in l^1 such that l^1/H is isomorphic to l^2. No weakly compact subset M in l^1 has the closed unit ball of l^1/H as canonical image $K(M)$. Hence K' is a weak monomorphism of $H^\perp$ into l^∞ but not a $\mathfrak{T}_k$-monomorphism since condition b) of § 32, 4.(3) is not satisfied.

On the other hand, the following converse of the question holds:

(5) *Let E, F be* (F)*-spaces and let $A \in \mathfrak{L}(E, F)$. If A' is a $\mathfrak{T}_k$-homomorphism, then A is a homomorphism.*

Since A' is a weak homomorphism, $A(E)$ is closed and the assertion follows from the BANACH–SCHAUDER theorem.

In § 21, 6. we defined on E' the topology $\mathfrak{T}_c$ of uniform convergence on the precompact subsets of E. This topology is more appropriate for studying duality relations for homomorphisms than the topologies $\mathfrak{T}_b$ or $\mathfrak{T}_k$, as the following theorem shows.

(6) *Let E, F be* (F)*-spaces. $A \in \mathfrak{L}(E, F)$ is a homomorphism if and only if A' is a $\mathfrak{T}_c$-homomorphism.*

Proof. $\mathfrak{T}_c$ is coarser than $\mathfrak{T}_k$ on E' and F' by § 21, 6.(1). Therefore, if A' is a $\mathfrak{T}_c$-homomorphism, A' is a weak homomorphism and A a homomorphism (BANACH–SCHAUDER theorem).

Conversely, let A be a homomorphism. Then $A(E)$ is closed and A' is $\mathfrak{T}_c$-continuous by § 32, 2.(4). To see that A' satisfies condition b) of § 32, 4.(3) it is enough to show that every compact subset M of $A(E)$ is the image of a compact set of E. To this end note that $\check{A}$ is an isomorphism, so $\check{A}^{(-1)}(M)$ is compact in $E/N[A]$. By § 22, 2.(7) there exists a compact set $M_1 \subset E$ such that $K(M_1) = \check{A}^{(-1)}(M)$ and therefore $M = \check{A}K(M_1) = A(M_1)$. This completes the proof.

$\mathfrak{T}_c$ denotes the topology on E of uniform convergence on the strongly compact sets of E'. This is the topology $\mathfrak{T}_b^\circ$ in the sense of § 21, 7.

(7) *Let E and F be* (B)*-spaces. $A \in \mathfrak{L}(E, F)$ is a homomorphism if and only if A is a $\mathfrak{T}_c$-homomorphism.*

If A is a $\mathfrak{T}_c$-homomorphism, then, since $\mathfrak{T}_c$ is weaker than the Mackey topology, A is a weak homomorphism and so by (3) a homomorphism. Conversely, if A is a homomorphism, then $A'(F')$ is weakly closed and A' is by 1.(3) a strong homomorphism. From § 22, 2.(7) it then follows that condition b) of § 32, 4.(3) is satisfied. By § 32, 2.(4) A'' is $\mathfrak{T}_c$-continuous and therefore A as the restriction of A'' to E is also $\mathfrak{T}_c$-continuous. Thus by § 32, 4.(3) A is a $\mathfrak{T}_c$-homomorphism.

3. Applications of the BANACH–DIEUDONNÉ theorem. In § 28, 3. we introduced the notion of local convergence (or Mackey convergence). Let N be a subset of a locally convex space E. We say that N is locally closed (or closed for the Mackey convergence) if the limit of every locally convergent sequence of elements of N belongs to N. We say that a linear mapping A of E in F is locally sequentially invertible if for each sequence $y_n \in A(E)$ which converges locally to o there exists a sequence $x_n \in E$ which converges locally to o and such that $Ax_n = y_n$.

We need the BANACH–DIEUDONNÉ theorem for the proof of the following result on homomorphisms.

(1) *Let E be an* (F)-*space, F a metrizable locally convex space. $A \in \mathfrak{L}(E, F)$ is a homomorphism if and only if $A'(F')$ is either locally closed in $E'[\mathfrak{T}_s(E)]$ or strongly sequentially closed.*

Proof. a) If A is a homomorphism, then $A'(F')$ is weakly closed, so $A'(F')$ is locally closed and strongly sequentially closed.

b) Assume $A'(F')$ locally closed. Let M be an absolutely convex, weakly closed, and weakly bounded subset of E'. Then M is weakly compact by § 21, 5.(4), hence $(E')_M = E'_M$ is a (B)-space by § 20, 11.(2). Since the set $N = A'(F') \cap M$ is closed in E'_M, it follows that E'_N is a (B)-space too.

We next prove that N is weakly compact. F' is the dual of a metrizable space and therefore the union of a sequence $C_1 \subset C_2 \subset \cdots$ of absolutely convex and weakly compact sets C_n. Each set $N_n = A'(C_n) \cap M$ is weakly compact and hence closed in E'_N. Obviously $E'_N = \bigcup_{n=1}^{\infty} nN_n$ and it follows from BAIRE's theorem that one of the sets nN_n contains a ball of E'_N. Therefore there exists $\rho > 1$ such that $N \subset \rho N_n$; hence $N = \rho N_n \cap M$, and N is weakly compact.

Since this is true for every M, it follows from § 21, 10.(5) that $A'(F')$ is weakly closed. Hence A is a weak homomorphism and by 2.(3) a homomorphism.

c) M is strongly bounded by the theorem of BANACH–MACKEY; therefore the norm convergence in E'_M is stronger than the strong convergence in E'. Hence $A'(F')$ is locally closed if it is strongly sequentially closed.

We can now generalize 1.(4) to

(2) *Let E be an* (F)-*space, F a metrizable locally convex space, $A \in \mathfrak{L}(E, F)$. If A' is a strong homomorphism, then A is a homomorphism.*

It is sufficient to show that $A'(F')$ is weakly closed in E', and by § 21, 10.(5) this will be true whenever every weakly bounded subset B of $A'(F')$ is contained in a weakly compact subset of $A'(F')$.

We have $A' = \hat{A}'K$, where $\hat{A}'$ is by assumption a monomorphism of $(F'/N[A'])[\hat{\mathfrak{T}}_b(F)]$ in $E'[\mathfrak{T}_b(E)]$ with range $A'(F')$. We remark that K and $\hat{A}'$ are also weakly continuous. $F'/N[A']$ is the dual of the metrizable space $\overline{A(E)}$ and $(F'/N[A'])[\mathfrak{T}_b(\overline{A(E)})]$ is the strong dual of $\overline{A(E)}$. By § 22, 2. the topology $\hat{\mathfrak{T}}_b(F)$ on $F'/N[A']$ is stronger than $\mathfrak{T}_b(\overline{A(E)})$, whereas $\mathfrak{T}_s(F)$ and $\mathfrak{T}_s(\overline{A(E)})$ always coincide.

Since E is complete, B is strongly bounded in E'; hence $\hat{A}'^{(-1)}(B)$ is bounded in $(F'/N[A'])[\hat{\mathfrak{T}}_b(F)]$ and therefore in $(F'/N[A'])[\mathfrak{T}_b(\overline{A(E)})]$. Since $\overline{A(E)}$ is metrizable, $\hat{A}'^{(-1)}(B)$ is relatively weakly compact by § 21, 5. and therefore contained in a weakly compact subset C of $F'/N[A']$. Hence $\hat{A}'(C)$ is weakly compact and contains B.

(3) *Let E and F be* (F)*-spaces, $A \in \mathfrak{L}(E, F)$. A is a homomorphism if A' is either* a) *locally sequentially invertible or* b) *strongly sequentially invertible or* c) *weakly sequentially invertible.*

a) We must show that if a sequence $u_n \in A'(F')$ converges locally to u_0, then u_0 belongs to $A'(F')$. Let v_n be elements in $F'/N[A']$ such that $\hat{A}'v_n = u_n$. Now v_n is weakly Cauchy; for if not there would exist $(n_i, m_i) \to (\infty, \infty)$ such that $v_{n_i} - v_{m_i} \not\to \mathfrak{o}$ weakly. But $u_{n_i} - u_{m_i} \to \mathfrak{o}$ locally so, by assumption, there exist $z_i \in F'$ with $A'z_i = u_{n_i} - u_{m_i}$ and $z_i \to \mathfrak{o}$ locally. Thus for $K: F' \to F'/N[A']$ we would have $Kz_i = v_{n_i} - v_{m_i}$ and, since K is weakly continuous, this would show $v_{n_i} - v_{m_i} \to \mathfrak{o}$ weakly, which is a contradiction.

The weak Cauchy sequence $v_n \in F'/N[A']$ is weakly bounded and, since $\overline{A(E)}$ is an (F)-space, relatively weakly compact, so v_n has a weak limit v_0 and finally $\hat{A}'v_0 = u_0$.

The same arguments give the proofs in the cases b) and c).

In the case a) we have also the converse result.

(4) *Let E and F be barrelled spaces, A a homomorphism of E into F. Then A' is locally sequentially invertible.*

Proof. Let $u_n \in A'(F')$ be a local null sequence. There exists an absolutely convex, weakly closed, and weakly bounded set $C \subset E'$ such that $u_n \in \rho_n C$, $\rho_n \to 0$, $\rho_n > 0$. Since A is a homomorphism, there exists by § 32, 4.(3) an absolutely convex, weakly closed, and weakly bounded subset $B \subset F'$ such that $A'(B) \supset C \cap A'(F')$. Since $u_n \in \rho_n C$, there exist $v_n \in \rho_n B$ such that $A'v_n = u_n$ and v_n is a local null sequence.

From (3) and (4) we infer

(5) *Let E and F be* (F)*-spaces, $A \in \mathfrak{L}(E, F)$. A is a homomorphism if and only if A' is locally sequentially invertible.*

As we will see in § 33, 5., the corresponding theorem for case c) in (3) will be true only for separable spaces.

Remark. If E, F are (B)-spaces, $A \in \mathfrak{L}(E, F)$, and A' a strong homomorphism, then A' is strongly sequentially invertible; hence A is a homomorphism by (3). Thus 1.(4) follows also immediately from (3).

The results of this section were proved by the author for echelon spaces in his paper [6].

4. Homomorphisms in (B)- and (F)-spaces. We collect some of the results of § 33, 2.–4. in two main theorems.

(1) *Homomorphism Theorem for* (B)*-Spaces: Let E and F be* (B)*-spaces, A a continuous linear mapping of E into F. The following properties of A are equivalent:*

a) *A is a homomorphism;*

b) *A is a weak homomorphism*;
c) *$A(E)$ is closed*;
d) *A' is a strong homomorphism*;
e) *A' is a weak homomorphism*;
f) *$A'(F')$ is weakly closed*;
g) *$A'(F')$ is strongly closed*;
h) *A' is a $\mathfrak{T}_c$-homomorphism*;
i) *A is a $\mathfrak{T}_c$-homomorphism.*

Proof. a) and b) are equivalent by 2.(3), a) and c) by the BANACH–SCHAUDER theorem, a) and d) by 1.(3) and 1.(4), c) and e) by § 32, 3.(2), b) and f) too, d) and g) by § 32, 2.(4) and the BANACH–SCHAUDER theorem for A', a) and i) by 2.(7), and finally a) and h) by 2.(6).

(2) *Homomorphism Theorem for* (F)-*Spaces: Let E and F be* (F)-*spaces, A a continuous linear mapping of E into F. The following properties of A are equivalent:*
a) *A is a homomorphism*;
b) *A is a weak homomorphism*;
c) *$A(E)$ is closed*;
d) *A' is a weak homomorphism*;
e) *$A'(F')$ is weakly closed*;
f) *$A'(F')$ is locally closed or strongly sequentially closed*;
g) *A' is a $\mathfrak{T}_c$-homomorphism*;
h) *A is sequentially invertible*;
i) *A' is locally sequentially invertible.*

The proofs for the equivalence of properties a) to e) and a) and g) are the same as in the case of Theorem (1). The equivalence of a) and f) follows from 3.(1); a) and h) are equivalent by 2.(1), a) and i) by 3.(5).

Remark. We define three further properties of A: α) every bounded set $M \subset A(E)$ is contained in the image $A(N)$ of a bounded set N of E; β) resp. γ) the same property for weakly compact resp. compact subsets of $A(E)$. We prove that A is a homomorphism if it has one of these properties; and for this it is sufficient to show that A has property h).

Let $y_n \in A(E)$, $y_n \to \mathfrak{o}$; then there exist $\rho_n \to \infty$ such that $\rho_n y_n \to \mathfrak{o}$ by § 28, 1.(5). The set consisting of the $\rho_n y_n$ and $\mathfrak{o}$ is bounded, weakly compact, and compact in $A(E)$. By α), β), or γ) there exist $z_n \in E$ such that $Az_n = \rho_n y_n$ and the z_n are bounded in E. But then $(1/\rho_n)z_n$ converges to $\mathfrak{o}$ and $A_n(1/\rho_n)z_n = y_n$. Hence A has property h).

Observe that homomorphisms of (F)-spaces need not always have property α) (cf. § 31, 5.) nor property β) (cf. § 22, 4.). On the contrary, a homomorphism always has property γ), as follows from § 22, 2.(7). So γ) could be added to our list in (2).

5. Separability. A theorem of SOBCZYK. If E is locally convex and separable and if N is a countable dense set in E, we know (§ 21, 3.(3) and (4)) that the topologies $\mathfrak{T}_s(E)$ and $\mathfrak{T}_s(N)$ coincide on every equicontinuous subset M of E' and that on M the weak topology is metrizable. Therefore M is weakly closed if it is weakly sequentially closed. Using the BANACH–DIEUDONNÉ theorem we proved in § 21, 10.(7) that if E is a separable (F)-space, every convex subset of E' is weakly closed if it is weakly sequentially closed.

Finally, we proved in § 32, 1.(11) that if E is a separable (F)-space and A is a weakly sequentially continuous linear mapping of E' into a locally convex space F, then A is weakly continuous.

Taking the scalar field K for F we see that every weakly sequentially continuous linear functional on E' is continuous. This is a special case of the following general result:

(1) *Let E be locally convex, separable, and complete. Then every weakly sequentially continuous linear functional on E' is weakly continuous.*

This is an immediate consequence of GROTHENDIECK's result (§ 21, 9.(4)) and the metrizability of the equicontinuous sets in E'.

We now prove the following "lifting" property of separable spaces:

(2) *Let E be locally convex and separable, H a closed subspace. If $\hat{u}_n$ is an equicontinuous sequence weakly convergent to $\circ$ in E'/H°, then there exist $u_n \in \hat{u}_n$ such that u_n is equicontinuous and weakly convergent to $\circ$ in E'.*

By § 22, 1. an equicontinuous set in E'/H° is contained in a set of the form $K(U^\circ)$, U an absolutely convex neighbourhood of $\circ$ in E, and K is the canonical mapping of E' onto E'/H°. We will show that if $\hat{u}_n \in K(U^\circ)$, then there exist $u_n \in 2U^\circ$ with the desired properties.

We remark first that there are $v_n \in \hat{u}_n$, $v_n \in U^\circ$. Assume the sequence $y_1, y_2, \ldots$ to be dense in E. We need the following

Lemma. Let $\varepsilon > 0$ and $y_1, \ldots, y_k$ be given. Then there exists $N(\varepsilon) > 0$ such that for every $n \geqq N(\varepsilon)$ there exists $w_n \in U^\circ \cap H^\circ$ and $|(v_n + w_n)y_i| \leqq \varepsilon$, $1 \leqq i \leqq k$.

Proof. We assume the contrary. Then there exists a subsequence v'_n of v_n such that

(3) $|(v'_n + w)y_j| > \varepsilon$ for all $w \in U^\circ \cap H^\circ$ and a certain integer $j \in [1, k]$. Since U° is weakly compact and metrizable, there exists a subsequence v''_n of v'_n converging weakly to $u_0 \in U^\circ$. Since by assumption $\hat{v}''_n$ converges

weakly to $\mathfrak{o}$, it follows that $\hat{u}_0 = \mathfrak{o}$. Therefore $u_0 \in U^\circ \cap H^\circ$ and, taking $w = -u_0$ in (3), we obtain a contradiction. The lemma is proved.

To prove (2) we determine now $N_1 < N_2 < \cdots$ such that for every $n \geqq N_k$ there exists $w_n^{(k)} \in U^\circ \cap H^\circ$ such that $|(v_n + w_n^{(k)})y_i| \leqq 1/k$ for $i \leqq k$. Taking $w_n = \mathfrak{o}$ for $1 \leqq n \leqq N_1$, $w_n = w_n^{(1)}$ for $N_1 < n \leqq N_2$, $w_n = w_n^{(2)}$ for $N_2 < n \leqq N_3$ and so on, we have

$$|(v_n + w_n)y_1| \leqq 1 \quad \text{for all } n > N_1,$$
$$|(v_n + w_n)y_1| \leqq \tfrac{1}{2} \quad \text{and} \quad |(v_n + w_n)y_2| \leqq \tfrac{1}{2} \quad \text{for all } n > N_2, \text{ and so on.}$$

This means that $u_n = v_n + w_n$ converges to zero in the weak topology generated by the dense sequence $y_1, y_2, \ldots$. Since $u_n \in \hat{u}_n$, $u_n \in 2U^\circ$ and $2U^\circ$ is weakly compact, this means u_n converges to $\mathfrak{o}$ in the sense of $\mathfrak{T}_s(E)$ and (2) is proved.

(4) *Corollary: Let E be a separable* (B)*-space, H a closed subspace. If $\hat{u}_n \in E'/H^\circ$ converges weakly to $\hat{\mathfrak{o}}$ and $\|\hat{u}_n\| \leqq r$, then there exist $u_n \in \hat{u}_n$, $\|u_n\| \leqq 2r$ such that u_n converges weakly to $\mathfrak{o}$ in E'.*

We are now able to prove the result mentioned at the end of 3.

(5) *Let E and F be* (F)*-spaces, F separable. $A \in \mathfrak{L}(E, F)$ is a homomorphism if and only if A' is weakly sequentially invertible.*

In view of 3.(3) we have to prove only that the condition is necessary. We may suppose that A' is a weak homomorphism of F' into E'. Suppose $u_n \in A'(F')$ converges weakly to $\mathfrak{o}$ in E'. Then $\hat{A}'^{(-1)}u_n = \hat{v}_n$ converges weakly to $\hat{\mathfrak{o}}$ in $F'/N[A']$. Since F is an (F)-space, the sequence $\hat{v}_n$ is equicontinuous; by (2) there exist therefore $v_n \in \hat{v}_n$ converging weakly to $\mathfrak{o}$ with $A'v_n = u_n$.

(2) is not true for nonseparable (B)-spaces, as the following example shows. Consider $E = l^\infty$ and $H = c_0$ and the sequence e_n of unit vectors in $E'/H^\circ = l^1$. The sequence e_n converges to $\mathfrak{o}$ weakly but not in the norm. By § 31, 2.(3) it is therefore not the canonical image of any sequence in E' converging weakly to $\mathfrak{o}$.

If J is the injection of c_0 into l^∞, then J' is a weak homomorphism but not weakly sequentially invertible.

We give an application of (4).

(6) *Let A be a monomorphism of c_0 into a separable* (B)*-space E and $H = A(c_0)$. Then A has a left inverse $B \in \mathfrak{L}(E, c_0)$ such that $\|B\| \leqq 2\|\check{A}^{-1}\|$.*

Proof. A' is a weak and strong homomorphism of E' onto l^1 with kernel H° and $\hat{A}'$ is a weak and strong isomorphism of E'/H° onto l^1. The sequence e_n of the unit vectors in l^1 converges weakly to $\mathfrak{o}$ and therefore also the sequence $\hat{u}_n = \hat{A}'^{(-1)}e_n$ converges weakly to $\mathfrak{o}$ in E'/H°. We have

$$\|\hat{u}_n\| \leqq \|\hat{A}'^{(-1)}\|\,\|e_n\| = \|\hat{A}'^{(-1)}\| = \|(\check{A})'^{(-1)}\| = \|(\check{A}^{-1})'\| = \|\check{A}^{-1}\| = r,$$

where $\hat{A}' = (\check{A})'$ follows from § 32, 3.(7) and $(\check{A})'^{(-1)} = (\check{A}^{-1})'$ from $(C^{-1})' = (C')^{-1}$ for any isomorphism (if $I = CC^{-1} = C^{-1}C$ then $I = (C^{-1})'C' = C'(C^{-1})'$).

From (4) follows the existence of a sequence $u_n \in \hat{u}_n$, $\|u_n\| \leqq 2r$, u_n weakly convergent to $\mathfrak{o}$ in E' and $A'u_n = e_n$.

If we now define Bx for every $x \in E$ as the sequence $(u_k x)$, then $Bx \in c_0$, $\|Bx\| = \sup_k |u_k x| \leqq 2r\|x\|$, and finally $BA = I$ since

$$BAe_n = (u_k Ae_n)_k = ((A'u_k)e_n)_k = (e_k e_n)_k = e_n \quad \text{for } n = 1, 2, \ldots.$$

The following result of Sobczyk [2] is now an easy consequence:

(7) *Let E be a separable* (B)*-space and H a closed subspace isomorphic to c_0. Then H has a topological complement.*

If H is norm isomorphic to c_0, there exists a projection P of norm $\leqq 2$ of E onto H.

Proof. Let A be the monomorphism of c_0 in E inducing the isomorphism $\check{A}$ on H. Let B be the left inverse with $\|B\| \leqq 2\|\check{A}^{-1}\|$. The product $P = AB$ is a continuous projection of E onto H: We have $P(E) \subset A(c_0) = H$ and, on the other hand, if $y_0 \in H$, then $y_0 = Ax_0$, $x_0 \in c_0$, and $Py_0 = A(BAx_0) = Ax_0 = y_0$; hence $P(E) = H$, $P^2 = P$. The kernel of P is a topological complement to H.

If A is a norm isomorphism, then $\|A\| = 1$, $\|\check{A}^{-1}\| = 1$, $\|P\| \leqq \|A\|\,\|B\| \leqq 2$.

(8) *The upper bound 2 for $\|P\|$ in* (7) *is sharp.*

We have $c = c_0 \oplus [e]$, where $e = (1, 1, \ldots)$. Let P be a projection of c onto c_0 and $Pe = y = (y_1, y_2, \ldots) \in c_0$. For every n we have $\left\|2\sum_1^n e_i - e\right\| = 1$, but since $P\left(2\sum_1^n e_i - e\right) = 2\sum_1^n e_i - y$ and $y_k \to 0$ we are able to choose n so large that $\left\|2\sum_1^n e_i - y\right\| \geqq 2 - \varepsilon$. From this follows $\|P\| \geqq 2 - \varepsilon$; hence $\|P\| \geqq 2$.

Theorem 2 is stated in Grothendieck [11]; the idea of the proof comes from Köthe [6]. For (6) and (7) compare Köthe [4']; for an even shorter proof see Veech [1'].

6. (FM)-spaces. Because of § 27, 2.(5) the results of 5. apply to (FM)-spaces which are separable. There is another important property of (FM)-spaces. Let E and F be (FM)-spaces and $A \in \mathfrak{L}(E, F)$. Contrary to the situation for arbitrary (F)-spaces, we show that the adjoint A' of a homomorphism A is always a strong homomorphism (as in the case of (B)-spaces).

In an (FM)-space E the bounded subsets coincide with the relatively compact subsets; hence $\mathfrak{T}_b(E)$ coincides with $\mathfrak{T}_c(E)$ on E'. Therefore in 4.(2) condition g) can be read as: A' is a strong homomorphism.

Without repeating the whole list of equivalent properties of 4.(2) for (FM)-spaces we record this result in

(1) *Let E and F be* (FM)*-spaces. $A \in \mathfrak{L}(E, F)$ is a homomorphism if and only if A' is a strong homomorphism.*

By a theorem of DIEUDONNÉ (§ 27, 2.(8)) a locally convex space E is the strong dual of an (FM)-space if and only if it is barrelled and has a countable fundamental system of absolutely convex compact subsets. These spaces are (M)-spaces, are reflexive, and their duals are (FM)-spaces. Hence the homomorphism theorem for (FM)-spaces can be interpreted as a theorem on homomorphisms of the dual spaces. From 4.(2) and (1) follows in particular a theorem of BANACH–SCHAUDER type.

(2) *Let E and F be barrelled spaces with countable fundamental systems of absolutely convex compact sets. A continuous linear mapping A of E into F is a homomorphism if and only if $A(E)$ is sequentially closed.*

This theorem applies, for example, to the spaces $H(\mathfrak{A})$ of locally holomorphic functions of § 27, 3. and 4. and to the co-echelon spaces of type (M) of § 30, 9.

The counterexamples in 2. show that the BANACH–SCHAUDER theorem is not true for continuous linear mappings of the strong duals of arbitrary (F)-spaces. GROTHENDIECK [10] constructed an example showing that (2) is false even for the strong duals of separable reflexive (F)-spaces. The example is closely related to the example in § 31, 5.

Let $\mathfrak{a}^{(k)}$ be the vector $(a_{ij}^{(k)})$ such that $a_{ij}^{(k)} = j^k$ for $i < k$ and $a_{ij}^{(k)} = i^k$ for $i \geqq k$, where $i, j, k = 1, 2, \ldots$. We may write $\mathfrak{a}^{(k)}$ in the form

$$\mathfrak{a}^{(k)} = (\overbrace{\mathfrak{b}_k, \ldots, \mathfrak{b}_k}^{k-1}, k^k \mathfrak{e}, (k+1)^k \mathfrak{e}, \ldots),$$

where $\mathfrak{b}_k = (1^k, 2^k, 3^k, \ldots)$, $\mathfrak{e} = (1, 1, 1, \ldots)$.

Let λ be the echelon space of pth order, $p > 1$, defined by the steps $\mathfrak{a}^{(k)}$, and $\lambda^\times$ the α-dual which is a co-echelon space of pth order (§ 30, 8.).

λ consists of all vectors $\mathfrak{x} = (x_{ij})$ such that

$$(1) \qquad \sum_{i,j=1}^{\infty} |a_{ij}^{(k)}|\, |x_{ij}|^p < \infty, \qquad k = 1, 2, \ldots.$$

$\lambda^\times$ consists of all $\mathfrak{u} = (u_{ij})$ such that

$$(2) \qquad \sum_{i,j=1}^{\infty} |a_{ij}^{(k)}|^{-q/p} |u_{ij}|^q < \infty$$

for some k, where $1/p + 1/q = 1$.

With the norms corresponding to (1) λ is an (F)-space and by § 30, 9.(1) even an (FM)-space. $\lambda^\times$ is the dual of λ and an (M)-space for the strong topology.

Let A be the linear mapping $A\mathfrak{x} = \left(\sum_{i=1}^{\infty} x_{i1}, \sum_{i=1}^{\infty} x_{i2}, \ldots\right)$ defined for every $\mathfrak{x} \in \lambda$. We prove first that $A\mathfrak{x}$ is an element of l^p.

Let $\mathfrak{v} = (v_1, v_2, \ldots)$ be an element of l^q and let $\tilde{\mathfrak{v}}$ be $(\mathfrak{v}, \mathfrak{v}, \ldots)$. We show that $\tilde{\mathfrak{v}}$ is an element of $\lambda^\times$. For this we choose k such that $kq/p > 1$. Then

$\sum_{i,j=1}^{\infty} |a_{ij}^{(k)}|^{-q/p} |v_j|^q < \infty$: For $i < k$ we have

$$\sum_j |a_{ij}^{(k)}|^{-q/p} |v_j|^q \leqq \sum_j |v_j|^q \quad \text{for } i \geqq k \sum_j |a_{ij}^{(k)}|^{-q/p} |v_j|^q = i^{-kq/p} \sum_j |v_j|^q.$$

Since $\mathfrak{v} \in \lambda^\times$, $\tilde{\mathfrak{v}}\,\mathfrak{x} = \sum_{i,j=1}^{\infty} v_j x_{ij}$ is absolutely convergent. Therefore

$$(3) \qquad \tilde{\mathfrak{v}}\,\mathfrak{x} = \sum_j v_j \sum_i x_{ij} = \mathfrak{v}(A\mathfrak{x}) = (A'\mathfrak{v})\mathfrak{x} < \infty$$

for all $\mathfrak{x} \in \lambda$ and all $\mathfrak{v} \in l^q$. Hence $A\mathfrak{x} \in l^p$.

By (3) we have $A'\mathfrak{v} = \tilde{\mathfrak{v}}$ and A' maps $l^q = (l^p)'$ in $\lambda^\times = \lambda'$; hence A is weakly continuous and therefore continuous from λ into l^p. A' is one-one and $A'(l^q)$ is the subspace of $\lambda^\times$ consisting of all $\tilde{\mathfrak{v}}$. If a vector $\tilde{\mathfrak{u}} = (\mathfrak{u}, \mathfrak{u}, \ldots)$ is an element of λ', then it must satisfy (2) for some k. But then we have $\sum_j |u_j|^q < \infty$ and $\tilde{\mathfrak{u}}$ is in $A'(l^q)$. Since a weak adherent point of $A'(l^q)$ must be of the form $(\mathfrak{u}, \mathfrak{u}, \ldots)$, we find that $A'(l^q)$ is weakly closed in λ'. But then *A is a homomorphism of λ onto l^p*. Now the unit ball in l^p is bounded but not compact; therefore it cannot be the image of a bounded set in λ, since these sets are all relatively compact. From § 32, 4.(3) it follows that A' is not a strong homomorphism of l^q in λ' though $A'(l^q)$ is weakly closed.

λ is an (FM)-space with l^p, a reflexive (B)-space, as quotient space, and there exist weakly compact sets in l^p which are not canonical images of weakly compact sets in λ.

§ 34. The theory of Pták

1. Nearly open mappings. For arbitrary locally convex spaces the homomorphism theorem (§ 32, 4.) gives necessary and sufficient conditions for a continuous linear mapping to be a homomorphism. In the case of

(F)-spaces we have the much stronger theorem of BANACH and SCHAUDER.

PTÁK [4] made the first successful attempt to extend this theorem to a larger class of spaces. We give here an exposition of his ideas and later developments which include generalizations of the closed-graph theorem.

PTÁK's starting point is an analysis of the classical proof of the BANACH–SCHAUDER theorem (cf. § 15, 12.). Assuming $A(E)$ to be not meagre in F, one proves first that $\overline{A(U)}$ is a neighbourhood of $\circ$ in $\overline{A(E)} = F$, where U is a neighbourhood of $\circ$ in E. Secondly, one shows that $A(U)$ itself is a neighbourhood of $\circ$ in F.

The following notion will help to describe the situation. Let E, F be locally convex and A a linear mapping of E into F. Call A nearly open if for every neighbourhood U of $\circ$ in E the closure $\overline{A(U)}$ in F is a neighbourhood of $\circ$ in $\overline{A(E)}$.

An equivalent definition is: A is nearly open if the closure of $A(U)$ in $A(E)$ is a neighbourhood of $\circ$ in $A(E)$ for every neighbourhood U of $\circ$ in E.

If we look again at the proof of the BANACH–SCHAUDER theorem, the following two questions are very natural:

α) What conditions on F assure that a linear continuous mapping A of a locally convex space E onto F is always nearly open?

β) Can one characterize the spaces E with the property that a linear continuous and nearly open mapping A of E into an arbitrary locally convex space F is always open?

The answer to question α) is easy:

(1) *The barrelled spaces F are characterized by the following property: Every linear (or every linear continuous) mapping A of an arbitrary locally convex space E onto F is nearly open.*

Proof. a) Let F be barrelled, A linear from E onto F. If U is an absolutely convex neighbourhood of $\circ$ in E, then $A(U)$ is absorbent and absolutely convex. Consequently, $\overline{A(U)}$ is a barrel and therefore a neighbourhood in F.

b) Let I be the identity mapping of $F[\mathfrak{T}_b(F')]$ onto $F[\mathfrak{T}]$; then I is continuous. We assume that I is nearly open. Then a barrel T in $F[\mathfrak{T}]$ is a strong neighbourhood of $\circ$ in F and, as $\overline{I(T)}$ is also a $\mathfrak{T}$-neighbourhood, so $F[\mathfrak{T}]$ is barrelled.

The first part of the proof of the BANACH–SCHAUDER theorem is a special case of (1) if we show that a nonmeagre subspace of an (F)-space is barrelled. This will be an immediate consequence of (2).

In § 4, 6. we called a subset M of a metric space E meagre in E if M is the union of countably many nowhere dense sets. This notion of meagreness is meaningful in topological vector spaces too, and we have

(2) *A nonmeagre linear subspace H of a locally convex space $E[\mathfrak{T}]$ is barrelled in the induced topology $\mathfrak{T}$ and dense in E.*

Let T be a barrel in H. Then $H = \bigcup_{n=1}^{\infty} nT$. Since H is not meagre in E, T is not nowhere dense in E; hence $\overline{T}$ is a closed neighbourhood of $\circ$ in E. Since T is closed in H, $T = \overline{T} \cap H$ is a neighbourhood of $\circ$ in $H[\mathfrak{T}]$ and H is therefore barrelled. That H is dense in E follows from $E = \bigcup_{n=1}^{\infty} n\overline{T}$.

A locally convex space which is not meagre in itself is called a Baire space. As a corollary of (2) we have

(3) *Every locally convex Baire space is barrelled.*

We now turn to question β) and begin with the dual characterization of continuous nearly open mappings.

(4) *Let $E[\mathfrak{T}_1]$ and $F[\mathfrak{T}_2]$ be locally convex spaces, $\mathfrak{M}_1$ resp. $\mathfrak{M}_2$ the class of equicontinuous sets in E' resp. F'.*

A linear continuous resp. weakly continuous mapping A of E into F is nearly open resp. continuous and nearly open if and only if $A'(\mathfrak{M}_2) \supset \mathfrak{M}_1 \cap A'(F')$ resp. $A'(\mathfrak{M}_2) = \mathfrak{M}_1 \cap A'(F')$.

The statement for weakly continuous mappings is a consequence of the statement for continuous mappings and § 32, 2.(1). So we need prove only the continuous case and then the theorem is obviously in close analogy to the homomorphism theorem (§ 32, 4.(3)).

Proof. a) Assume $A'(\mathfrak{M}_2) \supset \mathfrak{M}_1 \cap A'(F')$ and let U be an absolutely convex neighbourhood of $\circ$ in E. We must show that $\overline{A(U)}$ is a $\mathfrak{T}_2$-neighbourhood of $\circ$ in $\overline{A(E)}$. By the theorem of bipolars we have $\overline{A(U)} = A(U)^{\circ\circ}$. Now $A(U)^{\circ} = A'^{(-1)}(U^{\circ}) = A'^{(-1)}(U^{\circ} \cap A'(F'))$; since by assumption there exists $M_2 \in \mathfrak{M}_2$ absolutely convex such that $A'(M_2) \supset U^{\circ} \cap A'(F')$, we have $A(U)^{\circ} \subset A'^{(-1)}(A'(M_2)) = M_2 + N[A']$. Hence $\overline{A(U)} = A(U)^{\circ\circ} \supset (\overline{M_2 + N[A']})^{\circ} = M_2^{\circ} \cap \overline{A(E)}$ and this is a $\mathfrak{T}_2$-neighbourhood of $\circ$ in $\overline{A(E)}$.

b) The condition is necessary. Let A be continuous and nearly open and let $M_1 \in \mathfrak{M}_1 \cap A'(F')$ be absolutely convex and weakly closed in $A'(F')$. By assumption $\overline{A(M_1^{\circ})} \supset M_2^{\circ} \cap \overline{A(E)}$ for some absolutely convex and weakly closed $M_2 \in \mathfrak{M}_2$. Taking polars in F' we have

$$A(M_1^{\circ})^{\circ} = \overline{A(M_1^{\circ})}^{\circ} \subset (M_2^{\circ} \cap \overline{A(E)})^{\circ} = (M_2^{\circ} \cap N[A']^{\circ})^{\circ} = \overline{M_2 + N[A']}.$$

On the other hand, $A(M_1^{\circ})^{\circ} = A'^{(-1)}(M_1^{\circ\circ}) = A'^{(-1)}(M_1)$, since $M_1^{\circ\circ} \cap A'(F') = M_1$ because M_1 is weakly closed in $A'(F')$. Hence $A'^{(-1)}(M_1) \subset \overline{M_2 + N[A']}$. If we apply A' to this inequality, we get $A'(A'^{(-1)}(M_1)) = M_1$ on the left side and $A'(\overline{M_2 + N[A']}) \subset \overline{A'(M_2 + N[A'])} = \overline{A'(M_2)}$ on the right side. Therefore $M_1 \subset \overline{A'(M_2)} = A'(M_2)$ since M_2 is weakly compact.

Comparison of (4) with the homomorphism theorem gives

(5) *A linear continuous and nearly open mapping A of E into F, E and F locally convex, is open if and only if $A'(F')$ is weakly closed.*

Another important consequence of (4) is the following proposition:

(6) *Let $A \in \mathfrak{L}(E[\mathfrak{T}_1], F[\mathfrak{T}_2])$ be nearly open. Then for every weakly closed absolutely convex $\mathfrak{T}_1$-equicontinuous set M_1 in E' the set $M_1 \cap A'(F')$ is weakly closed in E'.*

By (4) $M_1 \cap A'(F')$ is contained in a weakly compact set $A'(M_2)$. But then $M_1 \cap A'(F') = M_1 \cap A'(M_2)$ is also weakly compact because it is the intersection of two such sets and therefore weakly closed in E'.

Remembering the definition of the topology $\mathfrak{T}^f$ and the characterization of $\mathfrak{T}^f$-closed sets in § 21, 8., we have the following equivalent formulation of (6):

(6') *If $A \in \mathfrak{L}(E[\mathfrak{T}_1], F[\mathfrak{T}_2])$ is nearly open, then $A'(F')$ is $\mathfrak{T}_1^f$-closed in E'.*

For the weak topology we have

(7) *Every weakly continuous linear mapping is nearly weakly open.*

This follows from (4) since $A'(\mathfrak{M}_2) \supset \mathfrak{M}_1 \cap A'(F')$ if $\mathfrak{M}_1$ and $\mathfrak{M}_2$ are the classes of all bounded subsets of finite dimension.

2. Pták spaces and the BANACH–SCHAUDER theorem. The results of 1. and question β) suggest the following definitions.

A locally convex space $E[\mathfrak{T}]$ is called a Pták space (or B-complete) if every $\mathfrak{T}^f$-closed linear subspace of E' is weakly closed. $E[\mathfrak{T}]$ is called an infra-Pták space (or B_r-complete) if every weakly dense $\mathfrak{T}^f$-closed linear subspace of E' coincides with E'.

Every Pták space is an infra-Pták space. It is an unsolved problem whether there exist infra-Pták spaces which are not Pták spaces.

(1) *Every infra-Pták space is complete.*

If H is a $\mathfrak{T}^f$-closed linear subspace of E' of co-dimension 1, it is either weakly closed or weakly dense in E'. The second case cannot happen by definition of an infra-Pták space. The assertion follows now from § 21, 9.(6).

We can now give a complete answer to question β) of 1. (cf. PTÁK [4]).

(2) a) *A locally convex space E is a Pták space if and only if every continuous linear and nearly open mapping A of E into an arbitrary locally convex space F is a homomorphism.*

b) *A locally convex space E is an infra-Pták space if and only if every one-one continuous linear and nearly open mapping A of E into an arbitrary locally convex space F is a monomorphism.*

Proof. i) Let E be a Pták space, A continuous linear and nearly open. $A'(F')$ is $\mathfrak{T}_1^f$-closed by 1.(6′) and therefore weakly closed. By 1.(5) A is a homomorphism.

Observing that in the case b) $A'(F')$ is weakly dense in E' because A is one-one, the same argument will show that A is open if E is an infra-Pták space.

ii) Let $E[\mathfrak{T}]$ be locally convex and $\mathfrak{M}$ the class of equicontinuous subsets of E'. We assume that E is not a Pták space; E' has therefore a $\mathfrak{T}^f$-closed but not weakly closed subspace H. By § 22, 2.(1) the topologies $\mathfrak{T}_s(E)$ and $\mathfrak{T}_s(H')$ coincide on H; hence the injection J of $H[\mathfrak{T}_s(H')]$ into $E'[\mathfrak{T}_s(E)]$ is a weak monomorphism. J' is the canonical homomorphism K of $E[\mathfrak{T}_s(E')]$ onto $H'[\mathfrak{T}_s(H)] = (E/H^\circ)[\mathfrak{T}_s(H)]$.

Since H is $\mathfrak{T}^f$-closed, $\mathfrak{M} \cap H$ is a saturated class of absolutely convex and $\mathfrak{T}_s(E/H^\circ)$-compact subsets of $H = (E/H^\circ)[\mathfrak{T}_s(H)]'$; therefore $\mathfrak{M} \cap H$ determines a topology $\mathfrak{T}'$ on E/H° which is compatible with the dual pair $\langle H, E/H^\circ\rangle$. It follows from $K' = J$ that $J(\mathfrak{M} \cap H) = \mathfrak{M} \cap H$ and from 1.(4) that K is a continuous and nearly open mapping of $E[\mathfrak{T}]$ onto $(E/H^\circ)[\mathfrak{T}']$. Since H is not weakly closed, K is not open.

In the same way one constructs a nearly open but not open continuous injection on E whenever E is not an infra-Pták space.

From (2) and 1.(1) follows immediately PTÁK's generalization of the BANACH–SCHAUDER theorem:

(3) *Every linear continuous mapping of a Pták space onto a barrelled space is a homomorphism.*

Every one-one linear continuous mapping of an infra-Pták space onto a barrelled space is an isomorphism.

The classical BANACH–SCHAUDER theorem follows from (3) in the following way. We proved in 1. that if $A(E)$ is not meagre in F, then $A(E)$ is barrelled. By § 21, 10.(5) every (F)-space is a Pták space; hence A is a homomorphism.

The Pták spaces E are characterized by the property that the linear continuous and nearly open mappings of E onto any locally convex space F are always open. It is natural to ask whether this remains valid for a larger class of spaces E by putting restrictions on the space F. We give two examples.

HUSAIN [1′] called a locally convex space $E[\mathfrak{T}]$ a $B(\mathscr{T})$-space if a $\mathfrak{T}^f$-closed subspace H of E' is weakly closed if in H all weakly bounded subsets are $\mathfrak{T}$-equicontinuous. HUSAIN proved the following proposition:

(4) *Every linear continuous mapping of a $B(\mathscr{T})$-space $E[\mathfrak{T}]$ onto a barrelled space is a homomorphism.*

The $B(\mathscr{T})$-spaces are characterized by this property.

Proof. Let $E[\mathfrak{T}_1]$ be a $B(\mathscr{T})$-space, $F[\mathfrak{T}_2]$ be barrelled, and $A \in \mathfrak{L}(E, F)$. By 1.(1) A is nearly open and therefore $A'(F')$ is $\mathfrak{T}_1^f$-closed. Now A' is a weak monomorphism. The class $\mathfrak{M}_2$ of equicontinuous subsets of F' consists of all weakly bounded subsets since F is barrelled. Therefore $A'(\mathfrak{M}_2)$ is the class of all weakly bounded subsets of $A'(F')$. We have $A'(\mathfrak{M}_2) = \mathfrak{M}_1 \cap A'(F')$ by 1.(4); hence the weakly bounded subsets of $A'(F')$ are equicontinuous. Since E is a $B(\mathscr{T})$-space, $A'(F')$ is weakly closed and A is a homomorphism by 1.(5).

If, conversely, $E[\mathfrak{T}]$ is not a $B(\mathscr{T})$-space, then there exists a $\mathfrak{T}^f$-closed but not weakly closed subspace of E' in which all weakly bounded subsets are equicontinuous. A repetition of the arguments in the second part of the proof of (2) leads to a nearly open but not open mapping of $E[\mathfrak{T}]$ onto the barrelled space $(E/H^\circ)[\mathfrak{T}_b(H)]$.

It is also possible to characterize the locally convex spaces $E[\mathfrak{T}]$ with the property that every linear continuous and nearly open mapping of E onto a locally convex space with the Mackey topology is a homomorphism. The characterizing property is the following: Let H be a $\mathfrak{T}^f$-closed subspace of E' such that the equicontinuous sets in H coincide with the weakly compact absolutely convex sets and their subsets; then H is weakly closed in E'. The proof is left to the reader.

3. Some results on Pták spaces. Our first result is a new characterization of infra-Pták spaces.

(1) *$E[\mathfrak{T}]$ is an infra-Pták space if and only if there is no strictly coarser locally convex topology $\mathfrak{T}_1$ on E with the property that the $\mathfrak{T}_1$-closure of a $\mathfrak{T}$-neighbourhood of $\circ$ is always a $\mathfrak{T}_1$-neighbourhood.*

Proof. The condition is necessary, since the identity map of $E[\mathfrak{T}]$ onto $E[\mathfrak{T}_1]$ is continuous and nearly open and hence an isomorphism by 2.(2) b).

Conversely, we suppose that the condition is satisfied and that A is a one-one continuous and nearly open mapping of E onto a locally convex space $F[\mathfrak{T}_1]$. We may identify E and F so that A is the identity map of $E[\mathfrak{T}]$ onto $E[\mathfrak{T}_1]$. Then for A to be nearly open means exactly that $\mathfrak{T}_1$ satisfies the condition of the theorem. $\mathfrak{T}_1$ therefore coincides with $\mathfrak{T}$ and A is an isomorphism. By 2.(2) b) $E[\mathfrak{T}]$ is an infra-Pták space since it is sufficient to consider only mappings onto locally convex spaces F.

We next investigate the hereditary properties of Pták spaces.

(2) *Every quotient of a Pták space is a Pták space.*
Conversely, a locally convex space is a Pták space if all its quotients are infra-Pták spaces.

Proof. a) Let E/H be a quotient of the Pták space E and A a continuous linear and nearly open mapping of E/H into the locally convex space F. If K is the canonical homomorphism of E onto E/H then the mapping AK of E into F is continuous. If U is an open neighbourhood of $\circ$ in E, then $K(U)$ is a neighbourhood of $\circ$ in E/H and $\overline{A(K(U))} = \overline{(AK)(U)}$ is, by assumption, a neighbourhood in F. Therefore AK is nearly open hence open since E is a Pták space. But then A too is a homomorphism and the first assertion follows from 2.(2).

b) The proof of the second statement is analogous. If A is linearly continuous and nearly open, then the map $\hat{A}$ of $E/N[A]$ into F is also nearly open; since $E/N[A]$ is an infra-Pták space $\hat{A}$ is open, and hence A is also.

Using 2.(1) we conclude

(3) *Every quotient of a Pták space is complete.*
Every homomorphism A of a Pták space into a locally convex space F has a range $A(E)$ which is a Pták space in the topology induced by F.

A complete locally convex space with a noncomplete quotient space is therefore never a Pták space. We saw in § 31, 6. that the locally convex direct sum of countable many spaces c_0 is an example for this situation. Hence there exist strict (LB)-spaces which are not Pták spaces and an inductive limit of a sequence of Pták spaces is in general not a Pták space.

(4) *Every closed subspace H of an infra-Pták space resp. Pták space is again an infra-Pták space resp. Pták space.*

Proof. a) Let E be a Pták space, H a closed subspace, J the injection of H in E. Then J' is the canonical homomorphism K of E' onto $E'/H^\circ = H'$. Let L be a subspace of E'/H° such that all subsets $(U \cap H)^\circ \cap L$ are $\mathfrak{T}_s(H)$-closed, where U is an absolutely convex neighbourhood of $\circ$ in E and the polar $(U \cap H)^\circ$ is taken in H'. We have to show that L itself is $\mathfrak{T}_s(H)$-closed in E'/H°.

We first consider the subspace $K^{(-1)}(L)$ of E' and prove that $U^\circ \cap K^{(-1)}(L)$ is always weakly closed. We have $J(U \cap H) \subset U, J(U \cap H)^\circ \supset U^\circ$, hence $U^\circ \cap K^{(-1)}(L) = U^\circ \cap J(U \cap H)^\circ \cap K^{(-1)}(L)$. From § 32, 1.(9) it follows that $J(U \cap H)^\circ = K^{(-1)}((U \cap H)^\circ)$; therefore $J(U \cap H)^\circ \cap K^{(-1)}(L) = K^{(-1)}((U \cap H)^\circ \cap L)$. By assumption $(U \cap H)^\circ \cap L$ is

$\mathfrak{T}_s(H)$-closed and, since $\hat{\mathfrak{T}}_s(H) = \mathfrak{T}_s(E)$ on E'/H°, it follows that $K^{(-1)}((U \cap H)^\circ \cap L)$ is $\mathfrak{T}_s(E)$-closed in E'. Hence $U^\circ \cap K^{(-1)}(L)$ is always weakly closed and, since E is a Pták space, $K^{(-1)}(L)$ is weakly closed in E'.

Finally, $L = K(K^{(-1)}(L))$ is weakly closed in E'/H° since K is a weak homomorphism, so the complement of $K^{(-1)}(L)$ in E' is weakly open and L is the complement of the weakly open image of the complement.

b) The proof for infra-Pták spaces is contained in a) if one considers only subspaces L which are weakly dense in E'/H°. It follows then that $K^{(-1)}(L)$ is also weakly dense in E': The polar L° of L in H is $\mathfrak{o}$. Every $y \in K^{(-1)}(L)^\circ$ lies in H since $K^{(-1)}(L) \supset H^\circ$ and $K^{(-1)}(L)^\circ \subset H^{\circ\circ} = H$. From $uy = 0$ for all $u \in K^{(-1)}(L)$ therefore follows $\hat{u}y = 0$ for all $\hat{u} \in L$. Hence $y \in L^\circ = \mathfrak{o}$ and $K^{(-1)}(L)^\circ = \mathfrak{o}$.

We know that all (F)-spaces are Pták spaces. PTÁK [4] proved

(5) *Let E be an* (F)*-space. Then $E'[\mathfrak{T}]$ is a Pták space for every locally convex topology $\mathfrak{T}$ between $\mathfrak{T}_c(E)$ and $\mathfrak{T}_k(E)$. The strong dual of a reflexive* (F)*-space is a Pták space.*

We have $(E'[\mathfrak{T}])' = E$. It is sufficient to prove that every $\mathfrak{T}^f$-closed subspace H of E is closed. Let z be in $\overline{H}$; then z is the limit of a bounded sequence $x_n \in H$. The x_n are contained in an absolutely convex compact and therefore $\mathfrak{T}$-equicontinuous closed subset M of E. Since $H \cap M$ is closed by assumption, z is in H and therefore H is closed.

From (5) it follows that a Pták space need not be quasi-barrelled.

(6) *If $E[\mathfrak{T}]$ is a complete locally convex space and if $\mathfrak{T}^f$ and $\mathfrak{T}^{lf}$ coincide on E', then $E[\mathfrak{T}]$ is a Pták space.*

Proof. Under our assumptions $\mathfrak{T}^f$ coincides with $\mathfrak{T}^\circ$ because of § 21, 9.(7), so $\mathfrak{T}^f$ is a topology compatible with the dual pair $\langle E', E\rangle$, and therefore a $\mathfrak{T}^f$-closed subspace of E' is weakly closed.

On the other hand, it follows from (6) that if a complete $E[\mathfrak{T}]$ is not a Pták space, then the topologies $\mathfrak{T}^f$ and $\mathfrak{T}^{lf}$ are different on E', so that $\mathfrak{T}^f$ is not a locally convex topology.

Every ω_d is a Pták space since every linear subspace of $\omega_d' = \varphi_d$ is weakly closed. It follows from (5) that φ is a Pták space since it is the strong dual of the reflexive (F)-space ω. In connection with (6) it is of interest to note that it was proved in § 21, 8.(2) that on $\varphi_d = \omega_d'$, $d \geqq 2^{\aleph_0}$, the topologies $\mathfrak{T}^f$ and $\mathfrak{T}^{lf}$ are different.

φ_d is complete and every quotient space is complete, but φ_d for $d \geqq 2^{\aleph_0}$ is even not an infra-Pták space. By (4) it will be sufficient to prove this for φ_d, $d = 2^{\aleph_0}$. By § 9, 5.(5) ω has the algebraic dimension $2^{\aleph_0}$; it is therefore possible to find a one-one linear mapping A of φ_d onto ω. Since the topology on φ_d is the finest locally convex topology, A is continuous. But A is not a monomorphism so that by 2.(3) φ_d is not an infra-Pták space.

We remark that it follows from (6) that the topologies $\mathfrak{T}^f$ and $\mathfrak{T}^{lf}$ on $\varphi'_d = \omega_d$ are different.

We give now an example of a topological product of two Pták spaces which is not a Pták space. We observed in § 27, 2. that the spaces $\varphi\omega$ and $\omega\varphi$ are (M)-spaces. They are dual to each other; therefore the absolutely convex and compact subsets form a fundamental set of equicontinuous sets. As was shown in HAGEMANN [1] and KÖTHE [2], a subspace of $\varphi\omega$ or $\omega\varphi$ is closed if it is sequentially closed. The same argument as in the proof of (5) shows now that a $\mathfrak{T}^f$-closed subspace of $\varphi\omega$ or $\omega\varphi$ is closed. Therefore $\varphi\omega$ and $\omega\varphi$ are Pták spaces. The product $\varphi\omega \oplus \omega\varphi$ is not a Pták space since it has quotients which are not complete (cf. § 23, 5.).

4. A theorem of KELLEY. Let $\mathfrak{C}(E)$ be the class of all absolutely convex closed subsets of a locally convex space $E[\mathfrak{T}]$. We define a uniform structure $\mathfrak{U}$ on $\mathfrak{C}(E)$ in the following way. Let U be an absolutely convex neighbourhood of $\mathfrak{o}$ in E and denote by N_U the class of all pairs (A, B), where A and B are in $\mathfrak{C}(E)$ and such that $A \subset B + U$ and $B \subset A + U$. These N_U are the vicinities of a base of a uniformity $\mathfrak{U}$ on $\mathfrak{C}(E)$. $\mathfrak{U}$ is Hausdorff since from $A \subset B + U$ for all U follows $A \subset B$, and therefore $A = B$ by symmetry.

$E[\mathfrak{T}]$ is called hypercomplete if the uniform space $\mathfrak{C}(E)$ is complete. KELLEY [2'] proved that this is the case if and only if E' has the KREIN–SMULIAN property, i.e., every absolutely convex subset C of E' is weakly closed whenever all $C \cap U^\circ$ are weakly closed in E'. A hypercomplete space is therefore a Pták space; the problem of the existence of a Pták space which is not hypercomplete seems to be open. But there is a closely related characterization of Pták spaces.

We consider decreasing nets C_α, $\alpha \in \mathsf{A}$, in $\mathfrak{C}(E)$, so that $C_\alpha \supset C_\beta$ if $\alpha \leqq \beta$. Such a net is called scalar if with each C_α of the net, every ρC_α, $\rho > 0$, is also a member of the net. We say that $\mathfrak{C}(E)$ is scalarly complete if every decreasing scalar Cauchy net has a limit in $\mathfrak{C}(E)$.

If the decreasing net C_α, $\alpha \in \mathsf{A}$, has a limit C in $\mathfrak{C}(E)$ and if U is an absolutely convex neighbourhood of $\mathfrak{o}$ in E, then we have $C \subset C_\beta + U$ for all $\beta \geqq \beta_0(U)$. From this follows

$$C \subset \bigcap_U \bigcap_\alpha (C_\alpha + U) = \bigcap_\alpha \bigcap_U (C_\alpha + U) = \bigcap_\alpha C_\alpha.$$

On the other hand, $C_\alpha \subset C + U$ for sufficiently large α and hence $\bigcap_\alpha C_\alpha \subset C$. Hence if a decreasing net C_α has a limit C in $\mathfrak{C}(E)$, then

$C = \bigcap_\alpha C_\alpha$. In particular, if C_α is scalar, then $C = \bigcap_\alpha C_\alpha$ is a closed subspace of E.

We prove now the theorem of KELLEY:

(1) *A locally convex space* $E[\mathfrak{T}]$ *is a Pták space if and only if* $\mathfrak{C}(E)$ *is scalarly complete.*

$E[\mathfrak{T}]$ *is an infra-Pták space if and only if every decreasing scalar Cauchy net* C_α *such that* $\bigcap_\alpha C_\alpha = \mathfrak{o}$ *has the limit* $\mathfrak{o}$ *in* $\mathfrak{C}(E)$.

Proof. a) Let E be a Pták space and C_α, $\alpha \in \mathsf{A}$, be a decreasing scalar Cauchy net in $\mathfrak{C}(E)$ with $\bigcap_\alpha C_\alpha = C$. We take polars C_α° in E' and set $H = \bigcup_\alpha C_\alpha^\circ$. Since $(\rho C_\alpha)^\circ = (1/\rho) C_\alpha^\circ$, we see that H is a subspace of E' and $H^\circ = C$.

Let U be a closed absolutely convex neighbourhood of $\mathfrak{o}$ in E. Since C_α is a Cauchy net, there exists $\beta \in \mathsf{A}$ such that $C_\beta \subset C_\alpha + U$ for all $\alpha \geqq \beta$. Since C_α is decreasing, this is true for all $\alpha \in \mathsf{A}$; therefore $(C_\alpha + U)^\circ \subset C_\beta^\circ$ for all α. It follows that $\frac{1}{2}(C_\alpha^\circ \cap U^\circ) \subset (C_\alpha + U)^\circ \subset C_\beta^\circ \subset H$; hence $H \cap U^\circ \subset 2C_\beta^\circ \subset H$. Since C_β° is weakly closed, the weak closure of $H \cap U^\circ$ is contained in H and $H \cap U^\circ$ itself is weakly closed. Since E is a Pták space, H is weakly closed and $H = C^\circ$.

Finally, it follows from $H \cap U^\circ \subset 2C_\beta^\circ$ that $C_\beta \subset 2(H \cap U^\circ)^\circ = \overline{2(H^\circ + U)} \subset H^\circ + 3U = C + 3U$, and this means that C_α converges to C.

b) If E is an infra-Pták space, we have to consider only Cauchy nets C_α such that $\bigcap_\alpha C_\alpha = \mathfrak{o}$. Then $H = \bigcup_\alpha C_\alpha^\circ$ is weakly dense in E' and it follows as in a) that $H = E'$ and from $U^\circ = H \cap U^\circ \subset 2C_\beta^\circ$ it follows that $C_\beta \subset 2U$; this is the convergence of C_α to $\mathfrak{o}$.

c) We assume that every decreasing scalar Cauchy net converges in $\mathfrak{C}(E)$. Let H be a $\mathfrak{T}^f$-closed subspace of E'. We have to prove that H is weakly closed.

Let U be any absolutely convex closed neighbourhood of $\mathfrak{o}$ in E; then the sets $(U^\circ \cap H)^\circ$ form a decreasing scalar net in $\mathfrak{C}(E)$. Since $\bigcup_U (U^\circ \cap H) = H$, $\bigcap_U (U^\circ \cap H)^\circ = H^\circ$.

We prove first that this is a Cauchy net. Let V, W be two neighbourhoods of $\mathfrak{o}$; then using the fact that H is $\mathfrak{T}^f$-closed, we conclude that

$$(V^\circ \cap H)^\circ + 2W \supset \overline{(V^\circ \cap H)^\circ + W} \supset ((V^\circ \cap H)^{\circ\circ} \cap W^\circ)^\circ$$
$$= (V^\circ \cap H \cap W^\circ)^\circ \supset (W^\circ \cap H)^\circ.$$

For all $V, W \subset \frac{1}{2}U$ we therefore have $(W^\circ \cap H)^\circ \subset (V^\circ \cap H)^\circ + U$ and this is the Cauchy condition.

By assumption $(U^\circ \cap H)^\circ$ converges to H°; hence for any given U there exists V such that $(V^\circ \cap H)^\circ \subset H^\circ + U$. Therefore

$$\bar{H} \cap U^\circ = (H^\circ + U)^\circ \subset (V^\circ \cap H)^{\circ\circ} = V^\circ \cap H \subset H$$

and $\bigcup_U (\bar{H} \cap U^\circ) = \bar{H} \subset H$, showing H is weakly closed.

d) If we assume that every scalar Cauchy net C_α with $\bigcap_\alpha C_\alpha = \mathfrak{o}$ converges and if H is $\mathfrak{T}^f$-closed and weakly dense in E', then the net $(U^\circ \cap H)^\circ$ has the intersection $\mathfrak{o}$ and the proof in c) then shows that $H = E'$ so that E is an infra-Pták space.

(1) permits a simple proof of 3.(4). If H is a closed subspace of a Pták space, then every decreasing scalar Cauchy net C_α on H has the limit $\bigcap_\alpha C_\alpha$ in H; therefore H too is a Pták space.

We give also a new proof for the completeness of an infra-Pták space E based on (1). We assume that E is not complete and that z is an element of $\tilde{E} \sim E$. We consider the sets $[z] + U$, where $[z]$ is the line through $\mathfrak{o}$ and z and U any closed absolutely convex neighbourhood of $\mathfrak{o}$ in $\tilde{E}$.

We prove first that $[z] + U$ is closed in $\tilde{E}$. If $[z] \subset U$ this is trivial. Assume $\beta z \notin U$ for $|\beta| \geqq n_0$. By § 15, 6.(10) the sets $\{\alpha z + u; |\alpha| \leqq k, u \in U\}$ are closed. If t is in $\overline{[z] + U}$ but not in $[z] + U$, then $t + U$ must contain elements $\alpha z + u, u \in U$, $|\alpha|$ arbitrary large. But then $(\alpha_1 z + u_1) - (\alpha_2 z + u_2)$ would be in $2U$, and $(\alpha_1 - \alpha_2)z \in 4U$, which contradicts $\beta z \notin U$, $|\beta| \geqq n_0$, for suitably chosen α_1, α_2.

The sets $[z] + U$ obviously form a scalar Cauchy net in $\mathfrak{C}(\tilde{E})$ with limit $[z]$. We consider now the net consisting of the sets $([z] + U) \cap E$. It is a decreasing Cauchy net in $\mathfrak{C}(E)$ with intersection $\mathfrak{o}$ and our proof will be complete if we can show that $\mathfrak{o}$ is not a limit of this net. We choose U such that $\mathfrak{o} \notin z + U$ and V such that $(z + U) \cap V$ is empty. Let $W \subset U$; then there exists $y \in E$ such that $z - y \in W$, so $y \in (z + W) \cap E \subset z + U$. Therefore for no $W \subset U$ $([z] + W) \cap E$ is contained in $V \cap E = \mathfrak{o} + V \cap E$; hence $\mathfrak{o}$ cannot be the limit of the Cauchy net.

5. Closed linear mappings. As we saw in § 15, 12., the classical closed-graph theorem is a simple consequence of the BANACH–SCHAUDER theorem. So we can expect a generalization of the closed-graph theorem which is a counterpart to Theorem 2.(3). But it will not be a simple consequence of 2.(3).

We begin with a study of mappings with a closed graph. We note first that this notion generalizes that of a continuous mapping.

(1) *Every continuous mapping A of a Hausdorff topological space R_1 in a Hausdorff topological space R_2 has a closed graph.*

The graph $G(A)$ consists of all pairs $(x, Ax) \in R_1 \times R_2$, where $x \in R_1$. Assume $(y, z) \in \overline{G(A)}$ and $z \neq Ay$. There exist in R_2 neighbourhoods U of z and V of Ay such that $U \cap V$ is empty. Since A is continuous, there exists a neighbourhood W of y with $A(W) \subset V$. Then the neighbourhood $W \times U$ of (y, z) contains no point of $G(A)$ and this contradicts $(y, z) \in \overline{G(A)}$.

A mapping with a closed graph will be called for short a closed mapping in the following. This is unfortunately in contradiction with the use of the word defined in § 1, 7., but it will be always clear from the context in which sense the word is used.

A closed linear mapping A of E in F may be not weakly continuous, and then A' will not map every element of F' into an element of E'. If A is a linear mapping of E in F, then the set of all $v \in F'$ such that $A'v$ is an element of E' is a linear subspace of F' and is called the domain (of definition) $D[A']$ of A'. It is easy to determine $D[A']$:

(2) *Let E, F be locally convex, A a linear mapping of E into F. The domain $D[A']$ of A' in F' is the union $\bigcup_U A(U)^\circ$, where U is any absolutely convex neighbourhood of $\circ$ in E.*

Proof. If $v \in A(U)^\circ \subset F'$, then $\sup_{x \in U} |v(Ax)| = \sup |(A'v)x| \leqq 1$ and $A'v \in E'$; hence $A(U)^\circ \subset D[A']$.

Conversely, if $v \in D[A']$, $A'v \in E'$ and there exists an absolutely convex U such that $\sup_{x \in U} |(A'v)x| \leqq 1$. But this means $v \in A(U)^\circ$.

We can now establish the following useful characterization of closed linear mappings.

(3) *Let A be a linear mapping of the locally convex space $E[\mathfrak{T}_1]$ in the locally convex space $F[\mathfrak{T}_2]$. Then the following properties of A are equivalent:*

a) *A is closed;*

b) *$D[A']$ is weakly dense in F';*

c) *there exists a locally convex Hausdorff topology $\mathfrak{T}_2'$ on F which is coarser than $\mathfrak{T}_2$ and such that A is continuous from $E[\mathfrak{T}_1]$ in $F[\mathfrak{T}_2']$.*

Proof. i) b) follows from a). Let $G(A)$ be the closed graph of A. Let $z \neq \circ$ be an element of the polar $D[A']^\circ$ in F. Then $(\circ, z)$ is not an element of $G(A)$. Since $G(A)$ is a closed linear subspace of $E \times F$, it follows from the HAHN–BANACH theorem that there exists $(u, v) \in E' \times F'$ such that

$$\langle (u, v), (x, Ax) \rangle = ux + v(Ax) = 0 \quad \text{for all } x \in E \tag{4}$$

and

$$\langle (u, v), (\circ, z) \rangle = vz = 1. \tag{5}$$

From (4) follows $A'v = -u$; hence $v \in D[A']$. But this contradicts (5) because $z \in D[A']^\circ$. Therefore $D[A']^\circ = \mathfrak{o}$.

ii) c) follows from b). We assume that $H = D[A']$ is weakly dense in F'. If we equip F with the Hausdorff topology $\mathfrak{T}_s(H)$, then A is continuous from $E[\mathfrak{T}_s(E')]$ in $F[\mathfrak{T}_s(H)]$ by § 32, 1.(3) and therefore continuous from $E[\mathfrak{T}_1]$ in $F[\mathfrak{T}_s(H)]$.

iii) a) follows from c): Let A be a continuous linear mapping of $E[\mathfrak{T}_1]$ in $F[\mathfrak{T}_2']$, $\mathfrak{T}_2'$ coarser than $\mathfrak{T}_2$ but still Hausdorff. Then $G(A)$ is $\mathfrak{T}_1 \times \mathfrak{T}_2'$-closed, hence $\mathfrak{T}_1 \times \mathfrak{T}_2$-closed.

We note some useful corollaries.

(6) *Let A be a closed linear mapping of the locally convex space $E[\mathfrak{T}_1]$ in the locally convex space $F[\mathfrak{T}_2]$. If we replace $\mathfrak{T}_2$ by a stronger locally convex topology $\mathfrak{T}_2$, then A remains closed.*

(7) *The kernel $N[A]$ of a closed linear mapping A is closed. The linear mapping A is closed if and only if $\hat{A}$ is closed.*

Proof. The first statement follows from (3) c), since $N[A]$ is the kernel of a continuous mapping.

If A is continuous as a mapping of $E[\mathfrak{T}_1]$ in $F[\mathfrak{T}_2']$, then $\hat{A}$ is continuous from $(E/N[A])[\hat{\mathfrak{T}}_1]$ in $F[\mathfrak{T}_2']$; hence $\hat{A}$ is closed as a mapping of $(E/N[A])[\mathfrak{T}_1]$ in $F[\mathfrak{T}_2]$. The converse follows in a similar way.

(8) *Let A be a continuous linear mapping of a dense subspace H of the locally convex space E in the complete locally convex space F. If A has a closed graph in $E \times F$, then $H = E$.*

Proof. Let x_0 be an adherent point of H in E and $x_\alpha \in H$ a net converging to x_0. Then Ax_α is a Cauchy net in F and has a limit y_0 since F is complete. Hence (x_α, Ax_α) has the limit (x_0, y_0) and is in $G(A)$. Therefore $x_0 \in H$ and $H = E$.

(9) *Let $\mathfrak{T}_1$ and $\mathfrak{T}_2$ be two locally convex topologies on E. Then the identity mapping I of $E[\mathfrak{T}_1]$ onto $E[\mathfrak{T}_2]$ is closed if and only if there exists on E a locally convex topology $\mathfrak{T}$ coarser than $\mathfrak{T}_1$ and $\mathfrak{T}_2$.*

I is closed by (3) c) if and only if there exists a locally convex topology $\mathfrak{T} \subset \mathfrak{T}_2$ on E such that I is continuous. But this means $\mathfrak{T} \subset \mathfrak{T}_1$ and $\mathfrak{T} \subset \mathfrak{T}_2$.

The following example shows that even under rather strong assumptions the closed-graph theorem will not be true. We constructed in 3. a one-one continuous and nearly open mapping A of φ_d, $d = 2^{\aleph_0}$, onto ω which is not a monomorphism. Its inverse has a closed graph since A has a closed graph, but $A^{(-1)}$ is not continuous. Both spaces have very nice properties: ω is a reflexive (F)-space; φ_d is complete, reflexive, barrelled, and bornological. $A^{(-1)}$ is even nearly continuous in the sense of the next section.

On the other hand, the following special case is a simple consequence of (3).

(10) *Every closed linear mapping A of a locally convex space into ω_d, d any cardinal, is continuous.*

The proof uses the fact that the topology $\mathfrak{T} = \mathfrak{T}_s(\varphi_d)$ on $E = \omega_d$ is minimal in the sense that there exists no strictly coarser locally convex topology on E: From $\mathfrak{T}_1 \subset \mathfrak{T}$ it follows that $E[\mathfrak{T}_1]'$ is weakly dense in $E[\mathfrak{T}]' = \varphi_d$. But in φ_d every linear subspace is weakly closed; hence $E[\mathfrak{T}_1]' = \varphi_d$ and $\mathfrak{T}_1 \supset \mathfrak{T}_s(\varphi_d)$.

Assume now A closed. Then by (3) there exists a locally convex topology $\mathfrak{T}_1 \subset \mathfrak{T}$ on ω_d such that A is continuous into $\omega_d[\mathfrak{T}_1]$. But $\mathfrak{T}_1 = \mathfrak{T}$ and A is continuous.

6. Nearly continuous mappings and the closed-graph theorem. Let E and F be locally convex. A linear mapping A of E in F is called nearly continuous if the closure $\overline{A^{(-1)}(V)}$ of the inverse image of every neighbourhood V of $\mathfrak{o}$ in F is a neighbourhood of $\mathfrak{o}$ in E.

This notion is closely connected with the notion "nearly open": If A is a one-one linear mapping of E onto F, then A is nearly continuous if and only if $A^{(-1)}$ is nearly open.

The following proposition is a counterpart to 1.(1).

(1) *The barrelled spaces E are characterized by the following property: Every linear mapping A of E into an arbitrary locally convex space F is nearly continuous.*

Proof. a) Let E be barrelled, V an absolutely convex neighbourhood of $\mathfrak{o}$ in F. Then $A^{(-1)}(V)$ is absolutely convex and absorbent, hence $\overline{A^{(-1)}(V)}$ is a barrel and a neighbourhood of $\mathfrak{o}$ in E.

b) The second half follows from the second half of the proof of 1.(1) by considering $I^{(-1)}$ instead of I.

We will need the following generalizations of § 32, 1.(9) and (10):

(2) *Let A be a linear mapping of the locally convex space E in the locally convex space F and let $H = D[A']$.*

Then for every absolutely convex neighbourhood U of $\mathfrak{o}$ in E

$$A(U)^\circ = A'^{(-1)}(U^\circ) \tag{3}$$

and for every absolutely convex neighbourhood V of $\mathfrak{o}$ in F

(4) $$A^{(-1)}(V)^\circ = A'(H \cap V^\circ).$$

The proof of (3) is almost trivial: $v \in A(U)^\circ$ means $|v(Ax)| \leqq 1$ or $|(A'v)x| \leqq 1$ for all $x \in U$ and this means $A'v \in U^\circ$ or $v \in A'^{(-1)}(U^\circ)$. We remark that $A(U)^\circ \subset H$ by 5.(2).

We prove now (4). Let $v \in H \cap V^\circ$ and $u = A'v$. For each $y \in V \cap A(E)$ and $x \in A^{(-1)}y$ we have

$$|vy| = |(A'v)x| = |ux| \leqq 1.$$

Hence $A'(H \cap V^\circ) \subset A^{(-1)}(V)^\circ$.

Conversely, let $u \in A^{(-1)}(V)^\circ$. This means $|ux| \leqq 1$ for all $x \in A^{(-1)}y$, where y runs through $V \cap A(E)$. If we put $vy = ux$, $x \in A^{(-1)}y$, then v is uniquely defined on $A(E)$ since u is zero on $A^{(-1)}(\mathfrak{o})$. By the Hahn–Banach theorem v has an extension onto F such that $|vy| \leqq 1$ for all $y \in V$. It follows from $v(Ax) = vy = ux$ for all $x \in E$ that $A'v = u$; hence $v \in H \cap V^\circ$, $u \in A'(H \cap V^\circ)$, and $A^{(-1)}(V)^\circ \subset A'(H \cap V^\circ)$.

We give now the dual characterization of nearly continuous maps.

(5) *Let E and F be locally convex, $\mathfrak{M}_1$ resp. $\mathfrak{M}_2$ the class of equicontinuous sets in E' resp. F'. A linear mapping A of E in F is nearly continuous if and only if $A'(H \cap \mathfrak{M}_2) \subset \mathfrak{M}_1$, where $H = D[A']$.*

That A is nearly continuous means that for every absolutely convex neighbourhood V of $\mathfrak{o}$ in F there exists a closed absolutely convex neighbourhood U of $\mathfrak{o}$ in E such that $\overline{A^{(-1)}(V)} \supset U$. This is equivalent to $A^{(-1)}(V)^\circ \subset U^\circ$ and this by (3) to $A'(H \cap V^\circ) \subset U^\circ$.

(6) *A nearly continuous mapping A is continuous if and only if $D[A'] = F'$.*

This is an immediate consequence of (5) and § 32, 2.(1).

(7) *If A is nearly continuous from $E[\mathfrak{T}_1]$ in $F[\mathfrak{T}_2]$, then $H = D[A']$ is $\mathfrak{T}_2^f$-closed in F'.*

We have to prove that $H \cap V^\circ$ is weakly closed in F' for every absolutely convex neighbourhood V of $\mathfrak{o}$ in F. By (5) we have $A'(H \cap V^\circ) \subset U^\circ$ for some absolutely convex closed neighbourhood U of $\mathfrak{o}$ in E. Hence $H \cap V^\circ \subset A'^{(-1)}A'(H \cap V^\circ) \subset A'^{(-1)}(U^\circ) = A(U)^\circ$ by (3). Now $A(U)^\circ \subset H$ by 5.(2); hence $H \cap V^\circ = A(U)^\circ \cap V^\circ$ and $H \cap V^\circ$ is weakly closed in F'.

We are now able to prove Pták's counterpart to 2.(2) b):

(8) *A locally convex space F is an infra-Pták space if and only if every linear nearly continuous and closed mapping A of a locally convex space E into F is continuous.*

Proof. a) Suppose that F is an infra-Pták space. If A satisfies the conditions, then $D[A']$ is $\mathfrak{T}_2^f$-closed in F' by (7) and weakly dense in F' by 5.(3). But then $D[A'] = F'$ and A is continuous by (6).

b) We assume that F' contains a $\mathfrak{T}_2^f$-closed weakly dense subspace $H \neq F'$. If $\mathfrak{T}'$ is the topology of uniform convergence on the sets $H \cap U^\circ$, U a $\mathfrak{T}_2$-neighbourhood of $\mathfrak{o}$ in F, then by 2.(2) b) the identity mapping I of $F[\mathfrak{T}_2]$ onto $F[\mathfrak{T}']$ is nearly open but not open. The inverse mapping $I^{(-1)}$ is nearly continuous but not continuous and $I^{(-1)}$ has a closed graph since I is continuous.

From (1) and (8) follows now the closed-graph theorem corresponding to 2.(3):

(9) *Every closed linear mapping of a barrelled space E in an infra-Pták space F is continuous.*

Remark. The topology of F may be replaced by any weaker locally convex topology since the graph remains closed also for the original topology on F.

7. Some consequences, the HELLINGER–TOEPLITZ theorem. Before giving some applications of the closed-graph theorem we will prove that in 6.(9) the class of barrelled spaces E cannot be replaced by a larger class (cf. MAHOWALD [1']).

(1) *If every closed linear mapping of the locally convex space $E[\mathfrak{T}]$ into an arbitrary* (B)*-space is continuous, then $E[\mathfrak{T}]$ is barrelled.*

First proof. i) Let T be a barrel in E; then its Minkowski functional $q(x)$ is a seminorm on E by § 16, 4.(6). The kernel $N_T = \bigcap_{\lambda > 0} \lambda T$ of q is closed in E. We denote by E_T the quotient E/N_T with norm $q(\hat{x}) = q(x)$ and by $\tilde{E}_T$ the completion of E_T; $\tilde{E}_T$ is a (B)-space.

ii) *The canonical mapping K of E onto E_T is a closed linear mapping of E into $\tilde{E}_T$.*

K is a homomorphism of E onto $(E/N_T)[\hat{\mathfrak{T}}]$. Since $T \supset N_T$ is closed in $E[\mathfrak{T}]$, $E \sim T$ is open and hence $\hat{T} = K(T)$ is $\hat{\mathfrak{T}}$-closed since it is the complement of $K(E \sim T)$ and thus $\hat{T}$ is a barrel in $(E/N_T)[\hat{\mathfrak{T}}]$. Using 5.(7) we see that we have only to consider the case $N_T = \{\mathfrak{o}\}$.

Then K is the identity map of $E[\mathfrak{T}]$ onto $E[\mathfrak{T}']$, $\mathfrak{T}'$ the norm topology defined by T as closed unit ball. Let $\mathfrak{U}$ be a $\mathfrak{o}$-neighbourhood base of $E[\mathfrak{T}]$ consisting of absolutely convex U; then $\{U + (1/n)T;\ U \in \mathfrak{U},\ n \in N\}$ defines the $\mathfrak{o}$-neighbourhood base of a locally convex topology $\mathfrak{T}_0$ on E which is Hausdorff since $\bigcap_{U,n} (U + (1/n)T) = \bigcap_n (1/n)T = \{\mathfrak{o}\}$. Clearly, $\mathfrak{T}_0 \subset \mathfrak{T}$ and $\mathfrak{T}_0 \subset \mathfrak{T}'$.

It follows from $\bigcap_{U,n} (T + U + (1/n)T) = \bigcap_n [(n+1)/n]T = T$ that T is $\mathfrak{T}_0$-closed and § 18, 4.(4) c) implies that the completion $\tilde{E}_T = \tilde{E}[\tilde{\mathfrak{T}}']$ is continuously embedded in the completion $\tilde{E}[\tilde{\mathfrak{T}}_0]$. Hence K is continuous as map of $E[\mathfrak{T}]$ in $\tilde{E}_T$ equipped with $\tilde{\mathfrak{T}}_0$ and K is closed as map of $E[\mathfrak{T}]$ in $\tilde{E}_T$ equipped with $\tilde{\mathfrak{T}}' \supset \tilde{\mathfrak{T}}_0$ by 5.(6).

iii) By ii) and our assumption the mapping K of E into $\tilde{E}_T$ is continuous and the inverse image T of $\hat{T} = K(T)$, the closed unit ball in E_T, is a neighbourhood of $\circ$ in E, $E[\mathfrak{T}]$ is barrelled.

Second proof (WILANSKY [1']). We begin with the remark that the space $C_b(R)$ of all bounded continuous functions f on a Hausdorff topological space R is a (B)-space for the norm $\|f\| = \sup_{t \in R} |f(t)|$. The proof given in § 14, 9.(1) for compact R is valid in the general case too.

Let now T be a barrel in E, T° its polar in E' equipped with $\mathfrak{T}_s(E)$, and let F be $C_b(T^\circ)$. For any $x \in E$ we define f_x by $f_x(x') = \langle x, x' \rangle$. Since T is absorbent one has $f_x \in F$ and $A: x \mapsto f_x$ is a linear map of E in F.

The topology $\mathfrak{T}_p$ of pointwise convergence on T° (§ 24, 5.) is weaker than the norm topology on F and A is continuous as a map of E in $F[\mathfrak{T}_p]$ since A is obviously continuous from $E[\mathfrak{T}_s(E')]$ in $F[\mathfrak{T}_p]$. It follows from 5.(3) c) that A is closed as a map of E in the (B)-space F; hence A is continuous by assumption. Therefore the inverse image of the unit ball in $C_b(T^\circ)$ is a neighbourhood of $\circ$ in E. This means that

$$\left\{x \in E, \|f_x\| = \sup_{x' \in T^\circ} |\langle x, x' \rangle| \leqq 1\right\} = T^{\circ\circ} = T$$

is a neighbourhood in E.

We remark that we will show in 9. that the class of infra-Pták spaces of 6.(9) is not the largest class for which the closed-graph theorem remains valid.

We now apply 6.(9) to a problem on complementary subspaces. We prove first the following proposition:

(2) *Let the locally convex space E be the algebraic direct sum $E = H_1 \oplus H_2$ of two closed subspaces. Then the projection P of E onto H_1 with kernel H_2 is a closed linear mapping of E into itself.*

Proof. $G(P)$ consists of all the elements $(x_1 + x_2, x_1)$, $x_1 \in H_1$, $x_2 \in H_2$. $(u_1, u_2) \in E' \times E'$ is in $G(P)^\circ$ if $u_1x_1 + u_1x_2 + u_2x_1 = 0$ for all $x_1 \in H_1$, $x_2 \in H_2$. For $x_1 = \circ$ we obtain $u_1 \in H_2^\circ$; hence $u_1x_1 + u_2x_1 = 0$ or $u_1 + u_2 \in H_1^\circ$ or $u_2 = -u_1 + v$, $v \in H_1^\circ$. Therefore $G(P)^\circ$ consists of all elements $(u_1, -u_1 + v)$, $u_1 \in H_2^\circ$, $v \in H_1^\circ$. Similarly, $G(P)^{\circ\circ} = G(P)$ since H_1 and H_2 are closed.

From (2) and 6.(9) follows immediately

(3) *Two algebraically complementary closed linear subspaces H_1 and H_2 of a barrelled infra-Pták space are topologically complementary.*

This includes the case of (F)-spaces, for which the result was proved in § 15, 12.(6).

The BANACH–SCHAUDER theorem 2.(3), even in a more general version, is also an easy consequence of 6.(9):

(4) *Every closed linear mapping A of a Pták space E onto a barrelled space F is open.*

Proof. By 5.(7) the mapping $\hat{A}$ of $E/N[A]$ onto F is closed. The inverse mapping $\hat{A}^{(-1)}$ of F onto the Pták space $E/N[A]$ is also closed and is continuous by 6.(9). Therefore $\hat{A}$ and $A = K\hat{A}$ are open.

Contrary to the classical case, the open-mapping theorem is here a corollary to the closed-graph theorem, but the converse is not true. We will find that this relation between the two theorems is true also in other cases.

Another consequence of the closed-graph theorem is a generalization of the HELLINGER–TOEPLITZ theorem.

(5) *Let E be barrelled, F infra-Pták, and A a linear mapping of E into F such that A' is defined on a total subset M of F'. Then A is continuous.*

Proof. $D[A']$ contains the linear hull of M; hence $D[A']$ is weakly dense in F'. Thus A is closed by 5.(3) and continuous by 6.(9).

HELLINGER and TOEPLITZ [1] proved the following: Let $\mathfrak{A} = (a_{ik})$ be an infinite matrix such that for every $\mathfrak{x} \in l^2$ the vector $\mathfrak{A}\mathfrak{x} = \left(\sum_{k=1}^{\infty} a_{1k}x_k, \sum_{k=1}^{\infty} a_{2k}x_k, \ldots\right)$ exists and is again in l^2. Then $\mathfrak{A}$ is the matrix representing a continuous endomorphism of l^2. This is a special case of (5), where $E = F = l^2$ and M is the set of the unit vectors $\mathfrak{e}_p$ since $\mathfrak{A}'\mathfrak{e}_p$ is the pth row $\mathfrak{a}^{(p)}$ of $\mathfrak{A}$, which lies in l^2 because $\sum_{k=1}^{\infty} a_{pk}x_k < \infty$ for all $\mathfrak{x} \in l^2$.

This classical version of the HELLINGER–TOEPLITZ theorem was extended to all normal sequence spaces in KÖTHE–TOEPLITZ [2] and leads to a similar result which is not included in (5).

Let λ, μ be two normal sequence spaces containing φ and $\lambda^\times$, $\mu^\times$ their α-duals (§ 30, 1.). Let A be a linear mapping of λ in μ continuous in the sense of $\mathfrak{T}_s(\lambda^\times)$ and $\mathfrak{T}_s(\mu^\times)$. Denote $A\mathfrak{e}_k$ by $\mathfrak{a}_k$ and let $\mathfrak{A}$ be the matrix with columns $\mathfrak{a}_k = (a_{ik})$. Obviously, $A\mathfrak{e}_k = \mathfrak{A}\mathfrak{e}_k = \mathfrak{a}_k$ if we write the vectors $\mathfrak{e}_k$, $\mathfrak{a}_k$ as columns. Every element $\mathfrak{x} \in \lambda$ is the weak limit of its sections $\mathfrak{x}_n = \sum_1^n x_k\mathfrak{e}_k$; hence A is the weak limit of the sequence $\mathfrak{A}\mathfrak{x}_n = \sum_1^n x_k\mathfrak{a}_k$. By

§ 30, 5.(1) $\mathfrak{A}\mathfrak{x}_n$ is coordinatewise convergent to $A\mathfrak{x} \in \mu$; the ith coordinate of $A\mathfrak{x}$ is equal to $\sum_{k=1}^{\infty} a_{ik}x_k$. Therefore $A\mathfrak{x}$ can be calculated as the product $\mathfrak{A}\mathfrak{x}$ of the matrix $\mathfrak{A} = (a_{ik})$ with the column $\mathfrak{x}$. Since λ is normal, it follows that $\sum_k |a_{ik}x_k| < \infty$ for every i.

(6) *Every weakly continuous linear mapping of λ into μ is represented by a uniquely determined infinite matrix $\mathfrak{A}$.*

The generalization of the HELLINGER–TOEPLITZ theorem to sequence spaces is the following converse to (6).

(7) *Let λ, μ be normal sequence spaces containing φ. Every infinite matrix $\mathfrak{A} = (a_{ik})$ such that $\mathfrak{A}\mathfrak{x} \in \mu$ for every $\mathfrak{x} \in \lambda$ defines a weakly and $\mathfrak{T}_k$-continuous mapping of λ in μ.*

Proof. By assumption $\sum_{k=1}^{\infty} a_{nk}x_k < \infty$ for every $\mathfrak{x} \in \lambda$; therefore the nth row $\mathfrak{a}^{(n)}$ of $\mathfrak{A}$ is an element of $\lambda^\times$ since λ is normal. Let $\mathfrak{v} = (v_1, v_2, \ldots)$ be an element of $\mu^\times$ and $\mathfrak{v}_n$ the nth section of $\mathfrak{v}$. Then $\mathfrak{u}^{(n)} = v_1\mathfrak{a}^{(1)} + \cdots + v_n\mathfrak{a}^{(n)}$ is in $\lambda^\times$ and $\mathfrak{v}(\mathfrak{A}\mathfrak{x}) = \lim_{n\to\infty} \mathfrak{v}_n(\mathfrak{A}\mathfrak{x}) = \lim \mathfrak{u}^{(n)}\mathfrak{x}$. The sequence $\mathfrak{u}^{(n)}$ is therefore weakly Cauchy in $\lambda^\times$ and has a limit $\mathfrak{u}$ by § 30, 5.(3). Since $\mathfrak{u}\mathfrak{x} = \mathfrak{v}(\mathfrak{A}\mathfrak{x})$ for all $\mathfrak{x} \in \lambda$, it follows that the adjoint to $\mathfrak{A}$ maps $\mu^\times$ in $\lambda^\times$. But then $\mathfrak{A}$ is weakly continuous and hence continuous in the sense of $\mathfrak{T}_k(\lambda^\times)$ and $\mathfrak{T}_k(\mu^\times)$.

8. The theorems of A. and W. ROBERTSON. The generalizations 2.(3) and 6.(9) of the BANACH–SCHAUDER theorem and the closed-graph theorem are rather asymmetric in their assumptions since, as we have seen, the class of barrelled spaces is large and the class of Pták spaces small. Therefore the usefulness of these theorems in analysis is rather limited. So the question arises whether it is possible to replace the barrelled spaces by a smaller class and the Pták spaces by a larger class.

The first theorems in this direction were given by A. and W. ROBERTSON [1′] as an application of PTÁK's theory. They proved the following closed-graph theorem:

(1) *Let E be a locally convex hull $\sum_\alpha A_\alpha(E_\alpha)$ of Baire spaces E_α, F a locally convex hull $\bigcup_{n=1}^{\infty} B_n(F_n)$ of a sequence of Pták spaces F_n. Then every closed linear mapping A of E in F is continuous.*

Remark. The hull topology on F may be replaced by any weaker locally convex topology since the graph of A remains closed in the hull topology.

Proof. i) B_n is continuous from F_n in F by definition of the hull topology. Hence the kernel $N[B_n]$ is closed in F_n and $F_n/N[B_n]$ is again a Pták space. We may therefore assume that B_n is one-one.

ii) We consider first the case that E itself is a Baire space. It then follows from $E = A^{(-1)}(F) = \bigcup_{n=1}^{\infty} A^{(-1)}(B_n(F_n))$ that one of the sets $A^{(-1)}(B_n(F_n)) = H$ is not meagre in E and therefore $\overline{H}$ coincides with E.

Let A_0 be the restriction of A to H. Then $C_n = B_n^{(-1)}A_0$ is a linear mapping of H in F_n. We define a continuous injection J of $E \times F_n$ into $E \times F$ by $J(x, y) = (x, B_n y)$. Because $G(A)$ is assumed to be closed, therefore, $J^{(-1)}(G(A)) = G(C_n)$ is also closed in $E \times F_n$. Since H is barrelled by 1.(2) and since F_n is a Pták space, it follows from 6.(9) that C_n is continuous from H into F_n. But then $H = E$ by 5.(8) and $C_n \in \mathfrak{L}(E, F_n)$, $A_0 = A$. Therefore as the product of two continuous mappings $A = B_n C_n$ is continuous.

iii) Consider now the general case $E = \sum_{\alpha} A_\alpha(E_\alpha)$. If J_α is the continuous mapping $J_\alpha(x, y) = (A_\alpha x, y)$ of $E_\alpha \times F$ into $E \times F$, then $J_\alpha^{(-1)}(G(A)) = G(AA_\alpha)$ and $G(AA_\alpha)$ is closed in $E_\alpha \times F$. By ii) every AA_α is continuous and hence A is also, by § 19, 1.(7).

We make the following remark:

(2) *Let E be a locally convex hull $\sum_{\alpha} A_\alpha(E_\alpha)$ and H a closed subspace of E. Then E/H is the locally convex hull $\sum_{\alpha} KA_\alpha(E_\alpha)$, where K is the canonical mapping of E onto E/H.*

We have $E/H = K(E) = \sum_{\alpha} KA_\alpha(E_\alpha)$. Since $K(\Gamma_\alpha A_\alpha(U_\alpha)) = \Gamma_\alpha KA_\alpha(U_\alpha)$, the quotient topology of E/H coincides with the hull topology.

The second theorem is the corresponding BANACH–SCHAUDER theorem.

(3) *Let E be the locally convex hull $\bigcup_{n=1}^{\infty} B_n(E_n)$ of a sequence of Pták spaces E_n, F the locally convex hull $\sum_{\alpha} A_\alpha(F_\alpha)$ of Baire spaces F_α.*

Then every closed resp. continuous linear mapping A of E onto F is open resp. a homomorphism.

Remark. The hull topology on E may be replaced by any weaker locally convex topology as in (1).

Proof. By (2) we may assume that A is one-one. Since $G(A)$ is closed, the graph $G(A^{(-1)})$ of the inverse mapping is closed too. It follows from (1) that $A^{(-1)}$ is continuous and A is open.

All the spaces $\sum_{\alpha} A_\alpha(E_\alpha)$, E_α Baire spaces, are barrelled; they constitute a subclass of the class of all barrelled spaces. On the other hand, the class of spaces $\bigcup_{n-1}^{\infty} B_n(F_n)$, F_n Pták spaces, contains all Pták spaces but also spaces which are not Pták spaces, as we have seen in 3.

We know that every (F)-space is a Baire space. There exist normed spaces which are Baire spaces but are not (B)-spaces. For an example compare BOURBAKI [6], Vol. II, p. 3, Ex. 6.

That the class of Baire spaces is rather large is shown by the fact that *a topological product* $E = \prod_{\alpha \in A} E_\alpha$ *of* (F)-*spaces* E_α *is a Baire space.* We indicate the proof: If M_i is a sequence of nowhere dense subsets of E, then there exists a sequence $x^{(k)} \in E$ and closed absolutely convex neighbourhoods $U^{(k)}$ of $\circ$ in E such that $x^{(k+1)} + U^{(k+1)} \subset x^{(k)} + U^{(k)}$ and $(x^{(k)} + U^{(k)}) \cap M_k = \varnothing$. Let $U_n^{(\alpha)}$, $n = 1, 2, \ldots$, be a fundamental sequence of absolutely convex and closed neighbourhoods of $\circ$ in E_α. We can construct the neighbourhoods $U^{(k)}$ in such a way that for a denumerable subset $B = \{\beta_1, \beta_2, \ldots\} \subset A$ we have

$$U^{(k)} \subset \prod_{i=1}^{n_k} U_k^{(\beta_i)} \prod_{j=n_k+1}^{\infty} E_{\beta_j} \prod_{\alpha \notin B} E_\alpha .$$

But then the projections $x_B^{(k)}$ of $x^{(k)}$ onto $\prod_{i=1}^{\infty} E_{\beta_i}$ form a Cauchy sequence with a limit $x_B^{(\circ)}$. If z is any element of $\prod_{\alpha \notin B} E_\alpha$, then $x^{(\circ)} = (x_B^{(\circ)}, z)$ is contained in all $x^{(k)} + U^{(k)}$ and is therefore not in $\bigcup_{i=1}^{\infty} M_i$; hence E is not meagre.

We note the following important special cases of (1) and (3). Every (LF)-space satisfies the assumption for E and for F in (1) and (3). Therefore

(4) a) *Every closed linear mapping of an* (LF)-*space into an* (LF)-*space is continuous.*

b) *Every linear continuous mapping of an* (LF)-*space onto an* (LF)-*space is a homomorphism.*

We remark that (4) b) was first proved for strict (LF)-spaces by DIEUDONNÉ and SCHWARTZ [1] and for (LF)-spaces by KÖTHE [9].

Similar to 7.(3) we have

(5) *Two algebraically complementary closed linear subspaces of an* (LF)-*space are topologically complementary.*

A locally convex space E is called ultrabornological if it can be represented as the locally convex hull $\sum_\alpha A_\alpha(E_\alpha)$ of (B)-spaces E_α. E always has a representation of the simpler form $E = \sum_\alpha F_\alpha$, where the F_α are again (B)-spaces. To see this one replaces $A_\alpha(E_\alpha)$ by $\hat{A}_\alpha(E_\alpha/N[A_\alpha])$; then $F_\alpha = E_\alpha/N[A_\alpha]$ is again a (B)-space and $\hat{A}_\alpha$ is the injection I_α of F_α in E. By

§ 28, 4.(1) every ultrabornological space is bornological. Conversely, every sequentially complete bornological space is ultrabornological (§ 28, 2.(2)).

The following theorem is included in (1) and (3):

(6) *Every closed linear mapping of an ultrabornological space into an* (LF)-*space is continuous.*

Every continuous linear mapping of an (LF)-*space onto an ultrabornological space is a homomorphism.*

Remark. This is Theorem B of GROTHENDIECK ([13], p. 17) except that in Theorem B the graph of the mapping is supposed to be only sequentially closed. We will come back to this question in § 35.

GROTHENDIECK constructed in [10] an example of a continuous linear mapping A of a reflexive (LF)-space onto a closed subspace of a reflexive (LF)-space such that A is not a homomorphism.

9. The closed-graph theorem of KŌMURA. There is a more direct approach to the closed-graph theorem for barrelled spaces by KŌMURA [1], which leads to its sharpest possible form, which was given by VALDIVIA UREÑA [1']. ADASCH developed these ideas further in his papers [2']–[5']. We give here a short exposition of some of their results.

Let E be locally convex. Then the strong topology $\mathfrak{T}_b(E')$ on E is in general not compatible with the dual pair $\langle E', E\rangle$. We give a necessary and sufficient condition for the compatibility.

We denote by $\bar{H}$ the $\mathfrak{T}_s(E)$-quasi-closure of a subspace H of E' in the algebraic dual E^*. $\bar{H}$ is identical with the weak quasi-completion of H since E^* is weakly complete (§ 23, 1.).

(1) *$\mathfrak{T}_b(E')$ is compatible with $\langle E', E\rangle$ if and only if E' is quasi-closed in $E^*[\mathfrak{T}_s(E)]$.*

This follows immediately from § 23, 1.(3) and § 23, 6.(4).

$E[\mathfrak{T}_b(\bar{E}')]$ is a barrelled space with dual $\bar{E}'$ by (1). We call it the barrelled space associated to $E[\mathfrak{T}]$ and denote $\mathfrak{T}_b(\bar{E}')$ by $\mathfrak{T}^t$. We remark that $\mathfrak{T}^t$ depends only on $\langle E', E\rangle$, not on $\mathfrak{T}$.

It is obvious from (1) that $\mathfrak{T}^t$ is the coarsest barrelled topology on E which is finer than $\mathfrak{T}$.

The topology $\mathfrak{T}^t$ can also be constructed in the following way. We define the topology $\mathfrak{T}_\alpha$ for every ordinal α: $\mathfrak{T}_1 = \mathfrak{T}$, $\mathfrak{T}_{\alpha+1}$ is the strong topology on $E[\mathfrak{T}_\alpha]$, and for a limit ordinal β $\mathfrak{T}_\beta$ is the union of all $\mathfrak{T}_\alpha$, $\alpha < \beta$. It follows by transfinite induction (§ 23, 2.(1)) that $E[\mathfrak{T}_\alpha]'$ is contained in $\bar{E}'$ for every α; hence every $\mathfrak{T}_\alpha$ is coarser than $\mathfrak{T}^t$. There exists a first ordinal γ such that $\mathfrak{T}_{\gamma+1} = \mathfrak{T}_\gamma$ and $E[\mathfrak{T}_\gamma]$ is then barrelled; therefore $\mathfrak{T}_\gamma = \mathfrak{T}^t$.

(2) *If A is a continuous linear mapping of a barrelled space E in a locally convex space $F[\mathfrak{T}]$, then it remains continuous if we replace $\mathfrak{T}$ by $\mathfrak{T}^t$.*

Proof. If T is a barrel in F, then $A^{(-1)}(T)$ is a barrel in E since A is continuous; A is therefore continuous from E in $F[\mathfrak{T}_b(F')] = F[\mathfrak{T}_2]$. Using the above construction of $\mathfrak{T}^t$ and transfinite induction the statement becomes obvious. (2) can be proved also in the following way: Use § 32, 2.(1), then extend A' to $\bar{F}'$ by § 23, 1.(4); then the adjoint of this mapping has the desired property, again by § 32, 2.(1).

A locally convex space E is called an (s)-space if $\bar{H} \cap E' = \bar{H}$ (the weak closure taken in E') for every subspace H of E'; E is called an infra-(s)-space if $\bar{H} \cap E' = E'$ for every H weakly dense in E'.

We have the following characterization of infra-(s)-spaces:

(3) *A locally convex space $E[\mathfrak{T}]$ is an infra-(s)-space if and only if $\mathfrak{T}_1^t = \mathfrak{T}^t$ for every locally convex topology $\mathfrak{T}_1 \subset \mathfrak{T}$ on E.*

Proof. If $\mathfrak{T}_1 \subset \mathfrak{T}$, then $H = E[\mathfrak{T}_1]'$ is a weakly dense subspace of E' and $E[\mathfrak{T}_1^t]' = \bar{H}$. If $E[\mathfrak{T}]$ is an infra-(s)-space, we have $\bar{H} \cap E' = E'$, hence $\bar{H} = \bar{E}'$; therefore $E[\mathfrak{T}^t]$ and $E[\mathfrak{T}_1^t]$ have the same duals. Since these spaces are barrelled, it follows that $\mathfrak{T}_1^t = \mathfrak{T}^t$. Assume, on the other hand, that $\mathfrak{T}_1^t = \mathfrak{T}^t$ for every $\mathfrak{T}_1 \subset \mathfrak{T}$. If H is weakly dense in E', then $H = E[\mathfrak{T}_s(H)]'$ and $\mathfrak{T}_s(H) = \mathfrak{T}_1 \subset \mathfrak{T}$. From $E[\mathfrak{T}_1^t] = E[\mathfrak{T}^t]$ it follows that $\bar{H} = \bar{E}'$; hence $\bar{H} \cap E' = E'$ and $E[\mathfrak{T}]$ is an infra-(s)-space.

We are now able to prove KŌMURA's version of the closed-graph theorem:

(4) a) *Every closed linear mapping A of a barrelled space E in an infra-(s)-space $F[\mathfrak{T}]$ is continuous.*

b) *The infra-(s)-spaces are characterized by this property.*

Proof. a) There exists a locally convex topology $\mathfrak{T}_1 \subset \mathfrak{T}$ such that A is continuous from E into $F[\mathfrak{T}_1]$. By (2) A is also continuous from E into $F[\mathfrak{T}_1^t]$. But by (3) $F[\mathfrak{T}_1^t] = F[\mathfrak{T}^t]$ and $\mathfrak{T} \subset \mathfrak{T}^t$; hence A is continuous into $F[\mathfrak{T}]$.

b) Let $F[\mathfrak{T}]$ be a space with the property given in a) and let $\mathfrak{T}_1$ be a locally convex topology on F such that $\mathfrak{T}_1 \subset \mathfrak{T}$. Then $\mathfrak{T}_1^t \subset \mathfrak{T}^t$. The identity mapping I of $F[\mathfrak{T}_1^t]$ onto $F[\mathfrak{T}]$ is closed since $\mathfrak{T}_1 \subset \mathfrak{T}_1^t$ and $\mathfrak{T}_1 \subset \mathfrak{T}$ (5.(9)). By assumption I is continuous and by (2) I is also continuous from $F[\mathfrak{T}_1^t]$ onto $F[\mathfrak{T}^t]$. Therefore $\mathfrak{T}_1^t \supset \mathfrak{T}^t$ and we conclude $\mathfrak{T}_1^t = \mathfrak{T}^t$; hence $F[\mathfrak{T}]$ is an infra-(s)-space by (3).

This shows that the class of infra-(s)-spaces is maximal for the closed-graph theorem for barrelled spaces.

We list some properties of infra-(s)-spaces and (s)-spaces.

(5) *If $E[\mathfrak{T}]$ is an infra-*(s)*-space and $\mathfrak{T}_1 \subset \mathfrak{T}$, $\mathfrak{T}_1$ locally convex on E, then $E[\mathfrak{T}_1]$ is an infra-*(s)*-space.*

This is a trivial consequence of (3).

(6) *Every closed subspace H of an infra-*(s)*-space F is an infra-*(s)*-space.*

Let E be barrelled, A linear from E in H, and $G(A)$ closed in $E \times H$; then $G(A)$ is closed in $E \times F$ and by (4) a) A is continuous. Hence H is an infra-(s)-space by (4) b).

(7) *Every Pták space is an* (s)*-space; every infra-Pták space is an infra-*(s)*-space.*

Let H be a subspace of E', $E[\mathfrak{T}]$ a Pták space. We show that $\overline{\overline{H}} \cap E'$ is $\mathfrak{T}^f$-closed. If U is an absolutely convex neighbourhood of $\mathfrak{o}$ in E, the set $\overline{\overline{H}} \cap U^\circ$ is weakly bounded in $\overline{\overline{H}}$. If the net $x_\alpha \in \overline{\overline{H}} \cap U^\circ$ converges weakly to $x_0 \in E'$, then $x_0 \in \overline{\overline{H}}$ and $x_0 \in U^\circ$ since U° is weakly compact. Therefore $x_0 \in \overline{\overline{H}} \cap U^\circ$ and $\overline{\overline{H}} \cap U^\circ$ is weakly closed. Since $E[\mathfrak{T}]$ is a Pták space, it follows that $\overline{\overline{H}} \cap E' = \overline{\overline{H}}$ and $E[\mathfrak{T}]$ is an (s)-space.

The same proof is valid for infra-Pták spaces, but this case is also a consequence of (4) and 6.(9).

(8) *Every barrelled* (s)*-space is a Pták space; every barrelled infra-*(s)*-space is an infra-Pták space.*

If E is barrelled and H a subspace of E', then H is $\mathfrak{T}^f$-closed if and only if H is weakly quasi-complete, $H = \overline{\overline{H}}$. If E is, moreover, an (s)-space, then $H = \overline{\overline{H}} = \overline{H}$; hence E is a Pták space.

The same argument settles the case of a barrelled infra-(s)-space. The associated barrelled space $E[\mathfrak{T}^t]$ of an infra-(s)-space $E[\mathfrak{T}]$ is not always infra-(s); EBERHARDT [3′] gave an example in which even $E[\mathfrak{T}_b(E')]$ is not infra-(s). ADASCH [5′] and EBERHARDT [1′] proved the following weaker result.

(9) *If $E[\mathfrak{T}]$ is an infra-*(s)*-space, then $E[\mathfrak{T}^t]$ is complete.*

Proof. By § 21, 9.(6) it is sufficient to prove that every $(\mathfrak{T}^t)^f$-closed subspace H of co-dimension 1 in $(E[\mathfrak{T}^t])' = \overline{\overline{E}}'$ is weakly closed in $\overline{\overline{E}}'$. In $\overline{\overline{E}}'$ all weakly closed and weakly bounded sets M are equicontinuous and therefore weakly compact. All $H \cap M$ are therefore weakly closed and weakly compact, $H = \overline{\overline{H}}$, and H is weakly quasi-complete.

Since H has co-dimension 1 it is weakly closed or weakly dense in $\overline{\overline{E}}'$. We assume that H is weakly dense in $\overline{\overline{E}}'$ and will reach a contradiction.

$H + E'$ is either H or $\overline{\overline{E}}'$. In the first case we would have $E' \subset H$, $\overline{\overline{E}}' \subset \overline{\overline{H}} = H$, which is impossible. Therefore $H + E' = \overline{\overline{E}}'$ and $\overline{\overline{E}}'/H = E'/E' \cap H$ by § 7, 6.(6), so that $H_1 = E' \cap H$ has co-dimension 1 in E'.

Therefore $E' = [v_0] \oplus H_1$ and we have also $\bar{E}' = [v_0] \oplus H$ by § 7, 6.(2).

Assume that H_1 is weakly dense in E'. Since $E[\mathfrak{T}]$ is an infra-(s)-space, it follows that $\bar{H}_1 \cap E' = E'$; hence $H_1 = H \cap E' = E'$, which is a contradiction. Therefore $H_1 = E' \cap H$ is weakly closed in E', since it has co-dimension 1.

Next we prove that $[v_0] \oplus \bar{H}_1$ is weakly quasi-complete. This is true for $[v_0]$ and $\bar{H}_1$. Let B be an absolutely convex weakly bounded subset of $[v_0] \oplus \bar{H}_1$ with elements $u = \rho v_0 + v$, $v \in \bar{H}_1$. Assume there exist $u_n = \rho_n v_0 + v_n$ in B such that $\rho_n \neq 0$, $|\rho_n| \to \infty$. Then $v_0 + v_n/\rho_n$ is weakly convergent to $\circ$ and v_0 is the weak limit of a sequence of elements of $\bar{H}_1$. But this is in contradiction to the weak quasi-completeness of $\bar{H}_1$. Hence every weakly bounded subset B is contained in a set $B_1 \oplus B_2$, B_1 compact in $[v_0]$, B_2 weakly compact in $\bar{H}_1$. Therefore $B_1 \oplus B_2$ is compact for the product topology, which is finer than $\mathfrak{T}_s(E)$ on $B_1 \oplus B_2$ and therefore coincides with this topology. Hence $[v_0] \oplus \bar{H}_1$ is weakly quasi-complete.

Since $E' \subset [v_0] \oplus \bar{H}_1$, $[v_0] \oplus \bar{H}_1$ must coincide with the quasi-completion $[v_0] \oplus H$ of E'. From $\bar{H}_1 \subset H$ it follows that $\bar{H}_1 = H$. Now H is weakly dense in $\bar{E}'$; therefore $v_0 \in \bar{H}$. But $\bar{H} = \bar{\bar{H}}_1 = \bar{H}_1$; therefore $v_0 \in \bar{H}_1$, the closure taken in $\bar{E}'$. But $v_0 \in E'$ and therefore $v_0 \in \bar{H}_1 \cap E' = H_1$, since H_1 is weakly closed in E'. This is a contradiction.

We remark that the topological product of two (s)-spaces need not be an (s)-space. This follows from the example at the end of 3., since $\varphi\omega$ and $\omega\varphi$ are barrelled Pták spaces and the barrelled space $\varphi\omega \oplus \omega\varphi$ is not a Pták space, therefore not an (s)-space by (8).

In KÖTHE [2], § 7, Satz 2, a weakly dense strict subspace H of $\varphi\omega \oplus \omega\varphi$ is constructed which is sequentially closed. Using the fact that a subset is weakly compact if it is weakly sequentially compact (§ 30, 6.(1)), one sees that H is $\mathfrak{T}^f$-closed and therefore $(\varphi\omega \oplus \omega\varphi)' = \omega\varphi \oplus \varphi\omega$ is not an infra-Pták space. Since $\omega\varphi \oplus \varphi\omega$ is barrelled, we see that the product of two infra-(s)-spaces may not be an infra-(s)-space again.

10. The open mapping theorem of ADASCH. We need two propositions on (s)-spaces.

(1) a) *Every quotient of an* (s)-*space is an* (s)-*space.*

b) *A locally convex space E is an* (s)-*space if and only if every quotient of E is an infra*-(s)-*space.*

Proof. a) Let L be a closed subspace of the (s)-space $E[\mathfrak{T}]$; then $(E/L)' = L^\circ \subset E'$ and $L^\circ \subset (E/L)^* \subset E^*$. For a subspace H of L° the weak quasi-completion $\bar{H}$ is the same in $\bar{L}^\circ$ and in $\bar{E}'$, since $\mathfrak{T}_s(E)$ and $\mathfrak{T}_s(E/L)$ coincide on $(E/L)^*$. Since E is an (s)-space, $\bar{H} \cap E' = \bar{H}$ and the weak closure $\overline{H}$ of H in E' is also the weak closure in L°. Finally, $\overline{H} \cap L^\circ = \overline{H} \cap E' \cap L^\circ = \overline{H}$ and E/L is an (s)-space.

b) Let H be a subspace of E'; then it is weakly dense in a weakly closed subspace of the form $L^\circ = (E/L)'$. Since E/L is an infra-(s)-space, $\overline{H} \cap L^\circ = L^\circ = \overline{H}$, where the weak closure is taken in E'. We have seen in the proof of (a) that $\overline{H}$ is also the weak quasi-completion of H in E' and, since $\overline{H} \cap E' \subset \overline{H}$, we have $\overline{H} \cap E' = \overline{H}$, so that E is an (s)-space.

(2) *Every closed subspace H of an* (s)-*space E is an* (s)-*space.*

Let L be a closed subspace of H. Then H/L is a closed subspace of E/L. Since E/L is an infra-(s)-space, H/L is an infra-(s)-space by 9.(6) and the statement follows from (1) b).

We now prove ADASCH's open mapping theorem for barrelled spaces.

(3) a) *Every closed linear mapping A of an* (s)-*space E onto a barrelled space F is open.*

b) *The* (s)-*spaces are characterized by this property.*

Proof. a) By (1) a) it is sufficient to consider closed mappings which are one-one. But then 9.(4) a) applies to $A^{(-1)}$ and from the continuity of $A^{(-1)}$ it follows that A is open.

b) We assume first that every one-one closed linear mapping A of $E[\mathfrak{T}]$ onto a barrelled space F is open. Let $\mathfrak{T}_1$ be a locally convex topology on E such that $\mathfrak{T}_1 \subset \mathfrak{T}$. Then the identity mapping I of $E[\mathfrak{T}_1^t]$ onto $E[\mathfrak{T}]$ is closed, so $I^{(-1)}$ is open by assumption; hence I is continuous. By 9.(2) I is continuous from $E[\mathfrak{T}_1^t]$ onto $E[\mathfrak{T}^t]$; hence $\mathfrak{T}_1^t \supset \mathfrak{T}^t$. On the other hand, $\mathfrak{T}_1^t \subset \mathfrak{T}^t$; therefore $\mathfrak{T}_1^t = \mathfrak{T}^t$ and $E[\mathfrak{T}]$ is an infra-(s)-space by 9.(3).

Let now E/H be a quotient and $\hat{A}$ a one-one closed linear mapping of E/H onto a barrelled space F. Let K be the canonical homomorphism of E onto E/H; then $A = \hat{A}K$ is by 5.(7) a closed mapping of E onto F and by assumption open. But then $\hat{A}$ is open too and therefore E/H is an infra-(s)-space. (3) follows now from (1) b).

(3) b) says that the class of (s)-spaces is the maximal class of spaces E for which the open mapping theorem for closed linear mappings of E onto any barrelled space is true.

With the results of the last two sections it is also possible to improve Theorems 8.(1) and 8.(3) of ROBERTSON. We formulate the analogue to 8.(1) and indicate the proof; the theorem corresponding to 8.(3) is left to the reader.

(4) *Let E be a locally convex hull* $\sum_\alpha A_\alpha(E_\alpha)$ *of Baire spaces* E_α, *F a locally convex hull* $\bigcup_{n=1}^{\infty} B_n(F_n)$ *of a sequence of* (s)-*spaces* F_n. *Then every closed linear mapping A of E in F is continuous.*

The only change in the original proof occurs in ii). One proves again that $G(C_n)$ is closed in $E \times F_n$. But then C_n is continuous by 9.(4) and from 9.(2) it follows that C_n is continuous from H into $F_n[\mathfrak{T}_n^t]$. By 9.(9) this space is complete and it follows again from 5.(8) that $C_n \in \mathfrak{L}(E, F_n[\mathfrak{T}_n^t])$ and $A_0 = A$. But then C_n is also in $\mathfrak{L}(E, F_n)$ and the proof continues on the same lines as in 8.

VALDIVIA UREÑA [3'] has given different generalizations of (4) which are based on his results on subspaces of infra-(s)-spaces and (s)-spaces. A typical one is the following: If a subspace of finite or countable co-dimension of a locally convex space E is an infra-(s)-space, then E itself is an infra-(s)-space.

It follows from 9.(5) that there exist many infra-(s)-spaces which are not infra-Pták spaces; the weak dual of an (F)-space is an infra-(s)-space but not an infra-Pták space, since it is not weakly complete. But at the moment we know no example of an infra-(s)-space which is not an infra-Pták space for a stronger topology.

Again there is no example of an infra-(s)-space which is not an (s)-space. It follows from 9.(5) and (1) b) that an (s)-space remains an (s)-space if we replace the topology by a weaker locally convex topology and the class of (s)-spaces is strictly larger than the class of Pták spaces.

We recall the result 2.(4) of HUSAIN. He determined the class of $B(\mathscr{T})$-spaces as the maximal class of spaces E for which every continuous linear mapping of E onto a barrelled space is open. It follows immediately that every (s)-space is a $B(\mathscr{T})$-space. SULLEY [1'] gave an example of a $B(\mathscr{T})$-space which is not an infra-(s)-space.

Remark. Consider the dual pair $\langle F, E\rangle = \langle \varphi_d, \varphi_d\rangle$, where $d \geqq 2^{\aleph_0}$, and take $\mathfrak{T}_s(F)$ for the topology on E. Then E is a $B(\mathscr{T})$-space, since the only subspaces H of $E' = F$ in which every weakly bounded subset is equicontinuous are finite dimensional and therefore weakly closed. Now $\overline{F} = E^* = \omega_d$; therefore the associated barrelled topology on E is $\mathfrak{T}_b(\omega_d)$ and in this topology E is by 3. not an infra-Pták space. EBERHARDT [3'] has shown that $E[\mathfrak{T}_s(F)]$ is an infra-(s)-space.

ADASCH gave in his paper [4'] a generalization of PTÁK's theory (including the refinements treated in 8. and 9.) to general topological vector spaces. The barrelled spaces are replaced by the ultrabarrelled spaces which were introduced by W. ROBERTSON [2]. The notions of Pták spaces, infra-Pták spaces, (s)-spaces, and infra-(s)-spaces are easily generalized to topological vector spaces. The duality methods which we used in this exposition had to be replaced by new methods which go back to KELLEY's paper [1'] and are partly included in 4.

The paper [1'] of PERSSON was of some importance for ADASCH's work. For earlier results and references to the literature see HUSAIN's book [1'].

11. KALTON'S closed-graph theorems. We use the following notation. If F is a locally convex space, $\mathscr{C}^l(F)$ will denote the class of all locally convex spaces E for which every closed linear mapping of E into F is continuous. If $\mathscr{A}$ is a class of locally convex spaces F, $\mathscr{C}^l(\mathscr{A})$ is the intersection of all $\mathscr{C}^l(F)$, $F \in \mathscr{A}$.

We recall MAHOWALD's theorem (7.(1)); if $\mathscr{B}$ denotes the class of all (B)-spaces, then this theorem says that $\mathscr{C}^l(\mathscr{B})$ consists of all barrelled spaces. If $\mathscr{I}\mathscr{P}$ denotes the class of all infra-Pták spaces, then it follows from 6.(9) that $\mathscr{C}^l(\mathscr{I}\mathscr{P})$ is again the class of all barrelled spaces.

Let $\mathscr{B}_s$ resp. $\mathscr{I}\mathscr{P}_s$ denote the class of all separable (B)-spaces resp. all separable infra-Pták spaces. KALTON [1'] determined $\mathscr{C}^l(\mathscr{B}_s)$ and $\mathscr{C}^l(\mathscr{I}\mathscr{P}_s)$ and obtained new closed-graph theorems which have applications in summability theory.

We need some basic facts on separability and metrizable spaces.

If a (B)-space E is separable, then the weak topology on the closed unit ball of E' is metrizable (§ 21, 3.(4)). We prove the converse.

(1) *Let E be a normed space. If the closed unit ball of E' is metrizable in the weak topology, then E is separable.*

By assumption there exist finite dimensional bounded absolutely convex sets $A_1 \subset A_2 \subset \cdots$ in E such that $\bigcap_{n=1}^{\infty} A_n^\circ = \mathfrak{o}$. If $L(A)$ is the linear span of $A = \bigcup_{n=1}^{\infty} A_n$, then $L(A)^\circ = \mathfrak{o}$; hence $\overline{L(A)} = E$ and E is separable.

We need the following lemma of general topology.

(2) *Let M be a compact metrizable space and let the Hausdorff space N be the continuous image $f(M) = N$ of M. Then N is metrizable.*

Proof. $f(M)$ is compact and will be metrizable by § 7, 6.(3) if it has a countable base of open sets. If $O_1, O_2, \ldots$ is such a base on M, then all the finite unions $V = O_{n_1} \cup \cdots \cup O_{n_k}$ determine also a countable base on M. Since f is continuous and closed (§ 3, 2.(5)), the set $\tilde{V} = N \sim f(M \sim V) = \sim f(\sim V)$ is open in N and it will be sufficient to prove that the $\tilde{V}$ determine a base of open sets in N.

Assume G open in N and $p \in G$. Then $f^{(-1)}(p)$ is compact and $\subset f^{(-1)}(G)$ and there exists V such that $f^{(-1)}(p) \subset V \subset f^{(-1)}(G)$. We have only to verify that then $p \in \tilde{V} \subset G$.

Assume $p \notin \sim f(\sim V)$. Then $p \in f(\sim V)$ and $p = f(q)$, $q \notin V$, so $q \in f^{(-1)}(p)$, which contradicts $f^{(-1)}(p) \subset V$. Hence $p \in \tilde{V}$. From $V \subset f^{(-1)}(G)$ it follows that $\sim V \supset \sim f^{(-1)}(G)$, $f(\sim V) \supset f(\sim f^{(-1)}(G)) = \sim G$, and $\sim f(\sim V) \subset G$; hence $\tilde{V} \subset G$.

We recall that if M is an absolutely convex subset of a locally convex space $E[\mathfrak{T}]$, then $\mathfrak{T}$ induces on M a uniquely determined uniformity (§ 28, 5.(3)).

We note further:

(3) *A uniform space is metrizable if and only if its completion is metrizable.*

This is a consequence of § 6, 7.(1) and the definition of the vicinities of the completion.

(4) *Let E be locally convex. Every weak Cauchy sequence $u_n \in E'$ is contained in an absolutely convex weakly closed, weakly precompact, and weakly metrizable subset of E'.*

It is sufficient to prove for the $\mathfrak{T}_s(E)$-quasi-completion $\bar{E}'$ that every weakly convergent sequence u_n with limit u_0 is contained in an absolutely convex, weakly compact, and weakly metrizable subset of $\bar{E}'$. Introduce $v_0 = u_0$ and $v_n = u_n - u_0$, $n = 1, 2, \ldots$. Then v_n converges weakly to o in $\bar{E}'$ and $N = \overline{\Gamma\{v_0, v_1, \ldots\}}$ is the $\mathfrak{T}_s(c_0)$–$\mathfrak{T}_s(E)$-continuous image of the closed unit ball K of l^1 in $\bar{E}'$ (§ 20, 9.(6)). But K is weakly compact and $\mathfrak{T}_s(c_0)$-metrizable; therefore N is weakly compact and $\mathfrak{T}_s(E)$-metrizable by (2). Since $u_0, u_1, u_2, \ldots$ are elements of $2N$, the statement follows.

We will also use the following lemma.

(5) *Let E, F be locally convex, A a closed linear mapping of E in F. $D[A']$ is weakly sequentially closed in F' if E' is weakly sequentially complete.*

Let $v_n \in D[A']$ be weakly convergent to $v_0 \in F'$. Then $u_n = A'v_n$ is a weak Cauchy sequence in E' which has a limit $u_0 \in E'$ by assumption. For every $x \in E$

$$(A'v_0)x = v_0(Ax) = \lim v_n(Ax) = \lim\,(A'v_n)x = u_0x.$$

Consequently, $A'v_0 = u_0 \in E'$ and $v_0 \in D[A']$.

We now prove KALTON's first theorem.

(6) *Let E be locally convex. The following statements are equivalent: a) $E[\mathfrak{T}_k(E')] \in \mathscr{C}^l(\mathscr{I}\mathscr{P}_s)$; b) $E[\mathfrak{T}_k(E')] \in \mathscr{C}^l(c_0)$; c) E' is weakly sequentially complete.*

Proof. b) follows immediately from a).

We remark that c_0 is isomorphic to c and therefore $\mathscr{C}^l(c_0) = \mathscr{C}^l(c)$. We assume now that $E[\mathfrak{T}_k(E')] \in \mathscr{C}^l(c_0)$ and prove c).

Let u_n be a weak Cauchy sequence in E'. We define a linear mapping A of E in c by $Ax = (u_nx)$. If the net x_α converges in $E[\mathfrak{T}_k(E')]$ to x_0 and if Ax_α converges to y_0 in c, then $u_nx_\alpha \to u_nx_0$ for every n; hence $y_0 = Ax_0$ and A is closed. A is continuous from $E[\mathfrak{T}_k(E')]$ in c by assumption.

Recall that the continuous linear functionals e_n on c defined by $e_n y = y_n$ for $y \in c$ converge weakly to $v_0 \in c'$ defined by $v_0 y = \lim_{n\to\infty} y_n$ (§ 14, 7.). Consequently, $u_n = A'e_n$ converges weakly to $u_0 = A'v_0 \in E'$, which is defined by $u_0 x = \lim u_n x$. Hence E' is weakly sequentially complete.

The last step of the proof follows from the following closed-graph theorem:

(7) *Let E be locally convex and E' weakly sequentially complete. Every closed linear mapping A of E in a separable infra-Pták space F is continuous from $E[\mathfrak{T}_k(E')]$ in F.*

Proof. $D[A']$ is weakly dense and by (5) sequentially weakly closed in F'. If U is an absolutely convex neighbourhood of $\circ$ in F, then U° is weakly compact and weakly metrizable. The set $U^\circ \cap D[A']$ is weakly sequentially closed and therefore weakly closed. Since F is an infra-Pták space, it follows that $D[A'] = F'$ and A is weakly continuous and therefore continuous from $E[\mathfrak{T}_k(E')]$ in F.

We remark that by § 30, 5.(3) every perfect sequence space has property c) of (6).

The next theorem determines the whole class $\mathscr{C}^l(c_0)$.

(8) *$E[\mathfrak{T}]$ is in $\mathscr{C}^l(c_0)$ if and only if every weak Cauchy sequence in E' is equicontinuous.*

Proof. a) Assume $E[\mathfrak{T}] \in \mathscr{C}^l(c_0)$ and let u_n be a weak Cauchy sequence in E'. Define A by $Ax = (u_n x)$ as in the proof of (6). It follows again that A is continuous. Consequently, $\|Ax\| = \sup_n |u_n x| \leqq K < \infty$ for the elements x of some absolutely convex neighbourhood U of $\circ$ in E and this means the equicontinuity of the sequence u_n.

b) Conversely, if every weak Cauchy sequence in E' is equicontinuous, then E' is weakly sequentially complete and $E[\mathfrak{T}_k(E')]$ is in $\mathscr{C}^l(c_0)$ by (6). Hence, if A is a closed linear mapping of $E[\mathfrak{T}]$ in c_0, it is closed as a mapping of $E[\mathfrak{T}_k(E')]$ in c_0 and continuous by (6). Therefore A' maps the sequence $e_n \in l^1$ which converges weakly to $\circ$ into the sequence $A'e_n = u_n$ which again converges weakly to $\circ$. By assumption the set $M = \{u_1, u_2, \ldots\}$ is equicontinuous in E'. Therefore $\|Ax\| = \sup_n |e_n(Ax)| = \sup_n |u_n x| \leqq 1$ for $x \in M^\circ$ and A is continuous from $E[\mathfrak{T}]$ in c_0.

The separable analogue to MAHOWALD's theorem is the following theorem of KALTON.

(9) *The classes $\mathscr{C}^l(\mathscr{I}\mathscr{P}_s)$, $\mathscr{C}^l(C[0,1])$, and $\mathscr{C}^l(\mathscr{B}_s)$ coincide and consist of all locally convex spaces $E[\mathfrak{T}]$ with the property that every weakly bounded, weakly metrizable, absolutely convex set B in E' is equicontinuous.*

Proof. If $E \in \mathscr{C}^l(\mathscr{I}\mathscr{P}_s)$, then $E \in \mathscr{C}^l(C[0,1])$ and $E \in \mathscr{C}^l(\mathscr{B}_s)$, since every separable (B)-space is isometric to a subspace of $C[0,1]$ (§ 21, 3.(6)).

Now we assume that $E \in \mathscr{C}^l(\mathscr{B}_s)$ and that B is a weakly bounded absolutely convex subset of E' which is metrizable for $\mathfrak{T}_s(E)$. The set $B^\circ \subset E$ is a barrel and the space E_{B° constructed as in the proof of MAHOWALD's theorem (7.(1)) is a normed space and $\langle E'_B, E_{B^\circ}\rangle$ is a dual pair. In the sense of this duality the norm topology on E_{B° is the strong topology $\mathfrak{T}_b(E'_B)$. The dual $(E_{B^\circ})'$ is equal to $\bigcup_{n=1}^{\infty} nB^{\circ\circ}$, where $B^{\circ\circ}$ is the polar of B° in $(E_{B^\circ})^*$. Algebraically, $(E_{B^\circ})'$ is identical with $E/B^\perp = E/(E'_B)^\circ$ and on E'_B the topologies $\mathfrak{T}_s(E)$ and $\mathfrak{T}_s(E_{B^\circ}) = \mathfrak{T}_s(E/B^\perp)$ coincide. B is therefore metrizable for the topology $\mathfrak{T}_s(E_{B^\circ})$. Since $B^{\circ\circ}$ is the completion of B for $\mathfrak{T}_s(E_{B^\circ})$, it follows from (3) that $B^{\circ\circ}$ is weakly metrizable. Consequently, E_{B° is separable by (1).

The canonical mapping K of E onto E_{B° is closed as a mapping of E in $\tilde{E}_{B^\circ}$ by 7.(1). Since $\tilde{E}_{B^\circ}$ is a separable (B)-space, K is continuous by our assumption. Therefore B°, the inverse image of the unit ball $K(B^\circ)$, is a neighbourhood of o in E and B is equicontinuous.

The last step of the proof is the closed-graph theorem:

(10) *Let $E[\mathfrak{T}]$ be a locally convex space with the property that every weakly bounded and weakly metrizable absolutely convex set B in E' is equicontinuous and let F be a separable infra-Pták space. Then every closed linear mapping A of E in F is continuous.*

A weak Cauchy sequence $u_n \in E'$ is by (4) contained in a set B which is by assumption equicontinuous. It follows that E' is weakly sequentially complete and $E[\mathfrak{T}_k(E')] \in \mathscr{C}^l(\mathscr{I}\mathscr{P}_s)$ by (6). A closed linear mapping A of E in F is therefore continuous from $E[\mathfrak{T}_k(E')]$ in F. Let V be a closed absolutely convex neighbourhood of o in F. Then V° is weakly metrizable and so is $A'(V^\circ)$ by (2). By our assumption on E the set $A'(V^\circ)$ is equicontinuous; hence $A'(V^\circ)^\circ = A^{(-1)}(V^{\circ\circ}) = A^{(-1)}(V)$ is a neighbourhood of o in F and A is continuous.

For further results and examples see KALTON [1'].

§ 35. DE WILDE's theory

1. Webs in locally convex spaces. When in 1954 GROTHENDIECK stated his Theorem B (compare § 34, 8.(6)), he conjectured that this theorem should be true for a much larger class of spaces than the class of (LF)-spaces.

His conjecture said, in particular, the following: The class of spaces E is now the class of ultrabornological spaces, a subclass of the class of barrelled spaces, and we are looking for spaces F such that the closed-graph theorem for mappings from any E into F is true. This is the case for F a Banach space, as we know. The closed-graph theorem should remain valid if one performs any one of the following operations on Banach spaces F: taking closed subspaces, quotients, countable products, countable locally convex sums, and a finite number of iterations of these basic operations.

This conjecture was first solved in 1966 by RAÍKOW [1′] using ideas of SLOWIKOWSKI [1′], [2′]. At about the same time L. SCHWARTZ [2′] gave a new version of the closed-graph theorem, which was generalized by MARTINEAU [1′], [2′]; this new version included also a positive answer to GROTHENDIECK's conjecture. Finally, in 1967 DE WILDE [1′], [2′], [3′] gave a solution by a method which can be understood as a refinement of the classical methods of Banach. We give here an exposition of DE WILDE's approach and some of the consequences.

We start with the fundamental notion of a web in a locally convex space E. Let $\mathscr{W} = \{C_{n_1,\ldots,n_k}\}$ be a class of subsets $C_{n_1,\ldots,n_k}$ of E, where k and $n_1, \ldots, n_k$ run through all the natural numbers. $\mathscr{W}$ is called a web if it satisfies the relations

$$\text{(w)}\quad E = \bigcup_{n_1=1}^{\infty} C_{n_1}, \qquad C_{n_1,\ldots,n_{k-1}} = \bigcup_{n_1=1}^{\infty} C_{n_1,\ldots,n_k}$$

for $k > 1$ and all $n_1, \ldots, n_{k-1}$.

If all sets of a web are closed or absolutely convex, we say that the web is closed resp. absolutely convex.

A web $\mathscr{W}$ is a web of type $\mathscr{C}$ or a $\mathscr{C}$-web if the following condition is satisfied: For every fixed sequence n_k, $k = 1, 2, \ldots$, there exists a sequence of positive numbers ρ_k such that for all λ_k, $0 \leqq \lambda_k \leqq \rho_k$, and all $x_k \in C_{n_1,\ldots,n_k}$ the series $\sum_1^\infty \lambda_k x_k$ converges in E.

We remark that if this is the case, then $\sum_{k=1}^\infty \lambda_k x_k$ is convergent in E also under the weaker assumption that $|\lambda_k| \leqq \rho_k$ for all k. This follows for real λ_k by considering the sequences $\lambda_1^+, \lambda_2^+, \ldots$ and $\lambda_1^-, \lambda_2^-, \ldots$, and, moreover, the sequences $\Re\lambda_1, \Re\lambda_2, \ldots$ and $\Im\lambda_1, \Im\lambda_2, \ldots$ in the complex case (we use the usual definitions $\lambda^+ = \sup(\lambda, 0)$ and $\lambda^- = (-\lambda)^+$ for λ real).

It is obvious that the existence of a $\mathscr{C}$-web in E means a rather weak kind of sequential completeness of E. Conversely, we have

(1) *A web $\mathscr{W} = \{C_{n_1,\ldots,n_k}\}$ on E is a $\mathscr{C}$-web if for every fixed sequence n_k, $k = 1, 2, \ldots$, there exists a sequence $\mu_k > 0$ such that every sequence $\mu_k x_k$, where $x_k \in C_{n_1,\ldots,n_k}$, is contained in an absolutely convex bounded and sequentially complete subset M of E.*

Proof. We define $\rho_k = 2^{-k}\mu_k$. Then for $0 \leqq \lambda_k \leqq \rho_k$ we have $\lambda_k = \gamma_k\mu_k$ and $\sum_1^\infty \gamma_k \leqq 1$. Since $z_k = \mu_k x_k \in M$ and M is absolutely convex and sequentially complete, $\sum_1^\infty \gamma_k z_k$ converges in E and $\mathscr{W}$ is a $\mathscr{C}$-web.

We will also use another kind of web. A web $\mathscr{W}$ is called strict if it is absolutely convex and if for any sequence n_k, $k = 1, 2, \ldots$, there exists a sequence $\rho_k > 0$ such that for all $x_k \in C_{n_1,\ldots,n_k}$ and all λ_k, $0 \leqq \lambda_k \leqq \rho_k$, the series $\sum_1^\infty \lambda_k x_k$ converges in E and $\sum_{k_0}^\infty \lambda_k x_k$ is contained in $C_{n_1,\ldots,n_{k_0}}$ for all $k_0 = 1, 2, \ldots$.

Obviously, a strict web is a $\mathscr{C}$-web. Conversely, we have

(2) *If $\mathscr{W}$ is an absolutely convex and closed $\mathscr{C}$-web on E, then $\mathscr{W}$ is strict.*

By the definition of a $\mathscr{C}$-web we are able to choose the sequence ρ_k for the given sequence n_k such that $\sum_{k=1}^\infty \rho_k \leqq 1$. Then for $x_k \in C_{n_1,\ldots,n_k}$ and $0 \leqq \lambda_k \leqq \rho_k$ the sum $\sum_{k_0}^{k_0+N} \lambda_k x_k$ is always contained in $C_{n_1,\ldots,n_{k_0}}$, since this set is absolutely convex and $\sum_{k_0}^\infty \lambda_k x_k$ is in $C_{n_1,\ldots,n_{k_0}}$ because $C_{n_1,\ldots,n_{k_0}}$ is closed.

We remark further:

(3) *Let $\mathscr{W} = \{C_{n_1,\ldots,n_k}\}$ be a $\mathscr{C}$-web or a strict web on E. If ρ_k are the numbers corresponding to the sequence $C_{n_1,\ldots,n_k}$, $k = 1, 2, \ldots$, and if U is a neighbourhood of $\circ$ in E, then there exists k_0 such that $\rho_k C_{n_1,\ldots,n_k} \subset U$ for $k \geqq k_0$.*

Proof. Assume this is not true for a given U. Then there exist infinitely many k_i and $x_{k_i} \in C_{n_1,\ldots,n_{k_i}}$ such that $\rho_{k_i} x_{k_i} \notin U$. If we define $\lambda_j = \rho_j$ for $j = k_i$ and $\lambda_j = 0$ and x_j arbitrary in $C_{n_1,\ldots,n_j}$ for $j \neq k_i$, then $\sum_{j=1}^\infty \lambda_j x_j$ converges in E, which contradicts the fact that $\lambda_{k_j} x_{k_j}$ does not converge to $\circ$.

We give a first example.

(4) *On every* (F)-*space E there exists a strict web.*

Let $U_1 \supset U_2 \supset \cdots$ be a fundamental sequence of absolutely convex and closed neighbourhoods of $\circ$ in E. We define $C_{n_1,\ldots n_k} = \bigcap_{j=1}^k n_j U_j$. Then condition (w) is obviously satisfied. We take $\rho_k = 1/(2^k n_k)$ for a given sequence n_k. Then for $0 \leqq \lambda_k \leqq \rho_k$ it follows that $\lambda_k x_k \in (1/2^k)U_k$ for every $x_k \in C_{n_1,\ldots,n_k}$; hence $\sum_{k_0}^{k_0+m} \lambda_k x_k \in U_{k_0}$ and $\sum_1^\infty \lambda_k x_k$ converges in E. Therefore $W = \{C_{n_1,\ldots,n_k}\}$ is an absolutely convex and closed $\mathscr{C}$-web, which is strict by (2).

We introduce the following terminology. A locally convex space $E[\mathfrak{T}]$ in which there exists a $\mathscr{C}$-web will be said to be a webbed space; if there exists a strict web on $E[\mathfrak{T}]$, then we say $E[\mathfrak{T}]$ is a strictly webbed space. The hereditary properties of these classes of spaces will be studied in detail in 3.

2. The closed-graph theorems of DE WILDE. In his work [3'] DE WILDE gave many versions of the closed-graph theorem. We will restrict our exposition to the cases which are the most important for applications.

Let E and F be locally convex; a linear mapping A of E into F is called sequentially closed if its graph $G(A)$ is sequentially closed in $E \times F$. In view of the applications it is certainly important to have the closed-graph theorem in the stronger form that A is continuous if it is only sequentially closed.

We now give the first theorem of this kind; in its proof the basic ideas of DE WILDE's approach become very clear.

(1) *Every sequentially closed linear mapping A of an* (F)*-space E into a webbed space F is continuous.*

Proof. i) Let $\mathscr{W} = \{C_{n_1,\dots,n_k}\}$ be a $\mathscr{C}$-web on F. From 1.(w) it follows that

$$E = \bigcup_{n_1=1}^{\infty} A^{(-1)}(C_{n_1}), \qquad A^{(-1)}(C_{n_1,\dots,n_{k-1}}) = \bigcup_{n_k=1}^{\infty} A^{(-1)}(C_{n_1,\dots,n_k})$$

for $k > 1$ and all $n_1, \dots, n_{k-1}$

Now E is an (F)-space. Using BAIRE's theorem we find n_1 such that $A^{(-1)}(C_{n_1})$ is not meagre in E, then n_2 such that $A^{(-1)}(C_{n_1,n_2})$ is not meagre in E, and so on. Therefore we have a sequence $n_1, n_2, \dots$ such that every $A^{(-1)}(C_{n_1,\dots,n_k})$ is not meagre in E.

Let V be an absolutely convex and closed neighbourhood of o in F. Since $A^{(-1)}(C_{n_1,\dots,n_k}) = \bigcup_{m=1}^{\infty} A^{(-1)}(C_{n_1,\dots,n_k} \cap mV)$, there exists m_k such that $A^{(-1)}(C_{n_1,\dots,n_k} \cap m_k V)$ is not meagre in E. Since $\mathscr{W}$ is a $\mathscr{C}$-web, there exist $\rho_k > 0$ such that $\sum_{k=1}^{\infty} \lambda_k z_k$ converges in F for all λ_k, $0 \leqq \lambda_k \leqq \rho_k$, and all $z_k \in C_{n_1,\dots n_k}$. We determine $\nu_k \in (0, \rho_k]$ such that, for a given $\varepsilon > 0$, $\sum_{k=1}^{\infty} \nu_k m_k \leqq \varepsilon$, and we define $M_k = A^{(-1)}(\nu_k C_{n_1,\dots,n_k} \cap \nu_k m_k V)$. M_k is again not meagre in E.

ii) We treat first the case of an absolutely convex $\mathscr{W}$ (this includes all strictly webbed spaces). The reader should be aware of the close analogy of this proof with the classical proof of the BANACH–SCHAUDER theorem in § 15, 12.

In our case the sets M_k are absolutely convex and $\overline{M}_k$ contains an interior point and therefore an absolutely convex neighbourhood $U^{(k)}$ of $\circ$. We may assume $U^{(k)} \subset U_k$, where $U_1 \supset U_2 \supset \cdots$ is a given fundamental sequence of absolutely convex neighbourhoods of $\circ$ in E.

From $E = \bigcup_{n=1}^{\infty} nA^{(-1)}(V)$ and BAIRE's theorem it follows again that $\overline{A^{(-1)}(V)}$ contains a neighbourhood of $\circ$ of E. Hence the continuity of A will follow from $\overline{A^{(-1)}(V)} \subset (1 + \varepsilon)A^{(-1)}(V)$.

To prove this assume $x_0 \in \overline{A^{(-1)}(V)}$. There exists $x_1 \in A^{(-1)}(V)$ such that $x_0 - x_1 \in U^{(1)} \subset \overline{M}_1$, there exists $x_2 \in M_1$ such that $x_0 - x_1 - x_2 \in U^{(2)} \subset \overline{M}_2$, and so on. Since $x_0 - \sum_{k=1}^{n} x_k \in U^{(n)} \subset U_n$, we have $x_0 = \sum_{k=1}^{\infty} x_k$. By construction $Ax_1 \in V$, $Ax_{k+1} \in A(M_k) \subset \nu_k m_k V$ for $k \geqq 1$ and also $Ax_{k+1} \in \nu_k C_{n_1,\ldots,n_k}$. Therefore $\sum_{k=1}^{\infty} Ax_k$ converges in F and its limit y_0 lies in $V + \left(\sum_{k=1}^{\infty} \nu_k m_k\right) V \subset (1 + \varepsilon)V$. Since A is sequentially closed, $Ax_0 = y_0$ and $x_0 \in (1 + \varepsilon)A^{(-1)}(V)$.

iii) The general case of an arbitrary $\mathscr{C}$-web is a little more complicated. In this case there exist $x_k \in M_k$ and absolutely convex neighbourhoods $U^{(k)} \subset U_k$ such that $\overline{M}_k \supset x_k + U^{(k)}$. Again it will be sufficient to prove that $\overline{A^{(-1)}(V)} \subset (1 + 2\varepsilon)A^{(-1)}(V)$.

Assume $x_0 \in \overline{A^{(-1)}(V)}$. There exists $y_1 \in A^{(-1)}(V)$ such that $x_0 - y_1 \in U^{(1)}$ and we have $x_0 - y_1 + x_1 \in \overline{M}_1$. Having constructed $y_1, \ldots, y_{k-1}$, we find $y_k \in M_{k-1}$ such that

$$x_0 - \sum_1^k y_i + \sum_1^{k-1} x_i \in U^{(k)} \subset U_k \quad \text{and} \quad x_0 - \sum_1^k y_i + \sum_1^k x_i \in \overline{M}_k.$$

By construction $Ax_i \in \nu_i C_{n_1,\ldots,n_i}$ for all $i \geqq 1$; therefore $\sum_{i=1}^{\infty} Ax_i$ converges in F. Similarly, $Ay_{i+1} \in \nu_i C_{n_1,\ldots,n_i}$ for all $i \geqq 1$ and $\sum_{i=1}^{\infty} Ay_i$ converges in F.

Furthermore, since $Ay_1 \in V$, $Ay_{i+1} \in \nu_i m_i V$, and $Ax_i \in \nu_i m_i V$ for all $i \geqq 1$, and since V is closed, $\sum_1^k Ay_i - \sum_1^{k-1} Ax_i$ converges to an element $y_0 \in (1 + 2\varepsilon)V$. But $\sum_1^k y_i - \sum_1^{k-1} x_i$ converges to x_0, so, since A is sequentially closed, we have again $Ax_0 = y_0$ and $\overline{A^{(-1)}(V)} \subset (1 + 2\varepsilon)A^{(-1)}(V)$.

As a corollary to (1) we obtain now DE WILDE's closed-graph theorem for ultrabornological spaces:

(2) *A sequentially closed linear mapping of an ultrabornological space E into a webbed space F is continuous.*

Let E be $\sum_\alpha A_\alpha(E_\alpha)$, E_α a (B)-space. By § 19, 1.(7) it is sufficient to prove that all mappings AA_α of E_α in F are continuous. By (1) this will be true if every AA_α is sequentially closed.

But this is trivial: Assume that $x_n \to x_0$ in E_α and that $AA_\alpha x_n \to y_0$ in F. Since A_α is continuous, we have $A_\alpha x_n \to A_\alpha x_0$ in E and, since A is sequentially closed, it follows $AA_\alpha x_n \to y_0 = AA_\alpha x_0$.

So far we treated the case of a sequentially closed mapping. If one supposes that the mapping is closed, Theorems (1) and (2) are valid in a more general setting. The analogue to (1) is

(3) *Let E be locally convex and F a webbed space. If A is a closed linear mapping defined on a nonmeagre subspace $D[A]$ of E and $A(D[A]) \subset F$, then A is continuous and $D[A] = E$.*

That A is closed means here that its graph $G(A)$ is closed in $E \times F$ (not only in $D[A] \times F$).

Proof. The proof proceeds as the proof of (1) with the difference that now $D[A] = \bigcup_{n_1=1}^{\infty} A^{(-1)}(C_{n_1})$. We find again the sets M_k and elements $x_k \in M_k$ such that $\overline{M}_k \supset x_k + U^{(k)}$ for some absolutely convex neighbourhood $U^{(k)}$ in E. Furthermore, we construct, as in iii), for a given $x_0 \in \overline{A^{(-1)}(V)}$ the elements $y_1 \in A^{(-1)}(V)$, $y_k \in M_{k-1}$, such that

$$x_0 - \sum_1^k y_i + \sum_1^{k-1} x_i \in U^{(k)} \quad \text{and} \quad x_0 - \sum_1^k y_i + \sum_1^k x_i \in \overline{M}_k.$$

As in iii), we prove $\sum_1^\infty Ay_k \in F$, $\sum_1^\infty Ax_i \in F$, and that $\sum_1^k Ay_i - \sum_1^{k-1} Ax_i$ converges to an element $y_0 \in (1 + 2\varepsilon)V$.

But we are not able to prove that $\sum_1^k y_i - \sum_1^{k-1} x_i$ converges to x_0, since in E we have no fundamental sequence of neighbourhoods of $\mathfrak{o}$. Instead of this we will show that $(x_0, y_0) \in \overline{G(A)} = G(A)$. From this it then follows again that $Ax_0 = y_0$ and $\overline{A^{(-1)}(V)} \subset (1 + 2\varepsilon)A^{(-1)}(V)$. Since $D[A] = \bigcup_{m=1}^{\infty} mA^{(-1)}(V)$ is not meagre in E, it follows from Baire's theorem that $\overline{A^{(-1)}(V)}$ contains a neighbourhood of $\mathfrak{o}$ of E. This is true also for $A^{(-1)}(V) \subset D[A]$; hence $D[A] = E$ and A is continuous.

We prove, finally, that $(x_0, y_0) \in \overline{G(A)}$. Let U, W be fixed absolutely convex neighbourhoods of $\mathfrak{o}$ in E resp. F. Since $x_0 - \sum_1^k y_i + \sum_1^k x_i \in \overline{M}_k \subset M_k + U$, there exists $t_k \in M_k$ such that $x_0 - \sum_1^k y_i + \sum_1^k x_i - t_k \in U$ for all

$k = 1, 2, \ldots$. From the definition of the M_k it follows that the sequences At_k and Ax_k converge to $\mathfrak{o}$ in F. Therefore

$$y_0 - A\left(\sum_1^k y_i - \sum_1^k x_i - t_k\right) \in W \quad \text{for } k \geqq k_0.$$

Hence

$$(x_0, y_0) - \left(\sum_1^k y_i + \sum_1^k x_i - t_k, A\left(\sum_1^k y_i - \sum_1^k x_i - t_k\right)\right) \in (U, W)$$
$$\text{for } k \geqq k_0,$$

so that $(x_0, y_0) \in \overline{G(A)}$.

We made no assumption on E in (3) but, if E has a nonmeagre subspace H, then E itself is a Baire space, since from $E = \bigcup_{i=1}^{\infty} M_i$, M_i nowhere dense in E, would follow $H = \bigcup_{i=1}^{\infty} (M_i \cap H)$, $M_i \cap H$ nowhere dense in E. Therefore (3) is a slight generalization of

(4) *Every closed linear mapping of a Baire space E into a webbed space F is continuous.*

The proof of the following closed-graph theorem is similar to the proof of (2):

(5) *A closed linear mapping A of a locally convex hull $E = \sum_\alpha A_\alpha(E_\alpha)$ of Baire spaces E_α in a webbed space F is always continuous.*

3. The corresponding open-mapping theorems. The first three theorems are easy consequences of 2.(2), 2.(3), and 2.(5). The first one is of the BANACH–SCHAUDER type.

(1) *Every continuous linear mapping A of a webbed space F onto an ultrabornological space E is a homomorphism.*

We will show in 4. that every quotient of a webbed space is again a webbed space (4.(3)). Therefore we may assume that A is one-one. But then $A^{(-1)}$ satisfies the conditions of 2.(2) and is therefore continuous from E onto F; hence A is an isomorphism.

An analogous proof using § 34, 5.(7) and 2.(3) leads to

(2) *If A is a closed linear mapping of the webbed space F onto a nonmeagre subspace $A(F)$ of a locally convex space E, then A is open and $A(F) = E$.*

In the same way one deduces from 2.(5)

(3) *Every closed linear mapping A of a webbed space F onto a locally convex hull $E = \sum_\alpha A_\alpha(E_\alpha)$ of Baire spaces E_α is open.*

We remark that (1) is a special case of (3) and that (1) is not the full counterpart to 2.(2) because in the case of an ultrabornological space one expects an open mapping theorem for an A which is only sequentially closed. To prove such a theorem it is necessary to use again DE WILDE's refinement of the classical method. We prove first

(4) *Let A be a linear mapping which is defined on a subspace D of a webbed space F and which maps D onto a nonmeagre subspace $A(D)$ of the* (F)*-space E. If $G(A)$ is sequentially closed in $F \times E$, then A is open and $A(D) = E$.*

Let $\mathscr{W} = \{C_{n_1,\ldots,n_k}\}$ be a $\mathscr{C}$-web on F. Since $A(D) = \bigcup_{n_1=1}^{\infty} A(C_{n_1} \cap D)$ is nonmeagre in E, we find again using BAIRE's theorem (recursively) a sequence $n_1, n_2, \ldots$ such that $A(C_{n_1,\ldots,n_k} \cap D)$ is nonmeagre in E. Let V be an absolutely convex neighbourhood of $\circ$ in F; then there exists m_k such that $A(C_{n_1,\ldots,n_k} \cap D \cap m_k V)$ is not meagre in E. Since $\mathscr{W}$ is a $\mathscr{C}$-web, there exist $\rho_k > 0$ such that $\sum_{k=1}^{\infty} \lambda_k y_k$ converges in F for all λ_k, $0 \leqq \lambda_k \leqq \rho_k$, and all $y_k \in C_{n_1,\ldots,n_k}$. We determine $\nu_k \in (0, \rho_k)$ such that $\sum_{k=1}^{\infty} m_k \nu_k \leqq \varepsilon$ for a given $\varepsilon > 0$ and we define $M_k = \nu_k(C_{n_1,\ldots,n_k} \cap D \cap m_k V)$. By construction $A(M_k)$ is not meagre in E. Hence there exist $x_k \in M_k$ and neighbourhoods $U^{(k)}$ of $\circ$ in E such that $\overline{A(M_k)} \supset Ax_k + U^{(k)}$. We suppose $U^{(k)} \subset U_k$, where U_k, $k = 1, 2, \ldots$, is a fundamental sequence of neighbourhoods of $\circ$ in E.

Since $A(D) = \bigcup_{n=1}^{\infty} nA(D \cap V)$ is nonmeagre in E, it follows from BAIRE's theorem that $\overline{A(D \cap V)}$ contains a neighbourhood of $\circ$ of E. It remains therefore to prove that $\overline{A(D \cap V)} \subset (1 + 2\varepsilon)A(D \cap V)$ because then $A(D) = E$ follows.

To do this we follow the method of part iii) of the proof of 2.(1). Assume $y_0 \in \overline{A(D \cap V)}$. There exists $z_1 \in D \cap V$ such that $y_0 - Az_1 \in U^{(1)}$ and then $y_0 - Az_1 + Ax_1 \in \overline{A(M_1)}$. Having constructed $z_1, \ldots, z_{k-1}$, we find $z_k \in M_{k-1}$ such that

$$y_0 - \sum_1^k Az_k + \sum_1^{k-1} Ax_k \in U^{(k)} \subset U_k \quad \text{and} \quad y_0 - \sum_1^k Az_k + \sum_1^k Ax_k \in \overline{A(M_k)},$$
$$k = 1, 2, \ldots.$$

Clearly, $\sum_1^k Az_k - \sum_1^{k-1} Ax_k$ converges to y_0 in E. Using the same arguments as in 2.(1) iii), one shows that $\sum_1^k z_k - \sum_1^{k-1} x_k \in D$ converges in F to an element $x_0 \in (1 + 2\varepsilon)V$. Since A is sequentially closed in $F \times E$, $x_0 \in D$, and $y_0 = Ax_0$, hence $\overline{A(D \cap V)} \subset (1 + 2\varepsilon)A(D \cap V)$.

Now we are able to prove the open mapping theorem for ultrabornological spaces.

(5) *A sequentially closed linear mapping A of a webbed space F onto an ultrabornological space E is open.*

We write E in the form $E = \sum_\alpha E_\alpha$, E_α a (B)-space (compare § 34, 8.). Since $A(F) = E$, we have algebraically $F = \sum_\alpha A^{(-1)}(E_\alpha) = \sum_\alpha F_\alpha$. The restriction A_α of A to F_α is sequentially closed in $F \times E_\alpha$ and maps F_α onto E_α. By (4) A_α is open. Let V be an absolutely convex and open neighbourhood of $\circ$ in F_α; then $V_\alpha = V \cap F_\alpha$ is a neighbourhood of $\circ$ in F_α and $A_\alpha(V_\alpha) \supset U_\alpha$, where U_α is a neighbourhood of $\circ$ in E_α. Since $A(V)$ is absolutely convex and since $A(V) \supset A_\alpha(V_\alpha)$, it follows that $A(V) \supset \Gamma_\alpha U_\alpha$ and this is a neighbourhood of $\circ$ in E. Hence A is open.

The same proof can be used for the more general version:

(6) *Let A be a linear mapping of the subspace D of the webbed space F onto the ultrabornological space E. If G(A) is sequentially closed in F × E, then A is open.*

4. Hereditary properties of webbed and strictly webbed spaces. Our aim is to prove that the classes of webbed resp. strictly webbed spaces are stable under the operations mentioned in GROTHENDIECK's conjecture and thus give a positive answer to this conjecture.

(1) *Every sequentially closed subspace H of a webbed resp. strictly webbed space E is webbed resp. strictly webbed.*

If $\mathscr{W} = \{C_{n_1,\dots,n_k}\}$ is a web on E, then $\mathscr{W}_H = \{C_{n_1,\dots,n_k} \cap H\}$ satisfies condition 1.(w). We remark that in the definition of a web it is not required that the sets $C_{n_1,\dots,n_k}$ have all to be nonempty. $\mathscr{W}_H$ satisfies the additional conditions for a $\mathscr{C}$-web resp. a strict web since H is sequentially closed.

(2) *If A is a sequentially continuous linear mapping of E in F and if* $\mathscr{W} = \{C_{n_1,\dots,n_k}\}$ *is a strict resp.* $\mathscr{C}$*-web on E, then* $\{A(C_{n_1,\dots,n_k})\}$ *is a strict resp.* $\mathscr{C}$*-web on A(E).*

The proof consists in the trivial verification of the definitions. (2) has the following corollaries:

(3) *Every quotient of a webbed resp. strictly webbed space is webbed resp. strictly webbed.*

(4) *If* $E[\mathfrak{T}]$ *is a webbed resp. strictly webbed space, this is true also for* $E[\mathfrak{T}']$, *where* $\mathfrak{T}'$ *is a locally convex topology weaker than* $\mathfrak{T}$.

A result in the other direction is

(5) *Let $E[\mathfrak{T}]$ be a webbed resp. strictly webbed space. Then the associated bornological space $E[\mathfrak{T}^{\times}]$ and the associated barrelled space $E[\mathfrak{T}^{t}]$ have the same property.*

This theorem is a particular case of a stronger theorem which we will prove later (cf. 8.(5)).

(6) *The topological product $E = \prod_{i=1}^{\infty} E_i$ of a sequence of webbed resp. strictly webbed spaces is again a space of this type.*

Proof. Let $\{C^{(p)}_{n_1,\ldots,n_k}\}$ be the web on E_p. We define a web on E in the following way: We set $D_{n_1} = C^{(1)}_{n_1} \times E_2 \times E_3 \times \cdots$ for all $n_1 = 1, 2, \ldots$. Then we set

$$D_{n_1,(n_2,n_1^{(1)})} = C^{(1)}_{n_1,n_2} \times C^{(2)}_{n_1^{(1)}} \times E_3 \times \cdots,$$

where the pairs $(n_2, n_1^{(1)})$ may be denoted by one index $\tilde{n}_2 = 1, 2, \ldots$. The next step is

$$D_{n_1,(n_2,n_1^{(1)})(n_3,n_2^{(1)},n_1^{(2)})} = C^{(1)}_{n_1,n_2,n_3} \times C^{(2)}_{n_1^{(1)},n_2^{(1)}} \times C^{(3)}_{n_1^{(2)}} \times E_4 \times \cdots,$$

where again $(n_3, n_2^{(1)}, n_1^{(2)})$ may be replaced by one index $\tilde{n}_3$, and so on. It is clear that condition 1.(w) is satisfied; hence $\{D_{n_1,\tilde{n}_2,\ldots,\tilde{n}_k}\}$ is a web on E.

It remains to prove that this web is of type $\mathscr{C}$ resp. strict when this is the case for all the webs on the E_i. We begin with the assumption that all the $\{C^{(p)}_{n_1,\ldots,n_k}\}$ are $\mathscr{C}$-webs.

Let $n_1, \tilde{n}_2 = (n_2, n_1^{(1)}), \ldots, \tilde{n}_k = (n_k, n_{k-1}^{(1)}, \ldots, n_1^{(k-1)}), \ldots$ be a fixed sequence of indices. There exist by assumption numbers $\rho_k > 0$ such that $\sum_{k=1}^{\infty} \lambda_k x_k^{(1)}$ converges in E_1 for $0 \leqq \lambda_k \leqq \rho_k$ and $x_k^{(1)} \in C^{(1)}_{n_1,\ldots,n_k}$; there exist, furthermore, numbers $\rho_k^{(1)} > 0$ such that $\sum_{k=1}^{\infty} \lambda_k x_k^{(2)}$ converges in E_2 for $0 \leqq \lambda_k \leqq \rho_k^{(1)}$ and $x_k^{(2)} \in C^{(2)}_{n_1^{(1)},\ldots,n_k^{(1)}}$, and so on.

We define $\tilde{\rho}_1 = \rho_1$, $\tilde{\rho}_2 = \inf(\rho_2, \rho_1^{(1)})$, $\tilde{\rho}_3 = \inf(\rho_3, \rho_2^{(1)}, \rho_1^{(2)}), \ldots$. Assume now $x_k = (x_k^{(1)}, x_k^{(2)}, \ldots) \in D_{n_1,\tilde{n}_2,\ldots,\tilde{n}_k}$ and $0 \leqq \lambda_k \leqq \tilde{\rho}_k$; then it follows by our construction that $\sum_{k=1}^{\infty} \lambda_k x_k^{(p)}$ converges in E_p for every p and this means that $\sum_{k=1}^{\infty} \lambda_k x_k$ converges in the topological product $\prod_{i=1}^{\infty} E_i$. This settles the case of a $\mathscr{C}$-web.

Assume now that all the webs $\{C^{(p)}_{n_1,\ldots,n_k}\}$ are strict. Then they are absolutely convex and $\{D_{n_1,\tilde{n}_2,\ldots,\tilde{n}_k}\}$ is also absolutely convex. By definition of the $\tilde{\rho}_j$ we have then $\sum_{k_0}^{\infty} \lambda_k x_k^{(1)} \in C^{(1)}_{n_1,\ldots,n_{k_0}}$, $\sum_{k_0}^{\infty} \lambda_k x_k^{(2)} \in C^{(2)}_{n_1^{(1)},\ldots,n_{k_0-1}^{(1)}}$, and so on. Therefore $\sum_{k_0}^{\infty} \lambda_k x_k$ is in $D_{n_1,\tilde{n}_2,\ldots,\tilde{n}_{k_0}}$ for all $k_0 = 1, 2, \ldots$. Hence the web $\{D_{n_1,\tilde{n}_2,\ldots,\tilde{n}_k}\}$ is strict.

By § 19, 10.(3) every topological projective limit is topologically isomorphic to a closed subspace of a topological product; hence it follows from (1) and (6) that

(7) *A topological projective limit* $E[\mathfrak{T}] = \varprojlim A_{nm}(E_m[\mathfrak{T}_m])$ *of countably many webbed resp. strictly webbed spaces* $E_m[\mathfrak{T}_m]$ *is of the same type.*

We now prove

(8) *The topological inductive limit* $E[\mathfrak{T}] = \varinjlim E_n[\mathfrak{T}_n]$ *of a sequence of webbed resp. strictly webbed spaces* $E_n[\mathfrak{T}_n]$ *is of the same type.*

By assumption E_p has a $\mathscr{C}$-web resp. strict web $\{C^{(p)}_{n_1,\ldots,n_k}\}$ and we have to construct a web on $E = \bigcup_{p=1}^{\infty} E_p$.

We define $D_{n_1} = E_{n_1}$ and $D_{n_1,\ldots,n_k} = C^{(n_1)}_{n_2,\ldots,n_k}$ for all natural numbers $n_1, n_2, \ldots$. Obviously, $\{D_{n_1,\ldots,n_k}\}$ is a web on E. Assume that all $\{C^{(p)}_{n_2,\ldots,n_k}\}$ are $\mathscr{C}$-webs. Then for a fixed sequence $n_1, n_2, n_3, \ldots$ there exist $\rho_2, \rho_3, \ldots$ such that $\sum_{2}^{\infty} \lambda_k x_k$ converges in E_{n_1} for all λ_k, $0 \leqq \lambda_k \leqq \rho_k$, and all $x_k \in C^{(n_1)}_{n_2,\ldots,n_k} \subset E_{n_1}$. Choosing $\rho_1 > 0$ and x_1 in E_{n_1} arbitrary, the convergence of $\sum_{k=1}^{\infty} \lambda_k x_k$ in E_{n_1} and therefore in E follows. Hence $\{D_{n_1,\ldots,n_k}\}$ is a $\mathscr{C}$-web.

If the $\{C^{(p)}_{n_2,\ldots,n_k}\}$ are all strict, then $\sum_{k_0}^{\infty} \lambda_k x_k \in C^{(n_1)}_{n_2,\ldots,n_{k_0}}$ for all $k_0 \geqq 2$ and for $k_0 = 1$ we have $D_{n_1} = E_{n_1}$, so nothing is to prove in this case. Hence $\{D_{n_1,\ldots,n_k}\}$ is strict.

From (6) and (8), applied to the case of a locally convex direct sum, and from (3) and § 19, 1.(3) it follows immediately that

(9) *The locally convex hull* $E[\mathfrak{T}] = \sum_{n=1}^{\infty} A_n(E_n[\mathfrak{T}_n])$ *of countably many webbed resp. strictly webbed spaces is again a webbed resp. strictly webbed space.*

De Wilde's closed-graph theorem 2.(2) for ultrabornological spaces and the hereditary properties for webbed and strictly webbed spaces we have proved so far give a complete solution of Grothendieck's conjecture.

For strictly webbed spaces we have an additional hereditary property.

(10) *Let E be a space with a strict web* $\mathscr{W} = \{C_{n_1,\ldots,n_k}\}$. *Then the linear hull* $L(C_{n_1,\ldots,n_p})$ *of any set* $C_{n_1,\ldots,n_p}$ *is again a strictly webbed space.*

Since $C_{n_1,\ldots,n_k}$ is absolutely convex, it is obvious that the sets

$$D_{m_1,\ldots,m_l} = m_1 C_{n_1,\ldots,n_p,m_l,\ldots,m_2}, \qquad m_1, m_2, \ldots = 1, 2, \ldots,$$

define an absolutely convex web on $L(C_{n_1,\ldots,n_p})$. Let the sequence $m_1, m_2, \ldots$ be given and let $\rho_k > 0$ be the sequence associated to the sequence $n_1, \ldots, n_p, m_2, m_3, \ldots$ in E. Then we define $\tilde{\rho}_k = \rho_{p+k-1}$ as the sequence associated with $m_1, m_2, \ldots$ in $L(C_{n_1,\ldots,n_p})$. If $x_k \in m_1 C_{n_1,\ldots,n_p,m_2,\ldots,m_k}$ and $0 \leqq \lambda_k \leqq \tilde{\rho}_k$, then $\sum_{k=1}^{\infty} \lambda_k x_k$ converges in E and, since $\mathscr{W}$ is strict, we have $\sum_{k=1}^{\infty} \lambda_k x_k \in m_1 C_{n_1,\ldots,n_p} \subset L(C_{n_1,\ldots,n_p})$ and $\sum_{k=k_0}^{\infty} \lambda_k x_k \in m_1 C_{n_1,\ldots,n_p,m_2,\ldots,m_{k_0}} = D_{m_1,\ldots,m_{k_0}}$ and the web $\{D_{m_1,\ldots,m_k}\}$ is strict.

The question whether the strong dual of a webbed space is again webbed seems to be open. But there are some results in this direction.

(11) *The strong dual of a metrizable space E is strictly webbed.*

Let $U_1 \supset U_2 \supset \cdots$ be a fundamental sequence of neighbourhoods of $\mathfrak{o}$ in E. Then $E' = \bigcup_{n=1}^{\infty} U_n^\circ$ and E' is complete for the strong topology by § 21, 6.(4). We define $C_{n_1,\ldots,n_k} = U_{n_1}^\circ$ for all $n_1, n_2, \ldots$. This gives an absolutely convex web on E'. We choose $\rho_k > 0$ such that $\sum_{k=1}^{\infty} \rho_k \leqq 1$. Then for $0 \leqq \lambda_k \leqq \rho_k$ and $x_k \in U_{n_1}^\circ$ we have $\sum_{k_0}^{\infty} \lambda_k x_k \in U_{n_1}^\circ$ since $U_{n_1}^\circ$ is strongly complete. Hence E' is a strictly webbed space.

We remark that by using (4) we may replace the strong topology on E' by any weaker locally convex topology.

A (DF)-space has by definition (§ 29, 3.) a fundamental sequence of bounded absolutely convex sets. Using the same arguments as in the proof of (11), we find

(12) *Every sequentially complete* (DF)-*space is a strictly webbed space.*

Finally, we prove the following very useful result:

(13) *The strong dual E' of a locally convex hull $E = \sum_{i=1}^{\infty} A_i(E_i)$ of a sequence of metrizable spaces E_i is strictly webbed.*

From § 19, 2.(3) and (4) it follows easily that E can be written as the locally convex hull of the metrizable spaces $E_i/N[A_i]$ injected in E. We assume therefore that E is given as a locally convex hull of the form $E = \sum_{i=1}^{\infty} E_i$, E_i metrizable.

Let $U_{n_i}^{(i)}$, $n_i = 1, 2, \ldots$ be a fundamental sequence of absolutely convex neighbourhoods of $\mathfrak{o}$ in E_i. Then we define $C_{n_1,\ldots,n_k} = \bigcap_{i=1}^{k} (U_{n_i}^{(i)})^\circ$, where the polar is taken in E'. Every $u \in E'$ is bounded on a neighbourhood of $\mathfrak{o}$ of E_1; it is therefore contained in some $(U_{n_1}^{(1)})^\circ$ and hence $E' = \bigcup_{n_1=1}^{\infty} C_{n_1}$.

Similarly, we have $C_{n_1,\ldots,n_{k-1}} = \bigcup_{n_k=1}^{\infty} C_{n_1,\ldots,n_k}$ and therefore $\mathscr{W} = \{C_{n_1,\ldots,n_k}\}$ is an absolutely convex and closed web on E'.

To see that $\mathscr{W}$ is a $\mathscr{C}$-web we use 1.(1) with $\mu_k = 1$: A sequence $u_k \in C_{n_1,\ldots,n_k}$, $k = 1, 2, \ldots$, is equicontinuous on every E_i since $u_k \in (U_{n_i}^{(i)})^\circ$ for $k \geqq i$; therefore the set $\{u_k\}$ of all u_k, $k = 1, 2, \ldots$, is contained in some $(U_{m_i}^{(i)})^\circ$. But then $\{u_k\} \subset \left(\Gamma_{i=1}^{\infty} U_{m_i}^{(i)}\right)^\circ$ and $\Gamma_{i=1}^{\infty} U_{m_i}^{(i)}$ is a neighbourhood of $\mathfrak{o}$ in E by definition of the hull topology. It follows that $\{u_k\}$ is equicontinuous in E' and is therefore contained in an absolutely convex and weakly compact set in E' which is sequentially complete. Hence $\mathscr{W}$ is a $\mathscr{C}$-web. From 1.(2) it follows that $\mathscr{W}$ is strict.

De Wilde proved in [4′] other hereditary properties for webbed spaces. Compare also De Wilde [6′].

5. A generalization of the open-mapping theorem. Kato proved in [1′] the following generalization in the case of two Banach spaces E and F: If A is a closed linear operator from $D \subset E$ onto a subspace $A(D)$ of finite co-dimension in F, then $A(D)$ is closed and A is open. Goldberg gave a more general version of this theorem in [1′] and I showed in [7′] that even for $A(D)$ of denumerable co-dimension the theorem remains true and it follows in this case that $A(D)$ always has finite co-dimension. De Wilde examined this question in [4′] in the frame of his theory. He proved

(1) *Let $E[\mathfrak{T}_1]$ be a webbed space, $F[\mathfrak{T}_2]$ ultrabornological resp. a locally convex hull of Baire spaces, and let A be a sequentially closed resp. closed linear mapping of a subspace D of E in $F[\mathfrak{T}_2]$.*

If $A(D)$ has the algebraic complement H in F and if $H[\mathfrak{T}']$ is a webbed space for a locally convex topology $\mathfrak{T}' \supset \mathfrak{T}_2$ on H, then A is open, $A(D)$ and H are closed and topologically complementary in F, and $\mathfrak{T}' = \mathfrak{T}_2$ on H.

Proof. The product $E[\mathfrak{T}_1] \times H[\mathfrak{T}']$ is a webbed space by 4.(6). We define a mapping $\tilde{A}$ of $D \times H$ onto F by setting $\tilde{A}(x, z) = Ax + z$ for all $x \in D$, $z \in H$.

$\tilde{A}$ is sequentially closed resp. closed in $E \times H \times F$. We prove this only for the first case. Assume that $(x, z, y) \in \overline{G(\tilde{A})}$. Then there exist $x_n \in D$, $z_n \in H$ such that $x_n \to x \in E$, $z_n \to z \in H$, and $Ax_n + z_n \to y \in F$. Therefore $\tilde{A}(x, z) = Ax + z = y$ and $\tilde{A}$ is sequentially closed. From 3.(6) resp. 3.(4) it follows now that $\tilde{A}$ is open. If U_1 is a $\mathfrak{T}_1$-neighbourhood of $\mathfrak{o}$ in D and U_2 a $\mathfrak{T}'$-neighbourhood of $\mathfrak{o}$ in H, there exists a $\mathfrak{T}_2$-neighbourhood V of $\mathfrak{o}$ in F such that $\tilde{A}(U_1 \times U_2) = A(U_1) + U_2 \supset V$. It follows that

$A(U_1) \supset V \cap A(D)$ and therefore A is open. Since $(A(U_1) + U_2) \cap H = U_2 \supset V \cap H$, $\mathfrak{T}'$ must be weaker than $\mathfrak{T}_2$ on H, and since $\mathfrak{T}' \supset \mathfrak{T}_2$ by assumption, we have $\mathfrak{T}' = \mathfrak{T}_2$ on H.

Finally, we prove that the projection P of F onto H with kernel $A(D)$ is continuous. Let U_2 be given in H and let $\tilde{A}(U_1 \times U_2) \supset V$; then $P(V) \subset P(\tilde{A}(U_1 \times U_2)) = P(A(U_1) + U_2) = U_2$, so P is continuous. Hence $H = P(F)$ is closed and $A(D) = (I - P)(F)$ is closed too and a topological complement for H in F.

We have the following special case:

(2) *Let E be a webbed space, F ultrabornological resp. a locally convex hull of Baire spaces and A a sequentially closed resp. closed linear mapping of a subspace D of E in F.*

If $A(D)$ has finite or denumerable co-dimension in F, then A is open, $A(D)$ is closed and every algebraic complement H of $A(D)$ is a topological complement of $A(D)$ and $\mathfrak{T}_2$ coincides on H with the strongest locally convex topology.

Proof. If $\mathfrak{T}'$ is the strongest locally convex topology on H, then $H[\mathfrak{T}']$ is topologically isomorphic to φ and this is a webbed space by 4.(8). The proposition follows immediately from (1).

The result of KATO is obviously contained in (2). The result that $A(D)$ never has denumerable co-dimension is a special case of

(3) *Let E be a webbed space, F ultrabornological and metrizable, and A a sequentially closed linear mapping of a subspace D of E into F. If $A(D)$ has at most countable co-dimension in F, then A is open and $A(D)$ closed and of finite co-dimension.*

This follows immediately from (2) since F has no subspace topologically isomorphic to φ since φ is not metrizable.

We note some consequences of (1) concerning the existence of complementary subspaces.

(4) *Let $E[\mathfrak{T}]$ be ultrabornological or a locally convex hull of Baire spaces. If E is the algebraic direct sum $E = H_1 \oplus H_2$ of two subspaces which are webbed spaces for the induced topology, then $H_1[\mathfrak{T}]$ and $H_2[\mathfrak{T}]$ are closed and topologically complementary.*

This follows by applying (1) to the continuous injection of H_1 into E.

Especially interesting is the following result of DE WILDE:

(5) *Let E be a webbed space which is also ultrabornological or a locally convex hull of Baire spaces.*

a) *If E is the algebraic direct sum $E = H_1 \oplus H_2$ of two sequentially* *closed subspaces, then H_1 and H_2 are closed and topologically complementary.*

b) *Every sequentially closed subspace H of E of at most denumerable co-dimension in E is closed and has a topological complement which is finite dimensional or isomorphic to φ.*

Proof. a) is a special case of (4) since H_1, H_2 are webbed spaces by 4.(1). Applying (2) to the injection of H into E gives b). (5) a) is a sharper result for (LF)-spaces than § 34, 8.(5), since (LF)-spaces are ultrabornological and webbed spaces by 4.(9). On the other hand, (5) a) does not include the corresponding result (§ 34, 7.(3)) for barrelled Pták spaces.

This raises the general question of the relation between PTÁK's and DE WILDE's results. We first remark that φ_d for $d = 2^{\aleph_0}$ is not a webbed space, since there exists a closed linear mapping of ω onto φ_d which is not continuous (compare § 34, 5.) and ω is certainly ultrabornological. Next we remember that $\varphi\omega$ and $\omega\varphi$ are Pták spaces, even (s)-spaces, but $\varphi\omega \oplus \omega\varphi$ is not even an infra-(s)-space (compare § 34, 9.). On the other hand, it follows from the hereditary properties in 4. that $\varphi\omega \oplus \omega\varphi$ is a webbed space. There exist therefore barrelled webbed spaces which are not infra-(s)-spaces. As we will see in 6., the Pták spaces ω_d, $d > \aleph_0$, are not webbed spaces; therefore theorem § 34, 8.(1) of A. and W. ROBERTSON is not contained as a special case in the otherwise much stronger result 2.(2) of DE WILDE. It follows that the class of webbed spaces does not contain all locally convex spaces F for which a closed linear mapping of an ultrabornological space E into F is always continuous.

Therefore the results of § 34 are not contained in the results of DE WILDE, but his theory gives more information in many problems. Since the topological product of two Pták spaces need not be a Pták space, the proof of (1) cannot be applied when E and H are Pták spaces. Nevertheless, for a Pták space E and $A(D)$ of finite co-dimension in a barrelled space F (2) was proved by KÖTHE [7′] using a different method.

6. The localization theorem for strictly webbed spaces. Let A be a sequentially closed linear mapping of an (F)-space E into a webbed space F. It is natural to ask whether something can be said of the relation of $A(E)$ or $A(B)$, B bounded in E, to a given web $\mathscr{W}$ in F. The answer is simple for strict webs and contained in the following localization theorem of DE WILDE.

(1) a) *Let E be an* (F)-*space and F a space with a strict web $\mathscr{W} = \{C_{n_1,\ldots,n_k}\}$, A a sequentially closed linear mapping of E into F.*

Then there exists a sequence n_k and a sequence $U^{(k)}$ of neighbourhoods of $\circ$ in E such that $A(U^{(k)}) \subset C_{n_1,\ldots,n_k}$ for every $k = 1, 2, \ldots$. Hence $A(E) \subset L(C_{n_1,\ldots,n_k})$ for every k and if B is a bounded subset of E there exist $\alpha_k > \circ$ such that $A(B) \subset \alpha_k C_{n_1,\ldots,n_k}$ for every k.

b) *The same statement holds if A is defined on a nonmeagre subspace D of a Baire space E and $G(A)$ is closed in $E \times F$.*

The proof of a) follows the lines of the proof of 2.(1) ii). There exists a sequence n_k such that $A^{(-1)}(C_{n_1,\dots,n_k})$ is not meagre for every k. Since $\mathscr{W}$ is strict, the $C_{n_1,\dots,n_k}$ are absolutely convex and there exist $\rho_k > 0$ such that $\sum_{k_0}^{\infty} \lambda_k y_k \in C_{n_1,\dots,n_{k_0}}$ for all $\lambda_k \in [0, \rho_k]$ and all $y_k \in C_{n_1,\dots,n_k}$.

We define $M_k = \rho_k A^{(-1)}(C_{n_1,\dots,n_k})$; then $\overline{M}_k$ contains a neighbourhood $U^{(k)}$ of 0. It will be sufficient to prove $\overline{M}_k \subset A^{(-1)}(C_{n_1,\dots,n_k})$, since then $A(U^{(k)}) \subset C_{n_1,\dots,n_k}$. Assume $x_0 \in \overline{M}_k$. Then there exists $x_k \in M_k$ such that $x_0 - x_k \in U^{(k+1)} \subset \overline{M}_{k+1}$. Next we find $x_{k+1} \in M_{k+1}$ such that $x_0 - x_k - x_{k+1} \in U^{(k+2)}$ and so on. We have $x_0 = \sum_{p=0}^{\infty} x_{k+p}$ if we assume $U^{(k)} \subset U_k$, where U_k is a fundamental sequence of neighbourhoods of $\circ$ in E. Since $Ax_{k+p} \in \rho_{k+p} C_{n_1,\dots,n_{k+p}}$, we have $y_0 = \sum_{p=0}^{\infty} Ax_{k+p} \in C_{n_1,\dots,n_k}$. Since A is sequentially closed, $y_0 = Ax_0$ and $x_0 \in A^{(-1)}(C_{n_1,\dots,n_k})$.

The proof of b) is analogous to the proof of 2.(3): The construction of x_{k+p} is the same as before, but in this case one shows that $(x_0, y_0) \in \overline{G(A)} = G(A)$. Let (U, W) be a given neighbourhood of $\circ$ in $E \times F$. We have $x_0 - \sum_{p=0}^{m} x_{k+p} \in \overline{M}_{k+m+1} \subset M_{k+m+1} + U$; therefore there exists $t_{k+m} \in M_{k+m+1}$ such that $x_0 - \sum_{p=0}^{m} x_{k+p} - t_{k+m} \in U$. Since At_{k+m} converges to $\circ$ with $m \to \infty$, we have $y_0 - \sum_{p=0}^{m} Ax_{k+p} - At_{k+m} \in W$ for $m \geqq n_0$. Hence $(x_0, y_0) - \left(\sum_{p=0}^{m} x_{k+p} + t_{k+m}, A\left(\sum_{p=0}^{m} x_{k+p} + t_{k+m}\right)\right) \in (U, W)$, and $(x_0, y_0) \in \overline{G(A)}$.

Let F be an (LF)-space $\bigcup_{n=1}^{\infty} F_n[\mathfrak{T}_n]$. In 4.(8) we constructed a strict web $\{D_{n_1,\dots,n_k}\}$ on F such that $D_n = F_n$. From (1) a) it follows that there exists n such that $A(E) \subset F_n$. Hence A is closed from E into F_n and therefore continuous. So we see that the localization theorem contains as a special case Theorem § 19, 5.(4) of GROTHENDIECK.

The statement in (1) concerning the images of bounded sets is true in the following very general version:

(2) a) *Let A be a sequentially closed linear mapping of the locally convex space E into the space F with the strict web $\mathscr{W} = \{C_{n_1,\dots,n_k}\}$. For every absolutely convex bounded and sequentially complete set B in E there exist a sequence n_k and a sequence $\alpha_k > 0$ such that $A(B) \subset \alpha_k C_{n_1,\dots,n_k}$ for $k = 1, 2, \dots$.*

b) *If F is a space with the strict web $\mathscr{W} = \{C_{n_1,\dots,n_k}\}$, then for every* *absolutely convex bounded and sequentially complete set B of F there exist sequences n_k and $\alpha_k > 0$ such that $B \subset \alpha_k C_{n_1,\ldots,n_k}$ for $k = 1, 2, \ldots$.*

Proof. The space $E_B = \bigcup_{n=1}^{\infty} nB$ is a Banach space and the restriction of A to E_B has a sequentially closed graph in $E_B \times F$, so a) follows from (1) a). For $E = F$ and $A = I$, b) follows from a).

Another application of the localization theorem is

(3) *A strictly webbed space E which is a Baire space is an* (F)*-space.*

If we apply (1) b) to the identity mapping of E into E, there follow the existence of a sequence n_k and a sequence $U^{(k)}$ of neighbourhoods of $\circ$ in E such that $U^{(k)} \subset C_{n_1,\ldots,n_k}$ for every $k = 1, 2, \ldots$. If U is any neighbourhood of $\circ$ in E, there exists $\rho_k > 0$ such that $\rho_k C_{n_1,\ldots,n_k} \subset U$ for some k by 1.(3); hence $\rho_k U^{(k)} \subset U$ and the topology of E is given by the multiples of the $U^{(k)}$, which means that E is metrizable.

If x_n is a Cauchy sequence in E, there exists a subsequence x_{n_j} such that $x_{n_{j+1}} - x_{n_j} \in \rho_k U^{(k)} \subset \rho_k C_{n_1,\ldots,n_k}$ and $x_{n_1} + \sum_{j=1}^{\infty} (x_{n_{j+1}} - x_{n_j})$ converges. The sum is the limit of the sequence x_{n_j} and therefore the limit of the sequence x_n; hence E is complete.

It follows that a topological product of more than countably many (B)- or (F)-spaces is never a strictly webbed space, since these spaces are Baire spaces (§ 34, 8.).

For $\mathscr{C}$-webs the situation is more complicated. One obtains in this case

(4) a) *Let E be an* (F)*-space and F a space with a $\mathscr{C}$-web $\mathscr{W} = \{C_{n_1,\ldots,n_k}\}$, A a sequentially closed linear mapping of E into F.*

Then there exists a sequence n_k and a sequence $U^{(k)}$ of neighbourhoods of $\circ$ in E such that $A(U^{(k)}) \subset \overline{\Gamma C_{n_1,\ldots,n_k}}$ for every $k = 1, 2, \ldots$.

b) *The same statement holds if A is defined on a nonmeagre subspace D of a Baire space E and $G(A)$ is closed in $E \times F$.*

The proof follows the same pattern as the proof of (1). One defines $M_k = \rho_k A^{(-1)}(C_{n_1,\ldots,n_k})$ and there exists $y_k \in M_k$ such that $\overline{M}_k \supset y_k + U^{(k)}$ for some $U^{(k)}$. Then one has to show that $\overline{A^{(-1)}(\overline{\Gamma C_{n_1,\ldots,n_k}})} \subset (1 + 2\varepsilon) \times A^{(-1)}(\overline{\Gamma C_{n_1,\ldots,n_k}})$. This is done in a similar way as in 2.(1) iii) resp. 2.(3). The details are left to the reader (compare De Wilde [3'], p. 48).

With (4) b) it is possible to repeat the first part of the proof of (3) and we obtain

(5) *A webbed space E which is a Baire space is metrizable.*

It follows, in particular, that ω_d for $d > \aleph_0$ is not a webbed space.

7. Ultrabornological spaces and fast convergence. We introduced ultrabornological spaces in § 34, 8. as spaces which have a representation as a locally convex hull $E[\mathfrak{T}] = \sum_{\alpha} E_\alpha[\mathfrak{T}_\alpha]$ of (B)-spaces E_α. An ultrabornological space is bornological and barrelled (§ 27, 1.(3)). Every sequentially complete bornological space is ultrabornological; therefore every (F)-space is ultrabornological. Our intention is to collect further information on ultrabornological spaces. We begin with some characterizations.

We say that an absolutely convex bounded subset B of a locally convex space $E[\mathfrak{T}]$ is a Banach disk if the normed space E_B is a (B)-space with B as closed unit ball (in § 20, 11. we said that B is complete in itself).

(1) *$E[\mathfrak{T}]$ is ultrabornological if and only if $E[\mathfrak{T}]$ is the locally convex hull $\sum_B E_B$, where B runs through the Banach disks in E.*

Let $E[\mathfrak{T}]$ be the locally convex hull $E = \sum_\alpha E_\alpha$ of Banach spaces E_α. The closed unit ball B_α of E_α is a Banach disk in E; hence $E = \sum_\alpha E_{B_\alpha}$.

Let B be any Banach disk in E. The injection $E_B \to E$ is continuous. Since the hull topology of $\sum_B E_B$ is the finest locally convex topology on E such that these injections are continuous, it follows that the identity mapping $\sum_B E_B \to E = \sum_\alpha E_{B_\alpha}$ is continuous. In the same way one shows that the identity mapping $\sum_\alpha E_{B_\alpha} \to \sum_B E_B$ is continuous. Hence $\sum_\alpha E_{B_\alpha}$ and $\sum_B E_B$ define the same locally convex space.

(2) *$E[\mathfrak{T}]$ is ultrabornological if and only if $E[\mathfrak{T}]$ is the locally convex hull $\sum_K E_K$, where K runs through the absolutely convex compact subsets of E.*

Every K is a Banach disk and therefore $\sum_K E_K$ is ultrabornological. Assume now that E is ultrabornological. Then $E[\mathfrak{T}] = \sum_B E_B$, B the Banach disks in E. Let $\mathfrak{T}'$ be the topology of $\sum_K E_K$. We have to prove that $\mathfrak{T}$ and $\mathfrak{T}'$ coincide. Obviously $\mathfrak{T}' \supset \mathfrak{T}$. It will be sufficient to prove that every absolutely convex set U which absorbs all K absorbs all B too. Assume that U does not absorb the Banach disk B. Then there exists a sequence $x_n \in B$ such that $x_n \notin n^2 U$. The set C consisting of all x_n/n and $\mathfrak{o}$ is compact in E_B and its absolutely convex and closed hull $\overline{\Gamma C}$ in E_B is a set K. There exists therefore $\alpha > 0$ such that $\overline{\Gamma C} \subset \alpha U$; hence $(x_n/n) \in \alpha U$, $x_n \in \alpha n U$ for all n, which is a contradiction.

We introduced in § 28, 3. the notion of local convergence. The following notion (De Wilde [3']) is a sharpened form of local convergence. A sequence x_n of elements of a locally convex space $E[\mathfrak{T}]$ is said to be fast convergent to x_0 in E if there is an absolutely convex compact set $K \subset E$ such that x_n and x_0 lie in E_K and such that x_n converges to x_0 with respect to the norm of E_K. A sequence x_n which is fast convergent to $\mathfrak{o}$ is called a fast convergent null sequence. It is obvious that every fast convergent sequence is locally convergent.

In complete analogy to the characterization (§ 28, 3.(2)) of bornological spaces one obtains

(3) *$E[\mathfrak{T}]$ is ultrabornological if and only if every absolutely convex set M which absorbs all the fast convergent null sequences is a $\mathfrak{T}$-neighbourhood of $\mathfrak{o}$ in E.*

We will need the following slight improvement of § 28, 3.(1).

(4) *In an* (F)*-space E every sequence x_n convergent to x_0 is fast convergent to x_0.*

If x_n is fast convergent to x_0 in the locally convex space F, there exist $\rho_n > 0$, $\lim \rho_n = \infty$, such that $\rho_n(x_n - x_0)$ is fast convergent to $\mathfrak{o}$.

Proof. If $y_n \to \mathfrak{o}$ in the (F)-space, there exist $\rho_n > 0$ such that $\lim \rho_n = \infty$ and $\rho_n y_n \to \mathfrak{o}$ by § 28, 1.(5). The closed absolutely convex hull K of all $\rho_n y_n$ is compact in E and obviously $\|y_n\|_K \to 0$ in the norm of E_K. This is the first statement for a sequence converging to $\mathfrak{o}$. If $x_n \to x_0$ in E, then there exist positive $\rho_n \to \infty$ such that all $\rho_n(x_n - x_0)$ lie in an absolutely convex compact set K and $\|x_n - x_0\|_K \to 0$. If K_1 is the absolutely convex hull of K and the set $\{\alpha x_0, |\alpha| \leqq 1\}$, then K_1 is compact by § 20, 6.(5) and all x_n and x_0 lie in $2K_1$ and again $\|x_n - x_0\|_{K_1} \to 0$.

The second statement follows from § 28, 1.(5).

We point out that the last part of our proof shows also that in a locally convex space a sequence x_n is fast convergent to x_0 if and only if $x_n - x_0$ is fast convergent to $\mathfrak{o}$.

Remark. It follows from the first part of (4) that x_n is fast convergent to x_0 in E if and only if there exists a Banach disk B in E such that x_n converges to x_0 in the (B)-space E_B. Hence the notion of fast convergence in E depends only on the dual system $\langle E', E\rangle$.

From (1), (2), and (3) follows

(5) *A locally convex space $E[\mathfrak{T}_k(E')]$ is ultrabornological if and only if every linear functional on E is continuous which is bounded*

a) *on every Banach disk of E, or*

b) *on every absolutely convex compact subset of E, or*

c) *on every fast convergent null sequence of E.*

For linear mappings we obtain

(6) *Let A be a linear mapping of an ultrabornological space E in a locally convex space F. A is continuous if and only if*

a) *A maps fast convergent null sequences in fast convergent null sequences, or*

b) *A maps fast convergent null sequences in bounded sequences.*

The conditions are necessary. We prove that b) is sufficient. If V is a neighbourhood of $\mathfrak{o}$ in F and x_n is a fast convergent null sequence in E, then V absorbs the bounded sequence Ax_n; hence $A^{(-1)}(V)$ absorbs x_n. By (3) $A^{(-1)}(V)$ is a neighbourhood of $\mathfrak{o}$ in E.

We conclude this section with some remarks on the hereditary properties of ultrabornological spaces.

(7) *The locally convex hull $E[\mathfrak{T}] = \sum_{\alpha} A_\alpha(E_\alpha[\mathfrak{T}_\alpha])$ of ultrabornological spaces $E_\alpha[\mathfrak{T}_\alpha]$ is an ultrabornological space.*

In particular, every quotient of an ultrabornological space is ultrabornological and the locally convex direct sum $\bigoplus_{\alpha} E_\alpha[\mathfrak{T}_\alpha]$ of ultrabornological spaces is ultrabornological.

The first statement follows from the definition of an ultrabornological space and § 19, 1.(6). Another proof uses (2) and follows the argument of the proof of § 28, 4.(1).

From (7) it follows too that the topological product of a finite number of ultrabornological spaces is ultrabornological. In the general case one proceeds as in the case of bornological spaces. One obtains the following version of the Mackey–Ulam theorem (§ 28, 8.(6)):

(8) *The topological product of d ultrabornological spaces is ultrabornological if d is smaller than the smallest strongly inaccessible cardinal.*

Proof. This follows from the fact that ω_d is ultrabornological for these d and the result corresponding to § 28, 8.(3) a), i.e., that a product $\prod_{\alpha} E_\alpha$ of d ultrabornological spaces is ultrabornological if ω_d is ultrabornological. The proof of this result is the same as for § 28, 8.(3) a) with the only difference that one uses instead of § 28, 8.(1) the following proposition:

If the topological product $E = \prod_{\alpha} E_\alpha$ of ultrabornological spaces E_α is not ultrabornological, there exists a discontinuous linear functional on E which is bounded on all Banach disks and vanishes on $\bigoplus_{\alpha} E_\alpha$.

For the proof we remark that by (5) a) there exists a discontinuous u which is bounded on all Banach disks. Next we prove that u vanishes on the direct sum of all but finitely many E_α. Assume the contrary. Then there exists a sequence $x_k \in \bigoplus_{\alpha} E_\alpha$ such that $ux_k = k$ and the elements x_k are contained in finite sums $\bigoplus_{j=1}^{n_k} E_{\alpha_j^{(k)}}$ with pairwise disjoint sets of indices $\{\alpha_1^{(k)}, j = 1, \ldots, n_k\}$. Then E contains the Banach disk B consisting of all elements $\sum_{k=1}^{\infty} \gamma_k x_k$, $\sum |\gamma_k| \leqq 1$, and u is not bounded on B, which is a contradiction.

The linear functional vanishing on $\bigoplus_{\alpha} E_\alpha$ is now constructed as in § 28, 8.(1). For a different proof of (8) compare DE WILDE [7′].

8. The associated ultrabornological space. Let $E[\mathfrak{T}]$ be locally convex. Let $\mathfrak{T}^u$ be the locally convex topology defined by the family of all absolutely convex sets which absorb all Banach disks B of E as a system of neighbourhoods of $\mathfrak{o}$. Evidently, $E[\mathfrak{T}^u]$ is the locally convex hull $\sum_B E_B$ and ultrabornological. It follows from 7.(2) that $E[\mathfrak{T}^u]$ is identical with $\sum_K E_K$, where K is any compact Banach disk in E.

We have $\mathfrak{T} \subset \mathfrak{T}^u$ and $\mathfrak{T}^u$ is the weakest ultrabornological topology on E which is stronger than $\mathfrak{T}$; $E[\mathfrak{T}^u]$ is called the u l t r a b o r n o l o g i c a l s p a c e a s s o c i a t e d w i t h $E[\mathfrak{T}]$. If $E' = (E[\mathfrak{T}])'$ and if $\mathfrak{T}_1$ is compatible with the dual system $\langle E', E \rangle$, then $\mathfrak{T}^u = \mathfrak{T}_1^u$.

Obviously $\mathfrak{T}^u \supset \mathfrak{T}^\times$, where $\mathfrak{T}^\times$ is the associated bornological topology on E, and $\mathfrak{T}^u \supset \mathfrak{T}^t$, where $\mathfrak{T}^t$ is the associated barrelled topology; since $E[\mathfrak{T}^u]$ is barrelled, $\mathfrak{T}^u \supset \mathfrak{T}$ and $\mathfrak{T}^t$ is the weakest barrelled topology with this property.

(1) *If A is a linear continuous mapping of an ultrabornological space E into the locally convex space $F[\mathfrak{T}]$, then A remains continuous if we replace $\mathfrak{T}$ by $\mathfrak{T}^u$.*

Let V be a $\mathfrak{T}^u$-neighbourhood of $\mathfrak{o}$ in F; then V absorbs all fast converging null sequences in F and by 7.(5) $A^{(-1)}(V)$ absorbs all fast convergent null sequences in E; hence $A^{(-1)}(V)$ is a neighbourhood of $\mathfrak{o}$ in E by 7.(5).

From the definition of $E[\mathfrak{T}^u]$ and 7.(5) follows

(2) *$(E[\mathfrak{T}^u])'$ consists of all linear functionals $u^{(1)}$ on E which are bounded on all fast convergent null sequences of E or which are bounded on all Banach disks of E.*

By $\mathfrak{T}_{cf}(E)$ we denote the topology of uniform convergence on all fast convergent null sequences of E. If C is the set consisting of the elements of a fast convergent null sequence, then its closed absolutely convex cover $\overline{\Gamma(C)}$ is a compact subset of some E_B, where B is a Banach disk in E. The sets $\overline{\Gamma(C)}$ and their subsets constitute the saturated class $\mathfrak{M}_0$ defining the topology $\mathfrak{T}_{cf}(E)$ on $E' = (E[\mathfrak{T}])'$.

(3) *Let* $E[\mathfrak{T}]$ *be locally convex and* E' *its dual. Then* $E[\mathfrak{T}^u]'$ *is the completion* $\widetilde{E}'$ *of* $E'[\mathfrak{T}_{cf}(E)]$.

Proof. By GROTHENDIECK's theorem (§ 21, 9.(2)) $\widetilde{E}'$ consists of all linear functionals $u^{(2)}$ on E such that the restrictions on every $\overline{\Gamma(C)}$ are weakly continuous. Such a $u^{(2)}$ is always a $u^{(1)}$ in the sense of (2).

Conversely, let $u^{(1)}$ be given and a $\overline{\Gamma(C)}$. $\overline{\Gamma(C)}$ is compact in some E_B, B a Banach disk. Now $u^{(1)}$ is bounded on B and therefore continuous on E_B in the sense of the norm topology $\mathfrak{T}_B$ of E_B. Since $\mathfrak{T}_B$ and $\mathfrak{T}_s(E')$ coincide on the compact set $\overline{\Gamma(C)}$, $u^{(1)}$ is weakly continuous on $\overline{\Gamma(C)}$ and therefore a $u^{(2)}$.

As a consequence we obtain a characterization of ultrabornological spaces in close analogy to that of bornological spaces (§ 28, 5.(4)).

(4) *A locally convex space* $E[\mathfrak{T}]$ *is ultrabornological if and only if* $\mathfrak{T}$ *is the Mackey topology and* E' *is* $\mathfrak{T}_{cf}(E)$*-complete.*

Proof. If E is ultrabornological, then E' is $\mathfrak{T}_{cf}$-complete by (2). If, conversely, E' is $\mathfrak{T}_{cf}$-complete, then, as in the proof of (3), E' consists of all $u^{(1)}$ in the sense of (2) and E is ultrabornological by 7.(5).

We now apply our results on ultrabornological spaces to webbed spaces. The following theorem is due to M. POWELL [1'].

(5) *If* $E[\mathfrak{T}]$ *is a webbed space, then its associated ultrabornological space* $E[\mathfrak{T}^u]$ *is again webbed.*

Proof. Let $E[\mathfrak{T}]$ have the $\mathscr{C}$-web $\mathscr{W} = \{C_{n_1,\ldots,n_k}\}$. Then for a fixed sequence $n_1, n_2, \ldots$ there exist real numbers $\rho_k > 0$ such that $\sum_{k=1}^{\infty} \lambda_k x_k$ converges in E for all $x_k \in C_{n_1,\ldots,n_k}$ and all λ_k such that $|\lambda_k| \leqq \rho_k$, $k = 1, 2, \ldots$ (compare the remarks in 1. following the definition of a $\mathscr{C}$-web).

Let $x_k \in C_{n_1,\ldots,n_k}$ be a given sequence. Then the sequence $\rho_k x_k$ converges to o in $E[\mathfrak{T}]$. Hence K, the absolutely convex closed cover of the sequence $\rho_k x_k$, is compact in E (it is weakly compact by an argument analogous to that in § 20, 9.(6), therefore complete and hence compact by § 20, 6.(3)).

Let $E[\mathfrak{T}_1]$ be the locally convex hull of all the spaces E_K. Then $E[\mathfrak{T}_1]$ is ultrabornological and $\mathfrak{T}_1 \supset \mathfrak{T}$, hence $\mathfrak{T}_1 \supset \mathfrak{T}^u$ and $E[\mathfrak{T}^u]$, is webbed if $E[\mathfrak{T}_1]$ is webbed.

But this is nearly obvious: Take the same web $\mathscr{W} = \{C_{n_1,\ldots,n_k}\}$ and the real numbers $\sigma_k = \rho_k/2^k$ instead of ρ_k. Then $\sum_{k=1}^{\infty} \mu_k x_k$, $x_k \in C_{n_1,\ldots,n_k}$ and $|\mu_k| \leqq \sigma_k$, converges in E_K and therefore in $E[\mathfrak{T}_1]$.

It follows that all webbed spaces can be obtained from the ultrabornological webbed spaces by weakening the topology.

Taking into account the remark preceding (1), it is now clear that (5) implies Theorem 4.(5).

Recalling the definition of fast convergence we realize that the last part of the proof of (5) includes also the following statement:

(6) *If $\mathscr{W} = \{C_{n_1,\ldots,n_k}\}$ is a $\mathscr{C}$-web of the webbed space $E[\mathfrak{T}]$ and $n_1, n_2, \ldots$ a fixed sequence, there exists a sequence $\sigma_k > 0$ such that every series $\sum_{k=1}^{\infty} \mu_k x_k$ is fast convergent in $E[\mathfrak{T}]$, where $x_k \in C_{n_1,\ldots,n_k}$ and $|\mu_k| \leqq \sigma_k$.*

This fact enabled De Wilde to give a sharper form to his closed-graph theorem for ultrabornological spaces.

We say that a linear mapping A is fast sequentially closed if the graph $G(A)$ is closed for fast convergence in E and in F.

We begin with a simple case.

(7) *A linear fast sequentially closed mapping A of an ultrabornological space E into an* (F)*-space F is always continuous.*

Proof. Let K be absolutely convex and compact in E. Then the restriction A_K of A to E_K is sequentially closed by 7.(4); therefore A_K is continuous from E_K into F. By 7.(2) E is the locally convex hull of the E_K and therefore A is continuous.

The general theorem is the following.

(8) *A fast sequentially closed linear mapping A of an ultrabornological space E into a webbed space F is continuous.*

As in the proof of (7), it is sufficient to prove this for a (B)-space E. We indicate the necessary changes in the proof of 2.(1) to arrive at this new version. One has only to replace the numbers ρ_k by the numbers σ_k determined in (6) and to realize in part ii) of the proof that then $\sum_{k=1}^{\infty} Ax_k$ is fast convergent in F. Since E is a (B)-space, the convergence of $\sum_1^{\infty} x_k$ is fast anyway.

The corresponding open-mapping theorem is

(9) *A fast sequentially closed linear mapping A of a webbed space F onto an ultrabornological space E is open.*

The details of the proof are left to the reader.

De Wilde showed in [5′] that the following characterization of ultrabornological spaces corresponds to the characterization (§ 34, 7.(1)) of barrelled spaces by Mahowald.

(10) *If every linear fast sequentially closed mapping of the locally convex space $E[\mathfrak{T}]$ into an arbitrary* (B)*-space is continuous, then $E[\mathfrak{T}]$ is ultrabornological.*

The converse is a special case of (7).

Proof. By 7.(3) it is sufficient to prove that an absolutely convex set U which absorbs all fast convergent null sequences is a $\mathfrak{T}$-neighbourhood of $\mathfrak{o}$ in E.

As in the proof of § 34, 7.(1), we introduce the normed space $E_U = E/N_U$ and its completion $\tilde{E}_U$. It will be sufficient to prove that the canonical mapping J of E into $\tilde{E}_U$ is continuous, because then the inverse image of the open unit ball in $\tilde{E}_U$ is open and contained in U.

By assumption J is continuous if its graph is fast sequentially closed in $E \times \tilde{E}_U$. Let $x_n \in E$ be fast convergent to x_0 and let Jx_n converge to y_0 in $\tilde{E}_U$. There exist $\rho_n > 0$, $\lim \rho_n = \infty$, such that $\rho_n(x_n - x_0)$ is fast convergent to $\mathfrak{o}$. But then U absorbs the sequence $\rho_n(x_n - x_0)$ and therefore Jx_n converges to Jx_0 in $\tilde{E}_U$. It follows that $y_0 = Jx_0$ and J is fast sequentially closed.

9. Infra-(u)-spaces. Let $\mathscr{A}$ be a class of locally convex spaces. In analogy with the notation introduced in § 34, 11., $\mathscr{C}^\tau(\mathscr{A})$ will denote the class of all locally convex spaces F for which every closed linear mapping of an $E \in \mathscr{A}$ into F is continuous. If $\mathscr{A}$ is the class $\mathscr{T}$ of all barrelled spaces, then KŌMURA's closed-graph theorem (§ 34, 9.(4)) says that $\mathscr{C}^\tau(\mathscr{T})$ is the class of all infra-(s)-spaces.

Let now $\mathscr{U}$ be the class of ultrabornological spaces. DE WILDE's results show that the class of webbed spaces is contained in the class $\mathscr{C}^\tau(\mathscr{U})$, but there exist spaces in $\mathscr{C}^\tau(\mathscr{U})$ which are not webbed spaces. We will try to determine $\mathscr{C}^\tau(\mathscr{U})$. At the same time we will solve the following problem: What is $\mathscr{C}^\tau(\mathscr{B})$, the maximal class of spaces F, for which the closed-graph theorem for mappings of any (B)-space into a space F holds?

These results have been found independently by EBERHARDT [2′], GRATHWOHL [1′], and POWELL [1′] in 1972. But they may also be developed as special cases of a more general theory sketched by KŌMURA in his paper [1] of 1962. We will not give here an exposition of his general theory; it is given in detail by POWELL [1′]. The existence of such a theory will become plausible to the reader by observing the close analogy between the following and § 34, 9. and 10.

(1) *The classes $\mathscr{C}^\tau(\mathscr{B})$ and $\mathscr{C}^\tau(\mathscr{U})$ coincide.*

We have only to prove that an $F \in \mathscr{C}^\tau(\mathscr{B})$ is also contained in $\mathscr{C}^\tau(\mathscr{U})$. Let E be ultrabornological, $E = \sum_\alpha I_\alpha(E_\alpha)$, the E_α (B)-spaces and I_α the canonical injection of E_α in E. Let A be a closed linear mapping of E into $F \in \mathscr{C}^\tau(\mathscr{B})$. Then every $A_\alpha = AI_\alpha$ is a closed mapping of E_α into F and continuous since $F \in \mathscr{C}^r(\mathscr{B})$. But then A is also continuous and our statement follows.

We say that the locally convex space $E[\mathfrak{T}]$ is an infra-(u)-space if $\mathfrak{T}_1^u = \mathfrak{T}^u$ for every locally convex topology $\mathfrak{T}_1$ on E such that $\mathfrak{T}_1 \subset \mathfrak{T}$.

Corresponding to KŌMURA's closed-graph theorem we obtain

(2) *The class* $\mathscr{C}^r(\mathscr{U})$ *consists of all infra*-(u)-*spaces.*

If we replace the associated barrelled topology $\mathfrak{T}^t$ by the associated ultrabornological topology $\mathfrak{T}^u$ and recall 8.(1), then the proof of § 34, 9.(4) changes into a proof of (2).

The open-mapping theorem connected with this closed-graph theorem can be easily obtained.

We say that a locally convex space is a (u)-space if all its quotients are infra-(u)-spaces. We leave it to the reader to verify that the proof of ADASCH's open-mapping theorem (§ 34, 10.(3)) may be used in our case too and that one obtains

(3) a) *Every closed linear mapping A of a* (u)-*space E onto an ultrabornological space F is open.*

b) *The* (u)-*spaces are characterized by this property.*

It follows from 3.(1) that every webbed space is a (u)-space.

We make some remarks on these new classes of spaces.

(4) *If $E[\mathfrak{T}]$ is an infra*-(u)-*space resp.* (u)-*space and if $\mathfrak{T}_1 \subset \mathfrak{T}$, then $E[\mathfrak{T}_1]$ is again such a space.*

This follows easily from the definitions.

(5) *Every closed linear subspace of an infra*-(u)-*space resp.* (u)-*space is again such a space.*

The proofs of § 34, 9.(6) and § 34, 10.(2) can be used also in this case.

(6) *Every quotient of a* (u)-*space is a* (u)-*space.*

Proof. Let E be a (u)-space and E/H a quotient. A closed subspace of E/H is of the form L/H, where L is closed in E, and by § 15, 4. the quotient $(E/H)/(L/H)$ is isomorphic to E/L, which is an infra-(u)-space by assumption. It follows that E/H is a (u)-space.

This is nearly all that is known on hereditary properties in contrast to the subclass of webbed spaces. The statement corresponding to § 34, 9.(9) is false since there exist (LB)-spaces which are not complete (§ 31, 6.) and such a space is a webbed space and therefore an ultrabornological infra-(u)-space.

If $E[\mathfrak{T}]$ is a webbed space, then $E[\mathfrak{T}^u]$ is again a webbed space by 8.(5). But there exist infra-(u)-spaces $E[\mathfrak{T}]$ such that $E[\mathfrak{T}^u]$ is not infra-(u) (see Eberhardt [3'], Section 1).

Our definition of infra-(u)-spaces corresponds to the characterization of infra-(s)-spaces given in § 34, 9.(3). These spaces were defined by using a property of the dual. Such a dual characterization is possible also for infra-(u)-spaces.

(7) *$E[\mathfrak{T}]$ is an infra-*(u)*-space if and only if for every weakly dense subspace H of E' the completion of $H[\mathfrak{T}_{cf}(E[\mathfrak{T}_s(H)])]$ coincides with $\widetilde{E}'$, the completion of $E'[\mathfrak{T}_{cf}(E[\mathfrak{T}])]$.*

This follows from the definition of an infra-(u)-space, from 8.(3), and from the fact that it is sufficient to consider only the $\mathfrak{T}_1 \subset \mathfrak{T}$ of the form $\mathfrak{T}_1 = \mathfrak{T}_s(H)$, where H is weakly dense in E'. We remark that the topology $\mathfrak{T}_{cf}$ depends not only on the vector space E but also on the topology on E.

Finally, we mention another case of Kōmura's general theory. Let $\mathscr{N}$ be the class of normed spaces and $\mathscr{BO}$ the class of bornological spaces. Then $\mathscr{C}^\tau(\mathscr{N}) = \mathscr{C}^\tau(\mathscr{BO})$ and this class is very small. It was thoroughly investigated by Eberhardt [2'].

10. Further results. We did not follow here the method employed by many authors (De Wilde [3'], Raíkow [1']) to prove "two-sided" closed-graph theorems or, what is the same, closed-graph theorems for linear relations. This method has the advantage that a closed-graph theorem and an open-mapping theorem are special cases of one single theorem for a linear relation. But since these theorems are rather abstract and seem to have no interesting applications, we preferred our elementary approach.

We note some additional results. In his paper [8'] De Wilde proves the following: Let E and F be topological vector spaces and H a subspace of E of finite co-dimension. If A is a linear mapping of H into F with a graph closed in $H \times F$, then A is the restriction to H of a closed linear mapping from E into F. From this theorem follows: If every closed linear mapping of E into the fixed space F is continuous, then this property holds also for the subspaces of E of finite co-dimension.

In [7'] De Wilde studies the following problem: Let the class $\mathscr{F}$ of spaces F be fixed; for example, $\mathscr{F}$ is the class of infra-Pták-spaces or the class of strictly webbed spaces. Let $\mathscr{C}^l(\mathscr{F})$ be again the class of all spaces E such that the closed-graph theorem is true for mappings of E in every F in $\mathscr{F}$. Does $\mathscr{C}^l(\mathscr{F})$ contain all the topological products of its elements? The answer is positive in both of these cases and in others.

MacIntosh [1′] gave a version of the closed-graph theorem which is not contained in the previous results.

(1) *Let* $E[\mathfrak{T}]$ *be a sequentially complete locally convex space,* $\mathfrak{T}$ *the Mackey topology, and let* $E'[\mathfrak{T}_b(E)]$ *be complete. Let F be a semi-reflexive webbed space.*

Then every sequentially closed linear mapping A from E in F is continuous.

Proof. The associated bornological space $E[\mathfrak{T}^\times] = E_1$ is ultrabornological and from the closed-graph theorem 2.(2) it follows that A is continuous from E_1 into F. Hence A' is continuous from $F'[\mathfrak{T}_b(F)]$ into $E_1'[\mathfrak{T}_b(E)]$ (the bounded sets in E and E_1 are the same).

Since $E'[\mathfrak{T}_b(E)]$ is complete, E' is a closed subspace of $E_1'[\mathfrak{T}_b(E)]$ and its inverse image $D[A'] = A'^{(-1)}(E')$ is therefore a closed subspace of $F'[\mathfrak{T}_b(F)]$.

Since F is semi-reflexive, $\mathfrak{T}_b(E)$ coincides with $\mathfrak{T}_k(F)$ on F' and $D[A']$ is therefore a weakly closed subspace of F'. Hence $D[A'] = F'$, since $D[A']$ is weakly dense in F' as the domain of definition of the adjoint of a closed mapping. Therefore A' is weakly continuous from F' into E' and A is weakly continuous from E into F. Since every weakly continuous mapping is continuous for the Mackey topologies, our statement is proved.

An interesting consequence of (1) is the following result (see De Wilde [3′], p. 99).

(2) *Let E and F be* (F)-*spaces and A a weakly sequentially closed linear mapping from* E' *into* F'. *Then A is weakly continuous.*

Proof. $E'[\mathfrak{T}_k(E)]$ is sequentially complete by § 21, 6.(4). Its strong dual is E, which is complete. The space $F'[\mathfrak{T}_s(F)]$ is a webbed space by 4.(11) and 4.(4). Since F is barrelled, $F'[\mathfrak{T}_s(F)]$ is semi-reflexive by § 23, 3.(1). A is sequentially closed for the topologies $\mathfrak{T}_k(E)$ on E' and $\mathfrak{T}_s(F)$ on F'; therefore (1) applies and A is continuous and also weakly continuous.

As we said at the beginning of this paragraph, we followed here mainly the ideas of De Wilde. A short exposition of the methods of Schwartz and Martineau is contained in an appendix to the book [1′] of Treves. A detailed version of this theory and the connections with the theory of De Wilde are given in De Wilde [3′] and [9′].

Recently W. Robertson [1′] developed a systematic theory based on the ideas of Kelley (compare § 34, 4.), which leads to a very general but rather abstract closed-graph theorem which contains many of the theorems of §§ 34 and 35 as special cases. This theory is valid also for non-locally convex spaces. Different kinds of webbed spaces are also considered in the frame of this theory.

§ 36. Arbitrary linear mappings

1. The singularity of a linear mapping. In our study of the properties of linear continuous mappings of locally convex spaces we were led more and more to consider noncontinuous linear mappings and to investigate their properties. Our intention was to show that many of these mappings are really continuous but we encountered also examples where this was not the case. Noncontinuous mappings play an important role in Hilbert space theory. So the question seems very natural: Is it possible to develop a systematic theory of arbitrary linear mappings of locally convex spaces? The following exposition will be based on papers of ADASCH [1′], BROWDER [1′], and myself [3′].

Let A be a linear mapping defined on a subspace $D[A]$ of the locally convex space $E[\mathfrak{T}_1]$ with the range $R[A] = A(D[A])$ in the locally convex space $F[\mathfrak{T}_2]$. If the domain of definition $D[A]$ is dense in $E[\mathfrak{T}_1]$, then we say A is dense or densely defined.

We do not assume that A is continuous. It is natural to describe the discontinuity of A at the point $\mathfrak{o}$ in the following way. Let $\mathfrak{U} = \{U\}$ be the filter of all $\mathfrak{T}$-neighbourhoods of $\mathfrak{o}$ in $D[A]$. The images $A(U)$ of all the U generate a filter $A(\mathfrak{U})$ in F. We say that the set of all adherent points of $A(\mathfrak{U})$ is the singularity $S[A]$ of A at the point $\mathfrak{o}$ or, for short, the singularity of A. If $\overline{A(U)}$ denotes the closure of $A(U)$ in F, then we have

$$S[A] = \bigcap_{U \in \mathfrak{U}} \overline{A(U)}. \tag{1}$$

(2) *$S[A]$ is a closed subspace of F.*

Proof. Since it is sufficient to take only the intersection of all absolutely convex $\overline{A(U)}$, $S[A]$ is absolutely convex and closed. If $x \in S[A]$, $\rho > 0$, then $\rho x \in \bigcap_U \overline{A(\rho U)} = S[A]$; hence $S[A]$ is a linear subspace.

Using nets instead of filters, we have (compare § 2, 4. and 5.)

(3) *$S[A]$ is the set of all $y \in F$ such that there exists a net $x_\alpha \in D[A]$ converging to $\mathfrak{o}$ and Ax_α converging to y.*

This is equivalent to

(4) *$S[A]$ is the set of all $y \in F$ such that $(\mathfrak{o}, y) \in \overline{G(A)}$, $G(A)$ the graph of A in $E \times F$.*

We say that a linear mapping A is regular if $S[A] = \mathfrak{o}$ and singular if $S[A] \neq \mathfrak{o}$. From (4) it follows that A is regular if and only if $\overline{G(A)}$ is the graph of a linear mapping $\bar{A}$. This mapping is obviously a closed linear mapping in our former terminology and $\bar{A}$ is the uniquely determined smallest closed extension of A: In Hilbert space theory a linear mapping which has a closed extension is called closable; hence "regular" and "closable" are equivalent notions and we will use "closable" for "regular" too.

If A is continuous, then the kernel $N[A]$ of A can be defined in the following way: Let $\mathfrak{V} = \{V\}$ be the neighbourhood filter of $\circ$ in $F[\mathfrak{T}_2]$. The inverse images $A^{(-1)}(V) \subset D[A]$ define the inverse filter $\overline{A^{(-1)}(\mathfrak{V})}$ in $D[A]$. Let $\overline{A^{(-1)}(V)}$ be the closure of $A^{(-1)}(V)$ in $D[A]$. Then $\bigcap_{V \in \mathfrak{V}} \overline{A^{(-1)}(V)}$ is the set of adherent points of $A^{(-1)}(\mathfrak{V})$ in $D[A]$ and this set is identical with $N[A]$ since $N[A]$ is closed in $D[A]$.

For an arbitrary A we define now $\bigcap_{V \in \mathfrak{V}} \overline{A^{(-1)}(V)} = Q(A)$ as the extended kernel of A. Analogously to (2) one has

(5) *$Q[A]$ is a closed linear subspace of $D[A][\mathfrak{T}_1]$.*

We have the following connection between $S[A]$ and $Q[A]$:

(6) $\quad Q[A] = A^{(-1)}(S[A]), \qquad A(Q[A]) = S[A] \cap R[A].$

Proof. We have $\overline{A(U)} = \bigcap_{V \in \mathfrak{V}} (A(U) + V)$ and therefore

$$A^{(-1)}(S[A]) = A^{(-1)}\Big(\bigcap_U \bigcap_V (A(U) + V)\Big)$$
$$= \bigcap_U \bigcap_V (U + N[A] + A^{(-1)}(V)).$$

Since $A^{(-1)}(V) \supset N[A]$, $A^{(-1)}(V) + N[A] = A^{(-1)}(V)$; hence

$$A^{(-1)}(S[A]) = \bigcap_V \bigcap_U (U + A^{(-1)}(V)) = \bigcap_V \overline{A^{(-1)}(V)} = Q[A].$$

The second formula in (5) follows immediately from the first.

By analogy to (4) one has

(7) *$Q[A]$ is the set of all $x \in D[A]$ such that $(x, \circ) \in \overline{G(A)}$.*

By (6), $x \in Q[A]$ if and only if $Ax \in S[A]$ and this is equivalent to $(\circ, Ax) \in \overline{G(A)}$. Since $(x, Ax) \in G(A)$, $(x, Ax) - (\circ, Ax) = (x, \circ) \in \overline{G(A)}$ if $(\circ, Ax) \in \overline{G(A)}$ and conversely.

Obviously, $Q[A] \supset N[A]$ and, since $Q[A]$ is closed in $D[A]$, we have also $Q[A] \supset \overline{N}[A]$, the closure of $N[A]$ in $D[A]$.

We say that A is weakly singular if $Q[A] = \overline{N}[A]$ and strongly singular if $\overline{N}[A] \neq Q[A]$. It follows from (5) that if A is regular, then $Q[A] = N[A]$; hence in our terminology a regular mapping is also weakly singular.

For $\bar{N}[A]$ we have the relations

$$(8) \qquad A(\bar{N}[A]) = \bigcap_{U \in \mathfrak{U}} A(U), \qquad \bar{N}[A] = A^{(-1)}\Big(\bigcap_U A(U)\Big).$$

It is sufficient to prove the second formula which follows from

$$A^{(-1)}\Big(\bigcap_U A(U)\Big) = \bigcap_U (U + N[A]) = \bar{N}[A].$$

Let us point out that $S[A]$ and $Q[A]$ are not really dependent on the topologies $\mathfrak{T}_1$, $\mathfrak{T}_2$ but only on the dual systems $\langle E', E\rangle$ and $\langle F', F\rangle$. This is obvious for $S[A]$ from (4), since the closure $\overline{G(A)}$ of $G(A)$ is the same for all admissible topologies on E and F. It follows for $Q[A]$ from (6).

2. Some examples. i) Let e_i, $i = 1, 2, \ldots$, be an orthonormal basis of l^2. We write this basis also as a double sequence e_{ik}, $i, k = 1, 2, \ldots$. Let H be the linear space of all finite linear combinations of the e_{ik}. On $H = D[A]$ we define A by $Ae_{ik} = e_i$ for all i and k. A is a dense linear transformation of $H \subset l^2$ into l^2. Clearly, $A(H) = H$.

We determine the singularity of A. Let U be the closed unit ball in H; then $A(U) = H$: Every $x = \sum_{k=1}^{N} e_{ik}\xi_k$ with $\sum_1^n |\xi_k|^2 \leqq 1$ is in U and $Ax = e_i \sum_1^N \xi_k$ is, for suitably chosen N, ξ_k—an arbitrary multiple of e_i. Since the same is true for every multiple of U, we have $\bigcap_U A(U) = H$. By 1.(8) we have $D[A] = H = \bar{N}[A]$; therefore $Q[A] = \bar{N}[A]$ since $Q[A] \subset D[A]$. Hence A is weakly singular. Since $\overline{A(U)} = l^2$, we have $S[A] = l^2$; in this case the singularity is the closure of the range of A.

We remark that A is an open mapping of $H \subset l^2$ into l^2 since $A(U) = H$ is a neighbourhood of $\circ$ in $A(H) = H$. We have here an example of a dense open weakly singular and not regular linear mapping of a (B)-space into itself. Obviously, $N[A] \neq \bar{N}[A]$.

ii) Let $D[A] = H \subset l^2$ as in i). We define $A^{(n)}$ on H by $A^{(n)}e_{kn+j} = e_j$ for $j > 0, 1, \ldots, n-1$. Then $A^{(n)}(H) = [e_1, \ldots, e_n]$ and by using similar arguments as in i) we find

$$S[A^{(n)}] = A^{(n)}(H) = [e_1, \ldots, e_n], \qquad Q[A^{(n)}] = \bar{N}[A^{(n)}] = H.$$

$A^{(n)}$ is open, weakly singular, not regular, and of n-dimensional range which coincides with the singularity. So this is an example where $S[A^{(n)}]$ is of finite dimension n.

iii) We give now examples of strongly singular linear mappings of (B)-spaces.

Let E be a (B)-space and let A be a one-one linear mapping of E onto itself which is not continuous. $G(A)$ is not closed by the closed-graph theorem; hence there exists $(\circ, y) \in \overline{G(A)}$, $y \neq \circ$. $S[A] \neq \circ$ follows now from 1.(4), $Q[A] \neq \circ$ by 1.(6). Since $N[A] = \circ$, $\bar{N}[A] = \circ$ and A is strongly singular.

We give a concrete example. Let H be the linear span of the unit vectors e_i, $i = 1, 2, \ldots$, in c_0 and define A on H by $Ae_i = (i + 1)e_i$. Let G be an algebraic complement to H, $c_0 = H \oplus G$. For the construction of G one has to use Zorn's lemma. It is possible to do it in such a way that $y^{(0)} = (1, 1/2^2, 1/3^2, \ldots)$ is an element of G. We define A on G by $Ay = y$ for every $y \in G$. Obviously, A is then defined on c_0, A is one-one and onto c_0 and noncontinuous.

We determine an element of $S[A]$. Let U be the closed unit ball in c_0. For every natural number N there exists $k(N)$ such that $\| y_k^{(0)} - y^{(0)} \| < 1/N$ for all $k \geqq k(N)$, where $y_k^{(0)} = (1, \ldots, 1/k^2, 0, 0, \ldots)$ is the kth section of $y^{(0)}$. For $k \geqq k(N)$ all elements $A(y_k^{(0)} - y^{(0)})$ are therefore contained in $(1/N)A(U)$. Since $y_k^{(0)} \in H$ and $y^{(0)} \in G$, one has

$$A(y_k^{(0)} - y^{(0)}) = Ay_k^{(0)} - y^{(0)} = (1, 1/2, \ldots, 1/k, 0, 0, \ldots) - (y^{(0)} - y_k^{(0)}).$$

It follows that $z^{(0)} = (1, 1/2, 1/3, \ldots)$ is contained in $(1/N)\overline{A(U)}$ for every N and therefore $z^{(0)} \in S[A]$. The element $A^{(-1)}z^{(0)}$ is $\neq \circ$ and in $Q[A]$.

iv) Let A_1 be the restriction of A in iii) to $D[A_1] = H \oplus [y^{(0)}]$. Then A_1 is densely defined in c_0, the range $R[A_1] = D[A_1]$. It follows from iii) that $S[A_1] \neq \circ$. Therefore A_1 is a singular mapping of $D[A_1] \subset c_0$ into c_0.

We prove now that $Q[A_1] = \circ$, which means that A_1 is weakly singular. Let U be the closed unit ball in c_0, $U_0 = U \cap D[A_1]$. Since $R[A_1] = D[A_1]$, it is sufficient to prove that no $x_0 \in D[A_1]$, $x_0 \neq \circ$, is contained in all $\overline{A^{(-1)}(\varepsilon U_0)}$, $\varepsilon > 0$.

We may assume x_0 to be of the form $x_0 = z_0 + y^{(0)}$, $z_0 \in H$. An element y of εU_0 has the form $y = z + \lambda y^{(0)}$, $z = (z_1, z_2, \ldots) \in H$, and we have $|z_n + (\lambda/n^2)| \leqq \varepsilon$ for all $n = 1, 2, \ldots$.

We have to show that there exists $\varepsilon > 0$ such that the inequality $\|x_0 - A(z + \lambda y^{(0)})\| < \delta$ has no solution $z + \lambda y^{(0)} \in \varepsilon U_0$ for some $\delta > 0$.

If N_0 is sufficiently great, the Nth coordinate of x_0 is equal to $1/N^2$ for $n \geqq N_0$ since $z_0 \in H$. We have therefore for all $N \geqq N_0$

$$\left| \frac{1}{N^2} - (N + 1)z_N - \frac{\lambda}{N^2} \right| < \delta \quad \text{and} \quad \left| z_N + \frac{\lambda}{N^2} \right| \leqq \varepsilon.$$

From the second inequality it follows that $z_N = -(\lambda/N^2) + z'_N$, $|z'_N| \leqq \varepsilon$; from the first we obtain then $|\lambda + (1/N)| < \delta N + N(N + 1)\varepsilon$. For $N = N_0$ and $N = N_0 + 1$ we obtain two inequalities for λ which have no common solution λ for sufficiently small ε and δ.

From $Q[A_1] = \circ$ and 1.(6) we conclude that $S[A_1] \cap R[A_1] = \circ$; no point $\neq \circ$ of the singularity lies in the range of A_1.

3. The adjoint mapping. The usefulness of $S[A]$ and $Q[A]$ will become clear at this point, when we investigate whether some kind of duality theory can be developed for arbitrary linear mappings A.

We make the additional assumption that $D[A]$ is dense in $E[\mathfrak{T}_1]$ and hence $(D[A])[\mathfrak{T}_1]' = E'$. The domain of definition of A' is then the set $D[A']$ of all $v \in F'$ such that $A'v$ is an element of E'. Interpreting A as a linear mapping of $D[A]$ in F and applying § 34, 5.(2) we obtain

$$(1) \qquad D[A'] = \bigcup_{U \in \mathfrak{U}} A(U)^\circ,$$

where $\mathfrak{U}$ *is a basis of absolutely convex neighbourhoods of* $\circ$ *in* $(D[A])[\mathfrak{T}_1]$ *and the polars* $A(U)^\circ$ *are taken in* F'.

An easy consequence is

(2) *If A is a dense linear mapping of $D[A] \subset E$ into F, then*

$$D[A']^\perp = S[A] \quad \text{and} \quad \overline{D[A']} = S[A]^\perp,$$

where $\overline{D[A']}$ is the weak closure of $D[A']$ in F' and orthogonality relates to the dual pair $\langle F', F\rangle$.

The first relation follows from (1) and

$$D[A']^\perp = \Big(\bigcup_U A(U)^\circ\Big)^\circ = \bigcap_U A(U)^{\circ\circ} = \bigcap_U \overline{A(U)} = S[A].$$

The second relation follows immediately from the first.

As example i) of 2. shows, $D[A']$ may consist only of the element $\circ$. For $S[A] = \circ$, (2) specializes to

(3) *A dense linear mapping of $D[A] \subset E$ into F is regular (closable) if and only if $D[A']$ is weakly dense in F'.*

The classical relations between kernels and ranges of A and A' in the continuous case are contained in the following theorem:

(4) *Let A be a dense linear mapping of $D[A] \subset E[\mathfrak{T}_1]$ into $F[\mathfrak{T}_2]$, A' the adjoint mapping of $D[A'] \subset F'$ into E'. Then*

$$\text{a)} \qquad R[A]^\perp = N[A'] \quad \textit{and} \quad \overline{R[A]} = N[A']^\perp,$$

where orthogonality is defined by $\langle F', F\rangle$ and $\overline{R[A]}$ is the closure in $F[\mathfrak{T}_2]$;

$$\text{b)} \qquad R[A']^\perp = Q[A] \quad \textit{and} \quad \overline{R[A']} = Q[A]^\perp.$$

In this case orthogonality is defined by $\langle E', D[A]\rangle$ and $\overline{R[A']}$ is the $\mathfrak{T}_s(D[A])$-closure of $R[A']$ in E'.

Proof. a) If $v \in R[A]^\perp$, then $v \in D[A']$ by (1) and even $v \in N[A']$ since $v(Ax) = (A'v)x = 0$ for all $x \in D[A]$. The converse is also true; hence $R[A]^\perp = N[A']$. The second statement in a) follows by polarity.

b) Assume $x \in D[A]$. If $x \in R[A']^\perp$, then $(A'v)x = 0$ for all $v \in D[A']$ or $v(Ax) = 0$ for all $v \in D[A']$; therefore $Ax \in D[A']^\perp$. By (2) $Ax \in S[A]$ and $x \in A^{(-1)}(S[A]) = Q[A]$. Conversely, if $x \in Q[A]$, then by reversing the argument it follows that $x \in R[A']^\perp$. Polarity gives the second equation b).

The following corollary is obvious:

(5) *A is weakly singular if and only if the polar to $R[A']$ in $D[A]$ coincides with $\overline{N}[A]$.*

The range of A is dense in $F[\mathfrak{T}_2]$ if and only if A' is one-one. The range of A' is $\mathfrak{T}_s(D[A])$-dense in E' if and only if A is one-one and weakly singular.

The relations (4) are special cases of the following proposition which corresponds to § 32, 1.(9):

(6) *Let A be a dense linear mapping of $D[A] \subset E[\mathfrak{T}_1]$ into $F[\mathfrak{T}_2]$. Then* a) *$D[A'] \cap A[M]^\circ = A'^{(-1)}(M^\circ)$ for every subset M of $D[A]$;* b) *$D[A] \cap A'(N)^\circ = A^{(-1)}(N^\circ)$ for every subset N of $D[A']$. In* a) *the polars are taken in F' resp. E'; in* b) *in E resp. F.*

We prove a): $v \in A(M)^\circ$ is equivalent to $\Re v(Ax) \leqq 1$ for all $x \in M$. If therefore $v \in D[A']$, then $\Re(A'v)x \leqq 1$ or $A'v \in M^\circ$ or $v \in A'^{(-1)}(M^\circ)$. By reversing the argument one obtains a). The proof of b) is analogous.

By applying § 34, 6.(4) to A as a mapping of $D[A]$ in F we obtain

(7) *Let A be a dense linear mapping of $D[A] \subset E[\mathfrak{T}_1]$ into $F[\mathfrak{T}_2]$. For every absolutely convex and closed neighbourhood V of $\circ$ in F one has*

$$A^{(-1)}(V)^\circ = A'(D[A'] \cap V^\circ),$$

where the first polar is taken in E'. The sets $A'(D[A'] \cap V^\circ)$ are therefore $\mathfrak{T}_s(D[A])$-closed in E'.

We have the following corollary on the structure of $R[A']$ which corresponds to (1):

(8) *Let A be a dense linear mapping of $D[A] \subset E[\mathfrak{T}_1]$ into $F[\mathfrak{T}_2]$. Then*

$$R[A'] = \bigcup_{V \in \mathfrak{B}} A^{(-1)}(V)^\circ,$$

where $\mathfrak{B}$ is a basis of absolutely convex neighbourhoods of $\circ$ in F.

This follows from $R[A'] = \bigcup_V A'(D[A'] \cap V^\circ)$.

4. The contraction of A. We recall that if A is a closed linear mapping of the locally convex space $E[\mathfrak{T}_1]$ into the locally space $F[\mathfrak{T}_2]$, there exists on F a weaker locally convex topology $\mathfrak{T}_2'$ such that A is continuous as a mapping from $E[\mathfrak{T}_1]$ into $F[\mathfrak{T}_2']$ (§ 34, 5.(3)). Is a similar statement true for arbitrary linear mappings?

Let A, as before, be a linear mapping defined on a linear subspace $D[A]$ of $E[\mathfrak{T}_1]$ with values in $F[\mathfrak{T}_2]$. Since $S[A]$ is closed, the quotient $F/S[A]$ exists and is locally convex for the quotient topology $\hat{\mathfrak{T}}_2$. If K is the canonical homomorphism of F onto $F/(S[A])$, then KA is a linear mapping of $D[A] \subset E[\mathfrak{T}_1]$ into $(F/S[A])[\hat{\mathfrak{T}}_2]$.

(1) *KA is regular and $N[KA] = Q[A]$.*

Proof. We have $S[KA] = \bigcap_U \overline{KA(U)}$. The canonical image of $F \sim \overline{A(U)}$ in $F/S[A]$ is open and has the complement $\overline{KA(U)}$ since $\overline{A(U)} \supset S[A]$. Therefore $\overline{KA(U)} = K\overline{A(U)}$. From this follows $S[KA] = \bigcap_U K(\overline{A(U)})$. Now if y is an element of the residue class $\hat{y} \in K(\overline{A(U)})$, then $y \in \overline{A(U)}$ since $\overline{A(U)} \supset S[A]$. Hence $\bigcap_U K(\overline{A(U)}) = K \bigcap_U \overline{A(U)} = K(S[A]) = \mathfrak{o}$.

The second statement follows by 1.(6) from

$$(KA)^{(-1)}(\mathfrak{o}) = A^{(-1)}(K^{(-1)}(\mathfrak{o})) = A^{(-1)}(S[A]) = Q[A].$$

Therefore we call KA the regular contraction of A. If A is regular $KA = A$.

We come back to our problem. Clearly, there exists a finest topology $\mathfrak{T}_0$ on F with a basis of absolutely convex neighbourhoods of $\mathfrak{o}$ such that $\mathfrak{T}_0 \subset \mathfrak{T}_2$ and A is continuous from $(D[A])[\mathfrak{T}_1]$ into $F[\mathfrak{T}_0]$. When is $\mathfrak{T}_0$ locally convex, i.e., Hausdorff?

Let W be an absolutely convex $\mathfrak{T}_0$-neighbourhood of $\mathfrak{o}$. By assumption there exists an absolutely convex $\mathfrak{T}_2$-neighbourhood V of $\mathfrak{o}$ such that $W \supset V$ and an absolutely convex $\mathfrak{T}_1$-neighbourhood U of $\mathfrak{o}$ in $D[A]$ such that $W \supset A(U)$. Hence $W \supset \Gamma(A(U) \cup V)$. The class of all these sets $\Gamma(A(U) \cup V)$ is then obviously a $\mathfrak{T}_0$-neighbourhood basis of $\mathfrak{o}$ in F.

Since $\frac{1}{2}(A(U) + V) \subset \Gamma(A(U) \cup V) \subset A(U) + V$ for U, V absolutely convex, the class of all sets $A(U) + V$ is also a $\mathfrak{T}_0$-neighbourhood basis of $\mathfrak{o}$ in F.

The topology $\mathfrak{T}_0$ is Hausdorff if and only if the intersection of all neighbourhoods of $\mathfrak{o}$ is $\mathfrak{o}$. But

$$\bigcap_U \bigcap_V (A(U) + V) = \bigcap_U \overline{A(U)} = S[A].$$

Therefore $\mathfrak{T}_0$ is Hausdorff if and only if A is regular. In the general case we obtain a locally convex topology precisely on $F/S[A]$, but this means we have to consider KA instead of A. Therefore we have

(2) *Let A be a linear mapping of $D[A] \subset E[\mathfrak{T}_1]$ into $F[\mathfrak{T}_2]$ and let K be the canonical homomorphism of F onto $F/S[A]$. Let $\mathfrak{T}_0$ be the topology on $F/S[A]$ defined by the neighbourhood basis of $\circ$ consisting of all sets $K(A(U) + V)$, where U and V are absolutely convex $\mathfrak{T}_1$- resp. $\mathfrak{T}_2$-neighbourhoods of $\circ$ in $D[A]$ resp. F. Then $\mathfrak{T}_0$ is the finest locally convex topology on $F/S[A]$ weaker than $\hat{\mathfrak{T}}_2$ and such that KA is continuous from $D[A] \subset E[\mathfrak{T}_1]$ into $(F/S[A])[\mathfrak{T}_0]$.*

Recall that KA, the regular contraction, is defined as the mapping from $D[A] \subset E[\mathfrak{T}_1]$ into $(F/S[A])[\hat{\mathfrak{T}}_2]$. To avoid misunderstandings we will write from now on JKA for the mapping in (2), where J is the identity mapping of $(F/S[A])[\hat{\mathfrak{T}}_2]$ onto $(F/S[A])[\mathfrak{T}_0]$, and will call JKA the continuous contraction of A.

As a special case we obtain

(3) *Let A be a linear mapping of $D[A] \subset E[\mathfrak{T}_1]$ into $F[\mathfrak{T}_2]$. Then A is regular if and only if there exists a locally convex topology $\mathfrak{T}_0 \subset \mathfrak{T}_2$ on F such that A is continuous from $(D[A])[\mathfrak{T}_1]$ into $F[\mathfrak{T}_0]$.*

This contains the corresponding result (§ 34, 5.(3)) for closed linear mappings and we remark that we have given a precise description of the finest topology $\mathfrak{T}_0$ with these properties.

5. The adjoint of the contraction. We make the additional assumption on A that $D[A]$ is dense in $E[\mathfrak{T}_1]$. The regular contraction KA is then a dense linear mapping of $D[A] \subset E[\mathfrak{T}_1]$ into $(F/S[A])[\hat{\mathfrak{T}}_2]$. The adjoint $(KA)'$ is therefore a linear mapping of $D[(KA)'] \subset (F/S[A])[\hat{\mathfrak{T}}_2]'$ into E'. Now the canonical homomorphism K of F onto $F/S[A]$ has as adjoint the natural injection I of $(F/S[A])'$ into F' (§ 22, 1.) and $I(F/S[A]') = S[A]^{\perp} = \overline{D[A']}$. Furthermore,

$$(1) \qquad I(D[(KA)']) = D[A'] \quad \textit{and} \quad (KA)' = A'I.$$

Proof. For all $v \in (F/S[A])'$ and all $x \in D[A]$ we have

$$\langle v, KAx\rangle = \langle Iv, Ax\rangle \quad \text{and} \quad \langle (KA)'v, x\rangle = \langle A'Iv, x\rangle.$$

Hence $(KA)'v$ is in E' if and only if $A'Iv$ is in E'. From this follows the statement.

If one treats I as an identification, then (1) becomes

(2) *A and the regular contraction KA have the same adjoint, $A' = (KA)'$.*

We determine now the adjoint of the continuous contraction JKA.

$(JKA)'$ is a mapping of $(F/S[A])[\mathfrak{T}_0]'$ into E'. J' is the canonical injection of $(F/S[A])[\mathfrak{T}_0]'$ into $(F/S[A])[\hat{\mathfrak{T}}_2]'$; hence IJ' is the canonical injection of $(F/S[A])[\mathfrak{T}_0]'$ into F'. We have the following proposition:

(3) *Let A be a dense linear mapping of $D[A] \subset E[\mathfrak{T}_1]$ in $F[\mathfrak{T}_2]$. Then*

$$IJ'((F/S[A])[\mathfrak{T}_0]') = IJ'(D[(JKA)']) = D[A'] \quad \text{and} \quad (JKA)' = A'IJ'.$$

If we identify $(F/S[A])[\mathfrak{T}_0]'$ with $D[A']$ by IJ', then A and its continuous contraction JKA have the same adjoint A'.

We have to prove only the first identity, because the second identity follows then from

$$\langle w, JKAx\rangle = \langle A'IJ'w, x\rangle$$

for all $x \in D[A]$ and all $w \in (F/S[A])[\mathfrak{T}_0]'$.

We have $D[A'] = \bigcup_{U \in \mathfrak{U}} A(U)^\circ$ by 3.(1). If $\mathfrak{V}$ is a basis of absolutely convex $\mathfrak{T}_0$-neighbourhoods of o in F, then it follows that

$$D[A'] = \bigcup_{U \in \mathfrak{U}} \bigcup_{V \in \mathfrak{V}} (A(U)^\circ \cap V^\circ).$$

By the definition of $\mathfrak{T}_0$ in 4., the sets $JK(\ulcorner(A(U) \cup V))$ constitute a basis of absolutely convex $\mathfrak{T}_0$-neighbourhoods of o in $(F/S[A])[\mathfrak{T}_0]$. Its dual space can be written therefore as

$$\bigcup_U \bigcup_V JK(\ulcorner(A(U) \cup V))^\circ = \bigcup_U \bigcup_V J'^{(-1)}K'^{(-1)}(A(U)^\circ \cap V^\circ).$$

If we now apply IJ', we obtain $\bigcup_U \bigcup_V (A(U)^\circ \cap V^\circ) = D[A']$.

(3) enables us to apply our previous results on the adjoints of continuous mappings to the general case.

(4) *Let A be a dense linear mapping of $D[A] \subset E[\mathfrak{T}_1]$ in $F[\mathfrak{T}_2]$. Then A' is a continuous linear mapping of $(D[A'])[\mathfrak{T}_s(F)]$ into $E'[\mathfrak{T}_s(D[A])]$.*

$A' = (JKA)'$ is by (3) weakly continuous from $(D[A'])[\mathfrak{T}_s(F/S[A])]$ in $E'[\mathfrak{T}_s(D[A])]$. Since $D[A'] \subset S[A]^\perp \subset F'$, the topology $\mathfrak{T}_s(F/S[A])$ coincides by § 22, 2.(1) on $D[A']$ with $\mathfrak{T}_s(F)$.

We consider now the case that A' is a homomorphism.

(5) *Let A be a dense linear mapping of $D[A] \subset E[\mathfrak{T}_1]$ in $F[\mathfrak{T}_2]$. Then A' is a homomorphism of $(D[A'])[\mathfrak{T}_s(F)]$ in $E'[\mathfrak{T}_s(D[A])]$ if and only if $R[A] + S[A]$ is closed in F.*

By § 32, 3.(2), $A' = (JKA)'$ is a homomorphism if and only if $R[JKA] = R[KA]$ is closed in $(F/S[A])[\mathfrak{T}_s(D[A'])]$. Assume $R[KA]$ closed in $F/S[A]$

for $\mathfrak{T}_s(D[A'])$. Then it is closed for $\mathfrak{T}_2$ and $K^{(-1)}(R[KA]) = R[A] + S[A]$ is closed in F.

Conversely, if $R[A] + S[A]$ is closed in F, then $K(R[A] + S[A]) = R[KA]$ is closed in $(F/S[A])[\mathfrak{T}_2]$. By 3.(4) a), $R[KA]$ is the polar in $F/S[A]$ of $N[(KA)'] = N[A'] \subset S[A]^{\perp}$. But $N[A'] \subset D[A']$, so that $R[KA]$ is also the polar of $N[A']$ in the sense of the dual system $\langle D[A'], F/S[A]\rangle$ and $R[KA]$ is therefore $\mathfrak{T}_s(D[A'])$-closed.

The next proposition considers the case that KA and JKA coincide.

(6) *Let A be a dense linear mapping of $D[A] \subset E[\mathfrak{T}_1]$ in $F[\mathfrak{T}_2]$. If the regular contraction KA is continuous, then $D[A']$ is weakly closed in F'. Conversely, if $D[A']$ is weakly closed in F', then KA is weakly continuous from $(D[A])[\mathfrak{T}_s(E')]$ in $(F/S[A])[\mathfrak{T}_s(F')]$.*

If E and F are metrizable, then $D[A']$ is weakly closed in F if and only if KA is continuous.

Proof. If KA is continuous, then $(KA)' = A'$ is defined on the dual of $(F/S[A])[\mathfrak{T}_2]$, which coincides with $S[A]^{\circ}$ in F'.

If, conversely, $D[A']$ is weakly closed in F', then $D[A'] = S[A]^{\circ}$ by 3.(2); hence $\mathfrak{T}_s(D[A]) = \mathfrak{T}_s(S[A]^{\circ}) = \mathfrak{T}_s(F')$ on $F/S[A]$ and the statement follows from (4).

Finally, if E and F are metrizable, the topologies on $D[A]$ and $F/S[A]$ are Mackey topologies and weak continuity of KA therefore implies continuity.

6. The second adjoint. We begin with some remarks on closed linear mappings. The situation is now more general than in § 34, 5., since we suppose only that A is defined on a linear subspace $D[A]$ of $E[\mathfrak{T}_1]$ and that its graph $G(A)$ is closed in $E[\mathfrak{T}_1] \times F[\mathfrak{T}_2]$.

The kernel of a regular (closable) mapping is closed in $D[A]$, as we have seen in 1. For closed mappings we have the sharper result

(1) *If A is a closed linear mapping of $D[A] \subset E[\mathfrak{T}_1]$ in $F[\mathfrak{T}_2]$, then $N[A]$ is closed in E.*

If $x_{\alpha} \in N[A] \subset D[A]$ and $x_{\alpha} \to x \in E$, then $Ax_{\alpha} \to o$ in F; therefore $(x, o) \in \overline{G(A)} = G(A)$.

By (1) A can be written as $\hat{A}K$, K the canonical homomorphism of E onto $E/N[A]$, and one has

(2) *$\hat{A}$ is a closed linear mapping of $D[\hat{A}] \subset (E/N[A])[\hat{\mathfrak{T}}_1]$ in $F[\mathfrak{T}_2]$.*

Proof. $H = N[A] \times o$ is closed in $E \times F$ and $G(\hat{A})$ is the image of $G(A)$ by the canonical mapping of $E \times F$ onto $(E/N[A]) \times F = (E \times F)/H$.

Since $G(A)$ is closed in $E \times F$ and $G(A) \supset H$, we have that $G(\hat{A})$ is closed in $(E/N[A]) \times F$.

We recall that if A is regular, then A has a closed extension $\bar{A}$ such that $G(\bar{A}) = \overline{G(A)}$.

We showed in 5.(4) that for a densely defined A the adjoint A' is continuous from $(D[A'])[\mathfrak{T}_s(F)]$ into $E'[\mathfrak{T}_s(D[A])]$. What happens when we replace $\mathfrak{T}_s(D[A])$ by the stronger topology $\mathfrak{T}_s(E)$?

We denote by ${}^iG(A)$ the subspace of $F \times E$ consisting of all (Ax, x), $x \in D[A]$.

(3) *Let A be a dense linear mapping of $D[A] \subset E[\mathfrak{T}_1]$ in $F[\mathfrak{T}_2]$. Then $G(A)^\perp = {}^iG(-A')$. Hence A' is a closed linear mapping of $D[A'] \subset F[\mathfrak{T}_s(F)]$ in $E'[\mathfrak{T}_s(E)]$.*

$G(A)^\perp$ consists of all $(u, v) \in E' \times F'$ such that $\langle (u, v), (x, Ax) \rangle = ux + v(Ax) = 0$ for all $x \in D[A]$. Then $u = -A'v$ and $v \in D[A']$. Conversely, $(-A'v, v) \in G(A)^\perp$ for all $v \in D[A']$; hence $G(A)^\perp = {}^iG(-A')$.

(4) *Let A be a dense regular linear mapping of $D[A] \subset E[\mathfrak{T}_1]$ in $F[\mathfrak{T}_2]$. Then A and its closed extension $\bar{A}$ have the same adjoint, $A' = \bar{A}'$.*

We have $G(\bar{A}) = \overline{G(A)}$ and by (3) ${}^iG(-A') = G(A)^\perp = G(\bar{A})^\perp = {}^iG(-\bar{A}')$.

From this and 5.(4) follows a slight improvement of 5.(4) for regular mappings.

(5) *Let A be a dense regular linear mapping of $D[A] \subset E[\mathfrak{T}_1]$ in $F[\mathfrak{T}_2]$. Then A' is continuous from $(D[A'])[\mathfrak{T}_s(F)]$ in $E'[\mathfrak{T}_s(D[\bar{A}])]$.*

These results show that there are different ways to define the second adjoint of a regular dense mapping A. In this case $D[A']$ is weakly dense in F' by 3.(3). If A' is considered as a continuous mapping from $D[A']\,[\mathfrak{T}_s(F)]$ in $E'[\mathfrak{T}_s(D[A])]$, as in 5.(4), then its adjoint A'' coincides with A, which maps $(E')' = D[A]$ in $D[A']' = F$.

If we consider A' as in (5) as a continuous mapping from $(D[A'])[\mathfrak{T}_s(F)]$ in $E'[\mathfrak{T}_s(D[\bar{A}])]$, then $(A')'$ coincides with $\bar{A}$, which maps $D[\bar{A}]$ in F.

Finally, we may consider A' as in (3) as a closed linear mapping of $D[A'] \subset F'[\mathfrak{T}_s(F)]$ in $E'[\mathfrak{T}_s(E)]$. Since $D[A']$ is weakly dense in F', we are able to apply (3) to A' and we find $G(-A')^\perp = {}^iG((A')')$. Since ${}^iG(-A') = G(A)^\perp$, we conclude that $G((A')') = G(A)^{\perp\perp} = \overline{G(A)} = G(\bar{A})$. Hence again $(A')' = \bar{A}$.

(6) *Let A be a dense regular mapping of $D[A] \subset E[\mathfrak{T}_1]$ in $F[\mathfrak{T}_2]$. Then A' is a dense closed linear mapping of $D[A'] \subset F'[\mathfrak{T}_s(F)]$ in $E'[\mathfrak{T}_s(E)]$. The adjoint of A' is $\bar{A}$.*

For dense closed linear mappings A we have full duality by (6). The interpretation of A' as a closed linear mapping is more symmetric than the interpretation as a weakly continuous mapping and we will use this approach to duality in the general case too.

Let A be now a dense linear mapping of $D[A] \subset E[\mathfrak{T}_1]$ in $F[\mathfrak{T}_2]$. Then $D[A']$ is weakly dense in $(F/S[A])' = S[A]^\perp$ by 3.(2). We recall that the regular contraction KA maps $D[A]$ in $F/S[A]$ and that $(KA)' = A'$. If we apply (6) to KA, we obtain

(7) *Let A be a dense linear mapping of $D[A] \subset E[\mathfrak{T}_1]$ in $F[\mathfrak{T}_2]$ with the singularity $S[A]$. Then A' is a closed linear mapping of $D[A'] \subset (S[A]^\perp)[\mathfrak{T}_s(F)]$ in $E'[\mathfrak{T}_s(E)]$ and the adjoint A'' coincides with $\overline{KA}$, the closed extension of KA.*

It is therefore natural to ask for a characterization of those A which have a closed regular contraction. We treat this problem next.

7. Maximal mappings. If A is regular, then $\overline{G(A)}$ is the graph $G(\bar{A})$ of the closed extension $\bar{A}$ of A. If A is not regular, then $\overline{G(A)}$ is not the graph of a linear mapping. But it is possible that A has an extension $\breve{A}$ such that $G(A) \subset G(\breve{A}) \subset \overline{G(A)}$. We say that $\breve{A}$ is a slight extension of A. A slight extension $\tilde{A}$ of A is maximal if it has no proper extension which is a slight extension of A.

If A has no proper slight extension, we say that A is maximal.

(1) *Every linear mapping A of $D[A] \subset E[\mathfrak{T}_1]$ in $F[\mathfrak{T}_2]$ has a maximal slight extension $\tilde{A}$.*

Proof. Let $\mathfrak{M}(A)$ be the class of all subspaces G of $E \times F$ such that $G(A) \subset G \subset \overline{G(A)}$, where G contains no (o, y), $y \neq o$. $\mathfrak{M}(A)$ ordered by set theoretic inclusion satisfies the assumption of Zorn's lemma (§ 2, 2.(2)). There exists therefore a maximal $\tilde{G}$ and if we define $\tilde{A}x = y$ for every $(x, y) \in \tilde{G}$, then $\tilde{A}$ is a maximal slight extension of $\tilde{A}$.

We denote by H_A the image of $\overline{G(A)}$ under the projection of $E \times F$ onto E. Then we have

(2) *If $\tilde{A}$ is a maximal slight extension of A, then $D[\tilde{A}] = H_A$. A is maximal if and only if $D[A] = H_A$.*

By definition of a slight extension, $D[\breve{A}] \subset H_A$. We assume there exists $x_0 \in H_A \sim D[\breve{A}]$. Let $y_0 \in F$ be such that $(x_0, y_0) \in \overline{G(A)}$. It is easy to see that $G(\breve{A}) \oplus [(x_0, y_0)]$ is in $\mathfrak{M}(A)$. So A is not maximal. For a maximal slight extension $\tilde{A}$ we have therefore $D[\tilde{A}] = H_A$.

(3) *If A is regular there exists only one maximal slight extension, the closed extension $\overline{A}$. If A is not regular and if $D[A]$ is a proper subspace of H_A there exist infinitely many different maximal slight extensions of A.*

Assume A to be regular. Since $D[\overline{A}] = H_A$, $\overline{A}$ is a maximal slight extension of A. If $\tilde{A}$ is another maximal slight extension, then $(\mathrm{o}, \overline{A}x - \tilde{A}x) \in \overline{G(A)}$ and hence $\overline{A}x = \tilde{A}x$.

If A is not regular, then $S[A] \neq \mathrm{o}$. If $x_0 \in H_A \sim D[A]$, there exists, as in the proof of (2), $(x_0, y_0) \in \overline{G(A)}$ and by ZORN's lemma one constructs a maximal slight extension $\tilde{A}$ such that $G(\tilde{A}) \supset G(A) \oplus [(x_0, y_0)]$. If $z \in S[A]$, $z \neq \mathrm{o}$, then $(\mathrm{o}, z) \in \overline{G(A)}$ and $(x_0, y_0 + \rho z) \in \overline{G(A)}$ for every $\rho \neq 0$. Using these elements instead of (x_0, y_0) one finds infinitely many different maximal slight extensions of A.

(4) $S[\tilde{A}] = S[A]$ *for every maximal slight extension $\tilde{A}$ of A.*

This follows immediately from 1.(4).

(5) *All maximal slight extensions have the same extended kernel $Q[\tilde{A}]$, which is the set of all $x \in E$ such that $(x, \mathrm{o}) \in \overline{G(A)}$. If A is maximal, $Q[A]$ is closed in E.*

By 1.(6), $Q[\tilde{A}]$ consists of all $x \in D[\tilde{A}]$ such that $(x, \mathrm{o}) \in \overline{G(A)}$. But $D[\tilde{A}] = H_A$, the projection of $\overline{G(A)}$ onto E. From this follows the first statement. The second is immediate, since the set of all $(x, \mathrm{o}) \in \overline{G(A)}$ is closed in E.

The relation between $Q[A]$ and $Q[\tilde{A}]$ is not obvious. Since $Q[\tilde{A}]$ is closed in E, the closure $\overline{Q[A]}$ of $Q[A]$ in E is contained in $Q[\tilde{A}]$; but $\overline{Q[A]}$ can be a proper subspace of $Q[\tilde{A}]$, as the example below will show.

We are now able to answer the question raised at the end of 6.

(6) *Let A be a linear mapping of $D[A] \subset E$ into F. The regular contraction KA is closed if and only if A is maximal.*

Proof. Let A be maximal. Then $D[A] = D[KA] = H_A$ by (2). Let H be the closed subspace $\mathrm{o} \times S[A]$ of $E \times F$. Then $\overline{G(A)}/H$ consists of all (x, Ky), where $(x, y) \in \overline{G(A)}$. It is therefore obvious that $\overline{G(A)}/H \subset \overline{G(KA)}$. But since $\overline{G(A)} \supset H$, $\overline{G(A)}/H$ is closed and it follows that $\overline{G(KA)} = \overline{G(A)}/H$. Hence $H_{KA} = H_A$ and KA is maximal by (2) and closed by (3).

Conversely, if KA is closed, KA is maximal, $D[KA] = H_{KA}$, and $H_{KA} = H_A$, as we have seen. From $D[KA] = D[A] = H_A$ it follows that A is maximal.

(7) *Let A be a linear mapping of $D[A] \subset E$ into F and $\tilde{A}$ a maximal slight extension of A. Then the regular contraction $K\tilde{A}$ of $\tilde{A}$ is the closed extension $\overline{KA}$ of KA, where in both cases K is the canonical homomorphism of F onto $F/S[A]$.*

That K is the same follows from (4). That $K\tilde{A}$ is a slight extension of KA follows from $\overline{G(K\tilde{A})} = \overline{G(\tilde{A})}/H = \overline{G(A)}/H = \overline{G(KA)}$. From (6) follows $K\tilde{A} = \overline{KA}$.

Let G_A be the image of $\overline{G(A)}$ under the projection of $E \times F$ onto F. Corresponding to (2) we have

(8) *If A is maximal, then $R[A] + S[A] = G_A$.*

This is trivial for closed A. If A is arbitrary, $x \in D[A]$, $z \in S[A]$, then $(x, Ax + z) = (x, Ax) + (o, z) \in \overline{G(A)}$; therefore $R[A] + S[A] \subset G_A$. Assume now A to be maximal and $(x, y) \in \overline{G(A)}$. Then $x \in H_A = D[A]$; hence Ax is defined and $(x, y) - (x, Ax) \in \overline{G(A)}$, $y - Ax \in S[A]$; therefore $y \in R[A] + S[A]$.

The following example of constructing maximal slight extensions is due to ADASCH. We recall the linear mapping A_1 of Example 2. iv). A_1 is defined on $D[A_1] = H \oplus [y^{(o)}] \subset c_0$, where $y^{(o)} = (1, 1/2^2, \ldots, 1/n^2, \ldots)$. Let $z^{(o)} = (1, 1/2, \ldots, 1/n, \ldots)$ and

$$y^{(1)} = \left(\frac{1}{2} - \frac{1}{1}, \ldots, \frac{1}{(n+1)n^2} - \frac{1}{n^2}, \ldots\right),$$

The space E resp. F is defined as the subspace of c_0 such that

$$E = D[A] \oplus [y^{(1)}] \quad \text{resp.} \quad F = D[A] \oplus [z^{(o)}].$$

We proved in 2. that $R[A_1] = D[A_1]$, $[z^{(o)}] \subset S[A_1]$, and $Q[A_1] = o$, so that A_1 is weakly singular.

Since $D[A_1]$ has co-dimension 1 in E, either A_1 is maximal or $D[\tilde{A}_1] = E$ for every maximal slight extension $\tilde{A}_1$ of A_1. We use (5) to determine $Q[\tilde{A}_1]$.

Let $y_k^{(2)}$ be the kth section of $y^{(2)} = (1/2, \ldots, 1/[(n+1)n^2], \ldots)$. Then $y_k^{(2)} \in H$ and $y_k^{(2)} - y^{(o)}$ converges to $y^{(1)}$. Since $A_1(y_k^{(2)} - y^{(o)})$ converges to o, the element $(y^{(1)}, o)$ is in $\overline{G(A_1)}$ and $y^{(1)} \in Q[\tilde{A}_1]$. If $t = z + \lambda y^{(1)}$ with $z \in D[A]$ is any element of $Q[\tilde{A}_1]$, then $t - \lambda y^{(1)} \in Q[\tilde{A}_1] \cap D[A] = \underline{Q[A]} = o$ and therefore $Q[\tilde{A}_1] = [y^{(1)}]$. Hence A_1 is not maximal and $Q[A_1]$ is a proper subspace of $Q[\tilde{A}_1]$. We define the extension $\tilde{A}_1$ of A_1 by $\tilde{A}_1 y^{(1)} = z^{(o)}$. Since $(y^{(1)}, o) \in \overline{G(A_1)}$ and $(o, z^{(1)}) \in S[A_1] \subset \overline{G(A_1)}$, we have $G(\tilde{A}_1) \subset \overline{G(A_1)}$ and $\tilde{A}_1$ is a maximal slight extension of A_1.

We have $N[\tilde{A}_1] = o$, $Q[A_1] = [y^{(1)}]$, $S[\tilde{A}_1] = S[A_1] = [z^{(o)}]$. The first statement is trivial. We proved the second statement. The third statement follows from $\tilde{A}_1(Q[\tilde{A}_1]) = S[\tilde{A}_1] \cap R[\tilde{A}_1] = S[\tilde{A}_1]$ using 1.(6).

Obviously, the maximal slight extension $\tilde{A}_1$ is strongly singular. If we define the extension $\tilde{\tilde{A}}_1$ by $\tilde{\tilde{A}}_1 y^{(-1)} = o$, then it is easy to see that $\tilde{\tilde{A}}_1$ is also a maximal slight extension of A_1. But in this case $Q[\tilde{\tilde{A}}_1] = [z^{(o)}] = N[\tilde{\tilde{A}}_1]$; hence $\tilde{\tilde{A}}_1$ is weakly singular.

Therefore a maximal slight extension of a weakly singular mapping may be weakly singular or strongly singular.

8. Dense maximal mappings. We now make the additional assumption that $D[A]$ is dense in $E[\mathfrak{T}_1]$. We know then that the adjoint A' exists and we study the consequences of maximality for the duality properties.

(1) *Let A be a dense linear mapping of $D[A] \subset E$ into F and $\tilde{A}$ a maximal slight extension of A. Then $\tilde{A}' = A'$.*

This follows from ${}^tG(-A') = G(A)^{\perp} = G(\tilde{A})^{\perp}$ using 6.(3).

Conversely, we have

(2) *An extension $\tilde{A}$ of a dense linear mapping A is a maximal slight extension of A if $\tilde{A}' = A'$ and $D[\tilde{A}] = H_A$.*

Since ${}^tG(-A') = G(A)^{\perp} = G(\tilde{A})^{\perp}$, we have $\overline{G(A)} = \overline{G(\tilde{A})}$ and the statement follows from 7.(2).

The duality properties of dense maximal mappings are listed in the following theorem:

(3) *Let A be a dense maximal linear mapping of $D[A] \subset E[\mathfrak{T}_1]$ in $F[\mathfrak{T}_2]$. Then A' is a closed linear mapping of $D[A'] \subset (S[A]^{\perp})[\mathfrak{T}_s(F)]$ in $E'[\mathfrak{T}_s(E)]$ and $(A')'$ coincides with the regular contraction KA of A.*

The following duality relations hold:

a) $D[A'] = \bigcup_{U \in \mathfrak{U}} A(U)^{\circ}$, *$\mathfrak{U}$ a $\mathfrak{T}_1$-neighbourhood base of $\circ$ in $D[A]$;*

b) $R[A'] = \bigcup_{V \in \mathfrak{V}} A^{(-1)}(V)^{\circ}$, *$\mathfrak{V}$ a $\mathfrak{T}_2$-neighbourhood base of $\circ$ in F;*

c) $D[A] = D[KA] = H_A = \bigcup_{W \in \mathfrak{W}} A'(W)^{\circ}$, *$\mathfrak{W}$ a $\mathfrak{T}_s(F)$-neighbourhood base of $\circ$ in $D[A']$;*

d) $R[KA] = \bigcup_{X \in \mathfrak{X}} A'^{(-1)}(X)^{\circ}$, *$\mathfrak{X}$ a $\mathfrak{T}_s(E)$-neighbourhood base of $\circ$ in E'.*

The polars in a), b), c), d) *are taken in F', E', E, $F/S[A]$ respectively. The relation* d) *can be replaced by*

d′) $R[A] + S[A] = G_A = \bigcup_{X \in \mathfrak{X}} A'^{(-1)}(X)^{\circ}$, *where the polars are now taken in F.*

Proof. The first statement is implied by 6.(7) and 7.(6); a) is 3.(1); b) is 3.(8). c) follows from a) applied to A' instead of A. Similarly, d) follows from b) applied to A' and from 7.(2). Finally, d′) follows from d) by applying $K^{(-1)}$ to both sides and from 7.(8).

The following proposition will be useful later:

(4) *Let A be dense and maximal. Then the $\mathfrak{T}_s(E)$-closure of $R[A']$ in E' coincides with the $\mathfrak{T}_s(D[A]')$-closure.*

Assume that this is not the case. Then there exists u_0 which is in the $\mathfrak{T}_s(D[A])$-closure but not in the $\mathfrak{T}_s(E)$-closure of $R[A']$. There exists $x_0 \in E \sim D[A]$ such that $u_0 x_0 \neq 0$ but $(A'v)x_0 = 0$ for all $v \in D[A']$. Hence

$$\langle(-A'v, v), (x_0, \mathfrak{o})\rangle = (-A'v)x_0 + v\mathfrak{o} = 0 \quad \text{for all } v \in D[A'].$$

Therefore $(x_0, \mathfrak{o}) \in {}^tG(-A')^{\perp} = \overline{G(A)}$ and $x_0 \in Q[A] \subset D[A]$ by 7.(5), which is a contradiction.

§ 37. The graph topology. Open mappings

1. The graph topology. So far we investigated arbitrary linear mappings A of $D[A] \subset E[\mathfrak{T}_1]$ in $F[\mathfrak{T}_2]$ by weakening the topology on F and introducing the continuous contraction to represent the continuity properties of A.

There is another way to study arbitrary mappings A. Instead of weakening the topology on F one introduces on $D[A]$ a stronger topology so that A becomes continuous. More precisely, we have the following proposition:

(1) *Let A be a linear mapping of $D[A] \subset E[\mathfrak{T}_1]$ in $F[\mathfrak{T}_2]$. If $\mathfrak{U}$ is a base of absolutely convex $\mathfrak{T}_1$-neighbourhoods U of $\mathfrak{o}$ in $D[A]$ and $\mathfrak{V}$ a base of absolutely convex $\mathfrak{T}_2$-neighbourhoods of $\mathfrak{o}$ in F, then the class of all $U \cap A^{(-1)}(V)$ is a neighbourhood base of a locally convex topology $\mathfrak{T}_A$ on $D[A]$.*
$\mathfrak{T}_A$ is the coarsest topology $\mathfrak{T}_A \supset \mathfrak{T}_1$ such that A is continuous from $(D[A])[\mathfrak{T}_A]$ into $F[\mathfrak{T}_2]$.

It is obvious that the class of all $U \cap A^{(-1)}(V)$ defines a locally convex topology $\mathfrak{T}_A$ on $D[A]$ which is stronger than $\mathfrak{T}_1$ and such that A is continuous.

Assume that $\mathfrak{T}'$ is a locally convex topology on $D[A]$ with these properties. Then there exist $\mathfrak{T}'$-neighbourhoods W_1, W_2 of $\mathfrak{o}$ such that $W_1 \subset U$ and $A(W_2) \subset V$. But then $W_1 \cap W_2 \subset U \cap A^{(-1)}(V)$; hence $\mathfrak{T}' \supset \mathfrak{T}_A$.

$\mathfrak{T}_A$ is called the graph topology on $D[A]$ since

(2) *$(D[A])[\mathfrak{T}_A]$ is isomorphic to $G(A)$.*

The isomorphism is defined by $P(x, Ax) = x$. That it is topological follows from $P((U \times V) \cap G(A)) = U \cap A^{(-1)}(V)$.

If $\mathfrak{T}_1$ is defined on E by the system of seminorms p_α, $\mathfrak{T}_2$ on F by the system q_β, then $\mathfrak{T}_A$ on $D[A]$ is given by the system $\max(p_\alpha(x), q_\beta(Ax))$ or $[p_\alpha(x)^r + q_\beta(Ax)^r]^{1/r}$, $1 \leqq r < \infty$.

If E and F are normed spaces, then A will be continuous for the norm $\|x\|_A = \|x\| + \|Ax\|$ on $D[A]$, the so-called graph norm. If E and F are metrizable locally convex spaces, then $\mathfrak{T}_A$ is a metrizable topology on $D[A]$.

We have the following corollary to (2):

(3) *Let A be a linear mapping of $D[A] \subset E$ in F, where E and F are complete locally convex spaces.*

Then A is closed if and only if $(D[A])[\mathfrak{T}_A]$ is complete. If A is regular (closable) and $\bar{A}$ its closed extension, then $D[\bar{A}]$ is the completion of $(D[A])[\mathfrak{T}_A]$.

Proof. If $G(A)$ is closed in $E \times F$, $G(A)$ is complete and by (2) $(D[A])[\mathfrak{T}_A]$ is complete and conversely. The second statement follows from $G(\bar{A}) = \overline{G(A)}$ and (2).

We denote by I_A the identity mapping of $(D[A])[\mathfrak{T}_A]$ onto $(D[A])[\mathfrak{T}_1]$. We say that AI_A is the continuous refinement of A and we have to distinguish between A and AI_A from now on.

It is easy to determine $\mathfrak{T}_A$ in Examples § 36, 2. i) and ii). We leave the details to the reader.

2. The adjoint of AI_A. Since $\mathfrak{T}_A$ is finer than $\mathfrak{T}_1$ on $D[A]$, the adjoint I'_A is the canonical injection of $D[A]' = (D[A])[\mathfrak{T}_1]'$ into $D[AI_A]' = (D[A])[\mathfrak{T}_A]'$. We give a more detailed characterization of $D[AI_A]'$.

(1) *Let A be a linear mapping of $D[A] \subset E[\mathfrak{T}_1]$ in F_2. Let $\mathfrak{U}$ be a base of absolutely convex and closed $\mathfrak{T}_1$-neighbourhoods U of $\circ$ in $D[A]$ and $\mathfrak{V}$ a base of absolutely convex and closed $\mathfrak{T}_2$-neighbourhoods V of $\circ$ in F. Then*

$$D[AI_A]' = \bigcup_U \bigcup_V (U^\circ + A^{(-1)}(V)^\circ), \tag{2}$$

where the polars are taken in $D[AI_A]'$ or in the algebraical dual $D[A]^*$.

The class of sets $U^\circ + A^{(-1)}(V)^\circ$ is a fundamental system of $\mathfrak{T}_A$-equicontinuous subsets of $D[AI_A]'$.

Proof. A fundamental system of $\mathfrak{T}_A$-equicontinuous sets is given by the polars $(U \cap A^{(-1)}(V))^\circ$ of the $\mathfrak{T}_A$-neighbourhoods $U \cap A^{(-1)}(V)$ of $\circ$ in $D[AI_A]$. Since U and $A^{(-1)}(V) = (AI_A)^{(-1)}(V)$ are closed absolutely convex sets in $D[AI_A]$, it follows from § 20, 8.(10) that $(U \cap A^{(-1)}(V))^\circ = \overline{\Gamma(U^\circ \cup A^{(-1)}(V)^\circ)}$, the $\mathfrak{T}_s(D[A])$-closure taken in $D[AI_A]'$.

We have $\frac{1}{2}(U^\circ + A^{(-1)}(V)^\circ) \subset \Gamma(U^\circ \cup A^{(-1)}(V)^\circ) \subset U^\circ + A^{(-1)}(V)^\circ$. Since U° is $\mathfrak{T}_s(D[A])$-compact and $A^{(-1)}(V)^\circ$ $\mathfrak{T}_s(D[A])$-closed, the set $U^\circ + A^{(-1)}(V)^\circ$ is $\mathfrak{T}_s(D[A])$-closed by § 15, 6.(10). It is therefore possible to replace $\Gamma(U^\circ \cup A^{(-1)}(V)^\circ)$ by $\overline{\Gamma(U^\circ \cup A^{(-1)}(V)^\circ)}$ in the inequality and it follows that

$$\tfrac{1}{2}(U^\circ + A^{(-1)}(V)^\circ) \subset (U \cap A^{(-1)}(V))^\circ \subset U^\circ + A^{(-1)}(V)^\circ.$$

This proves (2).

It is convenient to give our result in another form. For $v \in F'$ the expression $v(Ax)$ defines a linear functional on $D[A]$ which we denote by $v \circ A$. This is an element of the algebraic dual $D[A]^*$. Since AI_A is continuous and $(AI_A)'$ maps F' in $D[AI_A]'$, we have $v(Ax) = v(AI_A x) = ((AI_A)'v)x$; hence $v \circ A = (AI_A)'v$. By § 34, 6.(4) we have $A^{(-1)}(V)^\circ = (AI_A)^{(-1)}(V)^\circ = (AI_A)'(V^\circ)$. It follows therefore from (2) that

$$(3) \qquad D[AI_A]' = D[A]' + F' \circ A = D[A]' + R[(AI_A)'].$$

If A *is densely defined, then* $D[A]' = E'$.

We assume now that $D[A]$ is dense in $E[\mathfrak{T}_1]$ and investigate the relations between A' and $(AI_A)'$.

(4) A' *is the restriction of* $(AI_A)'$ *to* E' *and the following relations hold:*

$$N[A'] = N[(AI_A)']; \qquad R[A'] = R[(AI_A)'] \cap E'.$$

If $v \in D[A']$, then $(A'v)x = v(AI_A x) = ((AI_A)'v)x$ for all $x \in D[A]$; hence $A'v = (AI_A)'v$. If $(AI_A)'w = 0$, then $w((AI_A)x) = w(Ax) = 0$ for all $x \in D[A]$; hence $w \in D[A']$, $A'w = \mathfrak{o}$. Obviously, $R[A'] \subset R[(AI_A)'] \cap E'$. Conversely, let $u \in E'$, $u = (AI_A)'v$, $v \in F'$. Then $ux = ((AI_A)'v)x = v(Ax)$ for all $x \in D[A]$; therefore $v \in D[A']$, $u = A'v \in R[A']$.

(5) *Let* $\overline{R[(AI_A)']}$ *be the* $\mathfrak{T}_s(D[A])$*-closure of* $R[(AI_A)']$ *in* $D[AI_A]'$. *Then* $\overline{R[(AI_A)']} = N[A]^\circ + F' \circ A = N[A]^\perp + R[(AI_A)']$, *where* $N[A]^\circ$ *is the polar of* $N[A]$ *in* E'.

By § 32, 1.(6) $\overline{R[(AI_A)']} = N[AI_A]^\circ$, the polar taken in $D[AI_A]'$. Using (3) we find that $w = u + v \circ A$ is orthogonal to $N[A]$ if and only if $\langle u, N[A]\rangle + \langle v \circ A, N[A]\rangle = \langle u, N[A]\rangle = 0$, which is the case for every $u \in N[A]^\circ$ and every $v \in F'$.

These results and the following proposition are due to Adasch [1'].

(6) *Let* A *be a dense linear mapping of* $D[A] \subset E[\mathfrak{T}_1]$ *into* $F[\mathfrak{T}_2]$. *Then* $R[(AI_A)']$ *is* $\mathfrak{T}_s(D[A])$*-closed in* $D[AI_A]'$ *if and only if* $R[A']$ *is* $\mathfrak{T}_s(D[A])$*-closed in* E' *and* A *is weakly singular.*

If $R[(AI_A)']$ is closed, then $R[(AI_A)'] = N[A]^\circ + R[(AI_A)']$ by (5); hence $N[A]^\circ \subset R[(AI_A)'] \cap E' = R[A']$ by (4). We proved in § 36, 3.(4) that $\overline{R[A']} = Q[A]^\circ$, where the polar is taken in E'; therefore $R[A'] \supset N[A]^\circ \supset Q[A]^\circ = \overline{R[A']}$. Hence $R[A']$ is $\mathfrak{T}_s(D[A])$-closed in E' and $N[A]^\circ = Q[A]^\circ$. Taking polars in $D[A]$ we find $\overline{N[A]} = Q[A]$; A is weakly singular.

Conversely, let A be weakly singular and $R[A']$ $\mathfrak{T}_s(D[A])$-closed in E'. Then $R[A'] = N[A]^\circ$ and $\overline{R[(AI_A)']} = R[A'] + R[(AI_A)'] = R[(AI_A)']$ by (5).

3. Nearly open mappings. As a first application of our general theory of linear mappings we investigate nearly open mappings, which were introduced in § 34, 1., where the continuous case, in particular, was studied.

We repeat the definition. A linear mapping A of $D[A] \subset E[\mathfrak{T}_1]$ in $F[\mathfrak{T}_2]$ is nearly open if for every absolutely convex $\mathfrak{T}_1$-neighbourhood U of $\mathfrak{o}$ in $D[A]$ the closure $\overline{A(U)}$ in F is a $\mathfrak{T}_2$-neighbourhood of $\mathfrak{o}$ in $\overline{R[A]}$.

(1) *Let A be a linear mapping of $D[A] \subset E[\mathfrak{T}_1]$ in $F[\mathfrak{T}_2]$. The following statements are equivalent:*

a) *A is nearly open;*
b) *the regular contraction KA is nearly open;*
c) *the continuous contraction JKA is nearly open;*
d) *the continuous refinement AI_A is nearly open.*

Proof. From a) follows b). Let U be an absolutely convex neighbourhood of $\mathfrak{o}$ in $D[A]$. By assumption $\overline{A(U)} \supset V \cap R[A]$ for some open absolutely convex neighbourhood V of $\mathfrak{o}$ in F. KA will be nearly open if we show that

$$3\overline{KA(U)} \supset 3K(\overline{A(U)}) \supset K(V) \cap K(R[A]).$$

If $Ky \in K(V) \cap K(R[A])$ we are able to choose $y \in V$ such that $y = Ax_0 + z_0$, where $x_0 \in D[A]$, $z_0 \in S[A]$. Then $Ax_0 \in V + S[A]$. Since $S[A] \subset \overline{A(U)}$, we have $Ax_0 \in V + \overline{A(U)} \subset A(U) + 2V$. Hence $Ax_0 = Ax_1 + z_1, x_1 \in U, z_1 \in 2V$. Therefore $z_1 \in 2(V \cap R[A]) \subset 2\overline{A(U)}$. It follows that $Ax_0 = Ax_1 + z_1 \in 3\overline{A(U)}$ and $y = Ax_0 + z_0 \in 3\overline{A(U)} + S[A]$. Finally, $Ky \in 3K(\overline{A(U)})$.

From b) follows a). If U is given there exists by assumption V such that $\overline{KA(U)} \supset K(V \cap R[A])$. Now $\overline{A(U)} \supset S[A]$; therefore $K(\overline{A(U)})$ is closed in $F/S[A]$ and we have $K(\overline{A(U)}) \supset K(V \cap R[A])$. If $y \in V \cap R[A]$, there exists $z_1 \in S[A]$ such that $y = y_1 + z_1$, $y_1 \in \overline{A(U)}$. But $z_1 \in \overline{A(U)}$. Therefore $y \in 2\overline{A(U)}$ and $2\overline{A(U)} \supset V \cap R[A]$; A is nearly open.

From b) follows c). It is sufficient to assume that A is regular and nearly open. Let $\overline{A(U)} \supset V \cap R[A]$, U and V absolutely convex. Then

$$\overline{A(U)} \supset \ulcorner(A(U) \cup (V \cap R[A])) = \ulcorner(A(U) \cup V) \cap R[A].$$

But $\ulcorner(A(U) \cup V)$ is a $\mathfrak{T}_0$-neighbourhood in F; hence JA is nearly open.

From c) follows b). If JA is nearly open, then A is nearly open, since $\mathfrak{T}_0$ is weaker than $\mathfrak{T}_2$.

From d) follows a). This is obvious since $\mathfrak{T}_A$ is finer on $D[A]$ than $\mathfrak{T}_1$. From a) follows d). We prove first the following lemma.

(2) *Let U be an absolutely convex neighbourhood of $\circ$ in the locally convex space E and K an absolutely convex subset of E. Then $\overline{U \cap K} = \overline{U} \cap \overline{K}$.*

$\overline{U \cap K} \subset \overline{U} \cap \overline{K}$ is obvious. Let $x \in \overline{U} \cap \overline{K}$. Then x is the limit of a net $x_\alpha \in K$. For a given $\varepsilon > 0$, $x \in (1 + \varepsilon)U$ and we may assume that all $x_\alpha \in (1 + \varepsilon)U$. Then $x_\alpha \in (1 + \varepsilon)U \cap \overline{(1 + \varepsilon)K} = (1 + \varepsilon)(U \cap K)$; hence $x/(1 + \varepsilon) \in \overline{U \cap K}$. It follows that $x \in \overline{U \cap K}$, and hence $\overline{U \cap K} \supset \overline{U} \cap \overline{K}$.

Let now A be nearly open and let $U \cap A^{(-1)}(V)$ be a $\mathfrak{T}_0$-neighbourhood of $\circ$ with absolutely convex U and V. Then by (2)

$$\overline{AI_A(U \cap A^{(-1)}(V))} = \overline{A(U) \cap (V \cap R[A])} = \overline{A(U)} \cap \overline{V \cap R[A]}$$

and this is a $\mathfrak{T}_2$-neighbourhood of $\circ$ in $\overline{R[A]}$ since A is nearly open. Hence AI_A is nearly open.

The dual characterization of continuous nearly open mappings of § 34, 1.(4) is valid also in the general case:

(3) *Let $E[\mathfrak{T}_1]$ and $F[\mathfrak{T}_2]$ be locally convex, $\mathfrak{M}_1$ resp. $\mathfrak{M}_2$ the class of equicontinuous subsets of E' resp. F'.*

A dense linear mapping A of $D[A] \subset E[\mathfrak{T}_1]$ in $F[\mathfrak{T}_2]$ is nearly open if and only if

$$A'(D[A'] \cap \mathfrak{M}_2) \supset \mathfrak{M}_1 \cap R[A'] \tag{4}$$

holds.

Proof. It is sufficient to show that (4) is necessary and sufficient for JKA to be nearly open. Now JKA is continuous from $(D[A])[\mathfrak{T}_1]$ into $(F/S[A])[\mathfrak{T}_0]$. If we apply § 34, 1.(4) to JKA and remember that $(JKA)' = A'$ we find that

$$A'(\mathfrak{M}) \supset \mathfrak{M}_1 \cap R[A'] \tag{5}$$

is the condition for JKA to be nearly open, where $\mathfrak{M}$ is the class of equicontinuous subsets of $D[A'] = (F/S[A])[\mathfrak{T}_0]'$. By § 36, 4. $\mathfrak{M}$ is the class of all sets $A(U)^\circ \cap V^\circ$ and their subsets. Since $A(U)^\circ \subset D[A']$, we have $\mathfrak{M} \subset D[A'] \cap \mathfrak{M}_2$ and (4) is a consequence of (5).

Conversely, we assume that (4) is true. Assume $U^\circ \cap R[A'] \in \mathfrak{M}_1 \cap R[A']$. Then there exists $V^\circ \in \mathfrak{M}_2$ such that $A'(D[A'] \cap V^\circ) \supset U^\circ \cap R[A']$. Now by § 36, 3.(6), $A'(A(U)^\circ) = A'(A'^{(-1)}(U^\circ)) \supset U^\circ \cap R[A']$ and therefore $A'(A(U)^\circ \cap V^\circ) \supset U^\circ \cap R[A']$; hence (5) is satisfied.

We remark in generalization of § 34, 1.(7) that every dense linear mapping A of $D[A] \subset E$ in F is nearly open for the weak topologies on E and F. It follows that for a linear mapping A to be nearly open means no restriction on the singularity of A.

(6) *Every maximal slight extension $\tilde{A}$ of a nearly open dense linear mapping is again nearly open.*

This follows from $\tilde{A}' = A'$ and (3).

Proposition § 34, 1.(6') is true also in the general setting.

(7) *Let A be nearly open dense and linear from $D[A] \subset E[\mathfrak{T}_1]$ in $F[\mathfrak{T}_2]$. Then $R[A']$ is $\mathfrak{T}_1^f(D[A])$-closed in E', i.e., for every absolutely convex $U \supset \mathfrak{o}$ in $D[A]$ the set $R[A'] \cap U^\circ$ is $\mathfrak{T}_s(D[A])$-closed in E'.*

By (4) there exists V such that $R[A'] \cap U^\circ \subset A'(D[A'] \cap V^\circ)$. This set is $\mathfrak{T}_s(D[A])$-closed in E' by § 36, 3.(7). But then $R[A'] \cap U^\circ = A'(D[A'] \cap V^\circ) \cap U^\circ$ is $\mathfrak{T}_s(D[A])$-closed too.

Baker [1'] proved a partial converse of (7) which gives a new dual characterization of nearly open mappings in this special case.

(8) *Let A be a dense linear mapping of $D[A] \subset E[\mathfrak{T}_1]$ onto $F[\mathfrak{T}_2]$, where $\mathfrak{T}_2$ is the Mackey topology. Then A is nearly open if and only if $R[A']$ is $\mathfrak{T}_1^f(D[A])$-closed in E'.*

We assume that $R[A']$ is $\mathfrak{T}_1^f(D[A])$-closed. If we are able to show that $A(U)^\circ \subset D[A'] \subset F'$ is $\mathfrak{T}_s(F)$-compact, then $A(U)^{\circ\circ} = \overline{A(U)}$ is a $\mathfrak{T}_2$-neighbourhood of $\mathfrak{o}$ in F and A is nearly open.

Since $R[A] = F$, we have $N[A'] = \mathfrak{o}$ by § 36, 3.(4), and by § 36, 5.(5) A' is a topological isomorphism of $(D[A'])[\mathfrak{T}_s(F)]$ onto $(R[A'])[\mathfrak{T}_s(D[A])]$. Since $A(U)^\circ = A'^{(-1)}(U^\circ \cap R[A'])$ by § 36, 3.(6) and since $U^\circ \cap R[A']$ is $\mathfrak{T}_s(D[A])$-compact by assumption, it follows that $A(U)^\circ$ is $\mathfrak{T}_s(F)$-compact.

4. Open mappings. In contrast to nearly open mappings, an open mapping does not have an arbitrary singularity.

(1) *Let A be an open linear mapping of $D[A] \subset E[\mathfrak{T}_1]$ in $F[\mathfrak{T}_2]$. Then A is always weakly singular.*

If U is an absolutely convex $\mathfrak{T}_1$-neighbourhood of $\mathfrak{o}$ in $D[A]$, then $A(U) = V \cap R[A]$, where V is an absolutely convex $\mathfrak{T}_2$-neighbourhood of $\mathfrak{o}$ in F. Hence every set $U + N[A]$ is of the form $A^{(-1)}(V)$. Therefore

$$Q[A] = \bigcap_V \overline{A^{(-1)}(V)} \subset \bigcap_U \overline{(U + N[A])} \subset \bigcap_U (2U + N[A])$$
$$= \bigcap_U (U + N[A]) = \overline{N}[A].$$

Hence $Q[A] = \bar{N}[A]$.

As we have seen in § 36, 2., there exist open linear mappings which are not regular.

(2) *If A is open, then the regular contraction KA is open. Conversely, if A is weakly singular and KA is open, then A is open.*

Proof. If A is open and $A(U) \supset V \cap R[A]$, then $3KA(U) \supset K(V) \cap K(R[A])$, as in the proof of 3.(1); therefore KA is open.

Conversely, assume that A is weakly singular and KA open. For every U_0 we have by § 36, 1.(6) and (8)

$$S[A] \cap R[A] = A(\bar{N}[A]) = \bigcap_U A(U) \subset A(U_0).$$

Since KA is open, $KA(U_0) \supset K(V_0 \cap R[A])$ for some V_0. If $Ax_0 \in V_0 \cap R[A]$, then there exists $z_1 \in S[A]$ such that $Ax_0 = Ax_1 + z_1$, $x_1 \in U_0$. Hence $z_1 \in S[A] \cap R[A] \subset A(U_0)$ and therefore $2A(U_0) \supset V_0 \cap R[A]$, A is open.

The following example shows that the assumption in (2) that A is weakly singular is necessary. We recall Example § 36, 2. i). Let G be an algebraic complement to H, so that $l^2 = H \oplus G$. We extend A from H to l^2 by defining A_1 on G as a nonopen linear mapping of G into G and $A_1 = A$ on H. Then A_1 is not open and $S[A_1] \supset S[A] = l^2$; therefore $S[A_1] = l^2$. Therefore $KA_1 = KA$ is the open mapping of l^2 onto $\mathfrak{o}$. It follows from (2) that A_1 must be strongly singular.

(3) *If A is open, then JKA is a homomorphism of $(D[A])[\mathfrak{T}_1]$ in $(F/S[A])[\mathfrak{T}_0]$. Conversely, if A is weakly singular and JKA is a homomorphism, then A is open.*

If A is weakly singular and JKA is open, then KA is open; hence A is open by (2). Conversely, let A be open. Then using (2) we may assume that A is regular and we have to show that JA is open. This is done as in 3.(1) for nearly open mappings.

A more satisfactory result is the following:

(4) *A linear mapping A of $D[A] \subset E[\mathfrak{T}_1]$ in $F[\mathfrak{T}_2]$ is open if and only if the continuous refinement AI_A is a homomorphism of $(D[A])[\mathfrak{T}_A]$ in $F[\mathfrak{T}_2]$.*

A $\mathfrak{T}_A$-neighbourhood of $\mathfrak{o}$ in $D[A]$ is of the form $U \cap A^{(-1)}(V)$ and $AI_A(U \cap A^{(-1)}(V)) = A(U) \cap V$. Therefore if A is open, $A(U)$ is open and $A(U) \cap V$ too; hence AI_A is open. Conversely, if AI_A is open it follows that $A(U) \cap V$ is open; hence $A(U)$ is open.

In § 32, 4.(3) we proved the homomorphism theorem, a dual characterization of homomorphisms. This theorem is a special case of the following dual characterization of open mappings.

(5) *Let A be a dense linear mapping of $D[A] \subset E[\mathfrak{T}_1]$ in $F[\mathfrak{T}_2]$. A is open if and only if the following conditions are satisfied:*

a) *$R[A']$ is $\mathfrak{T}_s(D[A])$-closed in E';*

b) *$A'(D[A'] \cap \mathfrak{M}_2) \supset \mathfrak{M}_1 \cap R[A']$, where $\mathfrak{M}_1$ resp. $\mathfrak{M}_2$ is the class of equicontinuous subsets of E' resp. F';*

c) *A is weakly singular.*

If A is further maximal, then a) *can be replaced by:*

a') *$R[A']$ is $\mathfrak{T}_s(E)$-closed in E'.*

We remark that b) is by 3.(3) always equivalent to: b') A is nearly open.

Proof. By (3) a weakly singular A is open if and only if JKA is a homomorphism. By the homomorphism theorem this is the case if and only if a) is satisfied and $A'(\mathfrak{M}) \supset \mathfrak{M}_1 \cap R[A']$, where $\mathfrak{M}$ is the class of $\mathfrak{T}_0$-equicontinuous subsets of $D[A']$. We proved in 3.(3) that this condition is equivalent to b).

That a) can be replaced by a') for a maximal A follows from § 36, 8.(4).

Since b) is always satisfied for the weak topologies, we have the following corollary to (5):

(6) *Let A be a dense linear mapping of $D[A] \subset E[\mathfrak{T}_1]$ in $F[\mathfrak{T}_2]$. A is weakly open if and only if A is weakly singular and $R[A']$ is $\mathfrak{T}_s(D[A])$-closed in E'.*

If A is dense and open, then A is weakly open.

A closed dense linear mapping A is weakly open if and only if $R[A']$ is $\mathfrak{T}_s(E)$-closed in E'.

In generalization of § 32, 4.(5) we have

(7) *Let $\mathfrak{T}_2$ be $\mathfrak{T}_k(F')$ on F. A linear mapping A of $D[A] \subset E[\mathfrak{T}_1]$ onto $F[\mathfrak{T}_2]$ is open if and only if it is weakly open.*

Proof. The condition is necessary by (5) and (6). Conversely, if A is weakly open it is sufficient to prove that A is nearly open. But this follows from 3.(8) since $R[A']$ is $\mathfrak{T}_s(D[A])$-closed.

(8) *If A is dense and open, then every maximal slight extension $\tilde{A}$ is open.*

Proof. We have $(\tilde{A})' = A'$; hence condition (5) b) for $\tilde{A}$ is satisfied. Since $\mathfrak{T}_s(D[\tilde{A}])$ is stronger than $\mathfrak{T}_s(D[A])$, condition (5) a) for $\tilde{A}$ is satisfied. It remains to prove that $\tilde{A}$ is weakly singular. Since A is weakly singular, $R[A'] = \overline{N}[A]^\circ$, where $\overline{N}[A]$ is the closure of $N[A]$ in $D[A]$ (§ 36, 3.(5)). The polar of $R[A']$ in $D[\tilde{A}]$ is therefore $\overline{N}[A]^{\circ\circ} = \overline{\overline{N}[A]}$, the closure of $\overline{N}[A]$ in $D[\tilde{A}]$. On the other hand, the polar of $R[A'] = R[\tilde{A}']$ in $D[\tilde{A}]$ is $Q[\tilde{A}]$ by § 36, 3.(4). It follows that $Q[\tilde{A}] = \overline{\overline{N}[A]} \subset \overline{N}[\tilde{A}]$; hence $Q[\tilde{A}] = \overline{N}[\tilde{A}]$.

We conclude with two simple but useful results on the range of an open mapping.

(9) *Let A be a closed dense and one-one linear mapping of $D[A] \subset E[\mathfrak{T}_1]$ in $F[\mathfrak{T}_2]$. If A is open and E complete, then $R[A]$ is closed in F.*

$A^{(-1)}$ is continuous from $(R[A])[\mathfrak{T}_2]$ onto $(D[A])[\mathfrak{T}_1]$. Let x_α be a net in $D[A]$ such that Ax_α converges to y_0 in F. Then x_α is a Cauchy net in $D[A]$ with a limit $x_0 \in E$. Since $G(A)$ is closed, $x_0 \in D[A]$ and $Ax_0 = y_0 \in R[A]$.

(9) is a special case of

(10) *Let A be a dense maximal open linear mapping of $D[A] \subset E[\mathfrak{T}_1]$ in $F[\mathfrak{T}_2]$. If $E/\overline{N}[A]$ is complete, then $R[A] + S[A]$ is closed in F.*

By § 36, 7.(5), $\overline{N}[A]$ is closed in E, so $E/\overline{N}[A]$ exists. The regular contraction KA is closed and dense. By § 36, 6.(2), the mapping $\widehat{KA}$ of $D[\widehat{KA}] \subset E/N[KA] = E/\overline{N}[A]$ in $F/S[A]$ is closed. By (9), $R[\widehat{KA}] = R[KA]$ is closed in $F/S[A]$ and therefore $K^{(-1)}(R[KA]) = R[A] + S[A]$ is closed in F.

5. Pták spaces. Open mapping theorems. We generalize some of our previous results on Pták spaces.

A linear mapping A of $D[A] \subset E[\mathfrak{T}_1]$ in $F[\mathfrak{T}_2]$ is nearly continuous (§ 34, 6.) if for every $\mathfrak{T}_2$-neighbourhood V of $\mathfrak{o}$ in F the closure $\overline{A^{(-1)}(V)}$ of $A^{(-1)}(V)$ in $D[A]$ is a $\mathfrak{T}_1$-neighbourhood U of $\mathfrak{o}$ in $D[A]$.

If $D[A]$ is dense in $E[\mathfrak{T}_1]$, then the class of all equicontinuous subsets of $(D[A])[\mathfrak{T}_1]' = E'$ coincides with the class $\mathfrak{M}_1$ of all equicontinuous subsets of $E[\mathfrak{T}_1]'$. Hence Propositions § 34, 6.(5) and § 34, 6.(7) read as follows:

(1) *Let A be a dense linear mapping of $D[A] \subset E[\mathfrak{T}_1]$ in $F[\mathfrak{T}_2]$. A is nearly continuous if and only if $A'(D[A'] \cap \mathfrak{M}_2) \subset \mathfrak{M}_1$. If A is nearly continuous, $D[A']$ is $\mathfrak{T}_2^f$-closed in F'.*

For the weak topologies the condition in (1) is always satisfied, therefore

(2) *Every dense linear mapping of $D[A] \subset E$ in F is nearly continuous in the sense of the weak topologies on E and F.*

From (1) and $\tilde{A}' = A'$ follows

(3) *If A is dense and nearly continuous, then every maximal slight extension $\tilde{A}$ of A is nearly continuous.*

We generalize the characterization of homomorphisms of Pták spaces given in § 34, 2.(2) to open mappings.

(4) *Let E be a Pták space, F locally convex, and A a maximal weakly singular linear mapping of $D[A] \subset E$ in F.*

If A is nearly open, then A is open and $R[KA]$ is closed in $F/S[A]$.

We can assume that $D[A]$ is dense in E; otherwise we replace E by $\overline{D[A]}$, which is also a Pták space.

Since A is nearly open, condition b) of 4.(5) is satisfied. A will be open by 4.(5) if $R[A']$ is $\mathfrak{T}_s(E)$-closed in E'. By 3.(7) all sets $R[A'] \cap U^\circ$ are $\mathfrak{T}_s(D[A])$-closed and hence $\mathfrak{T}_s(E)$-closed. The U° are also the polars of neighbourhoods of E since $D[A]$ is dense in E. Since E is a Pták space, it follows that $R[A']$ is weakly closed in E'.

That $R[KA]$ is closed follows from 4.(10) since every quotient of a Pták space is complete.

By § 34, 1.(1) every linear mapping onto a barrelled space is nearly open; hence (4) contains as a special case the following open-mapping theorem:

(5) *Let E be a Pták space, F barrelled. Then every maximal weakly singular mapping of $D[A] \subset E$ onto F is open.*

This generalizes the open-mapping theorem § 34, 7.(4), which resulted as a corollary to the closed-graph theorem § 34, 6.(9), whereas (5) is a consequence of the dual characterization of open linear mappings.

We list two further corollaries to (4):

(6) *Let E be a Pták space, F locally convex, and A a closed linear mapping of $D[A] \subset E$ in F. Then $R[A]$ is meagre in F or A is open and $R[A] = F$.*

If U is an absolutely convex neighbourhood of o in $D[A]$, then $R[A] = \bigcup_{n=1}^{\infty} nA(U)$. If $R[A]$ is nonmeagre in F, then $\overline{A(U)}$ is a neighbourhood of o in F. But then A is nearly open and hence open by (4) and $R[A] = F$.

(7) *Let E be a Pták space, F locally convex, and A a dense and weakly singular maximal linear mapping of $D[A] \subset E[\mathfrak{T}_1]$ onto $F[\mathfrak{T}_k(F')]$. If $R[A']$ is $\mathfrak{T}_1^f$-closed in E', then A is open.*

The assumption $R[A']$ is $\mathfrak{T}_1^f$-closed means that every set $R[A'] \cap U^\circ$, where U is a $\mathfrak{T}_1$-neighbourhood in E, is $\mathfrak{T}_s(E)$-closed. Since $U^\circ = (U \cap D[A])^\circ$ and since on U° $\mathfrak{T}_s(E)$ and $\mathfrak{T}_s(D[A])$ coincide, every $R[A'] \cap U^\circ$ is $\mathfrak{T}_s(D[A])$-closed in E'. By 3.(8) A is nearly open and by (4) open.

We investigated in some detail the behaviour of arbitrary linear mappings with the intention of understanding better the meaning of the closed-graph theorems and the open-mapping theorems.

If one does not make any assumption on a linear mapping A, then possibly $S[A] \neq \text{o}$ and there is no way to prove that A is continuous except in the case where E is a space with the strongest locally convex topology. Hence the natural assumption will be that A is regular (closable) and this is a necessary condition. If $D[A] = E$, then this means that A is closed. If $D[A]$ is a subspace of E, then the assumption that A is regular is more general. But if we have a closed-graph theorem which says that every closed linear mapping A of $D[A] \subset E$ in F is continuous, then these theorems remain true if we replace "closed" by "regular." This is obvious since the closed extension $\bar{A}$ is continuous by the closed-graph theorem and then A is continuous as a restriction of $\bar{A}$. The situation for the open-mapping theorems is different. Here the intention is to prove that a linear mapping A is open. The assumptions made on A in the open-mapping theorems in § 34 and § 35 (compare § 34, 10.(3), § 35, 3.) are that A is closed or sequentially closed. But these assumptions are not necessary, as we have seen, since there exist open linear mappings which are not regular.

The natural and necessary assumption is obviously that A be weakly singular. We succeeded in (5) in generalizing the open-mapping theorem for Pták spaces and barrelled spaces (§ 34, 7.(4)) to weakly singular maximal linear mappings. We ask if a similar generalization is possible in other cases. We give two examples.

We recall ADASCH's closed-graph theorem (§ 34, 10.(3)) that a closed linear mapping A of an (s)-space E onto a barrelled space F is open. It can be generalized to

(8) *A weakly singular maximal linear mapping of an* (s)*-space E onto a barrelled space F is open.*

The regular contraction KA is by § 36, 7.(6) a closed linear mapping of E onto $F/S[A]$ which is again a barrelled space. By ADASCH's theorem KA is open and by 4.(2) A is open.

We recall DE WILDE's theorem (§ 35, 3.(6)): A sequentially closed linear mapping A of $D[A] \subset E$, E a webbed space, onto an ultrabornological space F is open.

We prove the following variant:

(9) *Let E be a webbed space, F ultrabornological. Then a weakly singular maximal linear mapping of $D[A] \subset E$ onto F is open.*

By § 36, 7.(6) the regular contraction KA is closed and by DE WILDE's theorem KA is open and hence A.

An example in 6. will show that in (4), (5), (8), and (9) it is not possible to drop the assumption that A is maximal.

For further results see BAKER [1'], [2'] and BROWDER [1'].

6. Linear mappings in metrizable spaces. In this case more information is available.

(1) *Let E and F be normed spaces and A a dense linear mapping of $D[A] \subset E$ in F.*

a) *A is nearly continuous if and only if A' is strongly continuous.*

b) *A is nearly open if and only if A' is strongly open.*

c) *If E is a* (B)*-space and A, moreover, weakly singular and maximal, then A is open if and only if A' is strongly open.*

Proof. a) By 5.(1) A is nearly continuous if and only if $A'(D[A'] \cap \mathfrak{M}_2) \subset \mathfrak{M}_1$. But this means that the image of the closed unit ball of $D[A']$ is contained in a multiple of the closed unit ball of E'.

b) By 3.(3) A is nearly open if and only if $A'(D[A'] \cap \mathfrak{M}_2) \supset \mathfrak{M}_1 \cap R[A']$ and this means that the image of the unit ball of $D[A']$ contains a multiple of the unit ball of $R[A']$.

c) follows from b) and 5.(4).

Let L be an infinite dimensional vector space on which there are defined two inequivalent norms $\|x\|_1 \geqq \|x\|$ such that L with norm $\|x\|$ is a (B)-space F and with norm $\|x\|_1$ a normed space E. Then the identity mapping I of E onto F is continuous but not an isomorphism. I is nearly open by § 34, 1.(1) and I' is strongly open by (1). Since I' is strongly continuous, I' is a strong isomorphism. This is the counterexample mentioned at the end of § 33, 1. It shows too that (1) c) is false if we assume E only to be normed.

We remark that the definition of a strictly finer norm on a Banach space F seems possible only by using ZORN's lemma. Let $F = l^2$ and $\{x_\alpha\}$, $\alpha \in \mathsf{A}$, be an algebraic base of l^2 such that $\|x_\alpha\| = 1$ for every α. For $x = \sum_{i=1}^{n} \xi_{\alpha_i} x_{\alpha_i}$ one defines $\|x\|_1 = \sum_{i=1}^{n} |\xi_{\alpha_i}|$. One verifies easily that $\|x\|_1 \geqq \|x\|$ and that these norms are inequivalent. For details cf. GOLDBERG [1'], II.1.10.

We use this example also to obtain the counterexample mentioned at the end of 5. Let $\tilde{E}$ be the completion of the normed space E. Then I is a dense continuous linear mapping of $D[I] = E \subset \tilde{E}$ onto F. Its closed extension $\bar{I}$ is a homomorphism of $\tilde{E}$ onto F. Hence I is not maximal but otherwise satisfies the assumptions of A in 5.(4), 5.(8) and (9) and is not open.

We give some conditions for the continuity of linear mappings.

(2) *Let E and F be* (F)*-spaces, A a dense closed linear mapping of $D[A] \subset E$ in F. The following properties of A are equivalent:*

a) *A is continuous;*

b) *A is nearly continuous;*

c) $D[A] = E$;
d) $D[A'] = F'$.

If E and F are (B)-*spaces, one has the further equivalent property:*
e) *A' is strongly continuous.*

Proof. If we consider A as a mapping of $D[A]$ in F and if we assume b), then A is continuous by § 34, 6.(8); hence a) and b) are equivalent. From § 34, 5.(8) it follows that a) implies c); conversely, c) implies a) by the closed-graph theorem. a) and c) imply d) and d) implies a) by § 36, 5.(6).

If E and F are (B)-spaces, then b) and e) are equivalent by (1) a).

We now investigate open mappings.

For normed spaces we have an elementary characterization of open mappings which generalizes § 33, 1.(2).

(3) *Let E, F be normed spaces and A a linear mapping of $D[A] \subset E$ in F. Then A is open if and only if there exists $m > 0$ such that*

$$\|Ax\| \geqq m\|\hat{x}\| \quad \text{for all } x \in D[A],$$

where $\hat{x}$ is the residue class of x in $D[A]/N[A]$.

We remark that $N[A]$ is not necessarily closed in $D[A]$, so $\|\hat{x}\| = \inf_{x' \in x + N[A]} \|x'\|$ may be 0 for $\hat{x} \neq \hat{o}$.

Proof. a) Assume A to be open. There exists $\rho > 0$ such that for $y \in A(D[A])$, $\|y\| = 1$, there exists $x \in D[A]$, $\|x\| \leqq \rho$, such that $y = Ax$. Hence $\|Ax\| \geqq (1/\rho)\|x\| \geqq (1/\rho)\|\hat{x}\|$ for all $x \in D[A]$.

b) Conversely, assume $\|Ax\| \geqq m\|\hat{x}\|$. Then if $y \in A(D[A])$, $\|y\| = 1$, there exists x such that $1 = \|Ax\| \geqq (m - \varepsilon)\|x\|$ for a given $\varepsilon > 0$; hence $\|x\| \leqq 1/(m - \varepsilon)$ and the image of the ball of radius $1/(m - \varepsilon)$ in $D[A]$ contains the unit ball in $A(D[A])$ and A is open.

(4) *Let E and F be metrizable locally convex spaces, A a linear mapping of $D[A] \subset E$ in F. A is open if and only if A is sequentially invertible.*

Proof. A is open if and only if the continuous refinement AI_A is a homomorphism of $(D[A])[\mathfrak{T}_A]$ in $F[\mathfrak{T}_2]$ (4.(4)). The space $(D[A])[\mathfrak{T}_A]$ is again metrizable by 1. Therefore A is open if and only if AI_A is sequentially invertible (§ 33, 2.(1)).

Assume AI_A to be sequentially invertible. Then it is clear that A is sequentially invertible. Conversely, assume A to be sequentially invertible. Let $y_n \in A(D[A])$ converge to o; then there exist $x_n \to o$ in $D[A]$ such that $Ax_n = y_n$. But then $(x_n, Ax_n) \to o$ in $E \times F$ and this means that x_n converges to o in the sense of $\mathfrak{T}_A$; hence AI_A is sequentially invertible.

We give three further characterizations of open mappings.

(5) *Let E, F be metrizable locally convex spaces, A a dense linear mapping of $D[A] \subset E$ in F. A is open if and only if A is weakly open.*

Proof. If A is open, then it is weakly open by 4.(6). Assume now that A is weakly open. We consider A as a mapping A_0 of $D[A] \subset E$ onto $H = (R[A])[\mathfrak{T}_2]$. Then A is open if and only if A_0 is open, and A is weakly open if and only if A_0 is weakly open, since $\mathfrak{T}_s(H')$ and $\mathfrak{T}_s(E')$ coincide on H. It follows from 4.(7) that A_0 is open, since $\mathfrak{T}_2$ induces on H the Mackey topology.

(6) *Let E be an* (F)*-space, F metrizable locally convex, A a dense, maximal, and weakly singular linear mapping of $D[A] \subset E$ in F. A is open if and only if $R[A'] = A'(D[A'])$ is locally closed in $E'[\mathfrak{T}_s(E)]$ or strongly sequentially closed in E'.*

The conditions are both necessary by 4.(5). It remains to prove that if $R[A']$ is locally closed, it is $\mathfrak{T}_s(E)$-closed, since then A is weakly open by 4.(5) and open by (5).

Let $U_1 \supset U_2 \supset \cdots$ and $V_1 \supset V_2 \supset \cdots$ be neighbourhood bases of $\mathfrak{o}$ in $D[A]$ resp. F. Then the sets $W_n = K \sqcap (A(U_n) \cup V_n)$ are a neighbourhood base of $\mathfrak{o}$ for the topology $\mathfrak{T}_0$ on $F/S[A]$ (cf. § 36, 4.) and $(F/S[A])[\mathfrak{T}_0]$ is metrizable. Let $(\widetilde{F/S[A]})[\tilde{\mathfrak{T}}_0]$ be the completion. The continuous contraction JKA of A has a continuous extension B which maps $E[\mathfrak{T}_1]$ in $(\widetilde{F/S[A]})[\tilde{\mathfrak{T}}_0]$. Now $(F/S[A])[\mathfrak{T}_0]' = (\widetilde{F/S[A]})[\tilde{\mathfrak{T}}_0]' = D[A']$ by § 36, 5.(3) and therefore $B' = (JKA)' = A'$. By assumption $A'(D[A']) = R[B']$ is locally closed as a subspace of $E'[\mathfrak{T}_s(E)]$; it follows therefore from § 33, 3.(1) that B is a homomorphism and $A'(D[A'])$ is $\mathfrak{T}_s(E)$-closed.

(7) *Let E and F be* (F)*-spaces, A a dense maximal and weakly singular linear mapping of $D[A] \subset E$ in F. A is open if and only if A' is $\mathfrak{T}_c$-open.*

Proof. a) A is open if and only if the regular contraction KA is open. The topologies for A' are $\mathfrak{T}_c(F)$ and $\mathfrak{T}_c(E)$, and the topologies for $(KA)' = A'$ are $\mathfrak{T}_c(F/S[A])$ and $\mathfrak{T}_c(E)$. But $\mathfrak{T}_c(F)$ and $\mathfrak{T}_c(F/S[A])$ coincide on $S[A]^\perp \supset D[A']$; it will therefore be sufficient to consider the case that A is closed.

b) Let A be closed and A' $\mathfrak{T}_c$-open. The $\mathfrak{T}_c$-topologies on E' and F' are weaker than the Mackey topologies; therefore $D[A']$ is $\mathfrak{T}_c(F)$-dense in F'. Therefore A' is weakly dense and weakly closed and A'' exists and coincides with A. It follows from 4.(5) that $R[A''] = R[A]$ is closed in F and from 5.(5) that A is open.

c) Conversely, let A be open and closed. Then A' is a dense and closed linear mapping of $D[A'] \subset F'[\mathfrak{T}_c(F)]$ in $E'[\mathfrak{T}_c(E)]$. We verify the conditions of 4.(5).

We have again $A'' = A$ and $R[A'']$ is $\mathfrak{T}_s(F)$-closed; therefore we have only to prove $A(D[A] \cap \mathfrak{M}_1) \supset \mathfrak{M}_2 \cap R[A]$, where $\mathfrak{M}_1$ resp. $\mathfrak{M}_2$ is the class of relatively compact subsets of E resp. F. Since $R[A]$ is closed, it is sufficient to show that every absolutely convex and compact subset M_2 of $R[A]$ is the image of a set $M_1 \subset E$ with the same properties. By § 21, 10.(3) M_2 is the closed absolutely convex cover of a null sequence y_n. Since A is sequentially invertible by (4), there exists a null sequence $x_n \in D[A]$ such that $Ax_n = y_n$. Since A is closed, the closed absolutely convex cover M_1 of the sequence x_n is a compact subset of $D[A]$ and $A(M_1) = M_2$.

7. Open mappings in (B)- and (F)-spaces. As in the continuous case in § 33, 4., we collect the main results on open mappings in two theorems.

(1) *Let E and F be* (B)*-spaces and A a dense closed linear mapping of $D[A] \subset E$ in F. The following properties of A are equivalent:*

a) *A is open;*
b) *A is nearly open;*
c) *A is weakly open;*
d) *$R[A]$ is closed;*
e) *A' is strongly open;*
f) *A' is weakly open;*
g) *A' is $\mathfrak{T}_c$-open;*
h) *$R[A']$ is weakly closed;*
i) *$R[A']$ is strongly sequentially closed;*
j) *there exists $m > 0$ such that $\|Ax\| \geqq m\|\hat{x}\|$, $\hat{x} \in E/N[A]$.*

If A is dense, maximal, and weakly singular, all this remains true if we replace d) *by*

d') *$R[A] + S[A]$ is closed.*

We prove the second statement, which includes the case of a closed A. a) and b) are equivalent by 5.(4), a) and c) by 6.(5), a) and e) by 6.(1), a) and g) by 6.(7), and a) and j) by 6.(3).

d') follows from a) by 5.(4). Conversely, if $R[A] + S[A]$ is closed, then $R[KA]$ is closed in $F/S[A]$ and KA is open by 5.(5) when we consider KA as a mapping on $R[KA]$. Finally, A is open by 4.(2); hence a) follows from d').

The equivalence of h) and i) with a) is a consequence of 6.(6).

We show, finally, the equivalence of d') and f): A' is open as a mapping of $(D[A'])[\mathfrak{T}_s(F)]$ in $E'[\mathfrak{T}_s(E)]$ if and only if A' is open as the mapping $(KA)'$ of $D[A'] \subset (S[A]^\circ)[\mathfrak{T}_s(F/S[A])]$. As such it is dense and closed as the adjoint of the contraction KA. Since KA is closed, we have $(KA)'' = KA$ and by 4.(6) $(KA)' = A'$ is weakly open if and only if $R[KA]$ is weakly closed in $F/S[A]$. But this means d').

(2) *Let E and F be* (F)*-spaces and A a dense closed linear mapping of $D[A] \subset E$ in F. The following properties of A are equivalent:*

a) *A is open*;

b) *A is nearly open*;

c) *A is weakly open*;

d) *$R[A]$ is closed*;

e) *A is sequentially invertible*;

f) *A' is weakly open*;

g) *A' is $\mathfrak{T}_c$-open*;

h) *$R[A']$ is weakly closed*;

i) *$R[A']$ is locally closed or strongly sequentially closed.*

If A is dense, maximal, and weakly singular, all this remains true if we replace d) *by*

d') *$R[A] + S[A]$ is closed.*

The proofs given for (1) are valid also in this case. The equivalence of a) and e) follows from 6.(4).

In the case of (FM)-spaces we have, by analogy to § 33, 6.(1),

(3) *Let E and F be* (FM)*-spaces and A a dense, maximal, and weakly singular linear mapping of E in F. Then A is open if and only if A' is strongly open.*

We omit the trivial proof.

8. Domains and ranges of closed mappings of (F)-spaces. Not every subspace of an (F)-space F is the range of a closed linear mapping of another (F)-space E in F. It is possible to give an intrinsic characterization of the subspaces which are the range of a suitable closed linear mapping.

We say that a locally convex space $E[\mathfrak{T}]$ is an $(\mathrm{F})^t$-space if the associated barrelled space $E[\mathfrak{T}^t]$ is an (F)-space (§ 34, 9.).

(1) *A subspace H of the* (F)*-space $F[\mathfrak{T}]$ is the range $A(E)$ of a continuous linear mapping of an* (F)*-space $E[\mathfrak{T}']$ in F if and only if $H[\hat{\mathfrak{T}}]$, $\hat{\mathfrak{T}}$ the topology induced on H by $\mathfrak{T}$, is an* $(\mathrm{F})^t$*-space.*

If $H[\hat{\mathfrak{T}}]$ is an $(\mathrm{F})^t$-space, then the injection of $H[\hat{\mathfrak{T}}^t]$ in $F[\mathfrak{T}]$ is continuous and has the range H.

Assume, conversely, that $H = A(E)$, $A \in \mathfrak{L}(E, F)$. Without loss of generality we may suppose that A is one-one. We consider A as a mapping of $E[\mathfrak{T}']$ on $H[\hat{\mathfrak{T}}]$. Since $E[\mathfrak{T}']$ is barrelled, A is also continuous as a mapping of $E[\mathfrak{T}']$ onto $H[\hat{\mathfrak{T}}^t]$ by § 34, 9.(2). But then it is a one-one continuous mapping of an infra-Pták space onto a barrelled space and therefore an isomorphism by § 34, 2.(3); hence $H[\hat{\mathfrak{T}}^t]$ is an (F)-space.

For the larger class of closed linear mappings the result is the same.

(2) *Let E, F be* (F)*-spaces and A a closed linear mapping of $D[A] \subset E$ in F. Then $R[A]$ is an* $(\mathrm{F})^t$*-space.*

The continuous refinement AI_A of A is by 1.(3) a continuous mapping of the (F)-space $(D[A])[\mathfrak{T}_A]$ in F and $R[AI_A] = R[A]$, and so $R[A]$ is an $(\mathrm{F})^t$-space by (1).

(3) *Let E and F be* (F)*-spaces. Let $D \subset E$ be the domain of definition of a closed linear mapping in F; then D is an* $(\mathrm{F})^t$*-space and the graph topology $\mathfrak{T}_A$ coincides with the associated barrelled topology of D.*

We know from 1.(3) that $D[\mathfrak{T}_A]$ is an (F)-space and $\mathfrak{T}_A \supset \widehat{\mathfrak{T}}_1$, the topology induced on D by the topology $\mathfrak{T}_1$ of E. It follows from § 34, 9. that the associated barrelled topology $\widehat{\mathfrak{T}}_1^t$ is weaker than $\mathfrak{T}_A$ on D. Therefore the identity mapping I of $(D[A])[\mathfrak{T}_A]$ onto $(D[A])[\widehat{\mathfrak{T}}_1^t]$ is continuous and it follows again from § 34, 2.(3) that I is an isomorphism, which proves the statement.

If D is a subspace of $E[\mathfrak{T}]$ which is an $(\mathrm{F})^t$-space, then the identity mapping I of $D \subset E[\mathfrak{T}]$ onto $D[\widehat{\mathfrak{T}}^t]$ is closed and D is a domain of definition of a closed linear mapping; hence the class of all these domains is again the class of all subspaces of E which are $(\mathrm{F})^t$-spaces.

A subspace which has a countable but not finite algebraic dimension is not an $(\mathrm{F})^t$-space.

The class of range spaces of linear operators in Hilbert spaces has been studied in detail (cf. Fillmore and Williams [1′]).

§ 38. Linear equations and inverse mappings

1. Solvability conditions. Let A be a linear mapping of the locally convex space E in the locally convex space F. For a given element $y_0 \in F$

(1) $$Ax = y_0$$

is called a linear equation and the problem is to find all solutions $x \in E$ which satisfy (1). If $y_0 = \mathfrak{o}$ we have the homogeneous case and $N[A]$ is the set of all solutions. In the inhomogeneous case $y_0 \neq \mathfrak{o}$ all solutions are given by $x_0 + z$, $z \in N[A]$, where x_0 is one solution of (1).

The first problem is therefore to give necessary and sufficient conditions for the existence of one solution of (1) or, equivalently, to give necessary and sufficient conditions for the given y_0 to belong to the range of A.

We use again duality arguments.

(2) *Let E and F be locally convex and A a dense regular linear mapping of $D[A] \subset E$ in F.* (1) *is solvable in $D[A]$ if and only if $l(A'v) = vy_0$, $v \in D[A']$, defines on $R[A']$ a $\mathfrak{T}_s(D[A])$-continuous linear functional.*

Proof. If $x_0 \in D[A]$ and $Ax_0 = y_0$, then $v(Ax_0) = (A'v)x_0 = vy_0$ for all $v \in D[A']$ and x_0 is a $\mathfrak{T}_s(D[A])$-continuous linear functional on E' and hence on $R[A']$.

Conversely, if l is uniquely defined and $\mathfrak{T}_s(D[A])$-continuous on $R[A']$, then it can be continuously extended on E' by the HAHN–BANACH theorem and is therefore generated by an $x_0 \in D[A]$. It follows from $(A'v)x_0 = v(Ax_0) = vy_0$ for all $v \in D[A']$ that $Ax_0 = y_0$ since $D[A']$ is weakly dense in D' by § 36, 3.(3).

We remark that we can replace in this proof $\mathfrak{T}_s(D[A])$ by any locally convex topology on E' which is compatible with the dual pair $\langle E', D[A]\rangle$.

We formulate our result for the case $\mathfrak{T}_k(D[A])$ in the following way:

(3) *Let E and F be locally convex spaces and A a dense regular linear mapping of $D[A] \subset E$ in F. The equation* (1) *is solvable in $D[A]$ if and only if there exists an absolutely convex and $\mathfrak{T}_s(E')$-compact subset K of $D[A]$ such that*

$$(4) \qquad |vy_0| \leqq \sup_{x \in K} |(A'v)x| \quad \textit{for all } v \in D[A'].$$

If (4) *is satisfied,* (1) *has a solution $x_0 \in K$.*

We have only to prove the last statement: $p(u) = \sup_{x \in K} |ux|$ is a semi-norm on E'. (4) says that $|l(u)| \leqq p(u)$ on $R[A']$. The extension ux_0 of $l(u)$ can be chosen such that $|ux_0| \leqq p(u)$ on E'. But then $x_0 \in K^{\circ\circ} = K$.

If E and F are reflexive (B)-spaces and $A \in \mathfrak{L}(E, F)$, $y_0 \in A(E)$ if and only if there exists $r > 0$ such that $|vy_0| \leqq r\|A'v\|$ for all $v \in F'$ and then (1) has a solution x_0 with $\|x_0\| \leqq r$.

The following solvability condition was given by CROSS [1']:

(5) *Let E and F be locally convex and A a dense regular linear mapping of $D[A] \subset E$ in F.* (1) *is solvable for $y_0 \neq \mathfrak{o}$ if and only if $\overline{A'(D[A'] \cap [y_0]^\circ)}$ is a strict subspace of $\overline{A'(D[A'])}$, where in both cases the closure is the $\mathfrak{T}_s(D[A])$-closure.*

Proof. a) Necessity. We assume that $Ax_0 = y_0$, $x_0 \in D[A]$, and that $\overline{A'(D[A'] \cap [y_0]^\circ)} = \overline{R[A']}$. For $v \in D[A'] \cap [y_0]^\circ$ one has $0 = vy_0 = v(Ax_0) = (A'v)x_0$. But then $ux_0 = 0$ also for all $u \in \overline{A'(D[A'] \cap [y_0]^\circ)}$; hence $ux_0 = 0$ for all $u \in \overline{R[A']}$. From $(A'v)x_0 = v(Ax_0) = 0$ for all $v \in D[A']$ follows $Ax_0 = \mathfrak{o}$, which is a contradiction.

b) Sufficiency. We assume that $\overline{A'(D[A'] \cap [y_0]^\circ)}$ is a strict subspace of $\overline{R[A']}$. There exists $v_0 \in D[A'] \sim [y_0]^\circ$ such that $A'v_0 \notin \overline{A'(D[A'] \cap [y_0]^\circ)}$. We assume $v_0 y_0 = 1$. Using the HAHN–BANACH theorem we obtain $x_0 \in D[A]$ such that $v_0(Ax_0) = (A'v_0)x_0 = 1$ and $v(Ax_0) = (A'v)x_0 = 0$ for all $v \in D[A'] \cap [y_0]^\circ$.

We prove that this x_0 is a solution of (1). We have

$$D[A'] = (D[A'] \cap [y_0]^\circ) \oplus [v_0].$$

On $D[A']$ we have $v_0(Ax_0) = 1 = v_0 y_0$ and $v(Ax_0) = 0 = vy_0$ for all $v \in D[A'] \cap [y_0]^\circ$. Since $D[A']$ is weakly dense in F', it follows that $Ax_0 = y_0$.

The following problem is closely related to the linear equation (1). Let E be a locally convex space and $M = \{u_\alpha\}$, $\alpha \in \mathrm{A}$, an infinite set of elements of E'. The problem is whether the system of equations

$$(6) \qquad u_\alpha x = \gamma_\alpha, \qquad \alpha \in \mathrm{A},$$

where γ_α are given real or complex numbers, has a solution $x \in E$. One has the following result:

(7) *The system of equations* (6) *is solvable if and only if there exists an absolutely convex weakly compact subset K in E such that*

$$(8) \qquad \left|\sum_{k=1}^{n} \beta_k \gamma_{\alpha_k}\right| \leqq \sup_{x \in K} \left|\left(\sum_{1}^{n} \beta_k u_{\alpha_k}\right) x\right|, \qquad n = 1, 2, \ldots,$$

where the β_k are arbitrary real resp. complex numbers and $u_{\alpha_1}, \ldots, u_{\alpha_n}$ are n arbitrary different u_α.

If we define $l\left(\sum_1^n \beta_k u_{\alpha_k}\right) = \sum_1^n \beta_k \gamma_{\alpha_k}$, then l is a uniquely defined and $\mathfrak{T}_k(E)$-continuous linear functional on the linear span of M if and only if (8) is satisfied. By extension of l onto E' we obtain a solution $x_0 \in K$ of (6). (7) includes as special cases two classical results of HAHN [2].

(9) *Let $\{\gamma_\alpha\}$, $\alpha \in \mathrm{A}$, be an infinite set of real resp. complex numbers, $\{x_\alpha\}$, $\alpha \in \mathrm{A}$, a set of elements of a* (B)*-space E. The system of equations*

$$u x_\alpha = \gamma_\alpha, \qquad \alpha \in \mathrm{A},$$

has a solution u_0 in E' such that $\|u_0\| \leqq M$ if and only if

$$\left|\sum_1^n \beta_k \gamma_{\alpha_k}\right| \leqq M \left\|\sum_1^n \beta_k x_{\alpha_k}\right\|, \qquad n = 1, 2, \ldots,$$

for arbitrary real resp. complex β_k and n arbitrary different x_α.

(10) *Let E and F be reflexive* (B)*-spaces. Then* (6) *has a solution $x_0 \in E$, $\|x_0\| \leqq M$ if and only if $\left|\sum_1^n \beta_k \gamma_{\alpha_k}\right| \leqq M \left\|\sum_1^n \beta_k u_{\alpha_k}\right\|$ for arbitrary β_k.*

If E is not reflexive (10) is not true in general. For a finite number of equations one has the following classical result of HELLY [2].

(11) *Let E be a normed space, $u_1, \ldots, u_n$ elements of E', $\gamma_1, \ldots, \gamma_n$ real resp. complex numbers. Then the system of linear equations*

$$u_k x = \gamma_k, \qquad k = 1, \ldots, n,$$

has for every $\varepsilon > 0$ a solution x_ε such that $\|x_\varepsilon\| \leqq M + \varepsilon$ if and only if

$$\left|\sum_1^n \beta_k \gamma_k\right| \leqq M \left\|\sum_1^n \beta_k u_k\right\|$$

for arbitrary real resp. complex numbers β_k.

Proof. Denote by H the polar of the linear space $[u_1, \ldots, u_n]$. Then E/H is of dimension n and $(E/H)'$ can be identified with $[u_1, \ldots, u_n]$. The system of equations

$$u_k \hat{x} = \gamma_i, \qquad i = 1, \ldots, n, \quad \hat{x} \in E/H,$$

has by (10) a solution $\hat{x}_0$, $\|\hat{x}_0\| \leqq M$, if and only if $\left|\sum_1^n \beta_k \gamma_k\right| \leqq M \left\|\sum_1^n \beta_k u_k\right\|$ for all β_k. If this condition is satisfied there exists $x_\varepsilon \in \hat{x}_0$ such that $\|x_\varepsilon\| \leqq \|\hat{x}_0\| + \varepsilon = M + \varepsilon$ and x_ε is a solution of the original system of equations.

2. Continuous left and right inverses. The results of 1. are rather general but not very satisfactory. The existence of a solution of an equation $Ax = y_0$ is proved by using the Hahn–Banach theorem and is therefore nonconstructive, and for a different y_0 one has to verify the conditions again. One would prefer to have an explicit formula which gives immediately a solution for every possible y_0.

Before treating this problem in generality we consider first the continuous case.

Let $E[\mathfrak{T}_1]$ and $F[\mathfrak{T}_2]$ be locally convex spaces and $A \in \mathfrak{L}(E, F)$. We remark that

(1) *$A \in \mathfrak{L}(E, F)$ is an isomorphism if and only if there exists $B \in \mathfrak{L}(F, E)$ such that*

$$BA = I_E, \qquad AB = I_F. \tag{2}$$

If A is an isomorphism, then $B = A^{(-1)}$ is continuous and satisfies (2), and if (2) is satisfied with A and B continuous, then A is an isomorphism by § 1, 7.(2).

We recall from § 8, 4. that B is uniquely determined by (2) and is called the inverse A^{-1} of A.

Hence, if A is an isomorphism, $A^{-1}y_0$ is the unique solution of $Ax = y_0$ for every $y_0 \in F$.

If $B \in \mathfrak{L}(F, E)$ satisfies the relation $BA = I_E$, then B is called a continuous left inverse of A; analogously $C \in \mathfrak{L}(F, E)$ is a continuous right inverse of A if $AC = I_F$. One has the following theorem:

(3) *Let E and F be locally convex and $A \in \mathfrak{L}(E, F)$.*

a) *A has a continuous left inverse if and only if A is a monomorphism and $A(E)$ has a topological complement in F.*

b) *A has a continuous right inverse if and only if A is a homomorphism of E onto F and $N[A]$ has a topological complement in E.*

Proof. a) Let $B \in \mathfrak{L}(F, E)$ be a left inverse to A, $BA = I_E$. It follows that A is one-one and $N[AB] = N[B]$. Since $(AB)^2 = A(BA)B = AB$, AB is a continuous projection. It follows from $A(E) \supset (AB)(F) \supset A(BA)(E) = A(E)$ that the range of AB is $A(E)$ and $A(E)$ is closed as the range of a continuous projection. The kernel $N[B]$ of AB is then a topological complement to $A(E)$. Finally, A is an isomorphism of E onto $A(E)$ by (1) since the restriction B_0 of B to $A(E)$ satisfies $B_0 A = I_E$ and $AB_0 = I_{A(E)}$.

The conditions are also necessary: Let P be a continuous projection of F onto $A(E)$ and $B = \breve{A}^{-1}P$. Then $B \in \mathfrak{L}(F, E)$ and $BA = I_E$.

b) Let $AC = I_F$, $C \in \mathfrak{L}(F, E)$. Then C is one-one and CA is, as before, a continuous projection of E onto $C(F)$ with kernel $N[A]$. Hence $C(F)$ is a topological complement to $N[A]$. The mapping A can be considered as a continuous mapping of $C(F)$ onto F. It has the continuous inverse C and is therefore an isomorphism of $C(F)$ onto F; finally, A is a homomorphism of E onto F.

The conditions are sufficient: Let P be a continuous projection of E with kernel $N[A]$. Identifying $E/N[A]$ with the subspace $P(E)$, we define $C \in \mathfrak{L}(F, E)$ by $C = P\hat{A}^{-1}$ and have $AC = I_F$.

If B is a continuous left inverse to A, then all continuous left inverses to A are given by $B + D$, $D \in \mathfrak{L}(F, E)$ and $N[D] \supset A(E)$. Similarly, if C is a continuous right inverse to A, then all right inverses to A are given by $C + G$, $G \in \mathfrak{L}(F, E)$, $G(F) \subset N[A]$.

The importance of the existence of left resp. right inverses for solving equations is obvious: If there exists B such that $BA = I_E$ and if $y_0 \in R[A]$, then $x_0 = By_0$ is the unique solution of $Ax = y_0$.

If $AC = I_F$, then $Ax = y_0$ is solvable for every $y_0 \in F$ and $x_0 = Cy_0$ is a solution. The solution is not uniquely determined if $N[A] \neq \mathfrak{o}$. A different right inverse will in general give a different solution.

If $A \in \mathfrak{L}(E, F)$ has a left inverse $B \in \mathfrak{L}(F, E)$, then $A'B' = I_{E'}$; hence B' is a right inverse to A' and the transposed equation $A'v = u$ has the solution $v = B'u \in F'$ for a given $u \in E'$. Similarly for right inverses C.

We remark that our problem was first stated and solved in Hilbert space by TOEPLITZ [1′]. Using § 33, 1.(2) we have in this case

(4) *$A \in \mathfrak{L}(l^2_d)$ has a continuous left inverse if and only if A is bounded from below.*
$A \in \mathfrak{L}(l^2_d)$ has a continuous right inverse if and only if $R[A] = l^2_d$.

l^2_d is a space with the property that every closed subspace has a topological complement. We listed in § 31, 4. all known locally convex spaces G with this property. EBERHARDT [4′] discovered some new spaces G. It follows from (3) that every monomorphism A of a locally convex space E with $R[A]$ closed in G has a continuous left inverse. Similarly, every homomorphism A of a space G onto a locally convex space F has a continuous right inverse.

We leave it to the reader to formulate the theorems for the different spaces G. We remark further: If in (3) $A(E)$ resp. $N[A]$ has finite dimension or finite co-dimension, the left resp. right inverse exists.

Fortunately, these are not the only cases where the existence of continuous left resp. right inverses can be proved, as we will see in 3.

3. Extension and lifting properties. The HAHN–BANACH theorem shows that it is always possible to find a continuous extension of a continuous linear functional defined on a subspace of a locally convex space to the whole space. Simple examples show that it is in general not possible to extend continuously a continuous linear mapping of a subspace $H \subset E$ in F to a mapping of E in F.

We prove that this problem is closely related to the problem of the existence of a continuous left inverse.

(1) *Let H be a closed linear subspace of a locally convex space E. Then the following properties are equivalent:*

a) *there exists a continuous projection P of E onto H;*

b) *every continuous mapping A of H in a locally convex space F has a continuous extension $\tilde{A}$ mapping E in F;*

c) *every monomorphism A_1 of a locally convex space X in E with range H has a continuous left inverse B_1.*

Proof. The equivalence of a) and c) is by 2.(3) a). If P exists, then $\tilde{A} = AP$ is a continuous extension of A. Finally, we assume b); let J be the identity mapping of $H \subset E$ onto $F = H$ and $\tilde{J}$ a continuous extension to E. Then $J^{-1}\tilde{J}$ is a continuous projection of E onto H since $\tilde{J}J^{-1} = I_H$.

If the spaces involved, E, F, X, are all (B)-spaces, we have, moreover,

(2) *The following properties are equivalent:*

a) *there exists a continuous projection P of norm $\|P\| \leqq \lambda$ of E onto H;*

b) *every A has an extension $\tilde{A}$ such that* $\|\tilde{A}\| \leqq \lambda \|A\|$;

c) *every monomorphism A_1 has a left inverse B such that* $\|B\| \leqq \lambda \|\check{A}_1^{-1}\|$.

Proof. That b) and c) follow from a) is trivial from the construction of $\tilde{A}$ resp. B_1 (cf. 2.(3)). a) follows from b) since $J^{-1}\tilde{J}$ is a projection on H and $\|J^{-1}\tilde{J}\| = \lambda$. Finally, let I be the injection of H in E, B its left inverse, $\|B\| \leqq \lambda$; then $P = IB$ is a projection on H and $\|P\| \leqq \lambda$. Hence a) follows from c).

It was the extension problem that attracted attention first and the following result was the starting point. It is essentially due to PHILLIPS [1].

(3) *Let H be a linear subspace of the locally convex space $E[\mathfrak{T}]$ and $A \in \mathfrak{L}(H, l_d^\infty)$. Then A has an extension $\tilde{A} \in \mathfrak{L}(E, l_d^\infty)$.*

If E is a normed space, then there exists $\tilde{A}$ such that $\|\tilde{A}\| = \|A\|$.

Proof. There exists an absolutely convex neighbourhood U of $\mathfrak{o}$ in E such that $\|Az\| \leqq 1$ for all $z \in U \cap H$. Now $Az = y$ is an element (y_α), $\alpha \in \mathsf{A}$, of l_d^∞. If we define $y_\alpha = u_\alpha z$, then u_α is a linear functional on $A(H)$ such that $|u_\alpha z| \leqq 1$ for $z \in U \cap H$. Every u_α has an extension u_α on E such that $|u_\alpha x| \leqq 1$ for $x \in U$. The linear mapping $\tilde{A}x = (u_\alpha x)$, $\alpha \in \mathsf{A}$, is an extension of A and $\|\tilde{A}x\| \leqq 1$ for $x \in U$.

The result for normed spaces is included in our proof.

We have the following corollary:

(4) *Let E be a locally convex space and H a closed subspace isomorphic to l_d^∞. Then there exists a continuous projection P of E onto H.*

If E is a (B)*-space and H a closed subspace norm isomorphic to l_d^∞, then there exists a projection P of E onto H such that* $\|P\| = 1$.

Proof. Let J be the isomorphism of $H \subset E$ onto l_d^∞ and $\tilde{J}$ its extension by (3). Then $J^{-1}\tilde{J}$ is a continuous projection of E onto H which has norm 1 when J is a norm isomorphism.

We leave it to the reader to formulate the equivalent statements b) and c) of (1) resp. (2) in our case.

This result motivated the following definition: A (B)-space E is called a P_λ-space if it has the following property: If X is a (B)-space which contains E as a closed subspace, then there exists a projection P of X on E with $\|P\| \leqq \lambda$. l_d^∞ is a P_1-space in this terminology.

(5) *The following conditions are equivalent:*

a) *E is a P_λ-space;*

b) *let F, X be* (B)*-spaces, $X \supset E$; then every $A \in \mathfrak{L}(E, F)$ has a continuous extension $\tilde{A} \in \mathfrak{L}(X, F)$ such that* $\|\tilde{A}\| \leqq \lambda \|A\|$;

c) *let F, Y be* (B)*-spaces, $Y \supset F$; then every $A \in \mathfrak{L}(F, E)$ has a continuous extension $\tilde{A} \in \mathfrak{L}(Y, E)$ such that* $\|\tilde{A}\| \leqq \lambda \|A\|$.

The equivalence of a) and b) follows from (2). We assume a) and $A \in \mathfrak{L}(F, E)$. Then there exists l_d^∞ such that $E \subset l_d^\infty$ and $A \in \mathfrak{L}(F, l_d^\infty)$. By (3) A has an extension $\bar{A} \in \mathfrak{L}(Y, l_d^\infty)$, $\|\bar{A}\| = \|A\|$. If P is a projection of l_d^∞ onto E, $\|P\| \leqq \lambda$, then $\tilde{A} = P\bar{A} \in \mathfrak{L}(Y, E)$, $\|\tilde{A}\| \leqq \lambda\|A\|$.

Conversely, we assume c). The identity I on E has an extension $\tilde{I} \in \mathfrak{L}(Y, E)$, $\|\tilde{I}\| \leqq \lambda$, $\tilde{I}^2 = \tilde{I}$; thus $\tilde{I}$ is a projection of Y on E.

The class of P_1-spaces was determined by GOODNER [1′], NACHBIN [3], and KELLEY [1′] in the real case and by HASUMI [1′] in the complex case. A (B)-space is a P_1-space if and only if it is norm isomorphic to a space $C(K)$ of all continuous functions on a compact Stonean space K; a Hausdorff topological space is Stonean if the closure of every open set is again open. A relatively short proof of this result was given by KAUFMAN [1′].

Intensive research has been done on P_λ-spaces and related extension problems. We refer the reader to the work of LINDENSTRAUSS [1′] and the expository papers of NACHBIN [1′] and KÖTHE [6′].

A previous result, the theorem of SOBCZYK (§ 33, 5.(7)) and § 33, 5.(6), is another example belonging to this kind of problem, which says: If H is a closed subspace of a separable (B)-space E and H is norm isomorphic to c_0, then there exists a projection of norm $\leqq 2$ of E onto H.

For other examples of (B)-spaces with this "separable extension property" see BAKER [3′].

Conversely, recently ZIPPIN [1′] proved that a separable infinite dimensional (B)-space which is complemented in every separable (B)-space containing it is isomorphic to c_0.

If we use only the structure of locally convex spaces and not the richer structure of (B)-spaces, we obtain analogous problems which can be formulated in the following way.

Let $\mathscr{A}$ be a class of locally convex spaces. A space $E \in \mathscr{A}$ is called $\mathscr{A}$-detachable if it has the following property: Let X be any space in $\mathscr{A}$ and H a subspace of X isomorphic to E; then H has a topological complement in X.

If $\mathscr{A}$ is the class $\mathscr{L}$ of all locally convex spaces, then l_d^∞ is $\mathscr{L}$-detachable by (4). For other results compare KÖTHE [6′].

We will now establish an analogous relation between the existence of a continuous right inverse and the lifting of certain linear mappings.

Let A be a continuous linear mapping of E into the quotient F/H. We call A liftable in F if there exists $B \in \mathfrak{L}(E, F)$ such that $A = KB$, where K is the canonical homomorphism of F onto F/H.

(6) *Let E be locally convex and E/H a quotient. Then the following properties are equivalent:*

a) *there exists a continuous projection P of E with kernel H;*

b) *every continuous linear mapping A of a locally convex space X in E/H is liftable in E;*

c) *every homomorphism A_1 of E with kernel H onto a locally convex space F has a continuous right inverse.*

Proof. Assume a). We have $P = \hat{P}K$, $\hat{P}$ an isomorphism of E/H onto $P(E)$. Furthermore, $K\hat{P}$ is the identity on E/H. Let A be a continuous linear mapping of X in E/H. Then $A = (K\hat{P})A = K(\hat{P}A)$ and $B = \hat{P}A$ is the lifted mapping, so b) is true. Assume now b). Let I be the identical mapping of E/H, J its lifting in E, $I = KJ$. Then $P = JK$ is a continuous projection of E with kernel H. The equivalence of a) and c) follows by 2.(3) b).

If all spaces E, X, F are (B)-spaces we have the sharper corollary

(7) *The following properties of E and its quotient E/H are equivalent:*

a) *there exists a continuous projection P of norm $\|P\| \leqq \lambda$ of E with kernel H;*

b) *every $A \in \mathfrak{L}(X, E/H)$ has a lifting B such that $\|B\| \leqq \lambda\|A\|$;*

c) *every homomorphism A_1 of E onto F with kernel H has a right inverse C such that $\|C\| \leqq \lambda\|\hat{A}_1^{-1}\|$.*

Proof. b) and c) follow from a) by the construction of B resp. C in (6) resp. 2.(3). a) follows from b) since $P = JK$, as in the proof of (6). Finally, let K be the canonical homomorphism of E onto E/H. By c) it has a right inverse C of norm $\leqq \lambda$. Then $P = CK$ is a projection of norm $\leqq \lambda$ of E with kernel H and a) follows from c).

A (B)-space E will be called an H_λ-space if it has the following property: Let X be a (B)-space with a quotient X/H norm isomorphic to E; then there exists a projection P of X with kernel H and $\|P\| \leqq \lambda$.

Using (7) one obtains other equivalent definitions of an H_λ-space.

Let $\mathscr{A}$ be a class of locally convex spaces. $E \in \mathscr{A}$ is called liftable in $\mathscr{A}$ if for every $X \in \mathscr{A}$ and a quotient X/H isomorphic to E there exists a continuous projection of X with kernel H.

The following simple result is essentially due to Köthe [1′].

(8) *Every $l_{\mathfrak{a}}^1$ is liftable in the class of* (B)*-spaces. Every $l_{\mathfrak{a}}^1$ is an $H_{1+\varepsilon}$-space for any $\varepsilon > 0$.*

Proof. Let I be an isomorphism of $l_{\mathfrak{a}}^1$ onto X/H, X a (B)-space. We construct a lifting J of I in the following way.

If e_α is a unit vector in l_d^1, then $Ie_\alpha = \hat{x}_\alpha \in X/H$. For a given $\varepsilon > 0$ there exists $x_\alpha \in \hat{x}_\alpha$ such that $\|x_\alpha\| \leqq \|\hat{x}_\alpha\| + \varepsilon = \|I\| + \varepsilon$. We define J by $Je_\alpha = x_\alpha$. J is a continuous linear mapping of l_d^1 in X since

$$\left\| J\left(\sum_\alpha c_\alpha e_\alpha\right) \right\| = \left\| \sum c_\alpha x_\alpha \right\| \leqq \left(\sum |c_\alpha|\right)(\|I\| + \varepsilon) = \left\| \sum c_\alpha e_\alpha \right\| (\|I\| + \varepsilon).$$

Therefore $\|J\| \leqq \|I\| + \varepsilon$ and $I = KJ$; J is a lifting of I.

$P = JI^{-1}K$ is a continuous projection of X with kernel H. This proves that l_d^1 is liftable in the class of (B)-spaces.

If I is, moreover, a norm isomorphism, $\|P\| \leqq 1 + \varepsilon$ and l_d^1 is an $H_{1+\varepsilon}$-space.

Grothendieck proved in [14] the following counterpart to the result on P_1-spaces. There exists no H_1-space. Every (B)-space which is a $H_{1+\varepsilon}$-space for every $\varepsilon > 0$ is norm isomorphic to an l_d^1.

Köthe proved in [5′] that every (B)-space which is liftable in the class of (B)-spaces is isomorphic to an l_d^1. The proof relies on results of Pełczynski [1′] on complemented subspaces in l^1.

For further results on liftable spaces compare Köthe [5′] and Lindenstrauss [1′].

4. Inverse mappings. Let E and F be locally convex and A a linear mapping of $D[A] \subset E$ in F. If $N[A] = \mathfrak{o}$, then the inverse mapping $A^{(-1)}$ defines a linear mapping of $R[A]$ onto $D[A]$. We will denote it by A^{-1}. So far we have used this notation only in the case where $D[A] = E$ and $R[A] = F$ (cf. 2.). Again we have

(1) $$A^{-1}A = I_{D[A]}, \qquad AA^{-1} = I_{R[A]}$$

and the inverse A^{-1} is uniquely determined by (1) if it exists. We note further that $G(A^{-1}) = {}^iG(A)$.

From § 36, 3.(5) it follows immediately that

(2) *A weakly singular dense linear mapping A of $D[A] \subset E$ in F has an inverse A^{-1} if and only if $R[A']$ is $\mathfrak{T}_s(D[A])$-dense in E'.*

It is of special interest to know when A^{-1} is continuous.

(3) *A linear mapping A of $D[A] \subset E[\mathfrak{T}_1]$ in $F[\mathfrak{T}_2]$ has an inverse A^{-1} which maps $(R[A])[\mathfrak{T}_2]$ continuously onto $(D[A])[\mathfrak{T}_1]$ if and only if A is one-one and open.*

This is trivial. (3) shows that the continuity of A^{-1} is equivalent to the openness of A, and this has been one of the problems we investigated thoroughly before. We will therefore make only some remarks on this subject and leave it to the reader to reformulate our previous results on open mappings as statements on the existence of a continuous inverse.

We list first some general facts.

(4) *Let A be a dense linear mapping of $D[A] \subset E[\mathfrak{T}_1]$ in $F[\mathfrak{T}_2]$ and $\mathfrak{M}_1$ resp. $\mathfrak{M}_2$ the class of equicontinuous subsets of E' resp. F'.*

A has a continuous inverse A^{-1} if and only if a) $R[A'] = E'$ *and* b) $A'(D[A'] \cap \mathfrak{M}_2) \supset \mathfrak{M}_1 \cap R[A']$.

In particular, A has a weakly continuous inverse A^{-1} if and only if $R[A'] = E'$.

By (3) we have to prove that these conditions are equivalent to the fact that A is one-one and open. But this follows immediately from § 37, 4.(5) since $R[A'] = E'$ implies by § 36, 3.(5) that A is weakly singular and, conversely, any open mapping is weakly singular. The last statement follows from § 37, 4.(6).

(5) *Let A be a dense linear mapping of $D[A] \subset E$, E locally convex, onto a locally convex space $F[\mathfrak{T}_k(F')]$. Then A has a continuous inverse A^{-1} if and only if $R[A'] = E'$.*

This follows from (4) and § 37, 4.(7).

(6) *A dense linear mapping A of a locally convex space E into a metrizable locally convex space F has a continuous inverse A^{-1} if and only if $R[A'] = E'$.*

This is a special case of (5) since $A(E)$ is metrizable and its topology is therefore $\mathfrak{T}_k(A(E)')$.

As another example let us mention De Wilde's open-mapping theorem, Theorem § 35, 3.(5). It says: A sequentially closed one-one linear mapping A of a webbed space F onto an ultrabornological space E has a continuous inverse A^{-1}.

For the existence of $(A')^{-1}$ we have the following condition:

(7) *Let A be a dense linear mapping of E in F, where E and F are locally convex spaces. The inverse $(A')^{-1}$ exists if and only if $R[A]$ is dense in F.*

This is an immediate consequence of § 36, 3.(4) a).

We remark that this includes the case where $D[A']$ consists only of the element $\mathfrak{o}$.

(8) *Let A be a dense linear mapping of E in F, both locally convex. If A and A' have inverses, then $(A^{-1})' = (A')^{-1}$.*

$R[A]$ is dense in F by (7); hence A^{-1} is a dense linear mapping of $R[A] \subset F$ in E. By § 36, 6.(3) we have $G((A^{-1})') = {}^{i}G(-A^{-1})^{\circ} = G(-A)^{\circ}$.

Again by § 36, 6.(3) $G(-A)^\circ = {}^iG(A')$. By assumption A' too has an inverse and therefore $G((A')^{-1}) = {}^iG(A')$; and, finally, $G((A^{-1})') = G((A')^{-1})$, which includes the statement.

By analogy to the second statement of (4) we have

(9) *Let A be a dense linear mapping of $D[A] \subset E$ in F. The mapping A' of $(D[A'])[\mathfrak{T}_s(F)]$ onto $(R[A'])[\mathfrak{T}_s(D[A])]$ has a continuous inverse $(A')^{-1}$ if and only if $R[A] + S[A] = F$.*

This is a special case of § 36, 5.(5).

If A is, moreover, maximal we have a better result:

(10) *Let A be a dense maximal linear mapping of $D[A] \subset E$ in F. Then $(A')^{-1}$ is weakly continuous if and only if $R[A] + S[A] = F$. In particular, if A is a dense closed linear mapping of $D[A] \subset E$ in F, then $(A')^{-1}$ is weakly continuous if and only if $R[A] = F$.*

Proof. The weak topologies are $\mathfrak{T}_s(E)$ on $R[A']$ and $\mathfrak{T}_s(F)$ on $D[A']$. Since $\mathfrak{T}_s(D[A])$ is weaker on $R[A']$ than $\mathfrak{T}_s(E)$, it follows from (9) that the condition $R[A] + S[A] = F$ is sufficient.

We assume that A' is weakly open. Since $(KA)' = A'$, this means that the closed linear mapping KA has an adjoint which is dense and closed as a mapping from $D[A'] \subset (S[A]^\perp)[\mathfrak{T}_s(F/S[A])$ in $E'[\mathfrak{T}_s(E)]$. It is also open. Since A is maximal, KA is closed and therefore $(KA)'' = KA$. By § 37, 4.(6) $R[KA]$ is $\mathfrak{T}_s(S[A]^\perp)$-closed in $F/S[A]$. It follows from (7) that $R[KA] = F$ and this means $R[A] + S[A] = F$.

We now investigate the strong continuity of $(A')^{-1}$, which is always to be understood in the sense of $\mathfrak{T}_b(E)$ resp. $\mathfrak{T}_b(F)$ on $R[A']$ resp. $D[A']$.

(11) *Let E be locally convex, F quasi-barrelled, and A a dense linear mapping of $D[A] \subset E$ in F. If $(A')^{-1}$ exists and is strongly continuous, then $\overline{R[A]} = F$ and A is nearly open.*

Proof. $\overline{R[A]} = F$ follows from (7). By § 37, 3.(3) we have to show that $A'(D[A'] \cap \mathfrak{M}_2) \supset \mathfrak{M}_1 \cap R[A']$. This is equivalent to $(A')^{-1}(\mathfrak{M}_1 \cap R[A']) \subset D[A'] \cap \mathfrak{M}_2$. Every $M \in \mathfrak{M}_1 \cap R[A']$ is relatively weakly compact in E' and therefore strongly bounded. Since $(A')^{-1}$ is strongly continuous, $(A')^{-1}(M)$ is strongly bounded in F'. Since F is quasi-barrelled, $(A')^{-1}(M) \in D[A'] \cap \mathfrak{M}_2$.

As a corollary we obtain the following generalization (Mochizuki [1']) of the theorem of Banach–Hausdorff:

(12) *Let E be a Pták space, F quasi-barrelled, and A a dense closed linear mapping of $D[A] \subset E$ in F. If $(A')^{-1}$ exists and is strongly continuous, then A is open and $R[A] = F$.*

This follows from (11) and § 36, 5.(4).

The converse of (12) is not true even for (F)-spaces E and F, as is shown by Counterexample § 33, 2. 1), where K is a homomorphism onto $E/N[A]$ but $(K')^{-1}$ is not strongly continuous.

Closely related to (12) is

(13) *Let E be semi-reflexive, F locally convex, and A a dense closed linear mapping of $D[A] \subset E$ in F. If $(A')^{-1}$ exists and is strongly continuous, then $R[A] = F$.*

By assumption $(A')^{-1}$ is continuous from $(R[A'])[\mathfrak{T}_k(E)]$ in $F'[\mathfrak{T}_b(F)]$. Every continuous linear mapping is weakly continuous. Since $(R[A'])[\mathfrak{T}_k(E)]' = E/R[A']^\circ$, the weak topology on $R[A']$ is $\mathfrak{T}_s(E)$ and the weak topology on $F'[\mathfrak{T}_b(F)]$ is $\mathfrak{T}_s(F'')$. Since $\mathfrak{T}_s(F'')$ is stronger than $\mathfrak{T}_s(F)$, it follows that $(A')^{-1}$ is weakly continuous. $R[A] = F$ follows now from (10).

The following two results of KRISHNAMURTHY and LOUSTAUNAU [1'] have some interest in connection with the remark after (12).

(14) *Let $E'[\mathfrak{T}_b(E)]$ be metrizable and F barrelled or sequentially complete. If A is a dense linear mapping of $D[A] \subset E$ onto F, then $(A')^{-1}$ exists and is strongly continuous.*

$(A')^{-1}$ exists by (7) and maps the metrizable space $(R[A'])[\mathfrak{T}_b(E)]$ onto $(D[A'])[\mathfrak{T}_b(F)]$. We assume that $(A')^{-1}$ is not continuous. Then there exists a sequence $A'v_n \in R[A']$ which converges strongly to $\mathfrak{o}$ such that the sequence v_n is not strongly bounded (§ 28, 3.(4)). From $R[A] = F$ and $(A'v_n)x = v_n(Ax) \to 0$ for all $x \in D[A]$ follows the weak convergence of v_n to $\mathfrak{o}$. But under the assumptions on F every weakly bounded set in F' is strongly bounded, and this is a contradiction.

(14) applies in particular to (DF)-spaces E.

(15) *Let A be a dense linear mapping of the distinguished space E into the locally convex space F. If the strong dual $F'[\mathfrak{T}_b(F)]$ is an infra-(s)-space and if $\overline{R[A]} = F$ and $R[A'] = E'$, then $(A')^{-1}$ exists and is strongly continuous.*

Since $\overline{R[A]} = F$, $(A')^{-1}$ exists and maps $R[A'] = E'$ onto $D[A'] \subset F'$. By § 23, 7.(1) $E'[\mathfrak{T}_b(E)]$ is barrelled. The graph $G(A')$ is weakly closed (§ 36, 6.(3)) and therefore $(A')^{-1}$ is a closed linear mapping of the barrelled space $E'[\mathfrak{T}_b(E)]$ in the infra-(s)-space $F'[\mathfrak{T}_b(F)]$. By KŌMURA's closed-graph theorem $(A')^{-1}$ is continuous.

(15) is true in particular for semi-reflexive spaces E.

We close with some results on normed spaces and (B)-spaces. It follows immediately from § 37, 7.(1) that for a dense closed linear mapping A of a (B)-space E into a (B)-space F, $(A')^{-1}$ exists and is strongly continuous if and only if $R[A] = F$.

The next result is a trivial consequence of § 37, 6.(1) b).

(16) *Let E and F be normed spaces and A a dense linear mapping of $D[A] \subset E$ in F. Then $(A')^{-1}$ exists and is strongly continuous if and only if A is nearly open and $\overline{R[A]} = F$.*

We remark that by (14) in the case that F is a (B)-space it is sufficient to assume for A that $R[A] = F$.

(17) *Let E and F be normed spaces and A a dense linear mapping of $D[A] \subset E$ in F. The following two statements are equivalent:*
a) $\overline{R[A]} = F$ *and* A^{-1} *exists and is continuous;*
b) $R[A'] = E'$ *and* $(A')^{-1}$ *exists and is strongly continuous.*

Proof. a) implies that A is open; hence by (16) $(A')^{-1}$ exists and is strongly continuous. Finally, $R[A'] = E'$ follows from (6).

Assume now b). (6) implies the existence and continuity of A^{-1} and $\overline{R[A]} = F$ follows from (7).

The results presented here may look rather unsystematic and accidental. But they are the main tools in the theory of state diagrams for linear mappings first developed by TAYLOR [2] for continuous mappings of Banach spaces, then by GOLDBERG [1'] for closed mappings of Banach spaces and by KRISHNAMURTHY [1'] and KRISHNAMURTHY and LOUSTAUNAU [1'] for mappings of locally convex spaces. The reader will find there a systematic theory of the connections between the properties of the range and the inverse of A and the same properties of A'.

5. Solvable pairs of mappings. The following problem has its origin in the theory of partial differential equations and was treated systematically by BROWDER [1']. We give here only one of his results and follow the method used by GOLDBERG [1'].

Let E, F be locally convex and A_0, A_1 two dense linear mappings of E in F. We assume that A_1 is an extension of A_0, $A_0 \subset A_1$, that A_0 is one-one, and that $R[A_1] = F$. We are looking for a linear mapping A such that $A_0 \subset A \subset A_1$, A is one-one, and $R[A] = F$.

We give a purely algebraic construction of such an A. Let

$$(1) \qquad E = N[A_1] \oplus E_0, \qquad F = R[A_0] \oplus F_0,$$

where E_0, F_0 are algebraic complements. We define $D[A]$ as

$$(2) \qquad D[A] = D[A_0] \oplus (E_0 \cap A_1^{(-1)}(F_0))$$

and A by

$$(3) \quad A(x_1 + x_2) = A_0 x_1 + A_1 x_2 \quad \text{for } x_1 \in D[A_0],\ x_2 \in E_0 \cap A_1^{(-1)}(F_0).$$

We remark that the sum in (2) is direct: For $x \in D[A_0] \cap (E_0 \cap A_1^{(-1)}(F_0))$ we have $A_0x = A_1x \in R[A_0] \cap F_0 = o$. Since A_0 is one-one, $x = o$.

Clearly, $A_0 \subset A \subset A_1$. We show that A is one-one. Let $A(x_1 + x_2) = A_0x_1 + A_1x_2$ be o; hence $A_0x_1 = -A_1x_2$. Since $A_0x_1 \in R[A_0]$ and $-A_1x_2 \in F_0$, it follows that $A_0x_1 = o$ and therefore $x_1 = o$. From $A_1x_2 = o$ and $x_2 \notin N[A_1]$ follows $x_2 = o$; hence A is one-one.

Finally, $R[A] = F$: Let y be an element of F, then $y = A_0x_1 + z$, $z \in F_0$. Since $R[A_1] = F$, there exists $x_2 \in D[A_1]$ such that $A_1x_2 = z$, $x_2 \in A_1^{(-1)}(F_0)$. There exists $x_3 \in N[A_1]$ such that $x_2 - x_3 \in E_0 \cap A_1^{(-1)}(F_0)$; hence $x_2 - x_3 \in D[A]$ and $A(x_2 - x_3) = Ax_2 = z$. Finally,

$$y = A(x_1 + (x_2 - x_3)) \quad \text{and} \quad y \in R[A].$$

The construction of A depends on E_0 and F_0, but if these complements are given, A is uniquely determined. We leave the details to the reader.

(4) *If A_0^{-1} is continuous, A_1 closed and the decompositions* (1) *topological, then A is closed.*

Assume $x_1^\alpha + x_2^\alpha \to x$, $x_1^\alpha \in D[A_0]$, $x_2^\alpha \in E_0 \cap A_1^{-1}(F_0)$, and $A(x_1^\alpha + x_2^\alpha) \to y$. We have to prove that $x \in D[A]$ and $y = Ax$. Since $A \subset A_1$ and A_1 is closed, $A_1x = y$. Now $Ax_1^\alpha \in R[A_0]$ and $Ax_2^\alpha \in F_0$ and it follows from (1) that $Ax_1^\alpha \to y_1 \in R[A_0]$ and $Ax_2^\alpha \to y_2 \in F_0$, $y = y_1 + y_2$. Since A_0^{-1} is continuous, x_1^α converges to an element $x_1 \in D[A_0]$, $Ax_1 = y_1$. Hence x_2^α converges to an element $x_2 \in E_0$ since E_0 is closed. Now A_1 is closed; hence $A_1x_2 = y_2$ and therefore $x_2 \in A_1^{-1}(F_0)$ and $x_2 \in D[A]$. But then $x \in D[A]$ and $Ax = y$.

The pair (A_0, A_1) is called s o l v a b l e if A has, moreover, a continuous inverse A^{-1}. Since A is one-one, it is sufficient to prove that A is open. If we apply § 37, 5.(6) we find the following rather large classes of solvable pairs:

(5) *Let E be a Pták space, F locally convex, and $A_0 \subset A_1$ two dense linear mappings of E in F such that A_0^{-1} exists and is continuous and such that A_1 is closed and $R[A_1] = F$. Assume, further, that there exists a topological decomposition* (1).

Then there exists a one-one linear mapping A such that $A_0 \subset A \subset A_1$, $R[A] = F$, and such that A^{-1} exists and is continuous.

6. Infinite systems of linear equations. We begin with some elementary facts. In particular, we present the results of EIDELHEIT [1′], [2′], which give a complete answer for a special type of systems of linear equations.

Let E be an (F)-space defined by the sequence of semi-norms $p_1(x) \leqq p_2(x) \leqq \cdots$. We consider the system of equations

(1) $\quad u_i x = c_i, \qquad u_i \in E'$, c_i real resp. complex, $i = 1, 2, \ldots.$

We say that (1) is f u l l y s o l v a b l e if (1) has a solution $x \in E$ for every sequence $c = (c_1, c_2, \dots)$. The problem is to find conditions for the u_i which are necessary and sufficient for the full solvability of (1).

If we define $Ax = (u_1 x, u_2 x, \dots)$, then A is a linear mapping of E into ω. Since the u_i are continuous, it is easy to see that A is a continuous linear mapping of E into ω. The full solvability of (1) is therefore equivalent to $A(E) = \omega$. Since E and ω are (F)-spaces, (1) is fully solvable if and only if A is a homomorphism of E onto ω.

The adjoint A' is the linear mapping of φ into E' defined by

$$A'\left(\sum_1^n v_i e_i\right) = \sum_1^n v_i u_i.$$

The homomorphism theorem for (F)-spaces (§ 33, 4.(2)) implies that (1) is fully solvable if and only if A' is one-one and locally sequentially invertible. Since the bounded sets in φ are finite dimensional, this means that a sequence $A'v^{(n)}$, $v^{(n)} \in \varphi$, converges locally to $\mathfrak{o}$ in E' if and only if the $v^{(n)}$ are uniformly bounded in length and converge coordinatewise to $\mathfrak{o}$.

EIDELHEIT introduces the o r d e r $n(u)$ of $u \in E'$ as the smallest n such that $|ux| \leqq M p_n(x)$ for all $x \in E$ and some $M > 0$. If U_n is the closed neighbourhood of $\mathfrak{o}$ defined by $p_n(x) \leqq 1$, then $n(u)$ is also the smallest n such that $u \in E'_{U_n^\circ}$ in our terminology.

We formulate now EIDELHEIT's theorem.

(2) *The system* (1) *is fully solvable if and only if the following conditions are satisfied:*

i) *the* u_i, $i = 1, 2, \dots$ *are linearly independent*;

ii) *for every natural number* m *there exists a natural number* r_m *with the following property: Let* $u = \sum_1^r v_i u_i$ *be any linear combination with* $v_r \neq \mathfrak{o}$. *If* $n(u) \leqq m$, *then* $r \leqq r_m$.

P r o o f. a) Necessity. If (1) is fully solvable, A' is one-one and hence i) is satisfied. Assume that ii) is false. Then there exists for some m a sequence $u^{(k)} = A'v^{(k)}$ such that $|u^{(k)}x| \leqq M_k p_m(x)$ and the $v^{(k)}$ are not bounded in length. But the multiples $(1/M_k k)u^{(k)}$ then converge locally to $\mathfrak{o}$, which leads to a contradiction since the corresponding sequence $(1/M_k k)v^{(k)}$ in φ has unbounded length.

b) Sufficiency. i) implies that A' is one-one. From ii) it follows that for every sequence $u^{(k)} = A'v^{(k)}$ which converges locally to $\mathfrak{o}$ in E' the $v^{(k)}$ are uniformly bounded in length and therefore coordinatewise convergent. Hence A' is locally sequentially invertible.

EIDELHEIT applied (2) to the infinite system of equations

$$(3) \qquad \sum_{k=1}^{\infty} a_{ik} x_k = c_k, \qquad i = 1, 2, \dots,$$

where the a_{ik} and c_k are given numbers. We assume in the following that for every k there exists at least one $a_{ik} \neq 0$. This is no restriction on the generality.

We are first looking for solutions $x = (x_1, x_2, \ldots)$ of (3) such that $\sum_{k=1}^{\infty} |a_{ik}x_k| < \infty$ for every i or solutions in the sense of absolute convergence. The space of all these x is the (F)-space E defined by the semi-norms $p_m(x) = \sum_{k=1}^{\infty} \left(\sum_{j=1}^{m} |a_{jk}| \right) |x_k|$, $m = 1, 2, \ldots$, and the $a_i x = \sum_{k=1}^{\infty} a_{ik}x_k$ define elements a_i of E'. So we have a special case of (2) and we obtain

(4) *The system* (3) *is fully solvable in the sense of absolute convergence if and only if the following conditions are satisfied:*

i) *the rows a_i of the matrix (a_{ik}) are linearly independent;*

ii) *for every natural number m there exists a natural number r_m with the following property: Let $\sum_1^r v_i a_i$ be any linear combination with $v_r \neq$ o. If for some $M > 0$*

$$\left| \sum_{i=1}^{r} v_i a_{ik} \right| \leqq M \sum_{j=1}^{m} |a_{jk}| \quad \text{for } k = 1, 2, \ldots,$$

then $r \leqq r_m$.

The proof is obvious; we remark only that the equivalence of $\left| \left(\sum_1^r v_i a_i \right) x \right| \leqq M p_m(x)$ to the system of inequalities in ii) can be seen by taking $x = e_k$, $k = 1, 2, \ldots$.

There is a second problem connected with (3). We now allow as solutions all sequences x such that the sums $\sum_{k=1}^{\infty} a_{ik}x_k$ are convergent for all $i = 1, 2, \ldots$ or solutions in the sense of conditional convergence. We will show that the space of all these x is again an (F)-space and that therefore (2) can be applied.

Let the sequence $b = (b_1, b_2, \ldots)$, all $b_k \neq 0$, be given and let μ be the space of all sequences $x = (x_1, x_2, \ldots)$ such that $\sum_{k=1}^{\infty} b_k x_k$ converges. We define the norm $\|x\| = \sup_n \left| \sum_1^n b_k x_k \right|$ on μ. If $x \in \mu$ and $s_n = \sum_1^n b_k x_k$, then $Jx = s$, $s = (s_1, s_2, \ldots)$ is a convergent sequence and J is a norm isomorphism of μ onto c and μ is a (B)-space.

We determine μ'. We saw in § 14, 7. that a continuous linear functional v on c is of the form $vs = \sum_{k=1}^{\infty} v_k s_k + v_0 \lim_{k \to \infty} s_k$ and $\|v\| \leqq 1$ if and only if $\sum_{k=0}^{\infty} |v_k| \leqq 1$. Since in μ every x is the limit of its sections and $\mu \supset \varphi$, every element of μ' is of the form $wx = \sum_{k=1}^{\infty} w_k x_k = \sum_{k=1}^{\infty} u_k b_k x_k$. The closed unit ball in μ' will consist of all $w = J'v$, $\|v\| \leqq 1$. We determine all $u = (u_1, u_2, \ldots)$ which correspond to the v, $\|v\| \leqq 1$.

If x is defined by $b_p x_p = 1$, $b_k x_k = 0$ for $k \neq p$, then Jx is the sequence s with $s_i = 0$ for $i < p$, $s_k = 1$ for $k \geqq p$ and we have

$$wx = u_p = vs = \sum_p^\infty v_k + v_0.$$

Consequently, $v_p = u_p - u_{p+1}$ for $p \geqq 1$. Hence $\sum_1^\infty v_p = u_1 - \lim_{n\to\infty} u_n$ and, since $u_1 = \sum_1^\infty v_p + v_0$, we conclude that $v_0 = \lim_{n\to\infty} u_n$. Therefore

(5) *μ' consists of all $w = (u_1 b_1, u_2 b_2, \ldots)$ such that $\sum_{k=1}^\infty |u_k - u_{k+1}| < \infty$. The closed unit ball in μ' consists of all w such that*

$$|u_1 - u_2| + |u_2 - u_3| + \cdots + |\lim u_n| \leqq 1.$$

We come back to (3). Let F be the space of all sequences x such that $\sum_{k=1}^\infty a_{ik} x_k$ converges for all $i = 1, 2, \ldots$. F is metrizable for the semi-norms $q_j(x) = \sup_n \left| \sum_{k=1}^n a_{jk} x_k \right|$, $j = 1, 2, \ldots$. Let $x^{(n)}$ be a Cauchy sequence in F. If we delete in $x \in F$ all coordinates x_k for which $a_{jk} = 0$, then the remaining sequence $x_{(j)}$ lies in a Banach space μ_j with norm $q_j(x_{(j)}) = q_j(x)$. The sequence $x_{(j)}^{(n)}$ has a limit $x_j^{(0)}$ which is also the coordinatewise limit. It follows that $x^{(n)}$ converges coordinatewise to a sequence $x^{(0)}$ which is the limit of $x^{(n)}$ in F. So F is an (F)-space.

It is obvious that $\sum_{k=1}^\infty a_{ik} x_k$ is a continuous linear functional on F for every $i = 1, 2, \ldots$. So we are in the situation of (2).

We introduce the increasing sequence of semi-norms $p_1 = q_1$, $p_2 = \max(p_1, q_2), \ldots$. If V_k is the closed unit ball corresponding to p_k, then $U_1 = V_1$, $U_2 = V_1 \cap V_2, \ldots$ are the closed unit balls corresponding to $p_1, p_2, \ldots$. The polars in F' are U_1°, $U_2^\circ = \ulcorner(V_1^\circ \cup V_2^\circ), \ldots$ (§ 20, 6.(5), § 20, 8.(10)). It follows now from (5) that $F'_{U_m^\circ}$ consists of all sequences of the form

(6) $(u_{11}a_{11}, u_{12}a_{12}, \ldots) + \cdots + (u_{m1}a_{m1}, u_{m2}a_{m2}, \ldots)$,

$$\sum_{k=1}^\infty |u_{jk} - u_{jk+1}| < \infty, \qquad j = 1, \ldots, m.$$

EIDELHEIT's theorem takes now the form

(7) *The system* (3) *is fully solvable in the sense of conditional convergence if and only if the following conditions are satisfied:*

i) *the rows a_i of the matrix (a_{ik}) are linearly independent;*

ii) *for every natural number m there exists a natural number r_m with the* *following property: Let* $\sum_{i=1}^{r} v_i a_i$ *be any linear combination with* $v_r \neq 0$. *If* $\sum_{i=1}^{r} v_i a_i$ *is of the form* (6), *then* $r \leqq r_m$.

There is quite a literature on these and related questions. For further references see my own paper [6] and especially NIETHAMMER and ZELLER [1′].

CHAPTER EIGHT

Spaces of Linear and Bilinear Mappings

The set $\mathfrak{L}(E, F)$ of all continuous linear mappings of E in F, where both E and F are locally convex, is a vector space. If $F = \mathsf{K}$, then $\mathfrak{L}(E, F) = E'$ and so it is obvious that there are many possibilities to define a locally convex topology on $\mathfrak{L}(E, F)$. This is done in § 39 and by adapting the methods of Volume I it is possible to obtain generalizations of some classical theorems as the BANACH–MACKEY theorem and the BANACH–STEINHAUS theorem. The relation between equicontinuous and weakly compact subsets of $\mathfrak{L}(E, F)$ is a little more complicated than in the case of dual spaces.

Bilinear mappings $B(x, y)$, $(x, y) \in E \times F$, and $B(x, y) \in G$ are studied in § 40. If $G = \mathsf{K}$, then we speak of bilinear forms on $E \times F$. These notions were introduced in § 15, 14.; § 40 contains now a systematic study. The notion of hypocontinuity which lies between separate continuity and continuity of a bilinear mapping is a very useful tool in studying bilinear mappings. The fundamental results are the continuity theorems in § 39, 2. To extend bilinear mappings continuously is quite a difficult task (§ 39, 3.).

In § 41 we investigate the projective tensor product $E \otimes_\pi F$ and its completion $E \tilde{\otimes}_\pi F$, E and F locally convex. We follow GROTHENDIECK's ideas and methods. Results for special classes of spaces are given and different properties are studied in detail. Some problems remain unsolved.

As a necessary preparation to the investigation of the approximation property we treat in § 42 compact mappings, in particular the subclass of nuclear mappings. § 42, 1. contains some basic properties of compact mappings; § 42, 2. is interested in weakly compact mappings. A lot of examples are given, including HILBERT–SCHMIDT mappings in Hilbert space. There exists a canonical mapping $\bar{\psi}$ of $E'_b \tilde{\otimes}_\pi F$ in $\mathfrak{L}(E, F)$ for (B)-spaces E and F. The $\bar{\psi}$-images in $\mathfrak{L}(E, F)$ are defined as the nuclear mappings of E in F. The space of all nuclear mappings $\mathfrak{N}(E, F)$ is a normed space relative to the nuclear norm $\| \;\|_\nu$. If $\bar{\psi}$ is one-one, $Z \in E'_b \tilde{\otimes}_\pi F$, then $\|\bar{\psi}(Z)\|_\nu = \|Z\|_\pi$. In § 42, 8. the method of factoring mappings is applied to compact mappings and we are led to the class of $\mathscr{L}^\infty$-spaces of LINDENSTRAUSS and PEŁCZYNSKI. § 42, 9. contains the fixed point theorem of SCHAUDER–TYCHONOFF and the theorem of LOMONOSOV on the existence of invariant subspaces.

§ 43 investigates the approximation property. Important equivalent formulations are given: a) This property of E and the property that $\bar{\psi}$ is one-one from $E'_b \tilde{\otimes}_\pi E$ in $\mathfrak{L}(E)$ are equivalent for (B)-spaces E; b) the ε-tensor product $E \otimes_\varepsilon F$ and the ε-product $E\varepsilon F$ are introduced in § 43, 3. and a (B)-space E has the approximation property if and only if $E \tilde{\otimes}_\varepsilon F = E\varepsilon F$ for every (B)-space F. Hereditary properties of the approximation property are obtained in § 43, 4. ENFLO's example of a separable (B)-space not having the approximation property is only mentioned; we suppose its existence and we give in §43, 9. JOHNSON's example of a separable (B)-space without the approximation property and therefore without a basis. A few results on the existence of a basis and some remarks on the bounded approximation theory give only some idea of a vast field of recent research.

ε-products and ε-tensor products of locally convex spaces are studied in detail in § 44. In contrast to the projective tensor product there seems to be little known in this case, most of it due to L. SCHWARTZ. I have tried to add some details to this picture but many questions remain open.

One question was investigated very carefully by GROTHENDIECK, namely, the determination of the dual space of $E \otimes_\varepsilon F$. This is the space $\mathfrak{J}(E, F)$ of integral bilinear forms on $E \times F$, which we study in § 45. To every such form corresponds a continuous mapping which is also called integral. In the case of Hilbert space SCHATTEN proved that $\mathfrak{J}(E, F)$ can be identified with $E' \tilde{\otimes}_\pi F'$; hence integral and nuclear mappings coincide in this case. Our last theorem in § 45 (Theorem § 45, 7.(6)) generalizes SCHATTEN's result to (B)-spaces, where E is arbitrary, F reflexive. This theorem of GROTHENDIECK relies on a deep theorem on vector measures which we state without proof.

§ 39. Spaces of linear mappings

1. Topologies on $\mathfrak{L}(E, F)$. Let E and F be locally convex and $\mathfrak{L}(E, F)$ be the vector space of all continuous linear mappings of E in F. In § 21, 1. we defined topologies on $E' = \mathfrak{L}(E, \mathsf{K})$ in a systematic way. We follow these methods in our more general case.

In the sequel $\mathfrak{M}$ will always be a class of bounded subsets M of E with the properties

a) $\mathfrak{M}$ is total in E, i.e., $\bigcup_{M \in \mathfrak{M}} M$ is total in E;

b) if M_1 and M_2 are in $\mathfrak{M}$, then $M_1 \cup M_2$ is in $\mathfrak{M}$.

We call $\mathfrak{M}$ saturated if it has the further properties

c) if $M \in \mathfrak{M}$, then $\rho M \in \mathfrak{M}$ for every $\rho > 0$;

d) if $M \in \mathfrak{M}$ and $N \subset M$, then $N \in \mathfrak{M}$;

e) if M_1, M_2 are in $\mathfrak{M}$, then $\overline{\Gamma(M_1 \cup M_2)}$ is in $\mathfrak{M}$.

The saturated cover of $\mathfrak{M}$ will be denoted by $\tilde{\mathfrak{M}}$.

Let $\mathfrak{L}(E, F)$ and $\mathfrak{M}$ be given. If $M \in \mathfrak{M}$ and V is a neighbourhood of $\mathfrak{o}$ in F, we define $U(M, V)$ as the set of all $A \in \mathfrak{L}(E, F)$ such that $A(M) \subset V$. As we will see immediately, these sets $U(M, V)$ are a neighbourhood base of $\mathfrak{o}$ of a vector space topology $\mathfrak{T}_\mathfrak{M}$ on $\mathfrak{L}(E, F)$ and we will write $\mathfrak{L}_\mathfrak{M}(E, F)$ for this topological vector space. $\mathfrak{T}_\mathfrak{M}$ is called the topology of $\mathfrak{M}$-convergence on $\mathfrak{L}(E, F)$.

(1) *$\mathfrak{L}_\mathfrak{M}(E, F)$ is locally convex.*

i) Every $U(M, V)$ is absorbing since, given $A \in \mathfrak{L}(E, F)$, one has $A(M) \subset \rho V$ for some $\rho > 0$ because M is bounded, and thus $(1/\rho)A \in U(M, V)$.

ii) If V is absolutely convex, A_1 and A_2 in $U(M, V)$, then $\alpha_1 A_1(M) + \alpha_2 A_2(M) \subset \alpha_1 V + \alpha_2 V \subset V$ for $|\alpha_1| + |\alpha_2| \leqq 1$; hence $U(M, V)$ is absolutely convex.

iii) $U(M_1, V_1) \cap U(M_2, V_2) \supset U(M_1 \cup M_2, V_1 \cap V_2)$; hence we have a filter base.

iv) $\rho U(M, V) = U(M, \rho V)$ for V absolutely convex and $\rho > 0$.

v) $\mathfrak{T}_{\mathfrak{M}}$ is Hausdorff: If $A \in \bigcap_{M,V} U(M, V)$, then $A(M) \subset V$ for all M, V; hence $A\left(\bigcup_M M\right) \subset \bigcap_V V = \circ$ and therefore $A = \circ$ since $\mathfrak{M}$ is total.

We have the following generalization of § 21, 1.(4):

(2) *The topologies $\mathfrak{T}_{\mathfrak{M}}$ and $\mathfrak{T}_{\tilde{\mathfrak{M}}}$ on $\mathfrak{L}(E, F)$ coincide. Two topologies $\mathfrak{T}_{\mathfrak{M}_1}$ and $\mathfrak{T}_{\mathfrak{M}_2}$ on $\mathfrak{L}(E, F)$ coincide if and only if $\tilde{\mathfrak{M}}_1 = \tilde{\mathfrak{M}}_2$.*

Proof. Enlarging $\mathfrak{M}$ so that c) and d) are satisfied evidently does not change the topology. The same is true for e), since for every closed absolutely convex V we obtain $U(\overline{\Gamma(M_1 \cup M_2)}, V) = U(M_1 \cup M_2, V)$.

The second statement of (2) follows from § 21, 1.(4) and the first half of the following lemma:

(2′) *$E'[\mathfrak{T}_{\mathfrak{M}}]$ is topologically isomorphic to a complemented subspace H_1 of $\mathfrak{L}_{\mathfrak{M}}(E, F)$.*

F is topologically isomorphic to a complemented subspace H_2 of $\mathfrak{L}_{\mathfrak{M}}(E, F)$.

Proof. a) We define the mapping J_1 of $E'[\mathfrak{T}_{\mathfrak{M}}]$ with range $H_1 \subset \mathfrak{L}_{\mathfrak{M}}(E, F)$ by $J_1 u = u \otimes y_0$, where $y_0 \neq \circ$ in F and $(u \otimes y_0)x = (ux)y_0$. We choose $v_0 \in F'$ such that $v_0 y_0 = 1$ and define the mapping K_1 of $\mathfrak{L}_{\mathfrak{M}}(E, F)$ into $E'[\mathfrak{T}_{\mathfrak{M}}]$ by $K_1 A = v_0 A$.

One has $K_1 J_1 u = K_1(u \otimes y_0) = (v_0 y_0)u = u$, or $K_1 J_1 = I_{E'}$, the identity on E'. Then $P_1 = J_1 K_1$ is a projection of $\mathfrak{L}_{\mathfrak{M}}(E, F)$ onto H_1.

We leave it to the reader to check that J_1 and K_1 are continuous. It follows that J_1 is a topological isomorphism of $E'[\mathfrak{T}_{\mathfrak{M}}]$ onto $H_1 \subset \mathfrak{L}_{\mathfrak{M}}(E, F)$.

b) By a similar argument one obtains the second statement of (2′) by using the mapping J_2 of F onto $H_2 \subset \mathfrak{L}_{\mathfrak{M}}(E, F)$ defined by $J_2 y = u_0 \otimes y$, $u_0 \neq \circ$, from E', and the mapping K_2 of $\mathfrak{L}_{\mathfrak{M}}(E, F)$ into F defined by $K_2 A = A x_0$, where $u_0 x_0 = 1$.

We remark that our notation is in accordance with § 9, 7.

$\mathfrak{T}_{\mathfrak{M}}$ can be described by semi-norms. If $\{p_\alpha(y)\}$, $\alpha \in \mathrm{A}$, is a system of semi-norms defining the topology of F, then $\mathfrak{T}_{\mathfrak{M}}$ is defined by the system of semi-norms

$$(3) \qquad p_{M,\alpha}(A) = \sup_{x \in M} p_\alpha(Ax), \qquad \alpha \in \mathrm{A},\ M \in \mathfrak{M},$$

or by the system

(4) $\quad p_{M,N}(A) = \sup_{\alpha \in M, v \in N} |v(Ax)|, \qquad M \in \mathfrak{M}, N$ equicontinuous in F'.

The set $\{A; p_{M,\alpha}(A) \leqq 1\}$ is obviously identical with $U(M, V)$, where $V = \{y; p_\alpha(y) \leqq 1\}$. Hence $U(M, V)$ is $\mathfrak{T}_{\mathfrak{M}}$-closed if V is closed in F.

We list some important particular cases of $\mathfrak{T}_{\mathfrak{M}}$. Define $\mathfrak{F}$, $\mathfrak{K}$, $\mathfrak{C}$, $\mathfrak{B}^*$, $\mathfrak{B}$, as in § 21 as, respectively, the classes of all finite, absolutely convex and weakly compact, precompact, strongly bounded, and bounded subsets of E; then the corresponding topologies $\mathfrak{T}_{\mathfrak{M}}$ will be denoted by $\mathfrak{T}_s$, $\mathfrak{T}_k$, $\mathfrak{T}_c$, $\mathfrak{T}_{b*}$, and $\mathfrak{T}_b$, and the corresponding spaces by $\mathfrak{L}_s(E, F)$, $\mathfrak{L}_k(E, F)$, $\mathfrak{L}_c(E, F)$, $\mathfrak{L}_{b*}(E, F)$, and $\mathfrak{L}_b(E, F)$.

It is usual to call $\mathfrak{T}_s$, $\mathfrak{T}_c$, $\mathfrak{T}_b$, respectively, the topologies of simple, precompact, and bounded convergence. These are the weak, precompact, and strong topologies in the special case $\mathfrak{L}(E, K) = E'$.

In the same way we will use the simpler notations E_s, E'_s, and so on for $E[\mathfrak{T}_s(E')]$, $E'[\mathfrak{T}_s(E)]$, and so on.

We say that $\mathfrak{M}$ covers E if $\bigcup_{M \in \mathfrak{M}} M = E$ or, equivalently, if $\tilde{\mathfrak{M}} \supset \mathfrak{F}$. The following remarks will be useful.

(5) *If $\mathfrak{M}$ covers E, then the mapping $A \rightarrow Ax_0$, where x_0 is a fixed element of E, is continuous from $\mathfrak{L}_{\mathfrak{M}}(E, F)$ in F.*

If V is a given neighbourhood of $\mathfrak{o}$ in F and if $x_0 \in M \in \mathfrak{M}$, then $Ax_0 \in V$ for all $A \in U(M, V)$.

(6) *If $\mathfrak{M}$ covers E and V is a closed neighbourhood of $\mathfrak{o}$ in F, then $U(M, V)$ is closed in $\mathfrak{L}_s(E, F)$ (simply closed).*

Let A_0 be an adherent point of $U(M, V)$ in $\mathfrak{L}_s(E, F)$. Then for each $x \in E$ A_0x is by (5) an adherent point of the Ax, $A \in U(M, V)$, in F. If x are in M, then the Ax are in V; hence $A_0x \in V$ and $A_0(M) \subset V$. But this is the statement to be shown.

We recall that we introduced in § 14, 6. in the case of normed spaces E, F the uniform norm topology on $\mathfrak{L}(E, F)$. Obviously, this is the topology of bounded convergence in the case of normed spaces.

Let E, F be locally convex. We denote by $L(E, F)$ the space of all linear (not necessarily continuous) mappings of E in F (in Volume I we used the notation $\mathfrak{S}(E, F)$). Obviously, $\mathfrak{L}(E, F) \subset L(E, F)$. We introduce now a topology $\mathfrak{T}_s$ on $L(E, F)$ such that this inclusion will become topological: $\mathfrak{L}_s(E, F) \subset L_s(E, F)$.

We define again the $\mathfrak{T}_s$-neighbourhoods $U(M, V)$ of $\mathfrak{o}$ as the sets of all $A \in L(E, F)$ such that $A(M) \subset V$, where M is a finite subset of E and V is a neighbourhood of $\mathfrak{o}$ in F. It is straightforward to verify that $L_s(E, F)$ is locally convex and that $\mathfrak{L}_s(E, F) \subset L_s(E, F)$.

It is rather obvious that for classes $\mathfrak{M}$ strictly larger than $\mathfrak{F}$ the space $L_{\mathfrak{M}}(E, F)$, defined in the same way, will no longer be Hausdorff.

We determine the structure of $L_s(E, F)$. Let $\{x_\alpha\}$, $\alpha \in \mathrm{A}$, be a linear base for E. For each x_α define Ax_α to be some element of F. Extending this map linearly we obtain a well-defined linear map A from E in F. In fact all linear maps from E in F arise in this way. Hence $L_s(E, F)$ can be identified algebraically with $F^{\mathrm{A}} = \prod_\alpha F_\alpha$, where $F_\alpha = F$. This identification is a topological isomorphism if we equip F^{A} with the product topology and the topology on F is the given locally convex topology.

We remark that the topology of E does not directly enter in the definition of $\mathfrak{T}_{\mathfrak{M}}$, but that the size of the space $\mathfrak{L}(E, F)$ depends on the topology of E. We recall that $\mathfrak{L}(E, F)$ is always a subspace of $\mathfrak{L}(E_s, F_s)$ and that this space coincides with $\mathfrak{L}(E_k, F)$, where F has any locally convex topology between $\mathfrak{T}_s$ and $\mathfrak{T}_k$. It follows that there are many topologies on the space $\mathfrak{L}(E_s, F_s)$ which depend not only on the class $\mathfrak{M}$ of subsets of E but also on the class $\mathfrak{N}$ of subsets of F' determining the topology on F. The following definitions take care of this situation.

Let $\langle E', E\rangle$ and $\langle F', F\rangle$ be two dual pairs and let $\mathfrak{L}(E_s, F_s)$ be the space of all weakly continuous linear mappings of E in F. Let $\mathfrak{M}$ be a total class of weakly bounded subsets of E as before and $\mathfrak{N}$ a total class of weakly bounded subsets of F'. Define again, for $M \in \mathfrak{M}$ and $N \in \mathfrak{N}$, $U(M, N^\circ)$ as the set of all $A \in \mathfrak{L}(E_s, F_s)$ such that $A(M) \subset N^\circ$. The topology defined by these neighbourhoods of $\mathfrak{o}$ on $\mathfrak{L}(E_s, F_s)$ will be denoted by $\mathfrak{T}_{\mathfrak{M},\mathfrak{N}}$ and $\mathfrak{L}_{\mathfrak{M},\mathfrak{N}}(E_s, F_s)$ will be the corresponding locally convex space, if it exists.

We will also use the notations $\mathfrak{T}_{b,b}$, $\mathfrak{L}_{b^*,b^*}(E, F)$, and so on, according to the conventions introduced after (4).

(7) *$\mathfrak{L}_{\mathfrak{M},\mathfrak{N}}(E_s, F_s)$ is locally convex if $\mathfrak{M}$ or $\mathfrak{N}$ contains only strongly bounded sets.*

The proof of (1) can be repeated except for the proof that $U(M, N^\circ)$ is absorbing. Assume M strongly bounded in E and N weakly bounded in F'. Let A be in $\mathfrak{L}(E_s, F_s)$. Since A is weakly and hence strongly continuous, $A(M) \subset \rho N^\circ$ for some $\rho > 0$ and $U(M, N^\circ)$ is absorbing.

Assume now M weakly bounded and N strongly bounded. Then $A(M)$ is weakly bounded and by the definition of a strongly bounded set (§ 20, 11.) there exists $\rho > 0$ such that $A(M) \subset \rho N^\circ$. Again $U(M, N^\circ)$ is absorbing.

The adjoint mappings A' to the $A \in \mathfrak{L}(E_s, F_s)$ determine the space $\mathfrak{L}(F'_s, E'_s)$ of all weakly continuous linear mappings of F' in E' and we have under the assumptions of (7)

(8) *The spaces $\mathfrak{L}_{\mathfrak{M},\mathfrak{N}}(E_s, F_s)$ and $\mathfrak{L}_{\mathfrak{N},\mathfrak{M}}(F'_s, E'_s)$ are isomorphic under the correspondence $A \to A'$.*

This follows from $p_{M,N}(A) = p_{N,M}(A')$, since the semi-norms (4) determine again the topology $\mathfrak{T}_{\mathfrak{M},\mathfrak{N}}$.

This generalizes the fact that $\|A\| = \|A'\|$ in the case of normed spaces.

2. The BANACH–MACKEY theorem. Let $\langle E_2, E_1\rangle$ be a dual system; then this theorem says that a Banach disk in E_1 resp. E_2 is always strongly bounded (§ 20, 11.(3)). It follows that in a locally convex space E weakly bounded and $\mathfrak{T}_k$-bounded sets coincide and that, if E is sequentially complete, even weakly bounded and strongly bounded sets in E resp. E' coincide.

It is easy to find the generalizations of these results for the spaces $\mathfrak{L}(E, F)$.

(1) *Let E, F be locally convex. Then every simply bounded subset P of $\mathfrak{L}(E, F)$ is $\mathfrak{T}_{b^*}$-bounded.*

If P is simply bounded, then for every $x \in E$ and every equicontinuous set $N \subset F'$

$$\sup_{A\in P, v\in N} |v(Ax)| = \sup |(A'v)x| < \infty.$$

Hence the set $\bigcup_{A\in P} A'(N)$ is weakly bounded in E'. If M is strongly bounded in E, it follows that

$$\sup_{A\in P, v\in N, x\in M} |(A'v)x| = \sup |v(Ax)| < \infty.$$

But this means $A(M) \subset \rho N^\circ$ for all $A \in P$; thus P is bounded in $\mathfrak{L}_{b^*}(E, F)$.

Using the BANACH–MACKEY theorem we obtain two corollaries.

(2) *Let E, F be locally convex. Then every simply bounded subset P of $\mathfrak{L}(E, F)$ is bounded for the uniform convergence on Banach disks.*

We say that a locally convex space E is l o c a l l y c o m p l e t e if every bounded subset is contained in a Banach disk. E is locally complete if it is sequentially complete.

(3) *If E is sequentially complete or locally complete, then every simply bounded subset P of $\mathfrak{L}(E, F)$ is $\mathfrak{T}_b$-bounded.*

As a special case of (3) we obtain the so-called "principle of uniform boundedness":

(4) *Let E be a* (B)*-space, F a normed space. A subset P of $\mathfrak{L}(E, F)$ is $\mathfrak{T}_b$-bounded, i.e., $\sup_{A\in P} \|A\| < \infty$, if and only if $\sup_{A\in P} \|Ax\| < \infty$ for every $x \in E$.*

This is nothing new; we proved this theorem even more directly in § 15, 13.(2′).

For a single continuous linear mapping the BANACH–MACKEY theorem implies

(5) *Let E or F be sequentially complete or locally complete and $A \in \mathfrak{L}(E, F)$. Then if B_1 is a bounded subset of E and B_2 any weakly bounded subset of F',*

$$\sup_{v \in B_2, x \in B_1} |v(Ax)| < \infty.$$

Proof. It follows from the assumptions that either B_1 (and therefore $A(B_1)$) is strongly bounded or that B_2 is strongly bounded.

We prove now a similar result for bounded sets of mappings which is sharper than (1) for the spaces $\mathfrak{L}(E_s, F_s)$.

(6) *Let P be a subset of $\mathfrak{L}(E_s, F_s)$. If*

$$\sup_{A \in P} |v(Ax)| = \rho(v, x) < \infty \quad \textit{for every } x \in E,\ v \in F',$$

then

$$\text{(7)} \qquad \sup_{A \in P, v \in N, x \in M} |v(Ax)| = \sigma(N, M) < \infty$$

for every strongly bounded set $N \subset F'$, $M \subset E$.

Or, equivalently, every simply bounded subset P of $\mathfrak{L}(E_s, F_s)$ is also bounded in $\mathfrak{L}_{b^*, b^*}(E_s, F_s)$.

Proof. It follows from (1) that a simply bounded subset $P \subset \mathfrak{L}(E_s, F_s)$ is $\mathfrak{T}_{b^*}$-bounded; hence $\bigcup_{A \in P} A(M)$ is (weakly) bounded in F for every strongly bounded $M \subset E$. But every bounded subset of F is also $\mathfrak{T}_{b^*}$-bounded in F; thus (7) is satisfied.

As a corollary we obtain

(8) *Let E and F be sequentially complete or locally complete. Then every simply bounded subset of $\mathfrak{L}(E_s, F_s)$ is bounded in $\mathfrak{L}_{b,b}(E_s, F_s)$, i.e.,*

$$\sup_{A \in P, v \in N, x \in M} |v(Ax)| < \infty,$$

where N is weakly bounded in F', M bounded in E.

3. Equicontinuous sets. We proved in § 15, 13.(1) that a set $H \subset \mathfrak{L}(E, F)$ is uniformly equicontinuous if it is equicontinuous at the point $\mathfrak{o}$; for simplicity we will call such a set "equicontinuous."

This means that for every neighbourhood V of $\mathfrak{o}$ in F there exists a neighbourhood U of $\mathfrak{o}$ in E such that $A(U) \subset V$ for all $A \in H$ or $H(U) = \bigcup_{A \in H} A(U) \subset V$.

We remark that the absolutely convex cover of an equicontinuous set is again equicontinuous.

(1) *Every equicontinuous subset* H *of* $\mathfrak{L}(E, F)$ *is bounded for every locally convex topology* $\mathfrak{T}_{\mathfrak{M}}$ *on* $\mathfrak{L}(E, F)$.

Proof. Let $U(M, V)$, M bounded, be a $\mathfrak{T}_{\mathfrak{M}}$-neighbourhood of o in $\mathfrak{L}_{\mathfrak{M}}(E, F)$. By assumption there exists in E a neighbourhood U of o such that $H(U) \subset V$. Since M is bounded in E, $M \subset \rho U$ for some $\rho > 0$ and it follows that $H(M) \subset \rho V$ or $H \subset \rho U(M, V)$; thus H is $\mathfrak{T}_{\mathfrak{M}}$-bounded.

We recall that the theorem of Banach (§ 15, 13.(2)) states, conversely, that a simply bounded set $H \subset \mathfrak{L}(E, F)$ is equicontinuous if E is complete metrizable and F is any topological vector space. We obtain now the following general theorem for the locally convex case.

(2) *Let E be barrelled, F locally convex. A subset H of $\mathfrak{L}(E, F)$ is equicontinuous if and only if it is simply bounded.*

The condition is necessary by (1). Assume now H to be simply bounded and $x \in E$. Then there exists $\rho = \rho(x, V) > 0$ such that $\rho A x \in V$ for all $A \in H$, V a given neighbourhood of o in F. Hence the set $B = \bigcap_{A \in H} A^{(-1)}(V)$ is absorbing in E. If V is absolutely convex and closed, then B has these properties too; thus B is a barrel in E. Hence B is a neighbourhood of o in E and it follows from $H(B) \subset V$ that H is equicontinuous.

We remark that the class of barrelled spaces is the maximal class of locally convex spaces E for which (2) remains true: Assume (2) to be true for the space E and all locally convex spaces F. If we take $F = \mathsf{K}$, then $\mathfrak{L}(E, \mathsf{K}) = E'$ and it follows from (2) that every weakly bounded subset of E' is equicontinuous. But then the topology on E is the strong topology and E is barrelled.

(3) *Let E be quasi-barrelled, F locally convex. A subset H of $\mathfrak{L}(E, F)$ is equicontinuous if and only if it is $\mathfrak{T}_b$-bounded.*

If H is $\mathfrak{T}_b$-bounded and M a bounded subset of E, then there exists $\rho = \rho(M, V) > 0$ such that $\rho A(M) \subset V$ for all $A \in H$. The set $B = \bigcap_{A \in H} A^{(-1)}(V)$ is for V absolutely convex and closed again a barrel which absorbs now all bounded sets; B is therefore a neighbourhood of o in E and H is equicontinuous since $H(B) \subset V$.

Since a locally convex space E is quasi-barrelled if and only if the equicontinuous sets in E' coincide with the strongly bounded subsets, the class of quasi-barrelled spaces is the maximal class of spaces E for which (3) is true. Recall that every bornological space is quasi-barrelled (§ 28. 1.(1)).

We give a dual characterization of equicontinuous sets. By H' we denote the set of all $A' \in \mathfrak{L}(F'_s, E'_s)$, where $A \in H$.

(4) *$H \subset \mathfrak{L}(E, F)$ is equicontinuous if and only if for every equicontinuous set $N \subset F'$ the set $H'(N)$ is equicontinuous in E'.*

Let V be an absolutely convex and closed neighbourhood of $\mathfrak{o}$ in F. If H is equicontinuous there exists an absolutely convex and closed neighbourhood U of $\mathfrak{o}$ in E such that $H(U) \subset V$. By polarity, using § 32, 1.(9), we obtain $H'(V^\circ) \subset U^\circ$; thus the condition is necessary. Conversely, if V° is the given equicontinuous set in F' and $H'(V^\circ) \subset U^\circ$, using polarity again we have $H(U) \subset V$; hence the condition is sufficient.

Again for arbitrary locally convex E, F we have

(5) *$H \subset \mathfrak{L}(E, F)$ is $\mathfrak{T}_b$-bounded if and only if H' is equicontinuous in $\mathfrak{L}(F'_b, E'_b)$.*

If H is $\mathfrak{T}_b$-bounded and B is a given absolutely convex and closed bounded subset of E, then $H(B) \subset C$, where C is an absolutely convex and closed bounded subset of F. By polarity we obtain $H'(C^\circ) \subset B^\circ$ and this is the equicontinuity of H'. Using polarity again we obtain $H(B) \subset C$; thus the condition is also sufficient.

We note the following corollaries.

(6) *Let E be quasi-barrelled, F locally convex. $H \subset \mathfrak{L}(E, F)$ is equicontinuous if and only if H' is equicontinuous in $\mathfrak{L}(F'_b, E'_b)$.*

This is a consequence of (3) and (5).

(7) *Let E be quasi-barrelled and sequentially complete or locally complete, F locally convex. $H \subset \mathfrak{L}(E, F)$ is simply bounded if and only if H' is equicontinuous in $\mathfrak{L}(F'_b, E'_b)$.*

This follows immediately from (3) and 2.(3).

4. Weak compactness. Metrizability. We recall that an equicontinuous subset of E' is always weakly relatively compact. This is the ALAOGLU–BOURBAKI theorem (§ 20, 9.(4)). We will see that this is no longer true for equicontinuous subsets H of a space $\mathfrak{L}(E, F)$.

We proved in § 21, 3.(3) that on an equicontinuous set M in E' the topologies $\mathfrak{T}_s(E)$ and $\mathfrak{T}_s(N)$ coincide, where N is a total subset of E. The proof uses the ALAOGLU–BOURBAKI theorem. Nevertheless, the result is true in general.

(1) *Let H be an equicontinuous subset of $\mathfrak{L}(E, F)$ and let N be a total subset of E. Then the topology $\mathfrak{T}_s$ of simple convergence on E and the topology $\mathfrak{T}_s(N)$ of simple convergence on N coincide on H.*

Proof. Since equicontinuous sets remain equicontinuous by translation, we may assume that $o \in H$, so that we have to compare only neighbourhoods of o in H.

Let M be a finite subset $\{x_1, \ldots, x_n\}$ of E, V an absolutely convex neighbourhood of o in F, and $U(M, V)$ the corresponding $\mathfrak{T}_s$-neighbourhood. It will be sufficient to determine a finite subset G of N and a $\rho > 0$ such that $H \cap U(G, \rho V) \subset U(M, V)$.

There exists $U \supset o$ in E such that $H(U) \subset V$. Since N is total in E, there exist linear combinations $z_i = \sum_{k=1}^{n_i} \alpha_{ik} y_{ik}$ of elements $y_{ik} \in N$ such that $x_i - z_i \in \frac{1}{2}U$ for $i = 1, \ldots, n$. We now define G to be the set of all y_{ik} and put $\rho = 1/2\sigma$, where $\sigma > 0$ is such that $z_i \in \sigma \Gamma G$ for all $i = 1, \ldots, n$. Then if $A \in U(G, \rho V)$ we obtain $Az_i \in \frac{1}{2}V$ and from $A \in H$ it follows that $A(x_i - z_i) \in \frac{1}{2}V$. Thus $Ax_i \in V$ for all $A \in H \cap U(G, \rho V)$ and (1) is proved.

Our next theorem is a straightforward generalization of § 21, 6.(2).

(2) *Let H be an equicontinuous subset of $\mathfrak{L}(E, F)$. Then the topologies $\mathfrak{T}_s$ and $\mathfrak{T}_c$ coincide on H.*

Again we assume $o \in H$. We consider a $\mathfrak{T}_c$-neighbourhood $U(C, V)$ of o, where C is a precompact subset of E and V is an absolutely convex neighbourhood of o in F. Let U be a neighbourhood of o in E such that $H(U) \subset V$. Since C is precompact, there exists $M = \{x_1, \ldots, x_n\} \subset E$ such that $C \subset \bigcup_{i=1}^{n} (x_i + \frac{1}{2}U)$. We define now the $\mathfrak{T}_s$-neighbourhood $U(M, \frac{1}{2}V)$ and the statement will follow from $U(C, V) \supset H \cap U(M, \frac{1}{2}V)$.

Let A be in $H \cap U(M, \frac{1}{2}V)$. An arbitrary element y of C has the form $y = x_k + z$, $z \in \frac{1}{2}U$, therefore $Ay = Ax_k + Az \in \frac{1}{2}V + \frac{1}{2}V = V$. Thus $A \in U(C, V)$.

We recall from 1. that $\mathfrak{L}_s(E, F)$ can be considered as a subspace of $L_s(E, F)$ which is isomorphic to a topological product F^A.

(3) *Let H be an equicontinuous subset of $\mathfrak{L}(E, F)$ and let $\overline{H}$ be the closure of H in $L_s(E, F)$. Then $\overline{H} \subset \mathfrak{L}(E, F)$ and $\overline{H}$ is again equicontinuous.*

Let A_0 be an adherent point of H in $L_s(E, F)$. If V is an absolutely convex and closed neighbourhood of o in F, then there exists a neighbourhood $U \supset o$ in E such that $H(U) \subset V$. For each fixed $x \in U$ the element A_0x is an adherent point of $H(x)$ in F. Since V is closed, $A_0x \in V$. Hence $A_0(U) \subset V$ and the result follows.

(4) *Let E, F be locally convex and F quasi-complete. Assume that $\mathfrak{M}$ covers E. Then every closed equicontinuous subset H of $\mathfrak{L}_{\mathfrak{M}}(E, F)$ is complete in $\mathfrak{L}_{\mathfrak{M}}(E, F)$.*

Proof. $L_s(E, F) = F^A$ is quasi-complete as a topological product of quasi-complete spaces. Hence the weak closure $\bar{H}$ of H in $L_s(E, F)$, which is bounded, is $\mathfrak{T}_s$-complete. By (3) $\bar{H} \subset \mathfrak{L}_s(E, F)$. From 1.(6) and § 18, 4.(4) it follows that $\bar{H}$ is $\mathfrak{T}_{\mathfrak{M}}$-complete. Since $H \subset \bar{H}$ is $\mathfrak{T}_{\mathfrak{M}}$-closed, H is $\mathfrak{T}_{\mathfrak{M}}$-complete.

We are now able to prove GROTHENDIECK's generalization of the ALAOGLU–BOURBAKI theorem.

(5) *Let E, F be locally convex. The following properties of $\mathfrak{L}(E, F)$ are equivalent:*

a) *every equicontinuous subset H of $\mathfrak{L}(E, F)$ is relatively $\mathfrak{T}_s$-compact;*

b) *every bounded subset of F is relatively compact.*

Proof. b) $\rightharpoonup$ a). The closure $\bar{H}$ of H in $\mathfrak{L}_s(E, F)$ is equicontinuous by (3) and complete by (4). Since every bounded subset of F^A is relatively $\mathfrak{T}_s$-compact, $\bar{H}$ is $\mathfrak{T}_s$-compact in $\mathfrak{L}(E, F)$.

a) $\rightharpoonup$ b). We use an indirect proof and assume that F contains a bounded but not relatively compact subset B. We will construct an equicontinuous subset $H \subset \mathfrak{L}(E, F)$ which is not relatively $\mathfrak{T}_s$-compact.

For every $y \in F$ let $u_0 \otimes y$ be the linear mapping of E in F defined by $(u_0 \otimes y)x = (u_0 x)y$, where $u_0 \neq \circ$ is a fixed element of E'. Let H be the set of all $u_0 \otimes y$, $y \in B$.

H is equicontinuous: Given a neighbourhood $V \supset \circ$ in F, there exists $\rho > 0$ such that $\alpha B \subset V$ for all α, $|\alpha| \leqq \rho$. If U is a neighbourhood of $\circ$ in E such that $|u_0 x| \leqq \rho$ for all $x \in U$, then $(u_0 \otimes y)(U) = u_0(U)y \subset V$ for all $y \in B$; thus H is equicontinuous.

H is not relatively $\mathfrak{T}_s$-compact: Choose $x_0 \in E$ such that $u_0 x_0 \neq 0$. The map $A \to Ax_0$ of $\mathfrak{L}_s(E, F)$ in F is continuous (1.(5)), so if H were relatively $\mathfrak{T}_s$-compact, then $H(x_0)$ would be relatively compact in F. But since $H(x_0) = (u_0 x_0)B$, this is not the case; thus H is not relatively $\mathfrak{T}_s$-compact.

We now prove some results on metrizability and separability.

(6) *Suppose that $\mathfrak{M}$ is saturated. Then $\mathfrak{L}_{\mathfrak{M}}(E, F)$ is metrizable if and only if F is metrizable and there exists a sequence $M_1 \subset M_2 \subset \cdots$, $M_k \in \mathfrak{M}$, such that every set $M \in \mathfrak{M}$ is contained in some M_k.*

Proof. If the conditions are satisfied and $V_1 \supset V_2 \supset \cdots$ is a neighbourhood base of $\circ$ in F, then the $U(M_i, V_i)$, $i = 1, 2, \ldots$, constitute a $\mathfrak{T}_{\mathfrak{M}}$-neighbourhood base of $\circ$ in $\mathfrak{L}_{\mathfrak{M}}(E, F)$.

Assume now that $\mathfrak{L}_{\mathfrak{M}}(E, F)$ is metrizable. Recall from the proof of 1.(2′) that $E'[\mathfrak{T}_{\mathfrak{M}}]$ is isomorphic to a subspace of $\mathfrak{L}_{\mathfrak{M}}(E, F)$. Hence $E'[\mathfrak{T}_{\mathfrak{M}}]$ is metrizable. Let $U_1 \supset U_2 \supset \cdots$ be a neighbourhood base of $\circ$ in $E'[\mathfrak{T}_{\mathfrak{M}}]$; then the polars $M_1 = U_1^\circ \subset M_2 = U_2^\circ \subset \cdots$ in E are in $\mathfrak{M}$ and every set $M \in \mathfrak{M}$ is contained in some M_k.

Similarly, F is isomorphic to a subspace of $\mathfrak{L}_{\mathfrak{M}}(E, F)$ and therefore metrizable: Let $u_0 \neq o$ be a fixed element of E'; then we define for every $y \in F$ the map $(u_0 \otimes y)x = u_0(x)y$. Let G be the subspace of $\mathfrak{L}_{\mathfrak{M}}(E, F)$ consisting of all $u_0 \otimes y$. The correspondence $u_0 \otimes y \to y$ is an algebraic isomorphism of G onto F. Let V be a closed absolutely convex neighbourhood of o in F and $M \in \mathfrak{M}$. Then $G \cap U(M, V)$ consists of all $u_0 \otimes y$ such that $\rho y \in V$, $\rho = \sup_{x \in M} |u_0 x|$. Hence the isomorphism is topological.

(7) *Let E be separable, F metrizable. Then the topology $\mathfrak{T}_s$ of simple convergence is metrizable on every equicontinuous subset H of $\mathfrak{L}(E, F)$.*

Let N be a countable set dense in E and $\mathfrak{T}_s(N)$ the topology of simple convergence on N. It follows from (6) that $\mathfrak{L}(E, F)$ is metrizable for $\mathfrak{T}_s(N)$ and from (1) that H is metrizable for $\mathfrak{T}_s$.

(8) *If E and F are separable and N is a countable dense set in E, then $\mathfrak{L}(E, F)$ is separable for the topology $\mathfrak{T}_s(N)$.*

Proof. Let P be a countable set dense in F. If $N_\alpha = \{x_1, \ldots, x_k\}$ is a finite subset of N and P_β a finite sequence $z_1, \ldots, z_k$ of elements of P, there exists a mapping $A_{\alpha\beta} \in \mathfrak{L}(E, F)$ such that $A_{\alpha\beta} x_i = z_i$, $i = 1, \ldots, k$, as can easily be seen. The set $\mathfrak{A}$ of all these $A_{\alpha\beta}$ is countable and $\mathfrak{T}_s(N)$-dense in $\mathfrak{L}(E, F)$: Let $A \in \mathfrak{L}(E, F)$ and $U(N_\alpha, V)$ be given, V a neighbourhood of o in F. If $Ax_i = y_i$ for $x_i \in N_\alpha$, there exists $z_i \in P$ such that $y_i - z_i \in V$. For P_β the sequence $z_1, \ldots, z_k$ it follows that $(A - A_{\alpha\beta})x_i = y_i - z_i \in V$ or that $A_{\alpha\beta} \in A + U(N_\alpha, V)$.

(9) *Let E be separable and F separable and metrizable. Then every equicontinuous subset H of $\mathfrak{L}(E, F)$ is separable and metrizable for the topology $\mathfrak{T}_s$.*

Let $N \subset E$ be countable and dense. By (8) $\mathfrak{L}(E, F)$ and hence H is $\mathfrak{T}_s(N)$-separable. But $\mathfrak{T}_s(N)$ and $\mathfrak{T}_s$ coincide on H by (1). $\mathfrak{T}_s$ is metrizable on H by (7).

5. The BANACH–STEINHAUS theorem. The classical theorem (§ 15, 13.(3)) considers sequences of continuous linear mappings of a complete metrizable space into a topological vector space. The general theorem for locally convex spaces is now an easy consequence of previous results.

(1) *Let E be barrelled and F locally convex. Let A_α, $\alpha \in \mathrm{A}$, be a net in $\mathfrak{L}(E, F)$ such that for every $x \in E$ the net $A_\alpha x$ is bounded in F and converges to an element $A_0 x \in F$. Then $A_0 \in \mathfrak{L}(E, F)$ and the convergence of A_α to A_0 is uniform on every precompact set in E, $A_\alpha \to A_0$ in $\mathfrak{L}_c(E, F)$.*

Proof. The set H of all A_α is equicontinuous by 3.(2). A_0 is, as the $\mathfrak{T}_s$-limit of the net A_α, in $\overline{H} \subset L(E, F)$ and by 4.(3) $A_0 \in \mathfrak{L}(E, F)$. Finally, by 4.(2) A_α converges to A_0 in the topology $\mathfrak{T}_c$.

We give a second version of the BANACH–STEINHAUS theorem.

(2) *Let E be barrelled, F locally convex and complete. Let A_α, $\alpha \in \mathrm{A}$, be a simply bounded net in $\mathfrak{L}(E, F)$ such that $A_\alpha x$ is a Cauchy net in F for all x of a total subset N of E. Then A_α $\mathfrak{T}_c$-converges to a mapping $A_0 \in \mathfrak{L}(E, F)$.*

The set H of all A_α is again equicontinuous and A_α is a $\mathfrak{T}_s(N)$-Cauchy net in $H \subset \mathfrak{L}(E, F)$. It follows from 4.(1) that A_α is a Cauchy net for the topology $\mathfrak{T}_s$. Since $\mathfrak{L}_s(E, F) \subset L_s(E, F) = F^{\mathrm{A}}$ and F^{A} is complete, the net A_α has a limit A_0 in $L_s(E, F)$. By 4.(3) A_0 is in $\mathfrak{L}(E, F)$ and, again by 4.(2), A_0 is the $\mathfrak{T}_c$-limit of A_α.

Remark 1. In (2) it is sufficient to suppose that F is sequentially complete if we consider a simply bounded sequence A_n which is $\mathfrak{T}_s(N)$-Cauchy.

Remark 2. If one supposes in (1) and (2) only that E is quasi-barrelled, then one has to assume that the net A_α is $\mathfrak{T}_b$-bounded in $\mathfrak{L}(E, F)$ to obtain similar results (compare 3.(3)). The exact formulation is left to the reader.

Remark 3. HUSAIN [2′] calls a locally convex space E countably barrelled if every weakly bounded subset of E' which is the union of countably many equicontinuous sets is itself equicontinuous. This is equivalent to the following property of E: Every barrel which is the intersection of countably many absolutely convex closed neighbourhoods of $\mathfrak{o}$ is itself a neighbourhood of $\mathfrak{o}$.

The BANACH–STEINHAUS theorem in both versions (1) and (2) is true for sequences $A_n \in \mathfrak{L}(E, F)$ if E is countably barrelled.

One has only to show that the set H of all A_n is equicontinuous; the rest of the proof remains the same. If we look at the proof of 3.(2), we see that $B = \bigcap_{n=1}^{\infty} A_n^{(-1)}(V)$ is a barrel which is the intersection of countably many absolutely convex closed neighbourhoods of $\mathfrak{o}$ in E and therefore a neighbourhood of $\mathfrak{o}$ in E; hence H is equicontinuous.

6. Completeness. If E is a normed space and F a (B)-space, then $\mathfrak{L}_b(E, F)$ is complete, as we proved in § 14, 6.(5). The completeness of $\mathfrak{L}_{\mathfrak{M}}(E, F)$, in general, does not even depend on the completeness of E.

(1) *Let E be locally convex, F locally convex and complete. Then $\mathfrak{L}_{\mathfrak{M}}(E, F)$ is isomorphic to $\mathfrak{L}_{\overline{\mathfrak{M}}}(\tilde{E}, F)$, where $\overline{\mathfrak{M}}$ denotes the class of subsets of $\tilde{E}$ consisting of all closures $\overline{M}$ in $\tilde{E}$ of the sets $M \in \mathfrak{M}$.*

Proof. Every $A \in \mathfrak{L}(E, F)$ has a uniquely determined continuous extension $\bar{A}$ from E to $\tilde{E}$ and the topologies $\mathfrak{T}_{\mathfrak{M}}$ and $\mathfrak{T}_{\bar{\mathfrak{M}}}$ on $\mathfrak{L}(\tilde{E}, F)$ obviously coincide.

In contrast to this we obtain

(2a) *Let E, F be locally convex. If $\mathfrak{L}_{\mathfrak{M}}(E, F)$ is complete, then F is complete.*

By 1.(2′) F is isomorphic to a complemented, hence closed, subspace of $\mathfrak{L}_{\mathfrak{M}}(E, F)$. This implies the completeness of F.

Therefore, if we are interested in complete spaces $\mathfrak{L}_{\mathfrak{M}}(E, F)$, we have to assume that F is complete.

One obtains a second necessary condition as a consequence of 1.(2′).

(2b) *Let E, F be locally convex. If $\mathfrak{L}_{\mathfrak{M}}(E, F)$ is complete, then $E'[\mathfrak{T}_{\mathfrak{M}}]$ is complete.*

These two necessary conditions are sufficient in many cases, as the following theorem of Grothendieck [11] shows.

(3) *Let E, F be locally convex and assume that the topology on E is the Mackey topology. If $\mathfrak{M}$ is a class of bounded subsets of E which covers E and if F and $E'[\mathfrak{T}_{\mathfrak{M}}]$ are complete, then $\mathfrak{L}_{\mathfrak{M}}(E, F)$ is complete.*

Proof. Let A_α, $\alpha \in \mathrm{A}$, be a Cauchy net in $\mathfrak{L}_{\mathfrak{M}}(E, F)$. Then $A_\alpha x$ is a Cauchy net in F by 1.(5) and has a limit $A_0 x$ since F is complete. Obviously, $A_0 \in L(E, F)$. Since A_α is a Cauchy net for $\mathfrak{T}_{\mathfrak{M}}$, the convergence $A_\alpha \to A_0$ is uniform on every $M \in \mathfrak{M}$. It will be sufficient to show that A_0 is weakly continuous, since weak continuity implies continuity because the topology on E is Mackey's topology. A_0 is weakly continuous if $A_0' v \in E'$ for every $v \in F'$. Since $A_\alpha \to A_0$ uniformly on every $M \in \mathfrak{M}$, it follows that $u_\alpha = A_\alpha' v \in E'$ converges uniformly on every $M \in \mathfrak{M}$ to $u_0 = A_0' v$. Since $E'[\mathfrak{T}_{\mathfrak{M}}]$ is complete, $u_0 \in E'$.

We remark that we may replace the assumption that E has Mackey's topology by the weaker assumption that $\mathfrak{L}(E, F)$ contains all weakly continuous linear mappings of E in F. The proof remains unchanged.

We remark further that if we assume F only to be quasi-complete in (3), then $\mathfrak{L}_{\mathfrak{M}}(E, F)$ is quasi-complete. The proof remains the same; we have only to consider Cauchy nets A_α which are bounded in E.

As a special case we obtain

(4) *If E is bornological, F is complete (quasi-complete) locally convex, and $\mathfrak{M}$ contains all sets consisting of local null sequences in E, then $\mathfrak{L}_{\mathfrak{M}}(E, F)$ is complete (quasi-complete).*

The topology of a bornological space E coincides always with $\mathfrak{T}_k(E')$ (§ 28, 1.) and $E'[\mathfrak{T}_{\mathfrak{M}}]$ is complete by § 28, 5.(1).

(4) can be proved directly in a simpler way: As in the proof of (3), one shows that $A_\alpha \to A_0$ uniformly on every $M \in \mathfrak{M}$; hence A_0 is continuous on every M, especially on every local null sequence, and from § 28, 3.(4) it follows that A_0 is continuous.

(5) *If E is barrelled, F is quasi-complete, and $\mathfrak{M}$ covers E, then $\mathfrak{L}_{\mathfrak{M}}(E, F)$ is quasi-complete.*

If E is quasi-barrelled, F is quasi-complete, then $\mathfrak{L}_b(E, F)$ is quasi-complete.

Proof. Let E be barrelled and H a closed bounded subset of $\mathfrak{L}_{\mathfrak{M}}(E, F)$. Then H is simply bounded and equicontinuous by 3.(2). H is complete by 4.(4).

The second result follows similarly from 3.(3).

(6) *If E is a* (DF)*-space and F complete, then $\mathfrak{L}_b(E, F)$ is complete.*

Proof. A Cauchy net A_α has a $\mathfrak{T}_s$-limit $A_0 \in L(E, F)$ which is continuous on the bounded sets of E. From § 29, 3.(7) it follows that A_0 is continuous on E and therefore in $\mathfrak{L}(E, F)$.

We are now interested in the completion of a space $\mathfrak{L}_{\mathfrak{M}}(E, F)$. For the simple topology we obtain

(7) *The completion of $\mathfrak{L}_s(E, F)$, F complete, is $L_s(E, F)$.*

As we have seen in 1., $\mathfrak{L}_s(E, F)$ is a subspace of $L_s(E, F)$ and $L_s(E, F) = F^{\mathsf{A}}$ is complete as a topological product of complete spaces. It remains to prove that $\mathfrak{L}(E, F)$ is $\mathfrak{T}_s$-dense in $L_s(E, F)$. But this follows easily from the fact that for $A_0 \in L(E, F)$ there always exists $A \in \mathfrak{L}(E, F)$ such that $Ax_i = A_0 x_i$ for a finite set of given elements $x_i \in E$.

For the general case we reproduce a construction of ADASCH [6′] which generalizes a result of A. and W. ROBERTSON [2′].

We assume again that F is complete and that $\mathfrak{M}$ covers E and we consider $\mathfrak{L}_{\mathfrak{M}}(E, F)$.

Let G be a subspace of $L = L(E, F)$. We denote by $G(M, V)$ the set of all $A \in G$ such that $A(M) \subset V$, where $M \in \mathfrak{M}$ and V is a neighbourhood of $\circ$ in F. We define

(8) $$H = \bigcap_{M \in \mathfrak{M}, V} (\mathfrak{L}(E, F) + L(M, V)).$$

Clearly, $\mathfrak{L}(E, F) = \mathfrak{L} \subset H \subset L$. It is easy to check that H is a linear subspace of L. We define the topology $\mathfrak{T}_{\mathfrak{M}}$ on H by taking all sets $H(M, V)$ as neighbourhoods of $\circ$ in H. By checking the proof of 1.(1) we see that $\mathfrak{T}_{\mathfrak{M}}$ is a locally convex topology on H if we show that every set $H(M, V)$ is absorbing in H.

Suppose $A_0 \in H$; then by (8) $A_0 = A_1 + A_2$, $A_1 \in \mathfrak{L}$, $A_2 \in L(M, V/2)$. Obviously, $A_2 \in H$ and $A_0 - A_2 = A_1$ is contained in $\rho H(M, V/2)$ for some $\rho \geqq 1$. Therefore $A_0 = A_1 + A_2 \in \rho(2H(M, V/2)) \subset \rho H(M, V)$; thus $H(M, V)$ is absorbing and $H[\mathfrak{T}_{\mathfrak{M}}]$ locally convex and contains $\mathfrak{L}_{\mathfrak{M}}(E, F)$ as a subspace.

(9) *$H[\mathfrak{T}_{\mathfrak{M}}]$ is the completion of $\mathfrak{L}_{\mathfrak{M}}(E, F)$.*

We prove first that $\mathfrak{L}$ is dense in H. Let $A_0 \in H$ and $H(M, V)$, V absolutely convex, be given. Again $A_0 = A_1 + A_2$, $A_1 \in \mathfrak{L}$, $A_2 \in H(M, V)$; hence $A_1 = A_0 - A_2 \in A_0 + H(M, V)$ and $\mathfrak{L}$ is dense in H.

Secondly, we have to show that $H[\mathfrak{T}_{\mathfrak{M}}]$ is complete. A $\mathfrak{T}_{\mathfrak{M}}$-Cauchy net A_α in H has a pointwise limit A_0 in $L(E, F)$. We have to prove that $A_0 \in H$ and that A_α converges to A_0 in the sense of $\mathfrak{T}_{\mathfrak{M}}$.

Let V be absolutely convex and closed. Then $L(M, V)$ is $\mathfrak{T}_s$-closed in $L(E, F)$. There exists α_0 such that $A_\alpha - A_\beta \in H(M, V) \subset L(M, V)$ for $\alpha, \beta \geqq \alpha_0$. For the pointwise limit A_0 it follows that $A_0 - A_\beta \in L(M, V)$ for $\beta \geqq \alpha_0$. Hence $A_0 \in A_\beta + L(M, V) = A_\beta^{(1)} + A_\beta^{(2)} + L(M, V)$, $A_\beta^{(1)} \in \mathfrak{L}$, $A_\beta^{(2)} \in L(M, V)$, and thus $A_0 \in \mathfrak{L} + L(M, 2V)$. This and (8) imply that $A_0 \in H$ and, since $A_0 - A_\beta \in H(M, V)$ for $\beta \geqq \alpha_0$, the net A_α $\mathfrak{T}_{\mathfrak{M}}$-converges to A_0.

The result of A. and W. ROBERTSON is the special case $\mathfrak{L}_{\mathfrak{M}}(E, F) = E'[\mathfrak{T}_{\mathfrak{M}}]$.

(10) *Let E be locally convex and suppose that $\mathfrak{M}$ covers E. Then the completion of $E'[\mathfrak{T}_{\mathfrak{M}}]$ is the intersection $\bigcap_{M \in \mathfrak{M}} (E' + M^\circ)$, where M° is the polar of M taken in the algebraic dual E^*.*

7. The dual of $\mathfrak{L}_s(E, F)$. We give a concrete representation of the dual of $\mathfrak{L}_s(E, F)$ for arbitrary locally convex spaces E, F.

For $x \in E$, $v \in F'$, $\langle w, A\rangle = v(Ax)$ defines a continuous linear functional w on $\mathfrak{L}_s(E, F)$, as follows immediately from the definition of $\mathfrak{T}_s$ on $\mathfrak{L}(E, F)$. More generally, all the expressions

$$\text{(1)} \qquad \langle w, A\rangle = \sum_1^n v_i(Ax_i), \qquad x_i \in E,\ v_i \in F',$$

define elements w of $(\mathfrak{L}_s(E, F))'$.

We prove now that every continuous linear functional has such a representation. The proof of 6.(7) shows that $\mathfrak{L}_s(E, F)$ is dense in $L_s(E, F)$; hence $\mathfrak{L}_s(E, F)' = L_s(E, F)'$. We saw in 1. that $L_s(E, F)$ is isomorphic to F^{A} in such a way that every $A \in L(E, F)$ is represented as an element $(Ax_\alpha)_{\alpha \in \mathsf{A}}$, where x_α is a linear base of E. The dual of F^{A} is $\bigoplus_{\alpha \in \mathsf{A}} F'_\alpha$, $F_\alpha = F$, and thus every $w \in (F^\wedge)'$ is of the form $\langle w, A\rangle = \sum_1^n v_{\alpha_i}(Ax_{\alpha_i})$, which proves our statement.

We have to use the tensor product to obtain from (1) the isomorphic representation of L'_s. The correspondence $(v, x) \to v(Ax)$ is a bilinear mapping of $F' \times E$ into L'_s. By § 9, 7.(2) it defines a linear mapping of $F' \otimes E$ into L'_s given by $\sum v_i \otimes x_i \to w$, where w is defined by (1). This mapping is onto, as we have proved.

We prove that it is also one-one. An element $\neq o$ of $F' \otimes E$ can be written as $\sum_{i=1}^r v_i \otimes x_i$, $r > 0$, where the $v_i \in F'$ and the $x_i \in E$ are linearly independent (§ 9, 6.(8)). Let y_k, $k = 1, \ldots, r$, be elements of F such that $v_i y_k = \delta_{ik}$ for all $i, k = 1, \ldots, r$ (§ 9, 2.(7a)). Since $L(E, F)$ consists of all linear mappings of E in F, there exists $A_0 \in L(E, F)$ such that $A_0 x_i = y_i$, $i = 1, \ldots, r$. It follows that $\langle w, A_0\rangle = \sum_{i=1}^r v_i(A_0 x_i) = r \neq 0$. This concludes the proof of

(2) *Let E, F be locally convex. Then the dual of $\mathfrak{L}_s(E, F)$ can be identified with $F' \otimes E$ if we define the linear functionals by the formula*

$$\left\langle \sum_{i=1}^n v_i \otimes x_i, A \right\rangle = \sum_{i=1}^n v_i(Ax_i)$$

$$\textit{for } \sum_{i=1}^n v_i \otimes x_i \in F' \otimes E \textit{ and } A \in \mathfrak{L}(E, F).$$

As a corollary we obtain

(3) *Let E, F be locally convex and $\mathfrak{M}$ a saturated class of relatively weakly compact subsets of E which covers E. Then the dual of $\mathfrak{L}_{\mathfrak{M}}(E, F_s)$ is $F' \otimes E$.*

Obviously, $\mathfrak{L}_{\mathfrak{M}}(E, F_s) = \mathfrak{L}_{\mathfrak{M}}(E_s, F_s)$ and this can be written as $\mathfrak{L}_{\mathfrak{M},s}(E_s, F_s)$ in the sense defined in 1. By 1.(8) $\mathfrak{L}_{\mathfrak{M},s}(E_s, F_s)$ is isomorphic to $\mathfrak{L}_{s,\mathfrak{M}}(F'_s, E'_s)$ by transposition. Now $\mathfrak{L}_s(F'_s, E'[\mathfrak{T}_{\mathfrak{M}}]) \subset \mathfrak{L}_{s,\mathfrak{M}}(F'_s, E'_s) \subset L_s(F'_s, E'[\mathfrak{T}_{\mathfrak{M}}])$; hence $\mathfrak{L}_s(F'_s, E'[\mathfrak{T}_{\mathfrak{M}}])$ is dense in $\mathfrak{L}_{s,\mathfrak{M}}(F'_s, E'_s)$ and both have the same dual. By (2) the dual to $\mathfrak{L}_s(F'_s, E'[\mathfrak{T}_{\mathfrak{M}}])$ is $E \otimes F'$, since $(E'[\mathfrak{T}_{\mathfrak{M}}])' = E$.

To the isomorphism $A \to A'$ of $\mathfrak{L}_{\mathfrak{M}}(E, F_s)$ onto $\mathfrak{L}_{s,\mathfrak{M}}(F'_s, E'_s)$ corresponds the isomorphism $w \to w'$ of the continuous linear functionals given by

$$\langle w', A'\rangle = \left\langle \sum x_i \otimes v_i, A' \right\rangle = \sum x_i(A'v_i) = \sum v_i(Ax_i)$$

$$= \left\langle \sum v_i \otimes x_i, A \right\rangle = \langle w, A\rangle,$$

which proves our statement.

This is our first example of the determination of the dual of a space of linear mappings, which is in general a rather difficult task. The notion of the tensor product of vector spaces arises quite naturally in the discussion and the later investigation of this notion has here its first motivation.

8. Some structure theorems. We studied in § 22, 7. the duals of locally convex hulls and kernels and their topologies. We consider now the more general situation of $\mathfrak{L}(E, F)$, where E or F is a locally convex hull or kernel.

Let E be the locally convex hull $E = \sum_{\alpha} A_\alpha(E_\alpha)$ and F locally convex. A linear mapping B of E in F is continuous if and only if all BA_α are continuous linear mappings of E_α in F (§ 19, 1.(7)).

If we define $T_\alpha B = BA_\alpha$, then T_α is a linear mapping of $\mathfrak{L}(E, F)$ in $\mathfrak{L}(E_\alpha, F)$ and $\mathfrak{L}(E, F)$ is, in the sense of § 19, 6., the kernel $\mathsf{K}_{\alpha} T_\alpha^{(-1)}(\mathfrak{L}(E_\alpha, F))$ of the $T_\alpha^{(-1)}(\mathfrak{L}(E_\alpha, F))$.

We assume now that the $\mathfrak{L}(E_\alpha, F)$ have topologies $\mathfrak{T}_{\mathfrak{M}_\alpha}$, where $\mathfrak{M}_\alpha$ is a class of bounded subsets of E_α. Then again in the sense of § 19, 6. $\mathfrak{L}(E, F)$ will be the locally convex kernel $\mathsf{K}_{\alpha} T_\alpha^{(-1)}(\mathfrak{L}_{\mathfrak{M}_\alpha}(E_\alpha, F))$ and the kernel topology $\mathfrak{T}$ of $\mathfrak{L}(E, F)$ can easily be determined: *A* $\mathfrak{T}_{\mathfrak{M}_\alpha}$-neighbourhood W_α of $\mathfrak{o}$ in $\mathfrak{L}_{\mathfrak{M}_\alpha}(E_\alpha, F)$ consists of all B_α such that $B_\alpha(M_\alpha) \subset V_\alpha$, where $M_\alpha \in \mathfrak{M}_\alpha$ and V_α is a neighbourhood of $\mathfrak{o}$ in F. Hence the corresponding $\mathfrak{T}$-neighbourhood $T_\alpha^{(-1)}(W_\alpha)$ consists of all B such that $B(A_\alpha(M_\alpha)) \subset V_\alpha$. The finite intersections of the $T_\alpha^{(-1)}(W_\alpha)$ determine a $\mathfrak{T}$-neighbourhood base of $\mathfrak{o}$ in $\mathfrak{L}(E, F)$.

From this follows

(1) *Let E be the locally convex hull $E = \sum_{\alpha} A_\alpha(E_\alpha)$, F locally convex. Let $\mathfrak{M}_\alpha$ be a class of bounded subsets of E_α for every α and let $\mathfrak{M}$ be the class of all finite unions of sets contained in $\bigcup_{\alpha} A_\alpha(\mathfrak{M}_\alpha)$. Then $\mathfrak{L}_{\mathfrak{M}}(E, F)$ is the locally convex kernel of all the $T_\alpha^{(-1)}(\mathfrak{L}_{\mathfrak{M}_\alpha}(E_\alpha, F))$, where $T_\alpha B = BA_\alpha$ for $B \in \mathfrak{L}(E, F)$.*

We remark that $\mathfrak{T}_{\mathfrak{M}} = \mathfrak{T}_s$ if all $\mathfrak{T}_{\mathfrak{M}_\alpha} = \mathfrak{T}_s$.

We state two corollaries.

(2) *If $E = \bigoplus_{\alpha} E_\alpha$ and F locally convex, then $\mathfrak{L}_{\mathfrak{M}}(E, F) = \prod_{\alpha} \mathfrak{L}_{\mathfrak{M}_\alpha}(E_\alpha, F)$, where $\mathfrak{M}$ is the class of all finite unions of sets contained in $\bigcup_{\alpha} \mathfrak{M}_\alpha$.*

For $\mathfrak{T}_{\mathfrak{M}_\alpha} = \mathfrak{T}_s, \mathfrak{T}_k, \mathfrak{T}_b$ for all α, we have $\mathfrak{T}_{\mathfrak{M}} = \mathfrak{T}_s, \mathfrak{T}_k, \mathfrak{T}_b$, respectively.

(3) *Let H be a closed subspace of E and K be the canonical mapping of E onto E/H. Then $\mathfrak{L}_{\hat{\mathfrak{M}}}(E/H, F) = \mathsf{K}\, T^{(-1)}(\mathfrak{L}_{\mathfrak{M}}(E, F))$, where $\hat{\mathfrak{M}} = K(\mathfrak{M})$ and $TB = BK$ for $B \in \mathfrak{L}(E/H, F)$.*

T is an isomorphic injection of $\mathfrak{L}_{\hat{\mathfrak{M}}}(E/H, F)$ onto the subspace of $\mathfrak{L}_{\mathfrak{M}}(E, F)$ consisting of all A vanishing on H.

The situation is more involved if we assume E to be a locally convex kernel. We suppose first that E is a topological product $\prod_\alpha E_\alpha$ and F locally convex. Let B be an element of $\mathfrak{L}(E, F)$, B_α the restriction of B to E_α, P_α the projection of E onto E_α. Then for $x = (x_\alpha) \in E$ we have $B_\alpha x_\alpha = (BP_\alpha)x$ and $B_\alpha P_\alpha = BP_\alpha$ is in $\mathfrak{L}(E, F)$. If only finitely many B_α are $\neq o$, then $\sum_{i=1}^n B_{\alpha_i} \in \bigoplus_\alpha \mathfrak{L}(E_\alpha, F)$ and $B = \sum_{i=1}^n B_{\alpha_i} P_{\alpha_i} \in \mathfrak{L}(E, F)$. Thus we have

$$\mathfrak{L}\left(\prod_\alpha E_\alpha, F\right) \supset \bigoplus_\alpha \mathfrak{L}(E_\alpha, F)P_\alpha. \tag{4}$$

But in general $\bigoplus_\alpha \mathfrak{L}(E_\alpha, F)P_\alpha$ is a proper subspace of $\mathfrak{L}\left(\prod_\alpha E_\alpha, F\right)$. This is illustrated by the following example.

Let $E = F = \omega$ and let us write ω as $\prod_{u=1}^\infty \mathsf{K}_n$, where K_n is the scalar field K. Then every $B \in \bigoplus_{n=1}^\infty \mathfrak{L}(\mathsf{K}_n, \omega)$ has finite dimensional range, but the identity mapping from $\mathfrak{L}(\omega, \omega)$ does not.

But there is a case in which we have equality in (4).

(5) *Let E be the topological product $\prod_\alpha E_\alpha$ of a class of locally convex spaces E_α and F a locally convex space with a fundamental sequence $C_1 \subset C_2 \subset \cdots$ of bounded subsets. Then*

$$\mathfrak{L}(E, F) = \mathfrak{L}\left(\prod_\alpha E_\alpha, F\right) = \bigoplus_\alpha \mathfrak{L}(E_\alpha, F)P_\alpha.$$

Assume that there exists a $B \in \mathfrak{L}(E, F)$ such that infinitely many restrictions B_{α_n}, $n = 1, 2, \ldots$, are different from o. Then there exist $x_{\alpha_n} \in E_{\alpha_n}$ such that $y_n = B_{\alpha_n} x_{\alpha_n} = Bx_{\alpha_n} \notin C_n$ for every n. The set N of all x_{α_n} is a bounded subset of $\prod_\alpha E_\alpha$, but $B(N)$, the set of all y_n, is by construction unbounded in F, which contradicts the continuity of B. Hence $\mathfrak{L}(E, F) \subset \bigoplus_\alpha \mathfrak{L}(E_\alpha, F)P_\alpha$ and (5) follows from (4).

The topological situation is even more complicated. If $\mathfrak{M}_\alpha$ is a class of bounded subsets of E_α defining the topology $\mathfrak{T}_{\mathfrak{M}_\alpha}$ on $\mathfrak{L}(E_\alpha, F)$, then it is natural to introduce on $E = \prod_\alpha E_\alpha$ the class $\mathfrak{M}$ of bounded subsets consisting of all products $M = \prod_\alpha M_\alpha$, $M_\alpha \in \mathfrak{M}_\alpha$. We denote the topology $\mathfrak{T}_\mathfrak{M}$ on $\mathfrak{L}(E, F)$ by $\mathfrak{T}$ and the hull topology on $\bigoplus_\alpha \mathfrak{L}_{\mathfrak{M}_\alpha}(E_\alpha, F)$ by $\mathfrak{T}'$.

(6) *Under the assumptions of* (5) *the topology $\mathfrak{T}'$ of $\bigoplus_\alpha \mathfrak{L}_{\mathfrak{M}_\alpha}(E_\alpha, F)$ is finer than the topology $\mathfrak{T}$ of $\mathfrak{L}_\mathfrak{M}(E, F)$.*

Let $W = W(M, V)$ be a $\mathfrak{T}$-neighbourhood of $\mathfrak{o}$ in $\mathfrak{L}_{\mathfrak{M}}(E, F)$, where $M = \prod_\alpha M_\alpha$, $M_\alpha \in \mathfrak{M}_\alpha$, and V a neighbourhood of $\mathfrak{o}$ in F. Let W_α be the neighbourhood $W(M_\alpha, V)$ in $\mathfrak{L}_{\mathfrak{M}_\alpha}(E_\alpha, V)$. Then $W' = \ulcorner_\alpha W_\alpha$ is a $\mathfrak{T}'$-neighbourhood of $\mathfrak{o}$ in $\bigoplus_\alpha \mathfrak{L}_{\mathfrak{M}_\alpha}(E_\alpha, F)$. It is sufficient to prove $W' \subset W$. An element of W' has the form $\sum_\alpha c_\alpha B_\alpha$, $B_\alpha \in W_\alpha$, $\sum |c_\alpha| \leqq 1$. Hence $\sum c_\alpha B_\alpha(M_\alpha) \subset V$. For the corresponding $B \in \mathfrak{L}(E, F)$ we obtain $B(M) = \sum c_\alpha B_\alpha P_\alpha(M) = \sum c_\alpha B_\alpha(M_\alpha) \subset V$; hence $W' \subset W$.

For our next proposition we need the following lemma on (DF)-spaces.

(7) *Let V_n, $n = 1, 2, \ldots$, be a sequence of absolutely convex closed neighbourhoods of $\mathfrak{o}$ in the* (DF)*-space F. Then there exist $\rho_n > 0$, $n = 1, 2, \ldots$, such that $V = \bigcap_{n=1}^{\infty} \rho_n V_n$ is a neighbourhood of $\mathfrak{o}$ in F.*

Using condition b) of the definition of a (DF)-space (§ 29, 3.), we see that it is sufficient to prove that V absorbs every set of a fundamental sequence $C_1 \subset C_2 \subset \cdots$ of bounded subsets of F.

We choose ρ_n such that $\rho_n V_n \supset C_n$ ($\supset C_{n-1} \supset \cdots \supset C_1$) for $n = 1, 2, \ldots$. Then there exists a σ_k, $0 < \sigma_k \leqq 1$, such that $\bigcap_{n=1}^{k-1} \rho_n V_n \supset \sigma_k C_k$ and it follows that $\bigcap_{n=1}^{\infty} \rho_n V_n = V \supset \sigma_k C_k$, which proves the statement.

(8) *Let E be the topological product $\prod_{n=1}^{\infty} E_n$ of countably many locally convex spaces E_n and F a* (DF)*-space. Then $\mathfrak{L}_{\mathfrak{M}}(E, F)$ and $\bigoplus_{n=1}^{\infty} \mathfrak{L}_{\mathfrak{M}_n}(E_n, F)$ are isomorphic.*

In particular, $\mathfrak{L}_b(E, F) = \bigoplus_{n=1}^{\infty} \mathfrak{L}_b(E_n, F)$ and $\mathfrak{L}_c(E, F) = \bigoplus_{n=1}^{\infty} \mathfrak{L}_c(E_n, F)$.

Since we proved in (6) that $\mathfrak{T}'$ is finer than $\mathfrak{T}$, we have to show that every $\mathfrak{T}'$-neighbourhood W' of $\mathfrak{o}$ contains a $\mathfrak{T}$-neighbourhood of $\mathfrak{o}$. Recalling § 18, 5.(8), we can assume that W' is of the form $\bigoplus_{n=1}^{\infty} W_n$, where $W_n = W(M_n, V_n)$, $M_n \in \mathfrak{M}_n$, and V_n is an absolutely convex and closed neighbourhood of $\mathfrak{o}$ in F.

Applying (7) to the sequence of neighbourhoods V_n, we obtain

$$W(M_n, V_n) = W(\rho_n M_n, \rho_n V_n) \supset W(\rho_n M_n, V)$$

and it follows that $W' \supset \bigoplus_{n=1}^{\infty} W(\rho_n M_n, V)$. Hence it is sufficient to consider $\mathfrak{T}'$-neighbourhoods of the form $W' = \bigoplus_{n=1}^{\infty} W(M_n, V)$.

Define $W = W(M, V)$, where $M = \prod_{n=1}^{\infty} M_n \subset E$. Then W is a $\mathfrak{T}$-neighbourhood of $\mathfrak{o}$ in $\mathfrak{L}(E, F)$. Let B be in W; then $P_n(M) = M_n \subset M$ and $BP_n(M) = B_n(M_n) \subset V$. Hence $B_n \subset W(M_n, V)$ and $\sum_{n=1}^{k} B_n$, the element corresponding to B, lies in W'. Thus $W \subset W'$.

We remember from § 19, 6.(4) that a subspace H of a locally convex space E can be written as a special case of a locally convex kernel, $H = \mathsf{K}\, J^{(-1)}(E)$, where J is the injection of H into E. From analogy one would expect the relation

$$\mathfrak{L}(H, F) = \mathfrak{L}(E, F)J \tag{9}$$

for any locally convex F.

We prove (9) for a complemented subspace H. Let E be $H \oplus H_1$ and P, P_1 be the projections of E onto H resp. H_1 vanishing on H_1 resp. H. Then from (1) follows $\mathfrak{L}(E, F) = \mathfrak{L}(H, F)P \oplus \mathfrak{L}(H_1, F)P_1$. Multiplying by J from the right proves (9).

Now we suppose only that H is a closed subspace of E. If (9) were true, every $A \in \mathfrak{L}(H, F)$ would have a representation $A = BJ$, where $B \in \mathfrak{L}(E, F)$ is a continuous extension of A from H to E. But such an extension exists for every F if and only if H is complemented in E (§ 38, 3.(1)).

On the other hand, if H is dense in E and if F is complete, then every $A = BJ$, where B is the uniquely determined extension of A from H to E, and in this case (9) is true and J is an isomorphism of $\mathfrak{L}(E, F)$ onto $\mathfrak{L}(H, F)$.

We are now well prepared for the negative result in the general case of a locally convex kernel $E = \mathsf{K}_{\alpha}\, A_{\alpha}^{(-1)}(E_{\alpha})$. It can be identified with a subspace $\hat{E}$ of the topological product $\prod_{\alpha} E_{\alpha}$ (§ 19, 6.) and for any locally convex F we have $\mathfrak{L}(E, F) \supset \bigoplus_{\alpha} \mathfrak{L}(E_{\alpha}, F)A_{\alpha}$. The elements of the direct sum are restrictions to $\hat{E}$ of the elements of $\mathfrak{L}(\prod E_{\alpha}, F)$, but in general there will be many other mappings contained in $\mathfrak{L}(E, F)$.

We turn now to the dual situation and begin with the case that F is a locally convex kernel $F = \mathsf{K}_{\alpha}\, A_{\alpha}^{(-1)}(F_{\alpha})$. A linear mapping B of a locally convex space E into F is continuous if and only if all mappings $A_{\alpha}B$ are continuous, i.e., $A_{\alpha}B \in \mathfrak{L}(E, F_{\alpha})$ (§ 19, 6.(6)). Hence we have linear mappings A_{α} such that $A_{\alpha}\mathfrak{L}(E, F) \subset \mathfrak{L}(E, F_{\alpha})$ and therefore $\mathfrak{L}(E, F) = \mathsf{K}_{\alpha}\, A_{\alpha}^{(-1)}(\mathfrak{L}(E, F_{\alpha}))$ algebraically. In particular, $\mathfrak{L}\left(E, \prod_{\alpha} F_{\alpha}\right) = \prod_{\alpha} \mathfrak{L}(E, F_{\alpha})$ since every $\prod_{\alpha} B_{\alpha}$, $B_{\alpha} \in \mathfrak{L}(E, F_{\alpha})$, obviously defines a continuous mapping of E in $\prod_{\alpha} F_{\alpha}$.

Now let $\mathfrak{M}$ be a class of bounded subsets of E defining the topology $\mathfrak{T}_{\mathfrak{M}}$ on $\mathfrak{L}(E, F)$ and the $\mathfrak{L}(E, F_{\alpha})$. We prove that the kernel topology $\mathfrak{T}$ on $\mathsf{K}_{\alpha}\, A_{\alpha}^{(-1)}(\mathfrak{L}_{\mathfrak{M}}(E, F_{\alpha}))$ coincides with the topology $\mathfrak{T}_{\mathfrak{M}}$ on $\mathfrak{L}(E, F)$.

A neighbourhood W_α of o in $\mathfrak{L}_{\mathfrak{M}}(E, F_\alpha)$ consists of all B_α such that $B_\alpha(M) \subset V_\alpha$, where $M \in \mathfrak{M}$ and V_α is a neighbourhood of o in F_α. The corresponding $\mathfrak{T}$-neighbourhood $A_\alpha^{(-1)}(W_\alpha)$ consists of all B such that $B(M) \subset A_\alpha^{(-1)}(V_\alpha)$; this is a $\mathfrak{T}_{\mathfrak{M}}$-neighbourhood of o in $\mathfrak{L}(E, F)$ and every $\mathfrak{T}_{\mathfrak{M}}$-neighbourhood of o in $\mathfrak{L}(E, F)$ contains a finite intersection of neighbourhoods of this type. This concludes the proof of

(10) *Let E be locally convex and F the locally convex kernel $\underset{\alpha}{\mathsf{K}}\, A_\alpha^{(-1)}(F_\alpha)$. Let $\mathfrak{M}$ be a class of bounded subsets of E. Then $\mathfrak{L}_{\mathfrak{M}}(E, F)$ is the locally convex kernel $\underset{\alpha}{\mathsf{K}}\, A_\alpha^{(-1)}(\mathfrak{L}_{\mathfrak{M}}(E, F_\alpha))$.*

In particular, $\mathfrak{L}_{\mathfrak{M}}\left(E, \prod_\alpha F_\alpha\right) = \prod_\alpha \mathfrak{L}_{\mathfrak{M}}(E, F_\alpha)$.

If H is a subspace of F, J the injection of H in F, then $\mathfrak{L}_{\mathfrak{M}}(E, H) = \mathsf{K}\, J^{(-1)}(\mathfrak{L}_{\mathfrak{M}}(E, F))$ and $J(\mathfrak{L}_{\mathfrak{M}}(E, H))$ is the subspace of $\mathfrak{L}_{\mathfrak{M}}(E, F)$ consisting of all B with range in H.

Next let F be a locally convex direct sum $\bigoplus_\alpha F_\alpha$. We have always

$$\mathfrak{L}\left(E, \bigoplus_\alpha F_\alpha\right) \supset \bigoplus_\alpha \mathfrak{L}(E, F_\alpha), \tag{11}$$

since for $B_{\alpha_i} \in \mathfrak{L}(E, F_{\alpha_i})$ the finite sum $\sum_{i=1}^{k} B_{\alpha_i}$ can be identified with an element of $\mathfrak{L}\left(E, \bigoplus_\alpha F_\alpha\right)$. But in general the equality sign in (11) is false. To see this take $E = F = \varphi$ and write $F = \varphi = \bigoplus_{n=1}^{\infty} \mathsf{K}_n$, $\mathsf{K}_n = \mathsf{K}$. The equality sign in (11) would imply that every $B \in \mathfrak{L}(\varphi, \varphi)$ has a finite dimensional range, but this is not true for B the identity.

Nevertheless, there is a counterpart to (8).

(12) *Let E be a metrizable locally convex space and F the locally convex direct sum $\bigoplus_{n=1}^{\infty} F_n$ of countably many locally convex spaces F_n.*

Then $\mathfrak{L}_b(E, F) \cong \bigoplus_{n=1}^{\infty} \mathfrak{L}_b(E, F_n)$ and $\mathfrak{L}_c(E, F) \cong \bigoplus_{n=1}^{\infty} \mathfrak{L}_c(E, F_n)$.

We prove first that $\mathfrak{L}(E, F) = \bigoplus_{n=1}^{\infty} \mathfrak{L}(E, F_n)$. Let $B \in \mathfrak{L}(E, F)$; then $B_n = P_n B \in \mathfrak{L}(E, F_n)$, where P_n is the projection of F onto F_n. We have to prove that only a finite number of the B_n are different from o. Assume that this is not the case. Then there exists a sequence $x_j \in E$ such that $B_{n_j} x_j \neq o$, where $n_j \to \infty$. Since E is metrizable, there exist $\rho_j > 0$ such that the set $C = \{\rho_1 x_1, \rho_2 x_2, \ldots\}$ is bounded in E. But $B(C)$ is not bounded in F by § 18, 5.(4), in contradiction to the continuity of B.

This settles the algebraic part and we remark that this proof remains valid also in the general case of any locally convex direct sum $\bigoplus_\alpha F_\alpha$.

For the second part of the proof we remark that if $M_1, M_2, \ldots$ is a sequence of bounded subsets of E, there exist $\rho_n > 0$ such that $M = \bigcup_{n=1}^{\infty} \rho_n M_n$ is again bounded. The analogous statement for precompact subsets M_n follows from § 21, 10.(3) in the same way.

Let $\mathfrak{M}$ be in the following the class of bounded resp. precompact subsets of E. Let $\mathfrak{T}$ be the topology of $\mathfrak{L}_{\mathfrak{M}}(E, F)$ and $\mathfrak{T}'$ the topology of $\bigoplus_{n=1}^{\infty} \mathfrak{L}_{\mathfrak{M}}(E, F_n)$. Let W be a $\mathfrak{T}$-neighbourhood of $\mathfrak{o}$ of the form $W(M, V)$, where $M \in \mathfrak{M}$ and $V = \bigoplus_{n=1}^{\infty} V_n$, V_n a neighbourhood of $\mathfrak{o}$ in F_n. It follows again from § 18, 5.(8) that these V define a neighbourhood base of $\mathfrak{o}$ in F. A $B \in W$ can be written as $B = \sum_{n=1}^{k} P_n B = \sum B_n$, $B_n \in \mathfrak{L}(E, F_n)$, and from $B(M) \subset \bigoplus_{n=1}^{\infty} V_n$ follows $B_n(M) \subset V_n$ or $B_n \in W(M, V_n)$. Since $\tilde{W} = \bigoplus_{n=1}^{\infty} W(M, V_n)$ is a $\mathfrak{T}'$-neighbourhood and every $\sum_{n=1}^{k} B_n \in \tilde{W}$ defines a $B \in W$, it follows that $\mathfrak{T}' \supset \mathfrak{T}$.

Conversely, let $W' = \bigoplus_{n=1}^{\infty} W(M_n, V_n)$ be a $\mathfrak{T}'$-neighbourhood (we use again § 18, 5.(8)). We determine the $\rho_n > 0$ in such a way that $\bigcup_{n=1}^{\infty} \rho_n M_n = M$ is in $\mathfrak{M}$ and let W be the $\mathfrak{T}$-neighbourhood $W\left(M, \bigoplus_{n=1}^{\infty} \rho_n V_n.\right)$ Then for $B \subset W$ we have $B_n(M) \subset \rho_n V_n$, $B_n(\rho_n M_n) \subset \rho_n V_n$, $B_n(M_n) \subset V_n$; hence $\sum_n B_n \in W'$ and $\mathfrak{T} \supset \mathfrak{T}'$.

We remark that if we assume E and the F_n in (11) to be (F)-spaces, then the statement that $\mathfrak{L}(E, F) = \bigoplus_{n=1}^{\infty} \mathfrak{L}(E, F_n)$ is a special case of Grothendieck's theorem, Theorem § 19, 5.(4). Its general form says that $\mathfrak{L}\left(E, \bigcup_{n=1}^{\infty} F_n\right) = \bigcup_{n=1}^{\infty} \mathfrak{L}(E, F_n)$, where $F = \bigcup_{n=1}^{\infty} F_n$ is the inductive limit of (F)-spaces F_n and E is an (F)-space.

For an arbitrary locally convex space E and a locally convex hull $F = \sum_\alpha A_\alpha(F_\alpha)$ we always have $\mathfrak{L}(E, F) \supset \bigoplus_\alpha A_\alpha(\mathfrak{L}(E, F_\alpha))$, but equality will be an exception.

We verify this statement in the simple case $F/H = KF$, where K is the canonical homomorphism of F onto F/H. $\mathfrak{L}(E, F/H) = K(\mathfrak{L}(E, F))$ would mean that every $A \in \mathfrak{L}(E, F/H)$ has a representation $A = KB$ with $B \in \mathfrak{L}(E, F)$ (A is liftable in F). And this is true for every E if and only if H is a complemented subspace of F (§ 38, 3.(6)).

We indicate some examples. We use the notations and results of § 13, 5. and § 23, 5. on spaces of countable degree, especially that these spaces are all reflexive, so that their topology is always the strong topology.

Using (10) we obtain

$$(13)\quad \mathfrak{L}_b(\omega, \omega) = \mathfrak{L}_b\left(\omega, \prod_{n=1}^{\infty} \mathsf{K}_n\right) \cong \prod_{n=1}^{\infty} \mathfrak{L}_b(\omega, \mathsf{K}_n) \cong \omega\varphi \qquad (\mathsf{K}_n = \mathsf{K}).$$

Using (2) we obtain

$$(14)\qquad \mathfrak{L}_b(\varphi, \varphi) = \mathfrak{L}_b\left(\bigoplus_{n=1}^{\infty} \mathsf{K}_n, \varphi\right) \cong \prod_{n=1}^{\infty} \mathfrak{L}_b(\mathsf{K}_n, \varphi) \cong \omega\varphi.$$

(14) is also an immediate consequence of (13) and 1.(8).

$$(15)\quad \mathfrak{L}_b(\varphi, \omega) = \mathfrak{L}_b\left(\bigoplus_{n=1}^{\infty} \mathsf{K}_n, \omega\right) = \prod_{n=1}^{\infty} \mathfrak{L}_b(\mathsf{K}_n, \omega) \cong \omega.$$

$$(16)\quad \mathfrak{L}_b(\omega, \varphi) = \mathfrak{L}_b\left(\omega, \bigoplus_{n=1}^{\infty} \mathsf{K}_n\right) = \bigoplus_{n=1}^{\infty} \mathfrak{L}_b(\omega, \mathsf{K}_n) \cong \varphi \quad \text{by (12).}$$

In the same way we obtain the following isomorphisms:

$$\mathfrak{L}_b(\varphi, \omega\varphi) \cong \omega\varphi, \qquad \mathfrak{L}_b(\varphi, \varphi\omega) \cong \omega\varphi\omega, \qquad \mathfrak{L}_b(\omega, \omega\varphi) \cong \omega\varphi,$$
$$\mathfrak{L}_b(\omega, \varphi\omega) \cong \varphi\omega\varphi \quad \text{and} \quad \mathfrak{L}_b(\varphi\omega, \varphi\omega) \cong \omega\varphi\omega\varphi,$$

and from this by 1.(8)

$$\mathfrak{L}_b(\omega\varphi, \omega\varphi) \cong \omega\varphi\omega\varphi.$$

We are also able to settle the case $\mathfrak{L}_b(\varphi\omega, \omega\varphi)$, since

$$\mathfrak{L}_b(\varphi\omega, \omega\varphi) \cong \prod_{n=1}^{\infty} \mathfrak{L}_b(\omega_n, \omega\varphi) \cong \prod_{n=1}^{\infty} (\omega\varphi)_n \cong \omega\varphi;$$

but our methods fail in the case $\mathfrak{L}_b(\omega\varphi, \varphi\omega)$, which remains undetermined.

§ 40. Bilinear mappings

1. Fundamental notions. Bilinear mappings and bilinear forms were briefly introduced in § 9, 7. and studied again for metrizable spaces in § 15, 14., where the important Theorem § 15, 14.(3) of BOURBAKI was proved.

We will now make a more systematic study of this topic which will become useful in the theory of tensor products. Nearly all results presented here are due to BOURBAKI and GROTHENDIECK.

Let E, F, G be locally convex. We denote by $B(E \times F, G)$ the vector space of all bilinear mappings B of $E \times F$ in G and by $B(E \times F)$ the vector space of all bilinear forms mapping $E \times F$ in K.

For $B \in B(E \times F, G)$ we define $B_x \in L(F, G)$ by $B_x(y) = B(x, y)$, $x \in E$, $y \in F$, and $B_y \in L(E, G)$ by $B_y(x) = B(x, y)$. Further, let $\tilde{B} \in L(E, L(F, G))$ be defined by $\tilde{B}x = B_x$ and $\tilde{\tilde{B}} \in L(F, L(E, G))$ by $\tilde{\tilde{B}}y = B_y$.

Conversely, if $\tilde{B} \in L(E, L(F, G))$, define $B \in B(E \times F, G)$ by

$$B(x, y) = (\tilde{B}x)(y) = B_x y$$

and, if $\tilde{\tilde{B}} \in L(F, L(E, G))$, define $B \in B(E \times F, G)$ by

$$B(x, y) = (\tilde{\tilde{B}}y)(x) = B_y x.$$

Hence the correspondences $B \to \tilde{B}$ and $B \to \tilde{\tilde{B}}$ are one-one and onto and we obtain the algebraic isomorphisms

$$(1) \qquad B(E \times F, G) \cong L(E, L(F, G)) \cong L(F, L(E, G)),$$

$$(1') \qquad B(E \times F) \cong L(E, F^*) \cong L(F, E^*).$$

We recall that a bilinear mapping B of $E \times F$ in G is continuous if it is continuous as a mapping of $E \times F$ in G or, as one says, if it is continuous in both variables simultaneously.

We denote the vector space of all continuous bilinear mappings of $E \times F$ in G by $\mathscr{B}(E \times F, G)$ and by $\mathscr{B}(E \times F)$ we denote the space of all continuous bilinear forms.

It is obvious what equicontinuity of a set H of bilinear mappings means and by § 15, 14.(1) it is only necessary to check continuity or equicontinuity of bilinear mappings at the point $(\mathrm{o}, \mathrm{o}) \in E \times F$.

$B \in B(E \times F, G)$ is separately continuous if B_x and B_y are continuous for all x, y, i.e., if $B_x \in \mathfrak{L}(F, G)$ and $B_y \in \mathfrak{L}(E, G)$ for all x, y. It is obvious that every continuous bilinear mapping is separately continuous.

We denote the vector space of all separately continuous bilinear mappings of $E \times F$ in G by $\mathfrak{B}(E \times F, G)$, and $\mathfrak{B}(E \times F)$ denotes the space of all separately continuous bilinear forms.

The correspondences $B \to \tilde{B} \to \tilde{\tilde{B}}$ generate the following algebraic isomorphisms:

$$(2) \qquad \mathfrak{B}(E \times F, G) \cong \mathfrak{L}(E, \mathfrak{L}_s(F, G)) \cong \mathfrak{L}(F, \mathfrak{L}_s(E, G));$$

$$(2') \qquad \mathfrak{B}(E \times F) \cong \mathfrak{L}(E, F'_s) \cong \mathfrak{L}(F, E'_s).$$

Proof. Since $\mathfrak{B}(E \times F, G)$ is symmetric in E and F, it will be sufficient to prove the first isomorphism in (2).

a) If $B \in \mathfrak{B}(E \times F, G)$, then $\tilde{B} \in L(E, \mathfrak{L}(F, G))$. We must show that $\tilde{B}$ is continuous from E in $\mathfrak{L}_s(F, G)$. Let $\mathscr{U}(M, W)$ be a given neighbourhood of o in $\mathfrak{L}_s(F, G)$, where M is a finite subset of F and W a neighbourhood of o in G. By separate continuity of B there exists a neighbourhood U of o in E such that $B(U, M) \subset W$. This means $\tilde{B}(U)(M) \subset W$ or, equivalently,

$B_x(M) \subset W$ for all $x \in U$. Hence $B_x \in \mathscr{U}(M, W)$ for all $x \in U$. Thus $\tilde{B}(U) \subset \mathscr{U}(M, W)$ and $\tilde{B} \in \mathfrak{L}(E, \mathfrak{L}_s(F, G))$.

b) If $\tilde{B} \in \mathfrak{L}(E, \mathfrak{L}_s(F, G))$, then $B \in B(E \times F, G)$, where B is defined by $B(x, y) = \tilde{B}(x)(y)$. We must show that B is separately continuous. Obviously, $B_x = \tilde{B}(x) \in \mathfrak{L}(F, G)$. Let y be an element of F and $\mathscr{U}(\{y\}, W)$ a neighbourhood of $\circ$ in $\mathfrak{L}_s(F, G)$. Since $\tilde{B}$ is continuous, there exists a neighbourhood U of $\circ$ in E such that $\tilde{B}(U) \subset \mathscr{U}(\{y\}, W)$. This means $B(x, y) \in W$ for all $x \in U$ or $B_y(U) \subset W$ and $B_y \in \mathfrak{L}(E, G)$.

Remark. For bilinear forms the isomorphism $\mathfrak{L}(E, F_s') \cong \mathfrak{L}(F, E_s')$ of (2′) consists in taking adjoints: By definition

$$B(x, y) = \langle y, \tilde{B}x \rangle = \langle \tilde{\tilde{B}}y, x \rangle$$

for all $x \in E$ and all $y \in F = (F_s')'$ thus $\tilde{\tilde{B}} = \tilde{B}'$.

The following example shows that in general a separately continuous bilinear form B will not be continuous.

Let E be locally convex and of infinite dimension and let E' be its dual. We consider the canonical bilinear form $B(u, x) = ux$ mapping $E_s' \times E$ on the scalar field K. B is separately continuous since $B_u = u \in E'$ is continuous on E and $B_x = x$ is weakly continuous on E'.

But B is not continuous: Let U be an absolutely convex weak neighbourhood of $\circ$ in E' and M a subset of E such that $|ux| \leqq 1$ for all $u \in U$ and all $x \in M$. It follows that M is contained in U° and therefore finite dimensional. Therefore M can never be a neighbourhood of $\circ$ in E and B is not continuous.

We now define a type of continuity for bilinear mappings which lies between separate continuity and continuity and was introduced by BOURBAKI [2].

Let $\mathfrak{M}$, $\mathfrak{N}$ be classes of bounded subsets of E resp. F with the properties a), b) of § 39, 1. A separately continuous bilinear mapping B of $E \times F$ in G is said to be $\mathfrak{M}$-hypocontinuous if for every $M \in \mathfrak{M}$ and every neighbourhood W of $\circ$ in G there exists a neighbourhood V of $\circ$ in F such that $B(M, V) \subset W$.

Note that B is $\mathfrak{M}$-hypocontinuous if and only if for every $M \in \mathfrak{M}$ the collection $\{B_x; x \in M\}$ is equicontinuous in $\mathfrak{L}(F, G)$.

Similarly, $\mathfrak{N}$-hypocontinuity of B means that the collection $\{B_y; y \in N\}$ is equicontinuous in $\mathfrak{L}(E, G)$ for every $N \in \mathfrak{N}$.

Finally, B is $(\mathfrak{M}, \mathfrak{N})$-hypocontinuous if it is both $\mathfrak{M}$- and $\mathfrak{N}$-hypocontinuous.

We remark that $\mathfrak{M}$ resp. $\mathfrak{N}$ can be replaced by its saturated cover $\tilde{\mathfrak{M}}$ resp. $\tilde{\mathfrak{N}}$ without changing the notion of hypocontinuity. This is an immediate consequence of the following proposition, (3) a) or b).

The strongest type of hypocontinuity occurs when $\mathfrak{M}$ and $\mathfrak{N}$ are the classes $\mathfrak{B}$ of all bounded subsets of E resp. F. In this case we say that B is hypocontinuous.

It is obvious that separate continuity and $(\mathfrak{F}, \mathfrak{F})$-hypocontinuity are equivalent, where $\mathfrak{F}$ is the class of finite subsets.

Every continuous bilinear mapping is $\mathfrak{M}$-hypocontinuous and $\mathfrak{N}$-hypocontinuous for every class $\mathfrak{M}$ resp. $\mathfrak{N}$.

We denote the space of all $\mathfrak{M}$-hypocontinuous resp. $(\mathfrak{M}, \mathfrak{N})$-hypocontinuous bilinear mappings of $E \times F$ in G by $\mathfrak{X}^{(\mathfrak{M})}(E \times F, G)$ resp. $\mathfrak{X}^{(\mathfrak{M},\mathfrak{N})}(E \times F, G)$ and by $\mathfrak{X}(E \times F, G)$ the space of all hypocontinuous bilinear mappings.

For bilinear forms we use the notations $\mathfrak{X}^{(\mathfrak{M})}(E \times F)$, $\mathfrak{X}^{(\mathfrak{M},\mathfrak{N})}(E \times F)$, and $\mathfrak{X}(E \times F)$.

In (1) and (2) we characterized the linear mappings $\tilde{B}$ and $\tilde{\tilde{B}}$ corresponding to a bilinear resp. separately continuous bilinear mapping B. We have similar results for hypocontinuous mappings.

(3) a) *$B \in \mathfrak{B}(E \times F, G)$ is $\mathfrak{M}$-hypocontinuous if and only if $\tilde{\tilde{B}}$ maps F continuously in $\mathfrak{L}_{\mathfrak{M}}(E, G)$. Therefore $\mathfrak{X}^{(\mathfrak{M})}(E \times F, G)$ and $\mathfrak{L}(F, \mathfrak{L}_{\mathfrak{M}}(E, G))$ are algebraically isomorphic.*

b) *$B \in \mathfrak{B}(E \times F, G)$ is $\mathfrak{M}$-hypocontinuous if and only if $\tilde{B}$ maps every $M \in \mathfrak{M}$ into an equicontinuous subset of $\mathfrak{L}(F, G)$.*

Proof. a) If $B \in \mathfrak{X}^{(\mathfrak{M})}(E \times F, G)$, then, given $M \in \mathfrak{M}$ and $W \ni \circ$ in G, there exists $V \ni \circ$ in F such that $B(M, V) \subset W$. So $\tilde{\tilde{B}}(V)(M) \subset W$ and thus $\tilde{\tilde{B}}(V)$ is contained in the neighbourhood $U(M, W)$ of $\circ$ of $\mathfrak{L}_{\mathfrak{M}}(E, G)$. Therefore $\tilde{\tilde{B}} \in \mathfrak{L}(F, \mathfrak{L}_{\mathfrak{M}}(E, G))$.

Conversely, if $\tilde{\tilde{B}} \in \mathfrak{L}(F, \mathfrak{L}_{\mathfrak{M}}(E, G))$, then, given $U(M, W)$, there exists $V \ni \circ$ such that $\tilde{\tilde{B}}(V) \subset U(M, W)$. So $\tilde{\tilde{B}}(V)(M) \subset W$ and $B(M, V) \subset W$. Therefore $B \in \mathfrak{X}^{(\mathfrak{M})}(E \times F, G)$.

b) $B(M, V) \subset W$ if and only if $\tilde{B}(M)(V) \subset W$. So $B \in \mathfrak{X}^{(\mathfrak{M})}(E \times F, G)$ if and only if $\tilde{B}(M)$ is an equicontinuous subset of $\mathfrak{L}(F, G)$ for every $M \in \mathfrak{M}$.

If we replace in the last argument M by a neighbourhood U of $\circ$ of E we obtain for bilinear forms

(4) *$B \in \mathfrak{B}(E \times F)$ is continuous if and only if $\tilde{B}$ maps some neighbourhood U of $\circ$ in E into an equicontinuous subset of F' or if and only if $\tilde{\tilde{B}}$ maps some neighbourhood V of $\circ$ in F into an equicontinuous subset of E'.*

An important property of hypocontinuity is stated in

(5) a) *If $B \in \mathfrak{X}^{(\mathfrak{M})}(E \times F, G)$, then B is continuous on every set $M \times F$, $M \in \mathfrak{M}$.*

b) *If* $B \in \mathfrak{X}^{(\mathfrak{M},\mathfrak{N})}(E \times F, G)$, *then* B *is uniformly continuous on every set* $M \times N$, $M \in \mathfrak{M}$, $N \in \mathfrak{N}$.

Proof. a) Let $(x_0, y_0) \in M \times F$. We must find neighbourhoods of $\circ$, U, V in E resp. F, such that $B(x, y) - B(x_0, y_0) \in W$ for all $(x, y) \in (M \times F) \cap ((x_0, y_0) + (U \times V))$. We use the identity

$$(6) \qquad B(x, y) - B(x_0, y_0) = B(x, y - y_0) + B(x - x_0, y_0).$$

Let W be a neighbourhood of $\circ$ in G. Since B is $\mathfrak{M}$-hypocontinuous, there exists $V \ni \circ$ in E such that $B(M, V) \subset \frac{1}{2}W$; hence $B(x, y - y_0) \in \frac{1}{2}W$ for $x \in M$, $y - y_0 \in V$.

Since B_{y_0} is continuous, there exists $U \ni \circ$ in E with $B(U, y_0) \subset \frac{1}{2}W$. Thus if $x \in M$, $x - x_0 \in U$, then $B(x - x_0, y_0) \in \frac{1}{2}W$ and from (6) follows $B(x, y) - B(x_0, y_0) \in W$ for $x \in M$, $x - x_0 \in U$, $y - y_0 \in V$, which proves a).

b) Let $x, \bar{x}$ be in M, $y, \bar{y}$ in N. There exist neighbourhoods U, V such that $B(U, N) \subset \frac{1}{2}W$ and $B(M, V) \subset \frac{1}{2}W$. Then it follows from (6) that for $x - \bar{x} \in U$ and $y - \bar{y} \in V$,

$$B(x, y) - B(\bar{x}, \bar{y}) \in B(M, V) + B(U, N) \subset W,$$

and this is the uniform continuity of B on $M \times N$.

Remark. We say that $B \in \mathfrak{B}(E \times F, G)$ is sequentially continuous if $x^{(n)} \to x^{(\circ)}$ in E and $y^{(n)} \to y^{(\circ)}$ in F implies always $B(x^{(n)}, y^{(n)}) \to B(x^{(\circ)}, y^{(\circ)})$ in G.

If the class $\mathfrak{M}$ of bounded subsets of E contains all convergent sequences (for instance, if $\mathfrak{M}$ is the class of all compact subsets of E), then it follows from (5) a) that every $\mathfrak{M}$-hypocontinuous bilinear mapping is sequentially continuous.

Examples. 1) Let E be locally convex and $B(u, x) = ux$ the canonical bilinear form on $E'_b \times E$. Then B is $(\mathfrak{M}, \mathfrak{N})$-hypocontinuous, where $\mathfrak{M}$ is the class of all equicontinuous subsets of E' and $\mathfrak{N}$ the class of all bounded subsets of E.

2) If E is barrelled, then ux is hypocontinuous on $E'_b \times E$. This is a special case of the previous example.

3) Let E be an (F)-space which is not a (B)-space. Then ux is hypocontinuous and sequentially continuous on $E'_b \times E$. But ux is not continuous: The neighbourhoods U of $\circ$ in E are unbounded and therefore $\sup_{u \in B^\circ, x \in U} |ux| = \infty$ for every U and every bounded subset B of E.

2. Continuity theorems for bilinear maps. We introduced different notions of continuity for bilinear mappings in 1. If we investigate not only one but a whole set of bilinear mappings, we will have to use the corresponding notions of equicontinuity.

Let H be a family of separately continuous bilinear mappings $B \in \mathfrak{B}(E \times F, G)$. We say that H is separately equicontinuous if the family $\{B_x, B \in H\}$ resp. $\{B_y, B \in H\}$ is equicontinuous in $\mathfrak{L}(F, G)$ resp. $\mathfrak{L}(E, G)$ for every $x \in E$, $y \in F$.

Let $\mathfrak{M}$ be a class of bounded subsets of E. H is $\mathfrak{M}$-equihypocontinuous if, given $M \in \mathfrak{M}$ and $W \ni \circ$ in G, there exists $V \supset \circ$ in F such that $B(M, V) \subset W$ for all $B \in H$.

If $\mathfrak{N}$ is a class of bounded subsets of F, then $\mathfrak{N}$-equihypocontinuity is similarly defined.

H is $(\mathfrak{M}, \mathfrak{N})$-equihypocontinuous if it is both $\mathfrak{M}$- and $\mathfrak{N}$-equihypocontinuous. H is equihypocontinuous if it is $(\mathfrak{B}, \mathfrak{B})$-equihypocontinuous, where $\mathfrak{B}$ is the class of all bounded subsets.

We recall the theorem of Bourbaki (§ 15, 14.(3)). It says in the locally convex case that a separately continuous bilinear mapping of a product of two (F)-spaces in a locally convex space is always continuous.

It is possible to weaken the assumptions a little and to arrive at our first continuity theorem:

(1) *Let E and F be metrizable barrelled spaces and let G be locally convex. Then*

a) *every $B \in \mathfrak{B}(E \times F, G)$ is continuous, and*

b) *a family $H \subset \mathfrak{B}(E \times F, G)$ is equicontinuous if and only if the set $H(x, y) = \{B(x, y); B \in H\}$ is bounded in G for each fixed $(x, y) \in E \times F$.*

The proof is the same as for § 15, 14.(3) with the only difference that the theorem of Banach is used in the form § 39, 3.(2).

A slightly different version of (1) is

(2) *Let E and F be metrizable barrelled spaces and G locally convex. A family $H \subset \mathfrak{B}(E \times F, G)$ is equicontinuous if and only if it is separately equicontinuous.*

This follows easily from (1) b): Let H be separately equicontinuous and $(x, y) \in E \times F$ given. Then to $W \ni \circ$ in G there exists $V \ni \circ$ in F such that $B(x, V) \subset W$ for all $B \in H$. If $y \in \rho V$, $\rho > 0$, then $H(x, y) \in \rho W$. Thus $H(x, y)$ is bounded in G and the condition of (1) b) is satisfied.

We remark that, conversely, if this condition is satisfied, separate equicontinuity of H follows from the theorem of Banach (§ 39, 3.(2)).

We will now drop the metrizability assumption and look at the case of general barrelled spaces.

(3) *Let F be barrelled and E, G locally convex. Then*

a) *every separately continuous bilinear mapping B of $E \times F$ in G is*

$\mathfrak{B}$-hypocontinuous, where $\mathfrak{B}$ is the class of all bounded subsets of E, and, more generally,

b) *every separately equicontinuous subset H of $\mathfrak{B}(E \times F, G)$ is $\mathfrak{B}$-equi-hypocontinuous.*

Proof. a) By 1.(2) the corresponding mapping $\tilde{B}$ of E in $\mathfrak{L}_s(F, G)$ is continuous; thus the image $\tilde{B}(M)$ of a bounded subset M of E is simply bounded in $\mathfrak{L}_s(F, G)$. By the theorem of BANACH (§ 39, 3.(2)) $\tilde{B}(M)$ is equicontinuous in $\mathfrak{L}(F, G)$. Hence there exists $V \ni \circ$ in F such that $\tilde{B}(M)(V) = B(M, V) \subset W$, where $W \ni \circ$ in G is given. But this is the $\mathfrak{B}$-hypocontinuity of B.

b) The same proof will work if we show that the set $\tilde{H}(M) = \{\tilde{B}x;\, B \in H, x \in M\}$ is simply bounded in $\mathfrak{L}_s(F, G)$. But this follows immediately from

(4) *Let H be a subset of $\mathfrak{B}(E \times F, G)$, E, F, G locally convex. If H is separately equicontinuous, then the corresponding set $\tilde{H}$ in $\mathfrak{L}(E, \mathfrak{L}_s(F, G))$ is equicontinuous.*

This can be proved by applying the arguments of part a) of the proof of 1.(2) to H instead of to $B \in \mathfrak{B}(E \times F, G)$.

As an immediate consequence of (3) we obtain the second continuity theorem:

(5) *Let E, F be barrelled, G locally convex. Then*

a) *every $B \in \mathfrak{B}(E \times F, G)$ is hypocontinuous, and*

b) *every separately equicontinuous subset H of $\mathfrak{B}(E \times F, G)$ is equi-hypocontinuous.*

If E is a reflexive (F)-space which is not a (B)-space and $F = E'_b$, then E'_b is barrelled by § 23, 3.(4); Example 1. 3) shows that the canonical bilinear form on $E'_b \times E$ is not continuous. Hence it is not possible to replace "hypocontinuity" by "continuity" in (5).

The exceptional character of a continuity theorem of type (1) is made clear by the following proposition, which is another version of 1.(4).

(6) *Let E, F be locally convex. The following statements are equivalent:*

a) *every separately continuous bilinear form on $E \times F$ is continuous;*

b) *for every $A \in \mathfrak{L}(E, F'_s)$ there exists a neighbourhood U of $\circ$ in E such that $A(U)$ is equicontinuous in F'.*

It is obvious that b) is a rather strong condition for the pair E, F.

The following method reduces the study of bilinear mappings to that of certain sets of bilinear forms.

Let E, F, G be locally convex and $B \in B(E \times F, G)$. If w is an element of G', then $wB(x, y) = \langle w, B(x, y)\rangle$ is a bilinear form wB on $E \times F$. If M

is a subset of G' and H a subset of $B(E \times F, G)$, we denote by MH the set of all wB, $w \in M$, $B \in H$. If A is a set of complex numbers, $|A|$ will denote $\sup_{\alpha \in A} |\alpha|$ or $+\infty$.

The method is now described by the following two lemmas.

(7) *$B \in B(E \times F, G)$ is separately continuous if and only if all sets MB, M an equicontinuous subset of G', are separately equicontinuous in $\mathfrak{B}(E \times F)$.*

Let x be a fixed element of E and W any absolutely convex and closed neighbourhood of o in G. Then the following statements are equivalent:

α) to W there exists a neighbourhood V of o in F such that $B(x, V) \subset W$;

β) to W there exists a neighbourhood V of o in F such that $|W^\circ B(x, V)| \leqq 1$.

From this and the corresponding equivalence for a fixed y in F follows (7).

(8) *$H \subset B(E \times F, G)$ is equicontinuous resp. $\mathfrak{M}$-equihypocontinuous if and only if all sets MH, M an equicontinuous subset of G', are equicontinuous resp. $\mathfrak{M}$-equihypocontinuous in $\mathfrak{B}(E \times F)$.*

The proof is similar to that of (7) using the following equivalences:

a) $B(U, V) \subset W$ for all $B \in H$ if and only if $|W^\circ B(U, V)| \leqq 1$ for all $B \in H$;

b) $B(M, V) \subset W$ for all $B \in H$ if and only if $|W^\circ B(M, V)| \leqq 1$ for all $B \in H$.

We relate now the results of GROTHENDIECK [10], [11] on bilinear mappings of (DF)-spaces.

A linear mapping A of E in F is called bounded if there exists a neighbourhood U of o in E such that $A(U)$ is a bounded subset of F. A class H of linear mappings is called equibounded if there exists $U \ni o$ such that $H(U) = \bigcup_{A \in H} A(U)$ is a bounded subset of F.

(9) *Let E be a (DF)-space, F metrizable locally convex. Then every $A \in \mathfrak{L}(E, F)$ is bounded and every equicontinuous subset H of $\mathfrak{L}(E, F)$ is equibounded.*

Let $V_1 \supset V_2 \supset \cdots$ be a neighbourhood base of o in F. Then every set $U_n = \bigcap_{A \in H} A^{(-1)}(V_n)$ is a neighbourhood of o in E. By § 39, 8.(7) there exists a neighbourhood U of o in E such that $U \subset \rho_n U_n$ for suitable $\rho_n > 0$. Hence $H(U) \subset \rho_n V_n$ and H is equibounded.

(10) *Let E, F be (DF)-spaces and G locally convex. A bilinear mapping $B \in \mathfrak{B}(E \times F, G)$ is continuous if and only if B is hypocontinuous. A set $H \subset \mathfrak{B}(E \times F, G)$ is equicontinuous if and only if it is equihypocontinuous.*

It follows from (8) that it is sufficient to prove this for a set $H \subset \mathfrak{B}(E \times F)$.

Let H be an equihypocontinuous subset of $\mathfrak{L}(E \times F)$. Then for every absolutely convex bounded subset M of F there exists a neighbourhood U of $\mathfrak{o}$ in E such that $|H(U, M)| \leqq 1$. From this follows $\tilde{H}(U) \subset M^\circ$, where $\tilde{H}$ is the set of all $\tilde{B}$, $B \in H$. Hence $\tilde{H}$ is an equicontinuous subset of $\mathfrak{L}(E, F_b')$. Since F_b' is an (F)-space, $\tilde{H}$ is equibounded by (9). Hence there exists an absolutely convex neighbourhood U_1 of $\mathfrak{o}$ in E such that $\tilde{H}(U_1)$ is bounded in F_b'. Since E is a (DF)-space, U_1 is the union of a sequence K_n of absolutely convex bounded subsets of E; thus $\tilde{H}(U_1) = \bigcup_{n=1}^{\infty} \tilde{H}(K_n)$.

Using again the equihypocontinuity of H, there exists $V_n \ni \mathfrak{o}$ in F such that $|H(K_n, V_n)| \leqq 1$; hence $\tilde{H}(K_n) \subset V_n^\circ$ and $\tilde{H}(K_n)$ is equicontinuous in F'. Therefore $\tilde{H}(U_1)$, as the strongly bounded union of a sequence of equicontinuous subsets of F', is itself equicontinuous (compare the definition of a (DF)-space in § 29, 3.). Hence H is equicontinuous in $\mathfrak{B}(E \times F)$.

For an analogous theorem see also § 45, 3.(3).

Combining (10) with (5) we obtain the following counterpart to (2):

(11) *Let E, F be barrelled* (DF)-*spaces, G locally convex. Then every $B \in \mathfrak{B}(E \times F, G)$ is continuous and every separately equicontinuous subset H of $\mathfrak{B}(E \times F, G)$ is equicontinuous.*

(11) applies to the case where E and F are the strong duals of distinguished (F)-spaces (§ 29, 4.(3)). For the strong duals of arbitrary (F)-spaces we have a weaker result which will follow from

(12) *Let E, F, G be locally convex. Then every separately weakly continuous bilinear mapping from $E' \times F'$ into G is strongly hypocontinuous.*

Moreover, every separately weakly equicontinuous set H of bilinear mappings from $E' \times F'$ into G is strongly equihypocontinuous.

By (8) we need only prove the assertion for bilinear forms. So let $H \in \mathfrak{B}(E_s' \times F_s')$ be separately equicontinuous. Then $\tilde{H}$ is an equicontinuous set in $\mathfrak{L}(E_s', F_s)$. Let M be an absolutely convex weakly bounded subset of E'. Then $\tilde{H}(M)$ is weakly bounded in F and therefore an equicontinuous subset for F_b'. Hence there exists a strong neighbourhood V of $\mathfrak{o}$ in F' such that $|\tilde{H}(M)(V)| \leqq 1$ or $|H(M, V)| \leqq 1$. But this implies the strong equihypocontinuity of H with regard to E'. Interchanging the roles of E' and F' completes the proof.

As a direct consequence of (10) and (12) we obtain

(13) *Let E, F be* (F)-*spaces and G locally convex. If H is a set of separately weakly equicontinuous bilinear mappings of $E' \times F'$ in G, then H is strongly equicontinuous.*

3. Extensions of bilinear mappings. We have seen (§ 39, 6.(1)) that a linear continuous mapping $A \in \mathfrak{L}(E, F)$ has a uniquely determined continuous extension $\tilde{A} \in \mathfrak{L}(\tilde{E}, F)$ if F is complete. We are interested in the corresponding questions for bilinear mappings.

(1) *Let E, F, G be locally convex and E_0 resp. F_0 a dense subspace of E resp. F. If B is a separately continuous bilinear mapping of $E \times F$ in G which vanishes on $E_0 \times F_0$, then B is identically $\circ$ on $E \times F$.*

Proof. For a fixed $x \in E_0$, $B_x(y) = \circ$ for all $y \in F_0$. Continuity of B_x implies $B_x(y) = \circ$ for all $y \in F$. Therefore $B_y(x) = \circ$ for all $x \in E_0$, $y \in F$. Now continuity of B_y implies $B_y(x) = \circ$ for all $x \in E$, $y \in F$.

(1) says that if an extension of B from $E_0 \times F_0$ to $E \times F$ exists, this extension is uniquely determined.

There is no difficulty with continuous bilinear mappings, as the following proposition shows.

(2) *If B is a continuous bilinear mapping of $E \times F$ into a complete space G, then there exists a uniquely determined continuous extension $\bar{B}$ to $\tilde{E} \times \tilde{F}$.*

If H is an equicontinuous set of bilinear mappings, then $\bar{H} = \{\bar{B}; B \in H\}$ is also equicontinuous.

Proof. The uniqueness follows from (1). We construct an extension in the following way. For a fixed $x \in E$, $B_x \in \mathfrak{L}(F, G)$. It has a continuous extension $\hat{B}_x \in \mathfrak{L}(\tilde{F}, G)$. This defines an extension of B to $\hat{B}$ defined on $E \times \tilde{F}$. We want to show that $\hat{B}$ is (i) bilinear and (ii) continuous.

(i) Bilinearity: $\hat{B}$ is linear in the second variable by definition. Linearity in the first variable follows from

$$\begin{aligned}\hat{B}(\alpha_1 x_1 + \alpha_2 x_2, y) &= \lim_\beta B(\alpha_1 x_1 + \alpha_2 x_2, y_\beta) \\ &= \alpha_1 \lim_\beta B(x_1, y_\beta) + \alpha_2 \lim_\beta B(x_2, y_\beta) \\ &= \alpha_1 \hat{B}(x_1, y) + \alpha_2 \hat{B}(x_2, y),\end{aligned}$$

where $y \in \tilde{F}$ and $y = \lim y_\beta$, $y_\beta \in F$.

(ii) Continuity: There exist neighbourhoods U of $\circ$, V of $\circ$ in E resp. F such that $B(U, V) \subset W$, where W is a given closed neighbourhood of $\circ$ in G. Let $\bar{V}$ be the closure of V in $\tilde{F}$; then by taking limits in $\tilde{F}$ we obtain $\hat{B}(U, \bar{V}) \subset W$.

$\hat{B}$ is a continuous bilinear mapping of $E \times \tilde{F}$ in G. Proceeding in the same way we extend $\hat{B}$ to a continuous bilinear mapping $\bar{B}$ of $\tilde{E} \times \tilde{F}$ in G.

The second statement of (2) is nearly obvious: If $B(U, V) \subset W$ for all

$B \in H$, then $\bar{B}(\bar{U}, \bar{V}) \subset W$ for a closed W by the first part of the proof and this is the equicontinuity of $\bar{H}$.

In the case of bilinear forms on arbitrary locally convex spaces E, F, (2) enables us to identify $\mathscr{B}(E \times F)$ and $\mathscr{B}(\tilde{E} \times \tilde{F})$ as vector spaces for which equicontinuous subsets are preserved.

The assertion of Proposition (2) is no longer true if we replace "continuous" by "separately continuous", as is shown by the following example.

Let $E = F = \varphi$ be endowed with the l^∞-norm and define $B(x, y) = \sum_{n=1}^{\infty} x_n y_n$ for $x = (x_n) \in \varphi$ and $y = (y_n) \in \varphi$. B is separately continuous, since B_x is continuous on φ: $\|B_x\| = \sum_{n=1}^{k} |x_n|$, if x_k is the last nonvanishing coordinate of x. Since φ is dense in c_0, the continuous extension $\hat{B}_x$ of B_x to $\varphi \times c_0$ is given by $\hat{B}_x(y) = \sum_{n=1}^{\infty} x_n y_n = \hat{B}(x, y)$, $y = (y_n) \in c_0$.

But $\hat{B}$ is not separately continuous since $\hat{B}_y$ is not continuous: The sequence $x^{(n)} = (1, \ldots, 1/\sqrt{n}, 0, 0, \ldots)$ is bounded in φ, since $\|x^{(n)}\| = 1$, but for $y = (1, 1/\sqrt{2}, 1/\sqrt{3}, \ldots) \in c_0$ we have $\hat{B}(x^{(n)}, y) = 1 + \cdots + (1/n) \to \infty$.

Nevertheless, it is possible to extend hypocontinuous bilinear mappings in a modest way, as was shown by Bourbaki [6].

(3) *Let E, F, G be locally convex and G quasi-complete. Let E_0 be a dense subspace of E and $\mathfrak{M}$ a class of bounded subsets M of E_0 which covers E_0 and with the property that the class $\bar{\mathfrak{M}}$ of the closures $\bar{M}$ in E covers E. Then every bilinear $\mathfrak{M}$-hypocontinuous mapping B of $E_0 \times F$ in G has a uniquely determined $\bar{\mathfrak{M}}$-hypocontinuous extension $\bar{B}$ on $E \times F$.*

A corresponding statement is true for $\mathfrak{M}$-equihypocontinuity.

Proof. By assumption every element of E is contained in a set $\bar{M}$ and E is therefore contained in the quasi-completion of E_0. It follows from § 23, 1.(4) that every B_y, $y \in F$, has a uniquely determined continuous extension $\bar{B}_y$ to E and thus $\bar{B}(x, y) = \bar{B}_y x$ is a mapping of $E \times F$ in G, linear on E by definition and linear on F as in the proof of (2).

That $\bar{B}$ is $\bar{\mathfrak{M}}$-hypocontinuous follows now easily. For given $M \in \mathfrak{M}$ and $W \subset G$ there exists $V \subset F$ such that $B(M, V) \subset W$. Taking W closed in G, it follows that $\bar{B}(\bar{M}, V) \subset W$. Thus $\bar{B}$ is separately continuous and $\bar{\mathfrak{M}}$-hypocontinuous.

We can go one step further.

(4) *Let E, F, G be locally convex, G quasi-complete. Let E_0 resp. F_0 be a dense subspace of E resp. F and $\mathfrak{M}$ resp. $\mathfrak{N}$ a class of bounded subsets of E_0 resp. F_0 which cover E_0 resp. F_0 and with the property that the class $\bar{\mathfrak{M}}$ resp. $\bar{\mathfrak{N}}$ of the closures $\bar{M}$ resp. $\bar{N}$ in E resp. F covers E resp. F.*

Then every bilinear $(\mathfrak{M}, \mathfrak{N})$-hypocontinuous bilinear mapping B of $E_0 \times F_0$ in G has a uniquely determined $(\overline{\mathfrak{M}}, \overline{\mathfrak{N}})$-hypocontinuous extension $\bar{B}$ on $E \times F$.

A corresponding statement is true for $(\mathfrak{M}, \mathfrak{N})$-equihypocontinuity.

Proof of (4). We apply (3) and obtain an $\overline{\mathfrak{M}}$-hypocontinuous extension $\bar{B}$ defined on $E \times F_0$. B is $\mathfrak{N}$-hypocontinuous; thus for $N \in \mathfrak{N}$ and a closed $W \supset \mathfrak{o}$ in G there exists $U \supset \mathfrak{o}$ in E_0 such that $\underline{B(U, N)} \subset W$. By taking limits of bounded nets in E_0 we obtain an $(\overline{\mathfrak{M}}, \overline{\mathfrak{N}})$-hypocontinuous extension $\bar{\bar{B}}$ of $\bar{B}$ which satisfies our statement.

If we extend B first to $E_0 \times F$ and then to $E \times F$, we obtain the same bilinear mapping, since by (1) the extension is uniquely determined.

We discuss a result similar to (3) which appears as Lemma C in GROTHENDIECK [13], p. 26. It concerns the extension of a separately weakly continuous bilinear form from $E \times F$ to $E'' \times F$.

The topology on the bidual E'' will be $\mathfrak{T}_n$, the topology of uniform convergence on the equicontinuous subsets of E' (§ 23, 4.); the weak topology on E'' will be $\mathfrak{T}_s(E')$.

(5) *Let E, F be locally convex. Then*

a) *a hypocontinuous bilinear form B on $E \times F$ is separately weakly continuous and has a uniquely determined separately weakly continuous extension $\hat{B}$ to $E'' \times F$,*

b) *if H is an equihypocontinuous set of bilinear forms on $E \times F$, then $\hat{H} = \{\hat{B}; B \in H\}$ is $(\overline{\mathfrak{B}}, \mathfrak{B})$-equihypocontinuous on $E'' \times F$, where $\overline{\mathfrak{B}}$ is the class of the weak closures in E'' of the bounded sets in E,*

c) *if H is an equicontinuous set of bilinear forms on $E \times F$, then $\hat{H}$ is equicontinuous on $E'' \times F$.*

Proof. If B is separately continuous on $E \times F$, then by 1.(2') $B_x \in F'$ and $B_y \in E'$ and thus B is separately weakly continuous on $E \times F$. Since E is $\mathfrak{T}_s(E')$-dense in E'', B_y has a $\mathfrak{T}_s(E')$-continuous extension $\hat{B}_y$ to E''. Every $z \in E''$ is the $\mathfrak{T}_s(E')$-limit of a bounded net $x_\alpha \in E$ (§ 23, 2.(3)); therefore $\hat{B}_y z = \lim_\alpha B_y x_\alpha$. Hence $\hat{B}(z, y)$ is defined by $\lim_\alpha B(x_\alpha, y)$ on $E'' \times F$ and is $\mathfrak{T}_s(E')$-continuous on E''.

Now we use hypocontinuity. For M bounded in E there exists $V \ni \mathfrak{o}$ in F such that $|B(M, V)| \leqq 1$. If $\overline{M}$ denotes the $\mathfrak{T}_s(E')$-closure of M in E'', it follows from the weak continuity on E'' that $|\hat{B}(\overline{M}, V)| \leqq 1$. If $z \in \overline{M}$, then $|\hat{B}(z, V)| \leqq 1$, $\hat{B}_z \in F'$; hence $\hat{B}_z$ is $\mathfrak{T}_s(F')$-continuous on F. Since the $\overline{M}$ cover E'' by § 23, 2.(3), $\hat{B}$ is separately weakly continuous. That $\hat{B}$ is uniquely determined follows from (1); hence a) is proved.

If H is equihypocontinuous, there exists $V \ni \mathfrak{o}$ in F such that $|H(M, V)| \leqq 1$ for a given bounded subset M of E and $U \ni \mathfrak{o}$ in E such that $|H(U, N)| \leqq 1$, N a given bounded subset of F. By a) follows $|\hat{H}(\overline{M}, V)| \leqq 1$ and $|\hat{H}(\overline{U}, N)| \leqq 1$, which proves b).

Finally, $|\hat{H}(\overline{U}, V)| \leqq 1$ follows from $|H(U, V)| \leqq 1$ by a). Since $\overline{U} = U^{\circ\circ}$ if U is absolutely convex in E, this proves the equicontinuity of $\hat{H}$ on $E'' \times F$, the topology on E'' being $\mathfrak{T}_n$.

It is in general not possible to extend $\hat{B}$ from $E'' \times F$ to $E'' \times F''$ so that it remains separately weakly continuous. We consider the following example (GROTHENDIECK).

Let B be the canonical bilinear form ux on $E' \times E$, E a (B)-space, E' its strong dual. B is continuous on $E' \times E$, hence separately weakly continuous in the sense of $\mathfrak{T}_s(E'')$ on E' and $\mathfrak{T}_s(E')$ on E. The extension $\hat{B}$ to $E' \times E''$ according to (5) coincides with the canonical bilinear form on $E' \times E''$ and is continuous on $E' \times E''$ and separately continuous on $E'[\mathfrak{T}_s(E'')] \times E''[\mathfrak{T}_s(E')]$.

The problem is now to extend $\hat{B}$ to $E''' \times E''$ in such a way that the extension is again separately weakly continuous in the sense of $\mathfrak{T}_s(E'')$ and $\mathfrak{T}_s(E')$. Since the $\mathfrak{T}_s(E'')$-continuous extension of $\hat{B}_z = z$ from E' to E''' is uniquely determined, the only possible extension of $\hat{B}$ is the canonical bilinear form $\hat{\hat{B}}(w, z) = wz$ on $E''' \times E''$. But $\hat{\hat{B}}_w = w \in E'''$ is $\mathfrak{T}_s(E')$-continuous on E'' if and only if $w \in E'$.

Hence our problem has a negative answer except in the trivial case where E is reflexive.

It is interesting to see what happens when we reverse the order of extensions. We obtain first by (5) the separately continuous extension $\check{B}(w, x)$ of $B(u, x) = ux$ to $E'''[\mathfrak{T}_s(E'')] \times E[\mathfrak{T}_s(E')]$. If $u_\alpha \in E'$ $\mathfrak{T}_s(E'')$ converges to w in E'''; then $\check{B}(w, x) = \lim_\alpha u_\alpha x = wx$. We recall (§ 31, 1.(10)) that $E''' = E' \oplus E^\perp$. Let P be the continuous projection of E''' onto E', then $wx = (Pw, x)$ and thus $\check{B}(w, x) = (Pw)x$. The only possible extension of $\check{B}$ to $E''' \times E''$ is then $\check{\check{B}}(w, z) = (Pw)z$. Obviously, $\check{\check{B}}$ coincides with $\hat{B}$ on $E' \times E''$ but $\check{\check{B}}$ is not continuous on $E'''[\mathfrak{T}_s(E'')] \times E''[\mathfrak{T}_s(E')]$, since there exists no such extension of $\hat{B}$ as we have seen.

Nevertheless, the extension from $E \times F$ to $E'' \times F''$ is possible for a large class of bilinear forms on (B)-spaces E, F. A mapping $A \in \mathfrak{L}(E, F)$ is called w e a k l y c o m p a c t if $A(K)$ is relatively weakly compact in F, where K is the closed unit ball of E. Let B be a continuous bilinear form on $E \times F$. Then $\tilde{B} \in \mathfrak{L}(E, F'_b)$. We say that B is w e a k l y c o m p a c t if $\tilde{B}$ is weakly compact. This means that $\tilde{B}(K)$ is relatively $\mathfrak{T}_s(F'')$-compact in F'.

(6) *A continuous and weakly compact bilinear form B on the product $E \times F$ of two* (B)*-spaces E, F has a uniquely determined extension $\hat{\hat{B}}$ which is separately continuous on* $E''[\mathfrak{T}_s(E')] \times F''[\mathfrak{T}_s(F')]$.

P r o o f. We have $B(x, y) = (\tilde{B}x)y$ for all $x \in E$, $y \in F$, where $\tilde{B} \in \mathfrak{L}(E, F'_b)$. By § 32, 2.(6) the double adjoint $\tilde{B}'' \in \mathfrak{L}(E'', F'''_b)$ is the uniquely determined $\mathfrak{T}_s(E')$–$\mathfrak{T}_s(F'')$-continuous extension of $\tilde{B}$ to E''.

We define $\hat{\hat{B}}(z, t) = (\tilde{B}''z)t$ for $z \in E''$, $t \in F''$. If z_α $\mathfrak{T}_s(E')$-converges to $z_0 \in E''$, then $\tilde{B}''z_\alpha$ $\mathfrak{T}_s(F'')$-converges to $\tilde{B}''z_0 \in F''$; hence $\hat{\hat{B}}(z, t)$ is $\mathfrak{T}_s(E')$-continuous in z for fixed $t \in F''$.

Let K be the closed unit ball in E; then by assumption $\tilde{B}(K) \subset C$, where C is $\mathfrak{T}_s(F'')$-compact in F'. The unit ball of E'' is the $\mathfrak{T}_s(E')$-closure $\bar{K}$ of K in E'' and so $\tilde{B}''(\bar{K}) \subset \overline{\tilde{B}(K)} \subset C$ and therefore $\tilde{B}''(E'') \subset F'$. Hence if $z \in E''$ is fixed, $\hat{\hat{B}}(z, t) = (\tilde{B}''z)t$ is $\mathfrak{T}_s(F')$-continuous in t.

The uniqueness of $\hat{\hat{B}}$ is obvious.

4. Locally convex spaces of bilinear mappings. We introduced different spaces of bilinear mappings. There exist natural topologies on these spaces which we will now consider.

Let E, F, G be locally convex and $\mathfrak{B}(E \times F, G)$ the space of separately continuous bilinear mappings of $E \times F$ in G. Let $\mathfrak{M}$ resp. $\mathfrak{N}$ be a class of bounded subsets of E resp. F which covers E resp. F and satisfies condition b) of § 39, 1. Let $M \in \mathfrak{M}$, $N \in \mathfrak{N}$, and let W be a neighbourhood of $\circ$ in G. Then we define $\mathscr{U}(M, N, W)$ as the set $\{B \in \mathfrak{B}(E \times F, G); B(M, N) \subset W\}$. Obviously, $\mathscr{U}(M, N, W)$ is absolutely convex if W is absolutely convex. The intersection $\mathscr{U}(M_1, N_1, W_1) \cap \mathscr{U}(M_2, N_2, W_2)$ contains $\mathscr{U}(M_1 \cup M_2, N_1 \cup N_2, W_1 \cup W_2)$. From this it follows easily that the class of all $\mathscr{U}(M, N, W)$ is a neighbourhood base of $\circ$ of a topology $\mathfrak{T}_{\mathfrak{M},\mathfrak{N}}$ defined on $\mathfrak{B}(E \times F, G)$ by endowing each $B_0 \in \mathfrak{B}(E \times F, G)$ with the neighbourhoods $B_0 + \mathscr{U}(M, N, W)$. $\mathfrak{T}_{\mathfrak{M},\mathfrak{N}}$ is Hausdorff: If $B_0 \neq \circ$, then there exists $(x, y) \in E \times F$ such that $B_0(x, y) = w \neq \circ$. If $w \notin W$, then $B_0 \notin \mathscr{U}(M, N, W)$ for an $M \ni x$ and an $N \ni y$.

We write $\mathfrak{B}_{\mathfrak{M},\mathfrak{N}}(E \times F, G)$ for $\mathfrak{B}(E \times F, G)[\mathfrak{T}_{\mathfrak{M},\mathfrak{N}}]$. If $\mathfrak{M}$ and $\mathfrak{N}$ are the classes $\mathfrak{F}$ of all finite subsets, we call the topology $\mathfrak{T}_{\mathfrak{F},\mathfrak{F}}$ the simple topology $\mathfrak{T}_s$ and write also $\mathfrak{B}_s(E \times F, G)$. If $\mathfrak{M}$ resp. $\mathfrak{N}$ is the class $\mathfrak{B}$ of all bounded subsets of E resp. F, then $\mathfrak{T}_{\mathfrak{B},\mathfrak{B}} = \mathfrak{T}_b$ is the bibounded topology and we write $\mathfrak{B}_b(E \times F, G)$.

We remark that $\mathfrak{T}_{\mathfrak{M},\mathfrak{N}}$ remains unchanged if we replace $\mathfrak{M}, \mathfrak{N}$ by their saturated covers $\tilde{\mathfrak{M}}, \tilde{\mathfrak{N}}$.

Unfortunately, $\mathfrak{T}_{\mathfrak{M},\mathfrak{N}}$ is in general not locally convex. By § 18, 1.(1) this depends on whether the sets $\mathscr{U}(M, N, W)$ are all absorbent or not. Assume that for $B_0 \in \mathfrak{B}(E \times F, G)$ the set $B_0(M, N)$ is not bounded in G; then there exists a neighbourhood $W \supset \circ$ in G such that $B_0(M, N)$ is not contained in any multiple of W; thus $\mathscr{U}(M, N, W)$ does not absorb B_0. Conversely, if $B_0(M, N)$ is bounded in G, $\mathscr{U}(M, N, W)$ contains a multiple of B_0. Therefore

(1) *A subspace $\mathfrak{X}$ of $\mathfrak{B}_{\mathfrak{M},\mathfrak{N}}(E \times F, G)$ is locally convex if and only if the sets $B(M, N)$ are bounded in G for every $B \in \mathfrak{X}$ and every $M \in \mathfrak{M}$ and $N \in \mathfrak{N}$.*

The problem can be reduced to bilinear forms.

(2) *$\mathfrak{B}_{\mathfrak{M},\mathfrak{N}}(E \times F, G)$ is locally convex if and only if $\mathfrak{B}_{\mathfrak{M},\mathfrak{N}}(E \times F)$ is locally convex.*

Proof. Assume that $\mathfrak{B}_{\mathfrak{M},\mathfrak{N}}(E \times F)$ is locally convex. Let B be any element of $\mathfrak{B}(E \times F, G)$; then $wB \in \mathfrak{B}(E \times F)$ for every $w \in G'$. The set $wB(M, N)$ is bounded in K by (1) for every $M \in \mathfrak{M}$, $N \in \mathfrak{N}$. Hence $B(M, N)$ is weakly bounded or bounded in G. Again by (1) $\mathfrak{B}_{\mathfrak{M},\mathfrak{N}}(E \times F, G)$ is locally convex.

Conversely, if $z_0 \neq o$ is a fixed element of G and $B \in \mathfrak{B}(E \times F)$, then the correspondence $B \to Bz_0$ maps $\mathfrak{B}_{\mathfrak{M},\mathfrak{N}}(E \times F)$ isomorphically onto a subspace of $\mathfrak{B}_{\mathfrak{M},\mathfrak{N}}(E \times F, G)$ and, if this space is locally convex, it follows that $\mathfrak{B}_{\mathfrak{M},\mathfrak{N}}(E \times F)$ is locally convex.

Using previous results on spaces of linear mappings we obtain sufficient conditions for local convexity.

(3) *$\mathfrak{B}_{\mathfrak{M},\mathfrak{N}}(E \times F, G)$ is locally convex if* a) *$\mathfrak{M}$ or $\mathfrak{N}$ consists only of strongly bounded subsets, or* b) *if the closed absolutely convex bounded subsets of $\tilde{\mathfrak{M}}$ or $\tilde{\mathfrak{N}}$ are Banach disks, or* c) *if E or F is locally or sequentially complete.*

We have only to show that a), the weakest condition, is sufficient. By (2) we need only consider $\mathfrak{B}_{\mathfrak{M},\mathfrak{N}}(E \times F)$. Now $\mathfrak{B}(E \times F)$ is by 1.(2′) algebraically isomorphic to $\mathfrak{L}(E, F'_s)$ and this space is identical to $\mathfrak{L}(E_s, F'_s)$. One checks immediately that by this isomorphism $B \to \tilde{B}$, the topology on $\mathfrak{L}(E_s, F'_s)$ corresponding to $\mathfrak{T}_{\mathfrak{M},\mathfrak{N}}$, is the topology $\mathfrak{T}_{\mathfrak{M},\mathfrak{N}}$ introduced in § 39, 1. From § 39, 1.(7) it follows now that a) is sufficient.

We consider a special case which will be needed later. We denote by $\mathfrak{E}$ the class of equicontinuous subsets of the dual E' of a locally convex space E. The corresponding equicontinuous resp. bi-equicontinuous topology $\mathfrak{T}_{\mathfrak{E}}$ resp. $\mathfrak{T}_{\mathfrak{E},\mathfrak{E}}$ will be denoted by $\mathfrak{T}_e$ and, correspondingly, we will write $\mathfrak{L}_e(E', F)$ resp. $\mathfrak{B}_e(E'_s \times F'_s, G)$ for the space endowed with this topology. From (3) we obtain immediately

(4) *Let E, F, G be locally convex. Then $\mathfrak{B}_e(E'_s \times F'_s, G)$ is locally convex.*

In the case of bilinear forms we have

(5) *Let E, F be locally convex. Then $\mathfrak{B}_e(E'_s \times F'_s)$ is topologically isomorphic to $\mathfrak{L}_e(E'_k, F)$ and $\mathfrak{B}_e(E'_s \times F'_s)$ is complete if and only if E and F are complete.*

Proof. $\mathfrak{B}(E'_s \times F'_s)$ is by 1.(2′) algebraically isomorphic to $\mathfrak{L}(E'_s, F_s)$ and this space is identical with $\mathfrak{L}(E'_k, F)$. That the topologies correspond follows from the equivalence of $|B(M, N)| \leqq 1$ and $\tilde{B}(M) \subset N^\circ$ for equicontinuous absolutely convex M, N.

The statement on completeness follows from § 39, 6.(2a), (2b), (3).

Let us now consider spaces of hypocontinuous and continuous bilinear mappings. Since these are subspaces of the spaces of separately continuous bilinear mappings, we expect better results.

(6) *Let E, F, G be locally convex and let $\mathfrak{M}$ resp. $\mathfrak{N}$ be a class of bounded subsets of E resp. F which covers E resp. F. Then the spaces*

$$\mathfrak{X}^{(\mathfrak{M})}_{\mathfrak{M},\mathfrak{N}}(E \times F, G), \qquad \mathfrak{X}_{\mathfrak{M},\mathfrak{N}}(E \times F, G), \qquad \mathscr{B}_{\mathfrak{M},\mathfrak{N}}(E \times F, G)$$

are always locally convex.

It is enough to prove the first case.

But by the definition of $\mathfrak{M}$-hypocontinuity the set $B(M, N)$, $M \in \mathfrak{M}$, $N \in \mathfrak{N}$, is bounded in G for every $B \in \mathfrak{X}^{(\mathfrak{M})}(E \times F, G)$ and by (1) $\mathfrak{X}^{(\mathfrak{M})}_{\mathfrak{M},\mathfrak{N}}(E \times F, G)$ is locally convex.

(7) $\mathfrak{X}^{(\mathfrak{M})}_{\mathfrak{M},\mathfrak{N}}(E \times F, G)$ *is topologically isomorphic to* $\mathfrak{L}_{\mathfrak{N}}(F, \mathfrak{L}_{\mathfrak{M}}(E, G))$.

We proved in 1.(3) a) that $\mathfrak{X}^{(\mathfrak{M})}(E \times F, G)$ and $\mathfrak{L}(F, \mathfrak{L}_{\mathfrak{M}}(E, G))$ are algebraically isomorphic by the correspondence $B \to \tilde{\tilde{B}}$. The neighbourhood $\mathscr{U}(M, N, W)$ of $\circ$ in $\mathfrak{X}^{(\mathfrak{M})}_{\mathfrak{M},\mathfrak{N}}(E \times F, G)$ consists of all B such that $B(M, N) \subset W$. The corresponding set in $\mathfrak{L}(F, \mathfrak{L}_{\mathfrak{M}}(E, G))$ consists of all $\tilde{\tilde{B}}$ such that $\tilde{\tilde{B}}(N)(M) \subset W$ and this is the neighbourhood $\mathscr{U}(N, U(M, W))$ of $\circ$ in $\mathfrak{L}_{\mathfrak{N}}(F, \mathfrak{L}_{\mathfrak{M}}(E, G))$. This implies the assertion.

Remark. By 1.(2) $\mathfrak{B}(E \times F, G)$ is algebraically isomorphic to $\mathfrak{L}(F, \mathfrak{L}_s(E, G))$ and by the same procedure the topology $\mathfrak{T}_{\mathfrak{M},\mathfrak{N}}$ can be carried over from $\mathfrak{B}(E \times F, G)$ to $\mathfrak{L}(F, \mathfrak{L}_s(E, G))$ and we denote it again by $\mathfrak{T}_{\mathfrak{M},\mathfrak{N}}$. In general, $\mathfrak{L}_{\mathfrak{N}}(F, \mathfrak{L}_{\mathfrak{M}}(E, G))$ will be a proper subspace of $\mathfrak{L}_{\mathfrak{M},\mathfrak{N}}(F, \mathfrak{L}_s(E, G))$.

As a consequence of (7) we obtain

(8) *Let E, F be barrelled, G quasi-complete. Then $\mathfrak{B}(E \times F, G) = \mathfrak{X}(E \times F, G)$ and $\mathfrak{B}_{\mathfrak{M},\mathfrak{N}}(E \times F, G)$ is locally convex and quasi-complete for $\mathfrak{M}$ resp. $\mathfrak{N}$ covering E resp. F.*

Proof. One has always $\mathfrak{B}(E \times F, G) \supset \mathfrak{X}^{(\mathfrak{M})}(E \times F, G) \supset \mathfrak{X}(E \times F, G)$. From 2.(5) it follows that the three spaces coincide for E and F barrelled. $\mathfrak{X}^{(\mathfrak{M})}_{\mathfrak{M},\mathfrak{N}}(E \times F, G)$ is isomorphic to $\mathfrak{L}_{\mathfrak{N}}(F, \mathfrak{L}_{\mathfrak{M}}(E, G))$ by (7). The assertion follows by applying twice § 39, 6.(5).

The BANACH–MACKEY theorem is true for bilinear mappings in the following version.

(9) *Let E and F be sequentially or locally complete and G locally convex. Then every simply bounded subset H of $\mathfrak{B}(E \times F, G)$ is $\mathfrak{T}_{\mathfrak{M},\mathfrak{N}}$-bounded for the arbitrary class $\mathfrak{M}$ resp. $\mathfrak{N}$ of bounded subsets of E resp. F which covers E resp. F.*

Proof. a) We remark that by (3) c) $\mathfrak{B}_{\mathfrak{M},\mathfrak{N}}(E \times F, G)$ is locally convex. We reduce the problem to the case of bilinear forms. If W is an absolutely convex closed neighbourhood of $\mathfrak{o}$ in G, then for given $M \in \mathfrak{M}$, $N \in \mathfrak{N}$ the statements $H(M, N) \subset W$ and $|W^\circ H(M, N)| \leqq 1$ are equivalent. This means that H is $\mathfrak{T}_{\mathfrak{M},\mathfrak{N}}$-bounded in $\mathfrak{B}(E \times F, G)$ if and only if all sets QH, Q equicontinuous in G', are $\mathfrak{T}_{\mathfrak{M},\mathfrak{N}}$-bounded in $\mathfrak{B}(E \times F)$.

Hence, if all $\mathfrak{T}_s$-bounded subsets of $\mathfrak{B}(E \times F)$ are $\mathfrak{T}_{\mathfrak{M},\mathfrak{N}}$-bounded, the same is true for the subsets of $\mathfrak{B}(E \times F, G)$.

b) Let H be simply bounded in $\mathfrak{B}(E \times F)$. Now $\mathfrak{B}(E \times F)$ is algebraically isomorphic to $\mathfrak{L}(E, F'_s) = \mathfrak{L}(E_s, F'_s)$ and $\tilde{H}$, the subset of $\mathfrak{L}(E_s, F'_s)$ corresponding to H, is simply bounded in $\mathfrak{L}(E_s, F'_s)$. As we saw in the proof of (3), $\mathfrak{B}_{\mathfrak{M},\mathfrak{N}}(E \times F)$ is topologically isomorphic to $\mathfrak{L}_{\mathfrak{M},\mathfrak{N}}(E_s, F'_s)$. Now § 39, 2.(8) implies that $\tilde{H}$, and therefore H, is $\mathfrak{T}_{\mathfrak{M},\mathfrak{N}}$-bounded.

We close with a result of GROTHENDIECK [10] on (DF)-spaces.

(10) *Let E, F be* (DF)*-spaces, G locally convex, H a subset of* $\mathfrak{B}(E \times F, G)$. *If H is the union of a sequence of equicontinuous sets H_n and if H is bounded for the bi-bounded topology, then H itself is equicontinuous.*

Proof. By using 2.(8) and part a) of the proof of (9) one reduces (10) to the case of bilinear forms. By 2.(10) it is sufficient to prove that H is equihypocontinuous.

Let $\tilde{H}_n$ be the corresponding subset of $\mathfrak{L}(E, F'_b)$. If M is bounded in E, then the equicontinuity of H_n implies that $\tilde{H}_n(M)$ is equicontinuous in F'. From the assumption that H is bounded in $\mathfrak{B}_b(E \times F)$ it follows that $\tilde{H}(M)$ is strongly bounded in F'. Hence, by the definition of a (DF)-space, $\tilde{H}(M) = \bigcup_{n=1}^{\infty} \tilde{H}_n(M)$ is equicontinuous in F'; thus there exists $V \ni \mathfrak{o}$ in F such that $|H(M, V)| \leqq 1$. Therefore H is $\mathfrak{B}$-equihypocontinuous in the first variable. By repeating this argument for the second variable we obtain that H is equihypocontinuous.

5. Applications. Locally convex algebras. So far we have considered $\mathfrak{L}(E) = \mathfrak{L}(E, E)$ only as a vector space and $\mathfrak{L}_{\mathfrak{M}}(E)$ as a locally convex space. If we introduce the product or composition BA of two mappings as a further operation in $\mathfrak{L}(E)$, then $\mathfrak{L}(E)$ becomes an algebra over K with unit element I as in the case of normed spaces (§ 14, 6.). Obviously, the composition BA can be considered as a bilinear mapping of $\mathfrak{L}(E) \times \mathfrak{L}(E)$ into $\mathfrak{L}(E)$. If E is a normed space, then this mapping is continuous on $\mathfrak{L}_b(E) \times \mathfrak{L}_b(E)$ (§ 14, 6.(7)). But this will not be true in the general case $\mathfrak{L}_{\mathfrak{M}}(E)$, E any locally convex space.

We introduce the following generalization of the notion of a normed algebra. A real or complex algebra A endowed with a topology $\mathfrak{T}$ is said to be locally convex if the underlying vector space $A[\mathfrak{T}]$ is locally convex and if the multiplication yx of two elements is separately continuous.

With this definition we obtain

(1) *Let E be locally convex and $\mathfrak{M}$ a saturated class of bounded subsets covering E such that $A(\mathfrak{M}) \subset \mathfrak{M}$ for every $A \in \mathfrak{L}(E)$. Then $\mathfrak{L}_{\mathfrak{M}}(E)$ is a locally convex algebra with unit element.*

We have to prove that the product BA is separately continuous. Let W be the neighbourhood $W(M, V)$ in $\mathfrak{L}_{\mathfrak{M}}(E)$, $M \in \mathfrak{M}$, and V a neighbourhood of $\mathfrak{o}$ in E. Let B be fixed and $U \ni \mathfrak{o}$ such that $B(U) \subset V$. If A is contained in $W(M, U)$ or $A(M) \subset U$, then $BA(M) \subset V$ or $BA \in W(M, V)$. Hence BA is continuous in the first variable.

Now let A be fixed and $A(M) = M_1 \in \mathfrak{M}$. If $B \in W(M_1, V)$, then $BA \in W(M, V)$ and this is the continuity in the second variable.

As a consequence of (1), we see that for every locally convex E the corresponding algebras $\mathfrak{L}_s(E)$, $\mathfrak{L}_k(E)$, $\mathfrak{L}_{b^*}(E)$, and $\mathfrak{L}_b(E)$ are locally convex (for the definitions compare § 39, 1.); the case $\mathfrak{L}_{b^*}(E)$ follows from § 32, 2.(3).

It is possible to develop spectral theory in a locally convex algebra $A[\mathfrak{T}]$ (compare Neubauer [1'], [2']; Waelbroeck [1'], [2']) if in A multiplication is bounded, i.e., if the product $MN = \{yx; y \in M, x \in N\}$ of two bounded subsets of $A[\mathfrak{T}]$ is always a bounded subset of $A[\mathfrak{T}]$. One has the following sufficient condition:

(2) *If the locally convex algebra $A[\mathfrak{T}]$ is locally or sequentially complete, then multiplication in $A[\mathfrak{T}]$ is bounded.*

Proof. From the assumption and 4.(3) c) it follows that $\mathfrak{B}_b(A \times A, A)$ is locally convex and from 4.(1) that for every separately continuous bilinear mapping B the sets $B(M, N)$ are bounded, where M and N are bounded subsets of A. This is true in particular for $B(y, x) = yx$.

Neubauer ([1'], 8.8) gives an example of a commutative locally convex algebra which contains a sequence x_n converging to $\mathfrak{o}$ such that x_n^2 is unbounded.

If the bilinear mapping yx is sequentially continuous, i.e., if $y_n \to \mathfrak{o}$ and $x_n \to \mathfrak{o}$ imply $y_n x_n \to \mathfrak{o}$ in $A[\mathfrak{T}]$, then multiplication in $A[\mathfrak{T}]$ is bounded. This follows immediately from § 15, 6.(3). Also one sees easily that the boundedness of multiplication is equivalent to: If $x_n \to \mathfrak{o}$ and $y_n \to \mathfrak{o}$, then $y_n x_n$ is bounded.

From 2.(1) and 2.(11) we obtain

(3) *If the locally convex algebra $A[\mathfrak{T}]$ is barrelled and metrizable or barrelled and a* (DF)*-space, then multiplication in $A[\mathfrak{T}]$ is continuous.*

In the general metrizable case one has

(4) *If $A[\mathfrak{T}]$ is a metrizable locally convex algebra with bounded multiplication, then multiplication in $A[\mathfrak{T}]$ is continuous.*

It is sufficient to prove that yx is sequentially continuous. We assume the converse. Then there exist $x_n \to o$, $y_n \to o$ such that $y_n x_n \notin V$, where V is some neighbourhood of o in A. By § 28, 3.(1) there exists a sequence $\rho_n \to \infty$, $\rho_n > 0$ such that $\rho_n x_n \to o$. But then $\rho_n y_n x_n \notin \rho_n V$ and this contradicts the assumption that multiplication in A is bounded.

After these remarks on locally convex algebras we come back to multiplication of mappings. One has the following generalization of (1):

(5) *Let E, F, G be locally convex. The product BA, $B \in \mathfrak{L}(F, G)$, $A \in \mathfrak{L}(E, F)$ is a separately continuous bilinear mapping of $\mathfrak{L}_{\mathfrak{M}_2}(F, G) \times \mathfrak{L}_{\mathfrak{M}_1}(E, F)$ in $\mathfrak{L}_{\mathfrak{M}_1}(E, G)$ if $A(\mathfrak{M}_1) \subset \mathfrak{M}_2$ for every $A \in \mathfrak{L}(E, F)$; $\mathfrak{M}_1$ resp. $\mathfrak{M}_2$ is a class of bounded sets covering E resp. F.*

The proof of (1) can be immediately adapted to this more general situation.

From (5) and the remarks after (1) it follows that in many important cases the multiplication BA is separately continuous.

We give an example to show that sequential continuity of BA is a more difficult problem.

Let l_s^2 be a Hilbert space in its weak topology. Then $\mathfrak{L}_s(l_s^2)$ is a locally convex algebra by (1). Let $C_{ik} = (c_{ik})$ be the infinite matrix with $c_{ik} = 1$ and $c_{jl} = 0$ for $(j, l) \neq (i, k)$. Then C_{1n} and C_{n1} converge to o in $\mathfrak{L}_s(l_s^2)$, but the product $C_{1n}C_{n1} = C_{11}$ does not.

As the remark at the end of 1. shows, sequential continuity of the product will follow from hypocontinuity properties of the bilinear mapping BA. We note the following general result (GROTHENDIECK [11]).

(6) *We assume the situation described in* (5). *Let $\mathfrak{P}$ be the class of all subsets H of $\mathfrak{L}(E, F)$ such that $H(M) \in \mathfrak{M}_2$ for every $M \in \mathfrak{M}_1$. Then BA is $(\mathfrak{E}, \mathfrak{P})$-hypocontinuous, where $\mathfrak{E}$ is the class of all equicontinuous subsets of $\mathfrak{L}(F, G)$.*

Proof. a) We show first that BA is $\mathfrak{E}$-hypocontinuous. Let $\mathscr{W}$ be a neighbourhood of o in $\mathfrak{L}_{\mathfrak{M}_1}(E, G)$ and let Q be an equicontinuous subset of $\mathfrak{L}(F, G)$. We must find a neighbourhood $\mathscr{U}$ of o in $\mathfrak{L}_{\mathfrak{M}_1}(E, F)$ such that $Q\mathscr{U} \subset \mathscr{W}$, that is, $BA \in \mathscr{W}$ for all $B \in Q$ and all $A \in \mathscr{U}$.

Let $\mathscr{W}$ be $\mathscr{W}(M, W)$, where $M \in \mathfrak{M}_1$ and W is a neighbourhood of o in G. Since Q is equicontinuous in $\mathfrak{L}(F, G)$, there exists $V \ni o$ in F such that $Q(V) \subset W$. Define $\mathscr{U} = \mathscr{U}(M, V)$. Then for all $A \in \mathscr{U}$ and all $B \in Q$ we have $B(A(M)) \subset B(V) \subset W$. So $BA \in \mathscr{W}$ and $Q\mathscr{U} \subset \mathscr{W}$.

b) BA is $\mathfrak{P}$-hypocontinuous in the second variable. Let $\mathscr{W}$ be defined as before and $H \in \mathfrak{P}$. We have to find a neighbourhood $\mathscr{V}$ of $\circ$ in $\mathfrak{L}_{\mathfrak{M}_2}(F, G)$ such that $\mathscr{V}H \subset \mathscr{W}$. Now $H(M) = N \in \mathfrak{M}_2$ and we define $\mathscr{V}$ as $\mathscr{V}(N, W)$. Then for $B \in \mathscr{V}$ and $A \in H$ we have $BA(M) \subset B(N) \subset W$, so $\mathscr{V}H \subset \mathscr{W}$.

We have the following corollaries:

(7) a) *If* $\mathfrak{M}_1 = \mathfrak{M}_2 = \mathfrak{F}$, *the class of all finite subsets, then* BA *is* $\mathfrak{E}$-*hypocontinuous from* $\mathfrak{L}_s(F, G) \times \mathfrak{L}_s(E, F)$ *in* $\mathfrak{L}_s(E, G)$;

b) *If* $\mathfrak{M}_1 = \mathfrak{M}_2 = \mathfrak{B}$, *the class of all bounded subsets, then* BA *is* $(\mathfrak{E}, \mathfrak{N})$-*hypocontinuous from* $\mathfrak{L}_b(F, G) \times \mathfrak{L}_b(E, F)$ *in* $\mathfrak{L}_b(E, G)$, *where* $\mathfrak{N}$ *is the class of all bounded subsets of* $\mathfrak{L}_b(E, F)$;

c) *If* $\mathfrak{M}_1 = \mathfrak{M}_2 = \mathfrak{C}$, *the class of all precompact resp. all compact subsets, then* BA *is* $(\mathfrak{E}, \mathfrak{N})$-*hypocontinuous from* $\mathfrak{L}_{\mathfrak{C}}(F, G) \times \mathfrak{L}_{\mathfrak{C}}(E, F)$ *in* $\mathfrak{L}_{\mathfrak{C}}(E, G)$, *where* $\mathfrak{N}$ *is the class of all precompact resp. compact subsets of* $\mathfrak{L}_{\mathfrak{C}}(E, F)$.

Proof. a) If H is a finite subset of $\mathfrak{L}(E, F)$ and M a finite subset of E, then $H(M)$ is a finite subset of F; thus $H(M) \in \mathfrak{M}_2$. The statement follows now from (6) since $\mathfrak{F} \subset \mathfrak{P}$.

b) If H is a bounded subset of $\mathfrak{L}_b(E, F)$, then for any bounded subset M of E the set $H(M) = \bigcup_{A \in H} A(M)$ is a bounded subset of F, so we have again a special case of (6).

c) If H is a precompact resp. compact subset of $\mathfrak{L}_{\mathfrak{C}}(E, F)$ and M is a precompact resp. compact subset of E, then $H \times M$ is precompact resp. compact. The bilinear mapping $(A, x) \to Ax$ of $\mathfrak{L}_{\mathfrak{C}}(E, F) \times E$ in F is $\mathfrak{C}$-hypocontinuous in the second variable. Therefore $(A, x) \to Ax$ is continuous on $\mathfrak{L}_{\mathfrak{C}}(E, F) \times M$ by 1.(5) a). Hence $H(M)$ is precompact resp. compact in F. Thus $\mathfrak{N} \subset \mathfrak{P}$ and (6) proves the statement.

We are now able to deduce sequential continuity of BA by using 1.(5) a). If Q is an equicontinuous subset of $\mathfrak{L}(F, G)$, then BA is continuous on $Q \times \mathfrak{L}_s(E, F)$ by (7) a). If F is barrelled, then every simply bounded subset of $\mathfrak{L}(F, G)$ is equicontinuous (§ 39, 3.(2)). Hence

(8) *Let* E, G *be locally convex,* F *barrelled. If* $A_n \to \circ$ *in* $\mathfrak{L}_s(E, F)$ *and* $B_n \to \circ$ *in* $\mathfrak{L}_s(F, G)$, *then* $B_n A_n \to \circ$ *in* $\mathfrak{L}_s(E, G)$.

Using (7) b) and the continuity of BA on $\mathfrak{L}_b(F, G) \times H$, H a bounded subset of $\mathfrak{L}_b(E, F)$, we obtain

(9) *Let* E, F, G *be locally convex. If* $A_n \to \circ$ *in* $\mathfrak{L}_b(E, F)$ *and* $B_n \to \circ$ *in* $\mathfrak{L}_b(F, G)$, *then* $B_n A_n \to \circ$ *in* $\mathfrak{L}_b(E, G)$.

Similarly, we obtain from (7) c)

(10) *Let E, F, G be locally convex. If* $A_n \to o$ *in* $\mathfrak{L}_{\mathfrak{C}}(E, F)$ *and* $B_n \to o$ *in* $\mathfrak{L}_{\mathfrak{C}}(F, G)$, *then* $B_n A_n \to o$ *in* $\mathfrak{L}_{\mathfrak{C}}(E, G)$.

There exist similar theorems for converging nets.

§ 41. Projective tensor products of locally convex spaces

1. Some complements on tensor products. We defined in § 9, 6. the tensor product $E \otimes F$ of two vector spaces E, F as the quotient Λ/Λ_0, where $\Lambda(E \times F)$ is the space of all formal linear combinations of elements (x, y) of $E \times F$ and Λ_0 is defined as the linear subspace generated by the elements of the form § 9, 6.(1).

We consider the canonical mapping of $\Lambda = \Lambda(E \times F)$ onto $\Lambda/\Lambda_0 = E \otimes F$. Its restriction to $E \times F$ is called the canonical bilinear mapping χ of $E \times F$ into $E \otimes F$ and we have $\chi((x, y)) = x \otimes y$. We remark that $\chi(E \times F)$ generates $E \otimes F$ in the sense that $E \otimes F$ consists of all finite sums of elements of $\chi(E \times F)$.

We have to change some of the notations used in § 9 to conform to the notations introduced in § 39 and § 40. We exemplify this in reformulating the fundamental relation § 9, 7.(2) between bilinear mappings and linear mappings of tensor products.

(1) *Let* E, F, G *be vector spaces,* $B \in B(E \times F, G)$, *the space of all bilinear mappings of* $E \times F$ *in* G. *Then* $B = \dot{B}\chi$, *where* $\dot{B} \in L(E \otimes F, G)$, *the space of all linear mappings of* $E \otimes F$ *in* G. *Conversely, if* $\dot{B} \in L(E \otimes F, G)$, *then* $B = \dot{B}\chi \in B(E \times F, G)$.

The correspondence $B \to \dot{B}$ *is an algebraic isomorphism of* $B(E \times F, G)$. *and* $L(E \otimes F, G)$.

If we combine (1) with § 40, 1.(1), we obtain the identities

$$\text{(2)} \quad \dot{B}(x \otimes y) = B(x, y) = (\tilde{B}x)y = (\tilde{\tilde{B}}y)x, \qquad x \in E,\ y \in F.$$

They define the algebraic isomorphisms

$$\text{(3)} \qquad L(E \otimes F, G) \cong B(E \times F, G) \cong L(E, L(F, G)) \cong L(F, L(E, G)).$$

In the case of bilinear forms we obtain

$$\text{(4)} \qquad (E \otimes F)^* \cong B(E \times F) \cong L(E, F^*) \cong L(F, E^*).$$

We see that the notion of tensor product gives a third possibility of considering a bilinear mapping as a linear mapping.

The tensor product can be characterized by the following universal mapping property.

(5) *Let E, F be fixed vector spaces and H a vector space with the following properties:*

a) *there exists a bilinear mapping χ_1 of $E \times F$ into H such that $\chi_1(E \times F)$ generates H;*

b) *if B is a bilinear mapping of $E \times F$ in a vector space G, then $B = B_1\chi_1$, where $B_1 \in L(H, G)$.*

Then there exists an isomorphism A of $E \otimes F$ onto H such that $\chi_1 = A\chi$, where χ is the canonical bilinear mapping of $E \times F$ in $E \otimes F$.

Proof. It follows from (1) that $E \otimes F$ has properties a) and b). In (1) take $G = H$ and $B = \chi_1$; then there exists $A \in L(E \otimes F, H)$ such that $\chi_1 = A\chi$.

Interchanging the roles of $E \otimes F$ and H, one finds from a), b) an $A_1 \in L(H, E \otimes F)$ such that $\chi = A_1\chi_1$. Therefore $\chi = A_1A\chi$ and $\chi_1 = AA_1\chi_1$. The first relation means that A_1A is the identity on $\chi(E \times F)$. Since $\chi(E \times F)$ generates $E \otimes F$, it follows that $A_1A = I_{E\otimes F}$. Similarly, $AA_1 = I_H$. Thus A and A_1 are isomorphisms and especially $\chi_1 = A\chi$.

In § 9, 7. we introduced the tensor product $A \otimes B$ of two linear mappings $A \in L(E, E_1)$, $B \in L(F, F_1)$. We repeat the definition: The mapping (A, B) defined by

$$(A, B)(x, y) = (Ax) \otimes (By), \qquad x \in E,\ y \in F,$$

is a bilinear mapping of $E \times F$ into $E_1 \otimes F_1$. We denote the corresponding linear mapping of $E \otimes F$ into $E_1 \otimes F_1$ by $A \otimes B$, thus:

$$A \otimes B(x \otimes y) = (Ax) \otimes (By), \qquad x \in E,\ y \in F. \tag{6}$$

We determine the structure of the kernel of $A \otimes B$. Let M be a subspace of E and N a subspace of N. We denote by $D[M, N]$ the subspace of $E \otimes F$ generated by all elements $x \otimes y$, where x is in M or y is in N; $D[M, N] = M \otimes F + E \otimes N$.

(7) *Let A, B be linear mappings of E onto E_1 and F onto F_1, respectively. Then $A \otimes B$ is a linear mapping of $E \otimes F$ onto $E_1 \otimes F_1$. The kernel $N[A \otimes B]$ is the space $D[N_1, N_2]$, where $N_1 = N[A]$, $N_2 = N[B]$.*

Proof. The first statement is trivial. It is also clear from (6) that $N = N[A \otimes B] \supset D[N_1, N_2] = D$. We prove now the converse.

Let K be the canonical mapping of $E \otimes F$ onto $(E \otimes F)/D$. Let x and x' be in the same residue class $\hat{x} \in E/N_1$; then $x \otimes y - x' \otimes y = (x - x') \otimes y \in D$. Thus $K(x \otimes y) = K(x' \otimes y)$. Analogously, $K(x \otimes y) = K(x \otimes y')$ if y and y' are in the same residue class $\hat{y} \in F/N_2$ and we conclude easily that K defines a bilinear mapping $K_0(\hat{x}, \hat{y})$ on $E/N_1 \times F/N_2$.

If $A = \hat{A}K_1$, where K_1 is the canonical mapping of E onto E/N_1, and similarly $B = \hat{B}K_2$, we obtain for $x_1 = \hat{A}\hat{x}$, $y_1 = \hat{B}\hat{y}$ that

$$K(x \otimes y) = K_0(\hat{x}, \hat{y}) = K_0(\hat{A}^{-1}x_1, \hat{B}^{-1}y_1) = B_0(x_1, y_1),$$

where B_0 is a bilinear mapping of $E_1 \times F_1$ onto $(E \otimes F)/D$. But then $B_0(x_1, y_1) = \dot{B}_0(x_1 \otimes y_1)$, $\dot{B}_0 \in L(E_1 \otimes F_1, (E \otimes F)/D)$. It follows that

$$\dot{B}_0(x_1 \otimes y_1) = \dot{B}_0((A \otimes B)(x \otimes y)) = K(x \otimes y).$$

Hence if $z \in N$, then $\dot{B}_0((A \otimes B)z) = \circ = Kz$ or $z \in D[N_1, N_2]$.

We give a second proof: There exist direct decompositions $E = G \oplus N[A]$, $F = H \oplus N[B]$, and for $x \in E$, $y \in F$ we have

$$\begin{aligned} x + y &= (x^1 + x^2) \otimes (y^1 + y^2) \qquad (x^1 \in G, x^2 \in N[A], y^1 \in H, y^2 \in N[B]) \\ &= x^1 \otimes y^1 + [x^1 \otimes y^2 + x^2 \otimes y^1 + x^2 \otimes y^2] \\ &= x^1 \otimes y^1 + t, \qquad t \in D = D[N[A], N[B]]. \end{aligned}$$

Hence $E \otimes F = G \otimes H + D$. This sum will be direct and we will have $N \subset D$ if we prove $(A \otimes B)z \neq \circ$ for every $z \in G \otimes H$, $z \neq \circ$.

z has a representation $z = \sum_{i=1}^{n} x_i \otimes y_i$, $n \geqq 1$, with linearly independent x_i in G and linearly independent y_i in H (§ 9, 6.(8)). Since A and B are one-one on G and H, respectively, the Ax_i in E_1 and the By_i in F_1 are linearly independent and so are the $Ax_i \otimes By_i$ in $E_1 \otimes F_1$ by § 9, 6.(5). Hence $(A \otimes B)z = \sum_{i=1}^{n} Ax_i \otimes By_i \neq \circ$.

We remark

(8) $N[A \otimes B] = D[N[A], N[B]]$ *is true for arbitrary linear mappings* A, B.

We have only to replace E_1, F_1 in (7) by the ranges $R[A]$ and $R[B]$; this does not affect the kernels.

We note the following corollary to (7):

(9) *Let K_1 resp. K_2 be the canonical mapping of E onto E/N_1 resp. F onto F/N_2, where N_1, N_2 are subspaces of E and F, respectively. Then $K_1 \otimes K_2$ is a linear mapping of $E \otimes F$ onto $E/N_1 \otimes F/N_2$ which induces the canonical isomorphism $\hat{z} \to (K_1 \otimes K_2)z$ of $(E \otimes F)/D[N_1, N_2]$ onto $E/N_1 \otimes F/N_2$.*

(10) *Let J_1, J_2 be isomorphisms of M, N into E and F, respectively. Then $J_1 \otimes J_2$ is an isomorphism of $M \otimes N$ into $E \otimes F$.*

This follows immediately from (8). (10) is also equivalent to § 9, 6.(7).

2. The projective tensor product. If E and F are locally convex spaces, the problem arises immediately how to define a locally convex topology on $E \otimes F$ in a natural way. If E and F are normed spaces, one is similarly interested in suitable norms on $E \otimes F$. It was this problem which was first investigated by VON NEUMANN for E and F Hilbert spaces and later by SCHATTEN for arbitrary normed spaces (SCHATTEN [1′]). The tensor products of locally convex spaces were treated in GROTHENDIECK's thesis [13], which is the main source of the following exposition. We refer the reader also to the expositions in SCHAEFER [1′], SCHWARTZ [1′], and TREVES [1′].

As we will see in § 44, there are many "natural" topologies on $E \otimes F$, but not all of the same importance in applications. We start with the projective topology on $E \otimes F$.

We will use the following notation. If A and B are subsets of E and F, respectively, then $A \otimes B$ will denote the set of all $a \otimes b$, $a \in A$, $b \in B$. This definition introduces a certain abuse of notation, since $E \otimes F$, where E and F are vector spaces, is not exhausted by the elements of the form $x \otimes y$, $x \in E$, $y \in F$; $E \otimes F$ contains all finite sums of these elements too.

Our idea is now to define a finest locally convex topology on $E \otimes F$ such that the canonical bilinear mapping χ of $E \times F$ into $E \otimes F$ is continuous. Let U, V be absolutely convex neighbourhoods of $\mathfrak{o}$ in E and F, respectively; then $U \times V$ is a neighbourhood of $\mathfrak{o}$ in $E \times F$. Thus the absolutely convex cover $\Gamma(U \otimes V)$ of $U \otimes V = \chi(U \times V)$ should be a neighbourhood of $\mathfrak{o}$ of this topology and one expects that the class of all these sets will provide a neighbourhood basis. We prove first

(1) *$\Gamma(U \otimes V)$ is absorbing. If p and q are the semi-norms corresponding to U and V, then $\Gamma(U \otimes V)$ defines the semi-norm*

$$p \otimes q(z) = \inf \sum_{i=1}^{n} p(x_i)q(y_i), \qquad z \in E \otimes F,$$

where the infimum is taken over all representations $z = \sum_{i=1}^{n} x_i \otimes y_i$ in $E \otimes F$.

Proof. We show first that $\Gamma(U \otimes V)$ is absorbing. Let $z = \sum_{i=1}^{n} x_i \otimes y_i$ be an element of $E \otimes F$. Observe that

$$x_i' = \frac{x_i}{p(x_i) + \delta} \in U, \qquad y_i' = \frac{y_i}{q(y_i) + \delta} \in V$$

for every $\delta > 0$ and hence $x_i' \otimes y_i' \in U \otimes V$.

Given $\varepsilon > 0$ we may choose δ sufficiently small such that

$$(2)\ z = \sum_{i=1}^{n} (p(x_i) + \delta)(q(y_i) + \delta)x_i' \otimes y_i' \in \left(\sum_{i=1}^{n} p(x_i)q(y_i) + \varepsilon\right)\Gamma(U \otimes V).$$

Thus $\Gamma(U \otimes V)$ is absorbing.

Since $\Gamma(U \otimes V)$ is absolutely convex, it defines a semi-norm $r(z)$ on $E \otimes F$. We will show that $r(z) = p \otimes q(z)$.

Now $r(z) = \inf_{\lambda > 0} \lambda$, $z \in \lambda\Gamma(U \otimes V)$. It follows from (2) that $r(z) \leqq \sum_{i=1}^{n} p(x_i)q(y_i)$. Since this is true for every representation of z as a sum of elements of the form $x \otimes y$, we obtain

$$\text{(3)} \qquad r(z) \leqq \inf \sum_{i=1}^{n} p(x_i)q(y_i) = p \otimes q(z).$$

Conversely, suppose $z \in \lambda\Gamma(U \otimes V)$. Then $z = \sum \alpha_k(x'_k \otimes y'_k)$, $p(x'_k) \leqq 1$, $q(y'_k) \leqq 1$, $\alpha_k \geqq 0$, $\sum \alpha_k \leqq \lambda$. For this particular representation we have $\sum p(\alpha_k x'_k)q(y'_k) \leqq \sum \alpha_k \leqq \lambda$ and hence $p \otimes q(z) = \inf \sum p(x_i)q(y_i) \leqq \lambda$. This is true for every λ with $z \in \lambda\Gamma(U \otimes V)$; thus $p \otimes q(z) \leqq r(z)$. From (3) follows, finally, $r(z) = p \otimes q(z)$.

(4) *There exists a finest locally convex topology* $\mathfrak{T}_\pi$ *on* $E \otimes F$ *for which the canonical bilinear mapping* χ *of* $E \times F$ *into* $E \otimes F$ *is continuous. The class of all sets* $\Gamma(U \otimes V)$, *where* U, V *are absolutely convex neighbourhoods of* o *in* E *and* F, *respectively, is a* $\mathfrak{T}_\pi$-*neighbourhood base of* o *in* $E \otimes F$.

$\mathfrak{T}_\pi$ or π is called the projective topology on $E \otimes F$ and $E \otimes F$ equipped with this topology will be denoted by $E \otimes_\pi F$ and called the projective tensor product of E and F. If $\mathfrak{T}_1$, $\mathfrak{T}_2$ are the topologies on E and F, respectively, one writes also $\mathfrak{T}_\pi = \mathfrak{T}_1 \otimes_\pi \mathfrak{T}_2$.

Proof of (4). By (1) the sets $\Gamma(U \otimes V)$ are absorbing and absolutely convex and $\Gamma(U_1 \otimes V_1) \cap \Gamma(U_2 \otimes V_2) \supset \Gamma((U_1 \cap U_2) \otimes (V_1 \cap V_2))$. Therefore $\{\Gamma(U \otimes V)\}$ is a filter base on $E \otimes F$ which defines a locally convex topology. That this topology is Hausdorff will follow from

(5) $(E \otimes_\pi F)' \supset E' \otimes F'$ *and* $\langle E' \otimes F', E \otimes F\rangle$ *is a dual pair.*

Proof. Let $u \in E'$, $v \in F'$. There exist continuous semi-norms $p(x)$, $q(y)$ such that $|u(x)| \leqq p(x)$, $|v(y)| \leqq q(y)$. By § 9, 7.(2) every element of $E^* \otimes F^*$ defines a linear functional on $E \otimes F$. In our case we have

$$|(u \otimes v)z| = \left|(u \otimes v) \sum_{i=1}^{n} x_i \otimes y_i\right| = \left|\sum (ux_i)(vy_i)\right| \leqq \sum p(x_i)q(y_i).$$

This is true for every representation of z; therefore

$$\text{(6)} \qquad |(u \otimes v)z| \leqq p \otimes q(z)$$

and this proves the first statement.

Now for the proof that $\mathfrak{T}_\pi$ is Hausdorff and the second statement of (5) it will be sufficient to construct for a given $z \neq o$ in $E \otimes F$ a $w \in E' \otimes F'$

such that $wz \neq o$. By § 9, 6.(8) there exists a representation $z = \sum_{i=1}^{n} x_i \otimes y_i$, where the x_i and the y_i are linearly independent. Choose $u \in E'$ with $u(x_1) = 1$, $u(x_i) = 0$ for $i \neq 1$ and $v \in F'$ with $v(y_1) = 1$. Set $w = u \otimes v$. Then $wz = 1$.

We remark that if we take not all the absolutely convex neighbourhoods of o in E and F but only neighbourhood bases U_α, $\alpha \in \mathrm{A}$, V_β, $\beta \in \mathrm{B}$, then the class of all $\Gamma(U_\alpha \otimes V_\beta)$ is a neighbourhood base of o in $E \otimes_\pi F$. Correspondingly, if $\{p_\alpha\}$, $\{q_\beta\}$ are directed systems of semi-norms determining the topologies of E and F, respectively, then the $p_\alpha \otimes q_\beta$ form a directed system of semi-norms determining $\mathfrak{T}_\pi$ on $E \otimes F$.

From this remark and (4) follows

(7) *If E and F are metrizable locally convex spaces with defining semi-norms $p_1 \leqq p_2 \leqq \cdots$ and $q_1 \leqq q_2 \leqq \cdots$, respectively, then $E \otimes_\pi F$ is metrizable with defining semi-norms $p_1 \otimes q_1 \leqq p_2 \otimes q_2 \leqq \cdots$.*

If E and F are normed spaces with norms p and q, respectively, then $E \otimes_\pi F$ is a normed space with norm $p \otimes q$.

This norm is called the projective norm or π-norm on $E \otimes F$ and will be denoted by $\| \ \|_\pi$.

The semi-norms $p \otimes q$ have the following properties.

(8) *Let E, F be locally convex and p, q continuous semi-norms on E and F, respectively. Then*

a) $p \otimes q(x \otimes y) = p(x)q(y)$ *for* $x \in E$, $y \in F$;

b) *Let N_1, N_2 be the kernels of p and q, respectively. Then the kernel N of $\pi = p \otimes q$ is $D[N_1, N_2]$, which is therefore a closed subspace of $E \otimes_\pi F$.*

If $\hat{p}$, $\hat{q}$, $\hat{\pi}$ are the quotient norms on E/N_1, F/N_2, $(E \otimes F)/D[N_1, N_2]$, then

$$\pi(z) = \hat{\pi}(\hat{z}) = \hat{p} \otimes \hat{q}(z') \tag{9}$$

for every $z \in E \otimes F$, where $\hat{z}$ is its residue class in $(E \otimes F)/D$ and z' is the corresponding element in $E/N_1 \otimes F/N_2$.

Proof. a) Choose linear functionals $u \in E'$, $v \in F'$ with $u(x) = p(x)$, $v(y) = q(y)$, and $|u(x')| \leqq p(x')$, $|v(y')| \leqq q(y')$ for all $x' \in E$, $y' \in F$ (HAHN–BANACH). Then

$$p \otimes q(x \otimes y) \leqq p(x)q(y) = u(x)v(y) = u \otimes v(x \otimes y) \leqq p \otimes q(x \otimes y),$$

the last inequality being a consequence of (6).

b) We proved in 1.(9) that $\hat{z} \to z' = (K_1 \otimes K_2)z$ is an algebraic isomorphism of $(E \otimes F)/D$ onto $E/N_1 \otimes F/N_2$. We have $\pi(t) = o$ for every $t \in D[N_1, N_2]$ since $\pi(x \otimes y) = o$ if $x \in N_1$ or $y \in N_2$. It follows that $\pi(z + t) = \pi(z) = \hat{\pi}(\hat{z})$ and $\hat{\pi}$ is a semi-norm on $(E \otimes F)/D$.

Now $\hat{p} \otimes \hat{q}$ is a norm on $E/N_1 \otimes F/N_2$ by (7) and, if $\hat{\pi}(\hat{z}) = \hat{p} \otimes \hat{q}(z')$ for every $z \in E \otimes F$, then $\hat{\pi}$ is also a norm and the kernel of π is D.

Thus we have only to prove that $\pi(z) = \hat{p} \otimes \hat{q}(z')$ for every $z \in E \otimes F$. If z has the representation $z = \sum x_i \otimes y_i$, then $z' = (K_1 \otimes K_2)z$ has the representation $z' = \sum \hat{x}_i \otimes \hat{y}_i$ in $E/N_1 \otimes F/N_2$. If, conversely, a representation $z' = \sum \hat{x}_i \otimes \hat{y}_i$ is given, then it follows by 1.(9) that $z = \sum x_i \otimes y_i + t$, where $x_i \in \hat{x}_i$, $y_i \in \hat{y}_i$, and $t \in D$. From this and $\sum p(x_i)q(y_i) = \sum \hat{p}(\hat{x}_i)\hat{q}(y_i)$ it follows that $\pi(z) = \hat{p} \otimes \hat{q}(z')$.

We will be interested also in the completion of a projective tensor product $E \otimes_\pi F$. We will denote this completion by $E \tilde{\otimes}_\pi F$. If E and F are normed spaces, then $E \tilde{\otimes}_\pi F$ is a (B)-space for the π-norm. If E and F are metrizable spaces, then $E \tilde{\otimes}_\pi F$ is an (F)-space by (7).

Let E, F be locally convex. The algebraic isomorphism $\sum x_i \otimes y_i \to \sum y_i \otimes x_i$ of $E \otimes F$ onto $F \otimes E$ (§ 9, 6.) generates a topological isomorphism of $E \otimes_\pi F$ onto $F \otimes_\pi E$ and of $E \tilde{\otimes}_\pi F$ onto $F \tilde{\otimes}_\pi E$.

If E and F are normed spaces, these isomorphisms are even norm isomorphisms.

In this sense the π-tensor product is commutative.

It is also associative: The natural algebraic isomorphism of $(E \otimes F) \otimes G$ and $E \otimes (F \otimes G)$ generates the natural topological isomorphisms

$$(E \otimes_\pi F) \otimes_\pi G \cong E \otimes_\pi (F \otimes_\pi G) \quad \text{and} \quad (E \tilde{\otimes}_\pi F) \tilde{\otimes}_\pi G \cong E \tilde{\otimes}_\pi (F \tilde{\otimes}_\pi G).$$

3. The dual space. Representations of $E \tilde{\otimes}_\pi F$. We recall from 1.(1) the algebraic isomorphism $\dot{B} \to B$ of $L(E \otimes F, G)$ and $B(E \times F, G)$. If E, F, G are locally convex, we are interested in the continuous linear mappings $\dot{B}$ of $E \otimes_\pi F$ in G. What are the corresponding bilinear mappings B of $E \times F$ in G? We obtain

(1) *A linear mapping $\dot{B}$ of $E \otimes_\pi F$ in G is continuous if and only if the corresponding bilinear mapping $B = \dot{B}\chi$ of $E \times F$ in G is continuous. Thus $\mathfrak{L}(E \otimes_\pi F, G)$ is algebraically isomorphic to $\mathscr{B}(E \times F, G)$.*

We remark that (1) is true also for $E \tilde{\otimes}_\pi F$ if G is complete, since then $\mathfrak{L}(E \otimes_\pi F, G) = \mathfrak{L}(E \tilde{\otimes}_\pi F, G)$.

Proof of (1). Suppose $\dot{B}$ continuous. Then the continuity of χ implies that $B = \dot{B}\chi$ is continuous. Conversely, assume B continuous. If W is an absolutely convex neighbourhood of $\circ$ in G, there exist U, V such that $B(U \times V) \subset W$. Thus $\dot{B}(U \otimes V) \subset W$ and this implies $\dot{B}(\Gamma(U \otimes V)) \subset W$ since W is absolutely convex.

Similar to 1.(5) is the following characterization of the projective tensor product.

(2) *Let E, F be fixed locally convex spaces and H a locally convex space with the following properties:*

a) *there exists a continuous bilinear mapping χ_1 of $E \times F$ into H such that $\chi_1(E \times F)$ generates H;*

b) *if B is a continuous bilinear mapping of $E \times F$ in a locally convex space G, then $B = B_1\chi_1$, where $B_1 \in \mathfrak{L}(H, G)$.*

Then there exists an isomorphism A of $E \otimes_\pi F$ onto H such that $\chi_1 = A\chi$, where χ is the canonical bilinear mapping of $E \times F$ into $E \otimes F$.

It is easy to adapt the proof of 1.(5) to the present situation.

As a special case of (1) we obtain

(3) *The dual of $E \otimes_\pi F$ and $E \tilde{\otimes}_\pi F$ can be identified with $\mathscr{B}(E \times F)$.*

The duality $\langle \mathscr{B}(E \times F), E \otimes_\pi F \rangle$ is expressed by

$$\langle \dot{B}, z \rangle = \left\langle \dot{B}, \sum x_i \otimes y_i \right\rangle = \sum B(x_i, y_i), \qquad z = \sum x_i \otimes y_i.$$

(4) *The equicontinuous subsets of $(E \otimes_\pi F)' = (E \tilde{\otimes}_\pi F)'$ are the equicontinuous sets of bilinear forms on $E \times F$.*

$\ulcorner(U \otimes V)^\circ$ consists of all $\dot{B}$ such that $|\langle \dot{B}, \ulcorner(U \otimes V)\rangle| \leqq 1$. The corresponding set of bilinear forms $B = \dot{B}\chi$ consists of all $B \in \mathscr{B}(E \times F)$ such that $|B(U \times V)| \leqq 1$. This implies the statement.

(5) *$E' \otimes F'$ is $\mathfrak{T}_s(E \otimes F)$-dense in $\mathscr{B}(E \times F)$.*

This follows from 2.(5).

We consider now the case that E, F, and G are normed spaces. A continuous bilinear mapping B of $E \times F$ in G has the natural norm $\|B\| = \sup_{\|x\| \leqq 1, \|y\| \leqq 1} \|B(x, y)\|$. This norm defines the bibounded topology $\mathfrak{T}_b$ on $\mathscr{B}(E \times F, G)$ (§ 40, 4.). Obviously, $\|\chi\| = 1$ for the canonical bilinear mapping of $E \times F$ into $E \otimes_\pi F$. (1) and (3) can be improved in the following way.

(6) *If E, F, G are normed spaces, then the correspondence $\dot{B} \to B = \dot{B}\chi$ defines a norm isomorphism of $\mathfrak{L}_b(E \otimes_\pi F, G)$ onto $\mathscr{B}_b(E \times F, G)$.*

The strong dual $(E \otimes_\pi F)'_b = (E \tilde{\otimes}_\pi F)'_b$ is norm isomorphic to $\mathscr{B}_b(E \times F)$ and to $\mathfrak{L}_b(E, F'_b)$.

Proof. Let $z = \sum x_i \otimes y_i$ be in $E \otimes_\pi F$. From $\|\dot{B}z\| = \left\|\sum_i B(x_i, y_i)\right\| \leqq \|B\| \sum \|x_i\| \|y_i\|$ follows $\|\dot{B}z\| \leqq \|B\| \|z\|_\pi$. Thus $\|\dot{B}\| \leqq \|B\|$.

Conversely, $\|B(x, y)\| = \|\dot{B}(x \otimes y)\| \leqq \|\dot{B}\| \|x \otimes y\|_\pi = \|\dot{B}\| \|x\| \|y\|$; hence $\|B\| \leqq \|\dot{B}\|$. The last statement follows from the fact that $|B(U, V)| \leqq 1$

is equivalent to $|\tilde{B}(U)V| \leqq 1$, where U and V are the unit balls in E and F, respectively, and $\tilde{B}$ is the linear mapping of E in F'_b corresponding to B.

Let E, F be locally convex. $E \otimes F$ can be considered as a space of continuous linear mappings: For $A = (x, y)$ we define $\tilde{A} \in \mathfrak{L}(E'_s, F)$ by $\tilde{A}u = (ux)y$, $u \in E'$. The bilinear mapping $A \to \tilde{A}$ generates a linear mapping $\psi(\dot{A}) = \tilde{A}$ of $E \otimes F$ in $\mathfrak{L}(E'_s, F)$. Every $\tilde{A}$ has finite rank. The mapping ψ is one-one, which becomes obvious when one uses linearly independent x_i and y_i in the representation of $\dot{A}$, $\dot{A} = \sum x_i \otimes y_i$.

Conversely, if $\tilde{A} \in \mathfrak{L}(E'_s, F)$ has finite rank, $\tilde{A}$ can be written as $\tilde{A}u = \sum_{i=1}^{n} \alpha_i(u) y_i$, where the y_i are linearly independent. There exist $u_i \in E'$ such that $\tilde{A}u_i = y_i$ and $v_i \in F'$ such that $v_i y_k = \delta_{ik}$. It follows that

$$\alpha_i(u) = \langle v_i, \tilde{A}u \rangle = \langle \tilde{A}' v_i, u \rangle = \langle x_i, u \rangle, \quad \text{where } x_i \in E.$$

Thus $\tilde{A}$ corresponds to $\dot{A} = \sum x_i \otimes y_i$.

(7) *Let E, F be locally convex, $\dot{A} = \sum x_i \otimes y_i \in E \otimes F$. Then*

$$\tilde{A}u = \sum (ux_i) y_i, \qquad u \in E',$$

defines an element $\psi(\dot{A}) = \tilde{A} \in \mathfrak{L}(E'_s, F)$ and ψ is an algebraic isomorphism of $E \otimes F$ with the subspace of all maps of finite rank in $\mathfrak{L}(E'_s, F)$.

Correspondingly, one obtains

(8) *Let E, F be locally convex, $\dot{A} = \sum u_i \otimes v_i \in E' \otimes F'$. Then*

$$\tilde{A}x = \sum (u_i x) v_i, \qquad x \in E,$$

defines an element $\psi(\dot{A}) = \tilde{A} \in \mathfrak{L}(E, F'_s)$ and ψ is an algebraic isomorphism of $E' \otimes F'$ with the subspace of all maps of finite rank in $\mathfrak{L}(E, F'_s)$.

In connection with (5) the following corollary to (8) is of interest.

(9) *If E or F is equipped with the weak topology, then*

$$(E \otimes_\pi F)' \cong \mathscr{B}(E \times F) \cong E' \otimes F'.$$

It is sufficient to prove that if $B \in \mathscr{B}(E \times F)$, the corresponding map $\tilde{B} \in \mathfrak{L}(E, F'_s)$ has finite rank. But by § 40, 1.(4) $\mathscr{B}(E \times F)$ is isomorphic to the subspace of $\mathfrak{L}(E, F'_s)$ consisting of all $\tilde{B}$ which map a neighbourhood of $\mathfrak{o}$ into an equicontinuous set in F'. If F has the weak topology, such an equicontinuous set is finite dimensional; thus $\tilde{B}$ has finite rank.

We note as a special case

(10) $(E'_s \tilde{\otimes}_\pi F'_k)' \cong \mathscr{B}(E'_s \times F'_k) \cong E \otimes F \cong \mathscr{B}(E'_s \times F'_s) \cong (E'_s \tilde{\otimes}_\pi F'_s)'$.

We obtained in (7) a straightforward interpretation of the elements of $E \otimes F$ as linear mappings. Is such an interpretation also possible for the elements of $E \tilde{\otimes}_\pi F$?

We investigate first the case where E, F are (B)-spaces. The injection $\psi(\dot{A}) = \tilde{A}$ of $E \otimes F$ in $\mathfrak{L}(E'_s, F)$ is a continuous injection of $E \otimes_\pi F$ into $\mathfrak{L}_b(E'_b, F)$. To prove this we remark first that an element of finite rank in $\mathfrak{L}(E'_s, F)$ is also continuous from E'_b into F. The continuity of ψ follows now from

$$\|\psi(\dot{A})\| = \|\tilde{A}\| = \sup_{\|u\| \leqq 1} \|\tilde{A}u\| = \sup \left\|\sum (ux_i) y_i\right\| \leqq \sum \|x_i\| \|y_i\|,$$

since this implies $\|\tilde{A}\| \leqq \|\dot{A}\|_\pi = \inf \sum \|x_i\| \|y_i\|$, the infimum being taken over all representations of $\dot{A}$.

$\mathfrak{L}_b(E'_b, F)$ is complete; thus ψ can be continuously extended to $E \tilde{\otimes}_\pi F$ and we obtain

(11) *If E and F are* (B)-*spaces, the canonical continuous injection ψ of $E \otimes_\pi F$ in $\mathfrak{L}_b(E'_b, F)$ has a uniquely determined extension to a continuous linear mapping $\bar{\psi}$ of $E \tilde{\otimes}_\pi F$ in $\mathfrak{L}_b(E'_b, F)$.*

We will see later that in some exceptional cases $\bar{\psi}$ may fail to be one-one.

Now let E, F be locally convex. By (10) $E \otimes F$ can be identified with $\mathscr{B}(E'_s \times F'_s)$, which is a subspace of $\mathfrak{B}(E'_s \times F'_s)$, the space of separately continuous bilinear forms. Is it possible to extend the injection of $E \otimes F$ in $\mathfrak{B}(E'_s \times F'_s)$ to a mapping of $E \tilde{\otimes}_\pi F$ in $\mathfrak{B}(E'_s \times F'_s)$?

We equip $\mathfrak{B}(E'_s \times F'_s)$ with the bi-equicontinuous topology $\mathfrak{T}_e$ (§ 40, 4.) and obtain first

(12) *The injection of $E \otimes_\pi F$ into $\mathfrak{B}_e(E'_s \times F'_s)$ is continuous.*

It is sufficient to prove that the canonical bilinear mapping $\chi_1: (x, y) \to B(u, v) = (ux)(vy)$ of $E \times F$ in $\mathfrak{B}_e$ is continuous. Let M, N be absolutely convex equicontinuous subsets of E' and F', respectively. One checks immediately that $\chi_1(M^\circ \times N^\circ)$ is contained in the $\mathfrak{T}_e$-neighbourhood $\{B; |B(M, N)| \leqq 1\}$ of $\mathfrak{o}$ in $\mathfrak{B}_e$.

By § 40, 4.(5) $\mathfrak{B}_e(E'_s \times F'_s)$ is topologically isomorphic to $\mathfrak{L}_e(E'_k, F)$; thus the canonical injection of $E \otimes_\pi F$ into $\mathfrak{L}_e(E'_k, F)$ is continuous too.

Again by § 40, 4.(5) $\mathfrak{B}_e(E'_s \times F'_s)$ and $\mathfrak{L}_e(E'_k, F)$ are complete if and only if E and F are complete. From this and (12) we obtain

(13) *Let E and F be complete locally convex spaces. Then the canonical injection ψ of $E \otimes_\pi F$ in $\mathfrak{B}_e(E'_s \times F'_s)$ resp. $\mathfrak{L}_e(E'_k, F)$ has a uniquely determined continuous extension $\bar{\psi}$ to $E \tilde{\otimes}_\pi F$.*

Again there is the problem whether $\bar{\psi}$ is an injection or not. We give the following criterion which is of interest compared with (5):

(14) *Let E and F be complete locally convex spaces. The canonical mapping $\bar{\psi}$ of $E \tilde{\otimes}_\pi F$ in $\mathfrak{B}_e(E'_s \times F'_s)$ is one-one if and only if $E' \otimes F'$ is $\mathfrak{T}_s(E \tilde{\otimes}_\pi F)$-dense in $\mathscr{B}(E \times F) = (E \tilde{\otimes}_\pi F)'$.*

We prove first

$$(15) \qquad \langle u \otimes v, \dot{B} \rangle = \bar{\psi}(\dot{B})(u, v) = B(u, v),$$

where $\dot{B} \in E \tilde{\otimes}_\pi F$, $u \in E'$, $v \in F'$, and $B = \bar{\psi}(\dot{B}) \in \mathfrak{B}(E'_s \times F'_s)$.

This is trivial for $\dot{B} \in E \otimes F$ because of the duality $\langle E' \otimes F', E \otimes F \rangle$. If $\dot{B} = \mathfrak{T}_\pi\text{-}\lim \dot{B}_\alpha$, $\dot{B}_\alpha \in E \otimes F$, then $\psi(\dot{B}_\alpha) \to \bar{\psi}(\dot{B})$ in $\mathfrak{B}_e$ and (15) follows by continuity.

Now $\bar{\psi}$ is not one-one if and only if there exists a $\dot{B} \neq \mathrm{o}$ such that $\bar{\psi}(\dot{B}) = \mathrm{o}$ or by (15) a $\dot{B} \neq \mathrm{o}$ which is orthogonal to $E' \otimes F'$ in the duality $\langle \mathscr{B}(E \times F), E \tilde{\otimes}_\pi F \rangle$, where $\mathscr{B}(E \times F) \supset E' \otimes F'$. This implies the statement.

4. The projective tensor product of metrizable and of (DF)-spaces. GROTHENDIECK [13] obtained a concrete representation of the elements of $E \tilde{\otimes}_\pi F$ for metrizable locally convex spaces E and F. His method was simplified by A. and W. ROBERTSON [2′] and A. PIETSCH [2′]. We start with a result on normed spaces.

(1) *Let E and F be normed spaces, z an element of $E \tilde{\otimes}_\pi F$. Then for every $\varepsilon > 0$, z has a representation*

$$(2) \qquad z = \sum_{n=1}^{\infty} \lambda_n (x_n \otimes y_n), \qquad x_n \in E,\ \|x_n\| \leqq 1,\ y_n \in F,\ \|y_n\| \leqq 1,$$

and $\sum_{n=1}^{\infty} |\lambda_n| \leqq \|z\|_\pi + \varepsilon.$

Proof. We choose $z_n \in E \otimes F$ with $\|z - z_n\|_\pi < (1/2^{n+2})\varepsilon$ for $n = 1, 2, \ldots$. Then $\|z_1\|_\pi < \|z\|_\pi + \varepsilon/2$ and $\|t_n\|_\pi < (1/2^{n+1})\varepsilon$ for $t_n = z_{n+1} - z_n$, $n = 1, 2, \ldots$. From the definition of the projective norm follows by induction the existence of representations:

$$z_1 = \sum_{i=1}^{l_1} \lambda_i (x_i \otimes y_i), \qquad \|x_i\| \leqq 1,\ \|y_i\| \leqq 1, \sum_{1}^{l_1} |\lambda_i| < \|z\|_\pi + \varepsilon/2,$$

$$t_n = \sum_{l_n+1}^{l_{n+1}} \lambda_i (x_i \otimes y_i), \qquad \|x_i\| \leqq 1,\ \|y_i\| \leqq 1, \sum_{l_n+1}^{l_{n+1}} |\lambda_i| < (1/2^{n+1})\varepsilon,$$

$$n = 1, 2, \ldots.$$

Clearly, $z = z_1 + t_1 + t_2 + \cdots$ and (2) follows easily.

We remark that $\sum_{i=1}^{\infty} \|\lambda_i (x_i \otimes y_i)\|_\pi \leqq \sum_{i=1}^{\infty} |\lambda_i|$; thus the series (2) is absolutely convergent. Conversely, every absolutely convergent series $\sum_{n=1}^{\infty} x'_n \otimes y'_n$ defines an element of $E \tilde{\otimes}_\pi F$.

For the metrizable case we need a generalization of § 21, 10.(3) due to A. and W. ROBERTSON. Their proof is elementary, whereas § 21, 10.(3) was obtained as a corollary of the BANACH–DIEUDONNÉ theorem.

(3) *Let F be a dense subspace of a metrizable locally convex space E and M a precompact subset of E. Then there exists a sequence x_n of elements of F converging to $\circ$ such that every element x of M has a representation $x = \sum_{n=1}^{\infty} \lambda_n x_n$ with $\sum |\lambda_n| \leqq 1$.*

Proof. Let $\| \ \|_1 \leqq \| \ \|_2 \leqq \cdots$ be a sequence of semi-norms defining the topology on E. Since M is precompact and F dense in E, there exists a finite set $N_1 \subset F$ such that for every $x \in M$, $\|x - x^{(1)}\|_1 \leqq (1/2)(1/2^3)$ for some $x^{(1)} \in N_1$. Recalling that a finite union of precompact sets is again precompact, one obtains successively finite sets $N_2, \ldots, N_n$, such that

$$\|x - x^{(1)} - \cdots - x^{(n)}\|_n < \frac{1}{n+1} \cdot \frac{1}{2^{n+2}}$$

for every $x \in M$ and suitable $x^{(i)} \in N_i$. We note that

$$(4) \quad \|x^{(n)}\|_{n-1} \leqq \|x - x^{(1)} - \cdots - x^{(n)}\|_{n-1} + \|x - x^{(1)} - \cdots - x^{(n-1)}\|_{n-1} < \frac{1}{n} \cdot \frac{1}{2^n}.$$

Let $y^{(n)} = 2^n x^{(n)}$ for every $x^{(n)} \in N_n$. The sequence consisting first of the elements $y^{(1)}$, followed by the elements $y^{(2)}$ and so on, converges to $\circ$ in F because of (4) and (3) follows from

$$x = x^{(1)} + x^{(2)} + \cdots = \frac{1}{2} y^{(1)} + \frac{1}{2^2} y^{(2)} + \cdots.$$

Remark. *If E is normed, then for every $\varepsilon > 0$ the x_n in (3) can be chosen so that $\|x_n\| \leqq (1 + \varepsilon)\gamma$, where $\gamma = \sup_{x \in M} \|x\|$.*

We use the inequalities

$$\|x - x^{(1)}\| < \frac{\varepsilon \cdot \gamma}{2 \cdot 2^3} \quad \text{and} \quad \|x - x^{(1)} - \cdots - x^{(n)}\| < \frac{\varepsilon}{n+1} \frac{\gamma}{2^{n+2}}$$

in the same construction.

(5) *Let M be a compact subset of $E \tilde{\otimes}_\pi F$, where E and F are metrizable locally convex. Then there exist null sequences $x_n \in E$, $y_n \in F$ such that every element z of M has a representation*

$$z = \sum_{n=1}^{\infty} \lambda_n (x_n \otimes y_n), \qquad \sum |\lambda_n| \leqq 1.$$

Proof. By (3) there exists a null sequence $z_n \in E \otimes_\pi F$ such that every $z \in M$ has a representation $z = \sum \mu_n z_n$, $\sum |\mu_n| \leqq 1$. Let $p_1 \otimes q_1 \leqq p_2 \otimes q_2 \leqq \cdots$ be a sequence of semi-norms defining the π-topology on $E \tilde{\otimes}_\pi F$. For $r = 1, 2, \ldots$ we determine k_r such that $p_r \otimes q_r(z_n) < 1/r$ for $n \geqq k_r$. For $n < k_1$, z_n can be written as a finite sum $z_n = \sum_i \nu_{ni}(x_{ni} \otimes y_{ni})$, $x_{ni} \in E$, $y_{ni} \in F$, $\sum_i |\nu_{ni}| \leqq 1$. For $k_r \leqq n < k_{r+1}$ there exists a representation $z_n = \sum_i \nu_{ni}(x_{ni} \otimes y_{ni})$, $p_r(x_{ni}) < 1/\sqrt{r}$, $q_r(y_{ni}) < 1/\sqrt{r}$, and $\sum_i |\nu_{ni}| \leqq 1$. Thus $z = \sum_{n=1}^{\infty} \mu_n \sum_i \nu_{ni}(x_{ni} \otimes y_{ni})$. Reindexing the x_{ni} and y_{ni} using dictionary order, we obtain sequences converging to o. Since $\sum_{n=1}^{\infty} \sum_i |\mu_n \nu_{ni}| \leqq 1$, z has the desired representation.

Remark. (5) shows that every compact set in $E \tilde{\otimes}_\pi F$, E and F metrizable, is contained in a compact set of the form $\overline{\Gamma(C_1 \otimes C_2)}$, where C_1 and C_2 consist of sequences converging to o.

As a special case of (4) we obtain

(6) *Let E and F be metrizable locally convex. Then every $z \in E \tilde{\otimes}_\pi F$ has a representation*

$$z = \sum_{n=1}^{\infty} \lambda_n (x_n \otimes y_n), \qquad \sum |\lambda_n| \leqq 1,$$

where x_n and y_n are null sequences in E and F, respectively.

Suppose that M and N are bounded subsets of the locally convex spaces E and F, respectively. Then $\overline{\Gamma(M \otimes N)}$ is bounded in $E \tilde{\otimes}_\pi F$. It is natural to ask if all bounded subsets of $E \tilde{\otimes}_\pi F$ are contained in subsets of this form.

Recall that the bibounded topology $\mathfrak{T}_{b,b}$ on $\mathscr{B}(E \times F) = (E \tilde{\otimes}_\pi F)'$ is given by the neighbourhoods of o,

$$\mathscr{U}(M, N, 1) = \{B \in \mathscr{B}(E \times F); |B(M, N)| \leqq 1\},$$

where M, N are bounded subsets of E and F, respectively. Now

$$|B(M, N)| \leqq 1, \qquad |\dot{B}(M \otimes N)| \leqq 1, \qquad |\dot{B}(\overline{\Gamma(M \otimes N)})| \leqq 1$$

are equivalent; therefore $\mathfrak{T}_{b,b}$ is defined by the polars of the sets

$$\overline{\Gamma(M \otimes N)} \subset E \tilde{\otimes}_\pi F.$$

Therefore our problem is equivalent to the question: Is the strong topology $\mathfrak{T}_b(E \tilde{\otimes}_\pi F)$ on $\mathscr{B}(E \times F)$ identical with $\mathfrak{T}_{bb}$?

The answer is yes in the case of normed spaces (3.(6)); for E and F both metrizable the question seems to be open; in general the answer is no, as we will see by an example in 6. For (DF)-spaces we have

(7) *If E and F are* (DF)-*spaces, then $E \otimes_\pi F$ and $E \tilde{\otimes}_\pi F$ are* (DF)-*spaces. If C_n and D_n are fundamental sequences of bounded sets in E and F, respectively, then $\overline{\Gamma(C_n \otimes D_n)}$ is a fundamental sequence of bounded sets in $E \tilde{\otimes}_\pi F$.*
The strong topology on $(E \tilde{\otimes}_\pi F)'$ and the bibounded topology on $\mathscr{B}(E \times F)$ coincide.

To prove the last statement we have to show that the canonical isomorphism of $\mathscr{B}_b(E \times F)$ onto $(E \tilde{\otimes}_\pi F)'_b$ is continuous. The bibounded topology on $\mathscr{B}(E \times F)$ has a neighbourhood base of $\circ$ of the form $\mathscr{U}(C_n, D_n, 1) = \{B; |B(C_n, D_n)| \leqq 1\}, n = 1, 2, \ldots$. Therefore $\mathscr{B}_b(E \times F)$ is metrizable and bornological. Thus it will be sufficient to show that every sequence B_n converging to $\circ$ in $\mathscr{B}_b(E \times F)$ is a bounded set in $(E \tilde{\otimes}_\pi F)'_b$. By § 40, 4.(10) the set $\{B_1, B_2, \ldots\}$ is equicontinuous in $\mathscr{B}(E \times F)$, hence equicontinuous in $(E \tilde{\otimes}_\pi F)'$ by 3.(4), thus bounded in $(E \tilde{\otimes}_\pi F)'_b$.

It follows that $E \otimes_\pi F$ and $E \tilde{\otimes}_\pi F$ have fundamental sequences of bounded subsets and by § 40, 4.(10) condition b) of the definition of a (DF)-space (§ 29, 3.) is satisfied too and (7) is proved.

From (7) we obtain further properties of the projective tensor product of two (DF)-spaces.

(8) *Let E and F be* (DF)-*spaces. Then*

a) *if E and F are barrelled, $E \otimes_\pi F$ and $E \tilde{\otimes}_\pi F$ are barrelled,*

b) *if E and F are quasi-barrelled, $E \otimes_\pi F$ is quasi-barrelled and $E \tilde{\otimes}_\pi F$ is barrelled,*

c) *if E and F are bornological, $E \otimes_\pi F$ is bornological,*

d) *if E and F are* (M)-*spaces, $E \tilde{\otimes}_\pi F$ is an* (M)-*space.*

Proof. a) Let H be a $\mathfrak{T}_s(E \otimes_\pi F)$-bounded subset of $\mathscr{B}(E \times F)$. By 3.(4) we have to show that H is an equicontinuous set of bilinear forms on $E \times F$. The assumption means that $|H(x, y)| < \infty$ for every $(x, y) \in E \times F$. Then for every $x \in E$ the set $\tilde{H}(x)$ is bounded in F' and therefore $\tilde{H}(x)$ is equicontinuous in F' since F is barrelled. Similarly, for every $y \in F$ the set $\tilde{\tilde{H}}(y)$ is equicontinuous in E'. Thus H is separately equicontinuous and by § 40, 2.(11) H is equicontinuous. This proves a) for $E \otimes_\pi F$. That $E \tilde{\otimes}_\pi F$ is also barrelled follows from § 27. 1.(2).

b) Now we have to show that a $\mathfrak{T}_b(E \otimes_\pi F)$-bounded subset H of $\mathscr{B}(E \times F)$ is equicontinuous. For M, N bounded subsets of E and F, respectively, $|H(M, N)| < \infty$. It follows that $\tilde{H}(M)$ is strongly bounded in F'. Since F is quasi-barrelled, $\tilde{H}(M)$ is equicontinuous in F' and there exists a neighbourhood V of $\circ$ in F such that $|H(M, V)| \leqq 1$. Similarly, there exists a neighbourhood U of $\circ$ in E such that $|H(U, N)| \leqq 1$. Thus H is equihypocontinuous and by § 40, 2.(10) equicontinuous.

c) Bornological spaces are quasi-barrelled; thus $E \otimes_\pi F$ is quasi-barrelled by b). Recalling § 28, 1.(3), we see that we have to prove that every locally bounded linear functional $\dot{B}$ on $E \otimes_\pi F$ is continuous. Let B be the bilinear functional on $E \times F$ corresponding to $\dot{B}$. Then $|B(M, N)| < \infty$ for every pair M, N of bounded subsets of E and F, respectively. The same argument as in b) proves that B is a continuous bilinear form; thus $\dot{B}$ is continuous on $E \otimes_\pi F$.

d) If E and F are (M)-spaces, then $E \otimes_\pi F$ is barrelled by a). We have to prove that every bounded subset N of $E \tilde{\otimes}_\pi F$ is relatively compact. If C_n, D_n are fundamental sequences of bounded subsets in E and F, respectively, then $\overline{\Gamma(C_n \otimes D_n)}$ is a fundamental sequence of bounded subsets in $E \tilde{\otimes}_\pi F$; thus $N \subset \overline{\Gamma(C_n \otimes D_n)}$ for some n. By assumption C_n, D_n are precompact; hence $C_n \otimes D_n = \chi(C_n \times D_n)$ is precompact and $\overline{\Gamma(C_n \otimes D_n)}$ is compact.

5. Tensor products of linear maps. Let E_1, E_2, F_1, F_2 be locally convex. For $A_1 \in \mathfrak{L}(E_1, F_1)$, $A_2 \in \mathfrak{L}(E_2, F_2)$ the mapping

$$(A_1, A_2)(x_1, x_2) = (A_1x_1) \otimes (A_2x_2), \qquad x_1 \in E_1, x_2 \in E_2,$$

is a bilinear continuous mapping of $E_1 \times E_2$ into $F_1 \otimes_\pi F_2$. The corresponding linear map of $E_1 \otimes_\pi E_2$ in $F_1 \otimes_\pi F_2$ defined by

$$A_1 \otimes A_2(x_1 \otimes x_2) = (A_1x_1) \otimes (A_2x_2)$$

is continuous by 3.(1). The kernel $N[A_1 \otimes A_2]$ is equal to $D[N[A_1], N[A_2]]$ by 1.(8).

(1) *$A_1 \otimes A_2$ has a uniquely determined continuous extension $A_1 \tilde{\otimes}_\pi A_2$ which maps $E_1 \tilde{\otimes}_\pi E_2$ in $F_1 \tilde{\otimes}_\pi F_2$.*

If E_1, E_2, F_1, F_2 are normed spaces, then

$$\text{(2)} \qquad \|A_1 \otimes A_2\| = \|A_1 \tilde{\otimes}_\pi A_2\| = \|A_1\|\,\|A_2\|,$$

where the norms of $A_1 \otimes A_2$ and $A_1 \tilde{\otimes}_\pi A_2$ are taken in $\mathfrak{L}(E_1 \otimes_\pi E_2, F_1 \otimes_\pi F_2)$ and $\mathfrak{L}(E_1 \tilde{\otimes}_\pi E_2, F_1 \tilde{\otimes}_\pi F_2)$, respectively. This is easy to see: The norm of the bilinear mapping (A_1, A_2) is $\|A_1\|\,\|A_2\|$ and by 3.(6) this norm is equal to $\|A_1 \otimes A_2\|$.

Our first result is

(3) *If A_1, A_2 are homomorphisms onto F_1 and F_2, respectively, then $A_1 \otimes A_2$ is a homomorphism of $E_1 \otimes_\pi E_2$ onto $F_1 \otimes_\pi F_2$ with kernel $D[N[A_1], N[A_2]]$.*

The range of $A_1 \otimes A_2$ is $F_1 \otimes F_2$. If $\Gamma(U_1 \otimes U_2)$ is a neighbourhood of $\circ$ in $E_1 \otimes_\pi E_2$, there exist neighbourhoods V_1, V_2 of $\circ$ in F_1 and F_2, respectively, such that $A_1(U_1) \supset V_1$, $A_2(U_2) \supset V_2$. Thus

$$A_1 \otimes A_2(\Gamma(U_1 \otimes U_2)) \supset \Gamma(V_1 \otimes V_2)$$

and $A_1 \otimes A_2$ is open.

As an immediate corollary we obtain

(4) *The π-product $E/M \otimes_\pi F/N$ of two quotients is isomorphic to the quotient $(E \otimes_\pi F)/D[M, N]$.*

Our next result states that the π-tensor product of two complemented subspaces is a complemented subspace of the π-tensor product. More precisely,

(5) *Let F_1, F_2 be locally convex spaces, P_1, P_2 continuous projections with ranges $P_1(F_1) = E_1$, $P_2(F_2) = E_2$, and kernels N_1, N_2. Then $P_1 \otimes P_2$ is a continuous projection of $F_1 \otimes_\pi F_2$ onto the subspace $E_1 \otimes_\pi E_2$ with kernel $D[N_1, N_2]$ and $P_1 \tilde{\otimes}_\pi P_2$ is a continuous projection of $F_1 \tilde{\otimes}_\pi F_2$ onto the subspace $E_1 \tilde{\otimes}_\pi E_2$ with kernel $\overline{D[N_1, N_2]}$, where the closure is taken in $F_1 \tilde{\otimes}_\pi F_2$.*

Proof. $P_1 \otimes P_2$ is continuous, has the range $E_1 \otimes E_2$, and reproduces the elements of $E_1 \otimes E_2$. Hence $P_1 \otimes P_2$ is a continuous projection of $F_1 \otimes_\pi F_2$ onto $E_1 \otimes E_2$. We denote $E_1 \otimes E_2$ equipped with the topology induced by $F_1 \otimes_\pi F_2$ by $(E_1 \otimes E_2)_\pi$ and show that $(E_1 \otimes E_2)_\pi$ is isomorphic to $E_1 \otimes_\pi E_2$. If J_1, J_2 are the injections of E_1, E_2 into F_1 and F_2, respectively, then $J_1 \otimes J_2$ is the continuous injection of $E_1 \otimes_\pi E_2$ in $F_1 \otimes_\pi F_2$; thus the topology of $E_1 \otimes_\pi E_2$ is finer than the topology of $(E_1 \otimes E_2)_\pi$. Conversely, if we consider P_1 and P_2 as elements of $\mathfrak{L}(F_1, E_1)$ and $\mathfrak{L}(F_2, E_2)$, respectively, then $P_1 \otimes P_2$ is a continuous map of $F_1 \otimes_\pi F_2$ onto $E_1 \otimes_\pi E_2$; hence its restriction to $(E_1 \otimes E_2)_\pi$ is continuous and the topology of $(E_1 \otimes E_2)_\pi$ is finer than the topology of $E_1 \otimes_\pi E_2$.

This proves the first statement and $F_1 \otimes_\pi F_2$ is the direct topological sum $E_1 \otimes_\pi E_2 \oplus D[N_1, N_2]$. Since the completion of such a sum is the sum of the completions, $F_1 \tilde{\otimes}_\pi F_2 = E_1 \tilde{\otimes}_\pi E_2 \oplus \overline{\tilde{D}[N_1, N_2]}$ and this implies the second statement.

In (5) we have an example of a result on projective tensor products which remains true for the completed tensor products. This is not the case for the result (3) on homomorphisms in general, as we will find out now.

We prove first

(6) *Let E, F be locally convex. Then*

a) *the injection J of $E \otimes_\pi F$ into $\tilde{E} \otimes_\pi \tilde{F}$ is a monomorphism on a dense subspace, and therefore $E \tilde{\otimes}_\pi F$ and $\tilde{E} \tilde{\otimes}_\pi \tilde{F}$ are isomorphic,*

b) *if* E_1 *is dense in* E, F_1 *dense in* F, *then the topology on* $E_1 \otimes F_1$ *induced by* $E \otimes_\pi F$ *coincides with the topology of* $E_1 \otimes_\pi F_1$,

c) *if* E *and* F *are normed spaces,* $\|J\| = 1$ *and* $E \tilde{\otimes}_\pi F$ *and* $\tilde{E} \tilde{\otimes}_\pi \tilde{F}$ *are norm isomorphic.*

Proof. a) We remark first that if a $B \in \mathscr{B}(\tilde{E} \times \tilde{F}) = (\tilde{E} \otimes_\pi \tilde{F})'$ vanishes on $E \otimes F$, it vanishes on $\tilde{E} \otimes \tilde{F}$; thus $E \otimes F$ is dense in $\tilde{E} \otimes_\pi \tilde{F}$. By the homomorphism theorem (§ 32, 4.(3)) and 3.(4) J is a monomorphism if and only if J' maps $\mathscr{B}(\tilde{E} \times \tilde{F})$ onto $\mathscr{B}(E \times F)$ and equicontinuous sets onto equicontinuous sets. If $B \in \mathscr{B}(\tilde{E} \times \tilde{F})$, then it follows from $\langle B, J(x \otimes y)\rangle = \langle J'B, x \otimes y\rangle$ that $J'B$ is the restriction of B to $E \times F$. From § 40, 3.(2) follow now the required properties of J'.

b) is an easy consequence of a), and c) follows from $\|J'\| = 1$.

We are now able to prove

(7) *Let* A_1, A_2 *be homomorphisms of* E_1, E_2 *onto dense subspaces of* F_1 *and* F_2, *respectively. Then* $A_1 \otimes A_2$ *and* $A_1 \tilde{\otimes}_\pi A_2$ *are homomorphisms of* $E_1 \otimes_\pi E_2$, $E_1 \tilde{\otimes}_\pi E_2$ *onto dense subspaces of* $F_1 \otimes_\pi F_2$ *and* $F_1 \tilde{\otimes}_\pi F_2$, *respectively.*

The kernel $N[A_1 \tilde{\otimes}_\pi A_2]$ *is the closure* $\bar{D}$ *of* $D[N[A_1], N[A_2]]$ *in* $E_1 \tilde{\otimes}_\pi E_2$.

If E_1, E_2 *are metrizable and* A_1 *and* A_2 *homomorphisms onto* F_1 *and* F_2, *respectively, then* $A_1 \tilde{\otimes}_\pi A_2$ *is a homomorphism onto* $F_1 \tilde{\otimes}_\pi F_2$.

Proof. a) By (3) $A = A_1 \otimes A_2$ is a homomorphism of $E_1 \otimes_\pi E_2$ onto $A_1(E_1) \otimes_\pi A_2(E_2)$. By (6) b) this is a dense subspace of $F_1 \otimes_\pi F_2$ equipped with the induced topology; thus $A_1 \otimes A_2$ is a homomorphism in $F_1 \otimes_\pi F_2$.

It follows from § 32, 5.(3) that $\tilde{A} = A_1 \tilde{\otimes}_\pi A_2$ is also a homomorphism of $E_1 \tilde{\otimes}_\pi E_2$ in $F_1 \tilde{\otimes}_\pi F_2$.

b) A and $\tilde{A}$ have the same adjoint A' and, since A is a homomorphism, the range of A' is $D^\perp$, D the kernel of A, and the kernel of $\tilde{A}$ is $D^{\perp\perp}$, which is $\bar{D}$ by the theorem of bipolars.

c) If E_1, E_2 are metrizable, $E_1 \tilde{\otimes}_\pi E_2$ is an (F)-space, $E_1 \tilde{\otimes}_\pi E_2/\bar{D}$ is complete, the range of $A_1 \tilde{\otimes}_\pi A_2$ is complete and dense in $F_1 \tilde{\otimes}_\pi F_2$ and coincides therefore with $F_1 \tilde{\otimes}_\pi F_2$.

It follows from the last argument that $A_1 \tilde{\otimes}_\pi A_2$, where A_1, A_2 are homomorphisms onto F_1 and F_2, respectively, will be a homomorphism onto $F_1 \tilde{\otimes}_\pi F_2$ if and only if the quotient $(E_1 \tilde{\otimes}_\pi E_2)/\bar{D}$ is complete.

We obtain the following corollary:

(8) *Let* E *and* F *be metrizable,* E/M *and* F/N *two quotients. Then* $E/M \tilde{\otimes}_\pi F/N$ *is isomorphic to* $(E \tilde{\otimes}_\pi F)/\overline{D[M, N]}$.

If E *and* F *are normed spaces, this isomorphism is a norm isomorphism.*

The first statement is an immediate consequence of (7) for A_1, A_2, the canonical maps onto the quotient spaces.

For normed spaces we write $A = A_1 \tilde{\otimes}_\pi A_2$ as $A = \hat{A}K$, where K is the canonical homomorphism of $E \tilde{\otimes}_\pi F$ onto $(E \tilde{\otimes}_\pi F)/\bar{D}$. We have to show that $\hat{A}$ is a norm isomorphism. We note that by 3.(6) the normed spaces $(E \tilde{\otimes}_\pi F)'_b$ and $\mathscr{B}_b(E \times F)$ can be identified.

Let $B \in \mathscr{B}(E/M \times F/N)$. We have

$$B(\hat{x}, \hat{y}) = \langle \dot{B}, \hat{x} \otimes \hat{y} \rangle = \langle \dot{B}, A(x \otimes y) \rangle = (A'\dot{B})(x \otimes y) = (A'B)(x, y);$$

hence $A'B$ has the same norm in $\mathscr{B}(E \times F)$ as B in $\mathscr{B}(E/M \times F/N)$. Therefore $A' = J\hat{A}'$, where $\hat{A}'$ is a norm isomorphism of $\mathscr{B}(E/M \times F/N)$ onto a subspace of $\mathscr{B}(E \times F)$ which is norm isomorphic to $((E \tilde{\otimes}_\pi F)/\bar{D})'$ by § 22, 3.(1) b) and J is the injection of this subspace into $\mathscr{B}(E \times F)$. Thus $\hat{A}$ is a norm isomorphism.

Another proof follows from the norm isomorphism of $(E \otimes_\pi F)/D$ and $E/M \otimes_\pi F/N$, which is a consequence of 2.(8) b).

Let now E_1, F_1 be subspaces of E, F and let J_1, J_2 be their injections into E and F, respectively. J_1 and J_2 are monomorphisms. Then $J = J_1 \otimes J_2$ is the injection of $E_1 \otimes_\pi F_1$ into $E \otimes_\pi F$ and J is continuous, but in general not a monomorphism, so that the topology of $E_1 \otimes_\pi F_1$ is in general strictly finer than the topology induced by $E \otimes_\pi F$.

We study this question in some detail. We denote $J_1 \tilde{\otimes}_\pi J_2$ by $\bar{J}$. Our first result is

(9) *J and $\bar{J}$ are monomorphisms if and only if every equicontinuous subset H_1 of $\mathscr{B}(E_1 \times F_1)$ is the set of restrictions to $E_1 \times F_1$ of an equicontinuous subset H of $\mathscr{B}(E \times F)$.*

If E and F are metrizable, it is necessary and sufficient that every $B_1 \in \mathscr{B}(E_1 \times F_1)$ be the restriction of a $B \in \mathscr{B}(E \times F)$.

Proof. As in the proof of (6), we use the homomorphism theorem (§ 32, 4.(3)) and see that J and $\bar{J}$ are monomorphisms if and only if every equicontinuous subset H_1 of $\mathscr{B}(E_1 \times F_1)$ is contained in the image $J'(H)$ of an equicontinuous subset H of $\mathscr{B}(E \times F)$. Let $B \in \mathscr{B}(E \times F)$. It follows from $\langle \dot{B}, J(x_1 \otimes x_2) \rangle = \langle J'\dot{B}, x_1 \otimes x_2 \rangle$ for $x_1 \in E_1$, $x_2 \in F_1$ that $B_1 \in \mathscr{B}(E_1 \times F_1)$ corresponding to $\dot{B}_1 = J'\dot{B}$ is the restriction of B to $E_1 \times F_1$.

In the metrizable case one has only to prove that J is a weak monomorphism (§ 33, 2.(3)), and this is the case if and only if $J'(\mathscr{B}(E \times F)) = \mathscr{B}(E_1 \times F_1)$.

Let us now assume that E and F are normed spaces. Then J_1 and J_2 are norm isomorphisms into E and F, respectively. In this case one has

(10) *J and $\bar{J}$ are norm isomorphisms if and only if every continuous bilinear form B_1 on $E_1 \times F_1$ is the restriction of a continuous bilinear form B on $E \times F$ such that $\|B\| = \|B_1\|$.*

Assume that J is a norm isomorphism and let $B_1 \in \mathscr{B}(E_1 \times F_1)$. Then $B_1 J^{-1}$ is defined on $J(E_1 \otimes_\pi F_1)$, $\|B_1\| = \|B_1 J^{-1}\|$. By HAHN–BANACH $B_1 J^{-1}$ has a linear extension B to $E \otimes_\pi F$ such that $\|B\| = \|B_1\|$. Thus B_1 is the restriction of $B \in \mathscr{B}(E \times F)$ to $E_1 \times F_1$ and $\|B_1\| = \|B\|$.

Conversely, if the condition is satisfied, then J is a norm isomorphism, since

$$\|Jz\|_\pi = \sup_{\|B\| \leqq 1} |\langle B, Jz\rangle| = \sup_{\|B\| \leqq 1} |\langle J'B, z\rangle| = \sup_{\|B_1\| \leqq 1} |\langle B_1, z\rangle| = \|z\|_\pi$$

(9) and (10) are the dual formulations of our problem. We give now some positive and some negative results.

(11) a) *Let E, F be locally convex and E'' be the bidual equipped with the natural topology $\mathfrak{T}_n$. Let J_1 be the canonical injection of E into E'' and I the identity on F.*

Then the injection $J = J_1 \otimes I$ of $E \otimes_\pi F$ in $E'' \otimes_\pi F$ and the mapping $\bar{J} = J_1 \tilde{\otimes}_\pi I$ of $E \tilde{\otimes}_\pi F$ in $E'' \tilde{\otimes}_\pi F$ are monomorphisms.

b) *If, moreover, E and F are normed spaces, then J and $\bar{J}$ are norm isomorphisms.*

We remark that if E is quasi-barrelled, then E'' is the strong bidual (§ 23, 4.(4)).

Proof. We recall § 40, 3.(5) c). Let H_1 be an equicontinuous subset of $\mathscr{B}(E \times F)$. Then $|H_1(U, V)| \leqq 1$ for suitable neighbourhoods $U \ni \mathfrak{o}$, $V \ni \mathfrak{o}$ in E and F, respectively. By using weak continuity in the first variable one extends every $B_1 \in H_1$ to a B defined on $E'' \times F$ such that $|B(U^{\circ\circ}, V)| \leqq 1$. a) follows now from (9).

b) follows in the same way from (10).

(12) *Let J_1, J_2 be monomorphisms. If $J_1(E_1)$ and $J_2(E_2)$ are complemented in E and F, respectively, then $J = J_1 \otimes J_2$ and $\bar{J} = J_1 \tilde{\otimes}_\pi J_2$ are monomorphisms.*

This is a corollary to (9), but also to (5).

In the case of normed spaces $E_1 \subset E, F$ our problem is equivalent to an extension problem for linear continuous mappings. We recall the norm isomorphism of $\mathscr{B}_b(E \times F)$ and $\mathfrak{L}_b(E, F'_b)$ of 3.(6). A $B_1 \in \mathscr{B}(E_1 \times F)$ will be the restriction of a $B \in \mathscr{B}(E \times F)$ if and only if the corresponding mapping $\tilde{B}_1$ of E_1 in F'_b has a continuous extension to a mapping $\tilde{B}$ of E in F'_b. This remark is useful in the proof of the following result of SCHATTEN [1'].

(13) *Let E, F be* (B)*-spaces such that E'_b is a subspace of F. Let J_1 be the injection of E'_b in F, I the identity on E. Then $J = J_1 \otimes I$ and $\tilde{J} = J_1 \tilde{\otimes}_\pi I$ are monomorphisms of $E'_b \otimes_\pi E$ in $F \otimes_\pi E$ resp. $E'_b \tilde{\otimes}_\pi E$ in $F \tilde{\otimes}_\pi E$ if and only if E'_b is complemented in F.*

The condition is sufficient by (12). Assume, conversely, that J is a monomorphism. Then by the preceding remark the identity $I \in \mathfrak{L}(E'_b, E'_b)$ has a continuous extension $\check{I} \in \mathfrak{L}(F, E'_b)$ and $J_1\check{I}$ is a continuous projection of F onto E'_b since $(J_1\check{I})^2 = J_1\check{I}$.

We remark that if J_1 is a norm isomorphism, then J and $\tilde{J}$ are norm isomorphisms if and only if there exists a projection of norm 1 of F onto E'_b.

Using the examples of § 31, 3., it is now easy to construct many counterexamples: A closed reflexive noncomplemented subspace of a (B)-space F can always be written as E'. Then by (13) the injection of $E' \otimes_\pi E$ into $F \otimes_\pi E$ is not a monomorphism.

6. Further hereditary properties. Let E be a locally convex kernel $\mathsf{K}_\alpha A_\alpha^{(-1)}(E_\alpha)$, where α belongs to a directed set of indices A and there exist for $\alpha < \alpha'$ linear mappings $A_{\alpha\alpha'} \in \mathfrak{L}(E_{\alpha'}, E_\alpha)$ such that

$$(1) \qquad A_\alpha = A_{\alpha\alpha'}A_{\alpha'}, \qquad A_{\alpha\alpha'}A_{\alpha'\alpha''} = A_{\alpha\alpha''} \quad \text{for } \alpha < \alpha' < \alpha''$$

is satisfied. We suppose, moreover, that $\mathsf{K}_\alpha A_\alpha^{(-1)}(E_\alpha)$ is reduced, i.e., $A_\alpha(E)$ is dense in E_α for every α.

By § 19, 8.(1) a neighbourhood base of $\mathfrak{o}$ of the kernel topology on E is given by the sets $A_\alpha^{(-1)}(U_\alpha)$, where U_α is a neighbourhood of $\mathfrak{o}$ in E_α.

Let F be similarly a reduced locally convex kernel $\mathsf{K}_\beta B_\beta^{(-1)}(F_\beta)$ with additional maps $B_{\beta\beta'} \in \mathfrak{L}(F_{\beta'}, F_\beta)$ satisfying the relations corresponding to (1). Then

(2) *$E \otimes_\pi F$ is identical with a reduced locally convex kernel*

$$\mathsf{K}_{\alpha,\beta} (A_\alpha \otimes B_\beta)^{(-1)}(E_\alpha \otimes_\pi F_\beta)$$

with the additional maps $A_{\alpha\alpha'} \otimes B_{\beta\beta'} \in \mathfrak{L}(E_{\alpha'} \otimes_\pi F_{\beta'}, E_\alpha \otimes_\pi F_\beta)$ satisfying the relations corresponding to (1).

Proof. It is obvious that $E \otimes F$ can be written as

$$\mathsf{K}_{\alpha,\beta} (A_\alpha \otimes B_\beta)^{(-1)} (E_\alpha \otimes F_\beta)$$

and that the mappings $A_{\alpha\alpha'} \otimes B_{\beta\beta'}$ satisfy the relations corresponding to (1). Since $A_\alpha(E)$ is dense in E_α and $B_\beta(F)$ is dense in F_β, it follows that $(A_\alpha \otimes B_\beta)(E \otimes F)$ is dense in $E_\alpha \otimes_\pi F_\beta$ and the locally convex kernel $\mathsf{K}_{\alpha,\beta} (A_\alpha \otimes B_\beta)^{(-1)}(E_\alpha \otimes_\pi F_\beta)$ is reduced.

We denote by $\mathfrak{T}_\pi$ the topology of $E \otimes_\pi F$ and by $\mathfrak{T}$ the topology of the locally convex kernel $\mathsf{K}_{\alpha,\beta}(A_\alpha \otimes B_\beta)^{(-1)}(E_\alpha \otimes_\pi F_\beta)$. We have to prove that $\mathfrak{T}_\pi$ and $\mathfrak{T}$ coincide. Let $\{U_\alpha\}$, $\{V_\beta\}$ be bases of absolutely convex open neighbourhoods of o in E_α and F_β, respectively. Then the sets $W = \Gamma(A_\alpha^{(-1)}(U_\alpha) \otimes B_\beta^{(-1)}(V_\beta))$ define a $\mathfrak{T}_\pi$-neighbourhood base of o on $E \otimes F$; similarly, the sets $W' = (A_\alpha \otimes B_\beta)^{(-1)}(\Gamma(U_\alpha \otimes V_\beta))$ define a $\mathfrak{T}$-neighbourhood base of o.

Any $z \in W$ has the form $z = \sum_{i=1}^{n} \lambda_i x_i \otimes y_i$, $\sum |\lambda_i| \leqq 1$, $x_i \in A_\alpha^{(-1)}(U_\alpha)$, $y_i \in B_\beta^{(-1)}(V_\beta)$; hence $(A_\alpha \otimes B_\beta)z \in \Gamma(U_\alpha \otimes V_\beta)$, $W \subset W'$ and $\mathfrak{T}_\pi$ is finer than $\mathfrak{T}$.

We prove the converse first for the particular case that $A_\alpha(E) = E_\alpha$, $B_\beta(F) = F_\beta$. Since U_α, V_β are absolutely convex and open, one has for any $z' \in W'$ that $(A_\alpha \otimes B_\beta)z' = \sum_{j=1}^{m} \mu_j u_j \otimes v_j$, $\mu = \sum |\mu_j| < 1$, $u_j \in U_\alpha$, $v_j \in V_\beta$. By assumption there exist $x_j' \in A_\alpha^{(-1)}(U_\alpha), y_j' \in B_\beta^{(-1)}(V_\beta)$ such that $(A_\alpha \otimes B_\beta)(z' - \sum \mu_j x_j' \otimes y_j') = \mathsf{o}$. Hence $z' = \sum \mu_j(x_j' \otimes y_j') + t$, where t is in the kernel of $A_\alpha \otimes B_\beta$. Now $\sum \mu_j(x_j' \otimes y_j') \in \mu W$ and it follows from 1.(8) that $t \in \varepsilon W$ for every $\varepsilon > 0$, in particular for $\varepsilon = 1 - \mu$. Thus $z' \in W$ and $W' \subset W$, $\mathfrak{T}$ is finer than $\mathfrak{T}_\pi$.

In the general case $A_\alpha(E)$ and $B_\beta(F)$ are dense in E_α and F_β, respectively, and $A_\alpha(E) \otimes B_\beta(F)$ is dense in $E_\alpha \otimes_\pi F_\beta$. We have to prove that every W' is contained in a W. But

$$\begin{aligned} W' &= (A_\alpha \otimes B_\beta)^{(-1)}(\Gamma(U_\alpha \otimes V_\beta)) \\ &= (A_\alpha \otimes B_\beta)^{(-1)}(\Gamma(U_\alpha \otimes V_\beta) \cap (A_\alpha(E) \otimes B_\beta(F))) \end{aligned}$$

and by 5.(6) b)

$$\Gamma(U_\alpha \otimes V_\beta) \cap (A_\alpha(E) \otimes B_\beta(F)) \subset \Gamma(U_\alpha' \otimes V_\beta'),$$

where U_α', V_β' are suitable open neighbourhoods of o in $A_\alpha(E)$ and $B_\beta(F)$, respectively. It follows now from the result in the particular case that $W' \subset (A_\alpha \otimes B_\beta)^{(-1)}(\Gamma(U_\alpha' \otimes V_\beta')) \subset \Gamma(A_\alpha^{(-1)}(U_\alpha) \otimes B_\beta^{(-1)}(V_\beta))$, which is easily seen to be a set of type W.

In the terminology of GROTHENDIECK and SCHWARTZ the locally convex kernels of the type considered above are called "projective limits" and the hereditary property (2) for "projective limits" was the reason for using the term "projective tensor product." We use the term "projective limit" in a more restricted sense, as we pointed out in § 19, 8., and the hereditary property (2) is not true for projective limits in our sense. We will give an example at the end of 6.

Our next result deals with completed π-tensor products of two locally convex kernels.

Let E and F be two reduced locally convex kernels with all the properties assumed above and, moreover, in reduced form. Then

(3) *$E \tilde{\otimes}_\pi F$ is isomorphic to the reduced projective limit*

$$\varprojlim (A_{\alpha\alpha'} \tilde{\otimes}_\pi B_{\beta\beta'})(E_{\alpha'} \tilde{\otimes}_\pi F_{\beta'}).$$

We prove this in two steps. a) By (2) we have

$$E \otimes_\pi F = \underset{\alpha,\beta}{\mathsf{K}} (A_\alpha \otimes B_\beta)^{(-1)}(E_\alpha \otimes_\pi F_\beta).$$

This space is, by § 19, 8.(1), topologically isomorphic to a subspace of the projective limit $G = \varprojlim (A_{\alpha\alpha'} \otimes B_{\beta\beta'})(E_{\alpha'} \otimes_\pi F_{\beta'})$, the isomorphism being defined by the mapping $z \to (z_{\alpha\beta}) = ((A_\alpha \otimes B_\beta)z)$ from $E \otimes_\pi F$ into $\prod_{\alpha\beta} (E_\alpha \otimes_\pi F_\beta)$. G is reduced, since $(A_\alpha \otimes B_\beta)(E \otimes F)$ is dense in $E_\alpha \otimes_\pi F_\beta$.

We prove that $E \otimes_\pi F$ is dense in G. Let z in G and $\Gamma(U_\alpha \otimes V_\beta)$ be a neighbourhood of o in $E_\alpha \otimes_\pi F_\beta$; then there exists by assumption a $z^{(o)} \in E \otimes_\pi F$ such that

$$z_{\alpha\beta} - z^{(o)}_{\alpha\beta} \in \Gamma(U_\alpha \otimes V_\beta) \text{ or } z - z^{(o)} \in (A_\alpha \otimes B_\beta)^{(-1)}(\Gamma(U_\alpha \otimes V_\beta)),$$

which by § 19, 8.(1) proves the assertion.

b) The second step is an immediate corollary of

(4) *Let $E = \varprojlim A_{\alpha\beta}(E_\beta)$ be a reduced projective limit; then $\tilde{E}$ is isomorphic to the reduced projective limit $\varprojlim \tilde{A}_{\alpha\beta}(\tilde{E}_\beta)$, where $\tilde{A}_{\alpha\beta}$ is the continuous extension of $A_{\alpha\beta} \in \mathfrak{L}(E_\beta, E_\alpha)$ to $\tilde{A}_{\alpha\beta} \in \mathfrak{L}(\tilde{E}_\beta, \tilde{E}_\alpha)$.*

We remark first that the relations $\tilde{A}_{\alpha\beta}\tilde{A}_{\beta\gamma} = \tilde{A}_{\alpha\gamma}$, $\alpha < \beta < \gamma$, are easy consequences of the relations $A_{\alpha\beta}A_{\beta\gamma} = A_{\alpha\gamma}$ (one uses the adjoints). Secondly, E is topologically isomorphic to a subspace of $\varprojlim \tilde{A}_{\alpha\beta}(\tilde{E}_\beta)$ and the same argument as in the proof of (3) shows that E is even a dense subspace. Finally, § 19, 10.(2) concludes the proof.

(5) *For arbitrary locally convex spaces E_α and F_β and arbitrary sets of indices $\mathsf{A} = \{\alpha\}$, $\mathsf{B} = \{\beta\}$ one has the isomorphism*

$$\left(\prod_{\alpha\in\mathsf{A}} E_\alpha\right) \tilde{\otimes}_\pi \left(\prod_{\beta\in\mathsf{B}} F_\beta\right) \cong \prod_{(\alpha,\beta)\in\mathsf{A}\times\mathsf{B}} E_\alpha \tilde{\otimes}_\pi F_\beta.$$

It is sufficient to prove $\left(\prod_\alpha E_\alpha\right) \tilde{\otimes}_\pi F \cong \prod_\alpha (E_\alpha \tilde{\otimes}_\pi F)$ for a locally convex F.

We consider $\bigoplus_\alpha E_\alpha$ as a subspace of $\prod_\alpha E_\alpha$ in the obvious way; then $H = \left(\bigoplus_\alpha E_\alpha\right) \otimes F$ is a $\mathfrak{T}$-dense subspace of $\left(\prod_\alpha E_\alpha\right) \tilde{\otimes}_\pi F$, where $\mathfrak{T}$ is the tensor product topology. H can be identified algebraically with $\bigoplus_\alpha (E_\alpha \otimes F)$, which is dense in $\prod_\alpha (E_\alpha \tilde{\otimes}_\pi F)$ in the sense of its topology $\mathfrak{T}'$. If we prove that $\mathfrak{T}$ and $\mathfrak{T}'$ coincide on H, then the statement follows from the completeness of both spaces.

Let U_α, V be absolutely convex neighbourhoods of o in E_α and F, respectively. As $\mathfrak{T}$-neighbourhoods of o on H we take the sets W consisting of all $z = \sum_{i=1}^m \lambda_i(x_i^{(\alpha_1)} + \cdots + x_i^{(\alpha_n)} + z_i) \otimes y_i$, $\sum |\lambda_i| \leq 1$, $x_i^{(\alpha_j)} \in U_{\alpha_j}$, $y_i \in V$, $z_i \in \bigoplus_{\substack{\beta \neq \alpha_j \\ j=1,\ldots,n}} E_\beta$, where n is fixed, m arbitrary. As $\mathfrak{T}'$-neighbourhoods of o on H we use the sets W' of all elements $t = \sum t^{(\alpha)}$, $t^{(\alpha_j)} \in \ulcorner(U^{\alpha_j} \otimes V)$, $t^{(\beta)} \in E_\beta \otimes F, \beta \neq \alpha_j, j = 1, \ldots, n$.

One has $W \subset W'$, since $z \in W$ can be written as $z = \sum_\alpha t^{(\alpha)}$, where $t^{(\alpha_j)} = \sum \lambda_i x_i^{(\alpha_j)} \otimes y_i \in \ulcorner(U_{\alpha_j} \otimes V)$. Conversely, $W' \subset (n + 1)W$ since for $t \in W'$, $t = \sum t^{(\alpha)}$ one has $t^{(\alpha_j)} \in \ulcorner(U_{\alpha_j} \otimes V) \subset W$, $j = 1, \ldots, n$, and $\sum_{\substack{\beta \neq \alpha_j \\ j=1,\ldots,n}} t^{(\beta)} \in W$. Hence $\mathfrak{T}$ and $\mathfrak{T}'$ coincide on H.

For locally convex hulls one does not have results of the same generality.

(6) a) *Let E be the locally convex hull $\sum_{\alpha \in A} E_\alpha$ of an arbitrary class of locally convex spaces E_α and let F be a normed space. Then $\left(\sum_\alpha E_\alpha\right) \otimes_\pi F$ is isomorphic to $\sum_\alpha (E_\alpha \otimes_\pi F)$.*

b) *Let $E = \sum_{n=1}^\infty E_n$, E_n locally convex, and let F be a* (DF)*-space. Then $\left(\sum_{n=1}^\infty E_n\right) \otimes_\pi F$ is isomorphic to $\sum_n (E_n \otimes_\pi F)$.*

a) It is easy to check that $H = \left(\sum_\alpha E_\alpha\right) \otimes F$ can be considered algebraically as the linear span $\sum_\alpha (E_\alpha \otimes F)$. We write $(\sum E_\alpha) \otimes_\pi F = H[\mathfrak{T}]$ and $\sum_\alpha (E_\alpha \otimes_\pi F) = H[\mathfrak{T}']$. The space $E_\alpha \otimes_\pi F$ is continuously imbedded in $H[\mathfrak{T}]$ and $H[\mathfrak{T}']$. Since $\mathfrak{T}'$ is the hull topology on H, it follows that $\mathfrak{T}' \supset \mathfrak{T}$.

It will therefore be sufficient to show that every $\mathfrak{T}'$-equicontinuous set $M \subset (H[\mathfrak{T}'])'$ is $\mathfrak{T}$-equicontinuous. Every $B \in M$ is a linear functional on H, hence a bilinear form on $\left(\sum_\alpha E_\alpha\right) \times F$. It follows from the remark after the proof of § 19, 1.(7) that the restrictions of the $B \in M$ to $E_\alpha \otimes_\pi F$ form an equicontinuous set M_α of bilinear forms. Thus, if V is the unit ball in F, there exists a neighbourhood U_α of $\circ$ in E_α such that $|M_\alpha(U_\alpha, V)| \leqq 1$. But then $|M(U, V)| \leqq 1$ for $U = \ulcorner_\alpha U_\alpha$ and M is $\mathfrak{T}$-equicontinuous.

b) The proof is analogous. We find U_n in E_n, V_n in F such that $|M_n(U_n, V_n)| \leqq 1$ for $n = 1, 2, \ldots$. By § 39, 8.(7) there exist $\rho_n > 0$ such that $\bigcap_{n=1}^\infty \rho_n V_n = V$ is a neighbourhood of $\circ$ in F. Hence

$$\left| M_n\left(\frac{1}{\rho_n} U_n, V\right) \right| \leqq 1 \quad \text{and} \quad |M(U, V)| \leqq 1 \quad \text{for } U = \ulcorner_{n=1}^{\infty} \frac{1}{\rho_n} U_n.$$

Since a locally convex direct sum $\bigoplus_\alpha E_\alpha$ is complete if and only if all E_α are complete (§ 18, 5.(3)), we obtain

(7) *Under the same assumptions as in* (6) a), *$\left(\bigoplus_\alpha E_\alpha\right) \tilde{\otimes}_\pi F$ and $\bigoplus_\alpha (E_\alpha \tilde{\otimes}_\pi F)$ are isomorphic, and under the same assumptions as in* (6) b),

$\left(\bigoplus_{n=1}^{\infty} E_n\right) \tilde{\otimes}_\pi F$ *and* $\bigoplus_{n=1}^{\infty} (E_n \tilde{\otimes}_\pi F)$ *are isomorphic.*

As an example we investigate $\omega \otimes_\pi \varphi$. By (5) we have

$$\omega \tilde{\otimes}_\pi \varphi = \left(\prod_{i=1}^{\infty} e_i \mathsf{K}\right) \tilde{\otimes}_\pi \varphi \cong \prod_{i=1}^{\infty} \left((e_i \mathsf{K}) \tilde{\otimes}_\pi \varphi\right).$$

If we denote by $f_1, f_2, \ldots$ the unit vectors in φ, we see that every element z of $\omega \tilde{\otimes}_\pi \varphi$ can be represented by a double sequence $z = (\xi_{ik} e_i \otimes f_k)$. $\omega \otimes_\pi \varphi$ contains only elements such that $\xi_{ik} = 0$ for $k \geqq k_0$ for some k_0 independent of i, whereas every z is in $\omega \tilde{\otimes}_\pi \varphi$ for which $\xi_{ik} = 0$ for $k \geqq k(i)$, where $k(i)$ depends on i. Hence $\omega \otimes_\pi \varphi$ is incomplete and we have an example of a π-product of two projective limits which is not a projective limit.

Similarly, if M, N are bounded (compact) sets in ω and φ, respectively, then $\Gamma(M \otimes N)$ contains only elements z such that $|\xi_{ik}| \leqq R_{ik}$ and $\xi_{ik} = 0$ for $k \geqq k_0$. On the other hand, every set of elements z with $|\xi_{ik}| \leqq R_{ik}$ and $\xi_{ik} = 0$ for $k \geqq k(i)$ is bounded and compact in $\omega \tilde{\otimes}_\pi \varphi$. Thus in $\omega \tilde{\otimes}_\pi \varphi$ the sets $\overline{\Gamma(M \otimes N)}$, M and N bounded (compact), do not contain all bounded (compact) subsets.

By 3.(6) $(\omega \otimes_\pi \varphi)' = \omega' \otimes \varphi' = \varphi \otimes \omega$. We leave it to the reader to verify that the strong dual $(\varphi \otimes \omega)_b$ is the topological direct sum of a sequence of spaces isomorphic to ω (a space isomorphic to $\varphi\omega$) and that the bibounded topology $\mathfrak{T}_{bb}$ on $\varphi \otimes \omega$ is the topology $\mathfrak{T}_\pi$, so that $((\varphi \otimes \omega)_b)' = \omega \tilde{\otimes}_\pi \varphi$ and $(\varphi \otimes \omega)[\mathfrak{T}_{bb}]' = \omega \otimes_\pi \varphi$.

We remark finally that (7) b) is no longer true if F is an (F)-space. We take again $\omega \otimes_\pi \varphi$ and write $\varphi = \bigoplus_{i=1}^{\infty} \mathsf{K}_i$, $\mathsf{K}_i \cong \mathsf{K}$. Then $\omega \tilde{\otimes}_\pi \left(\bigoplus_{i=1}^{\infty} \mathsf{K}_i\right)$ is a space of type $\omega\varphi$, whereas $\bigoplus_{i=1}^{\infty} (\mathsf{K}_i \tilde{\otimes}_\pi \omega)$ coincides with $\omega \otimes_\pi \varphi$ and is of type $\varphi\omega$. Not only are the topologies induced on $\omega \otimes \varphi$ different; even the dual spaces are different.

7. Some special cases. a) PIETSCH gave in [1′] a concrete representation of $\lambda \tilde{\otimes}_\pi F$, where λ is a perfect sequence space, F arbitrarily locally convex, which we reproduce here.

The topology on λ will be the normal topology, defined by the elements $\mathfrak{u}$ of $\lambda^\times$ and the corresponding semi-norms $p_\mathfrak{u}$ (§ 30, 2.). The topology on F is given by a neighbourhood base $\{U\}$ of $\circ$, U absolutely convex, q_U the associated semi-norm.

We define $\lambda\{F\}$ as the space of all sequences $y = (y_n)$, $y_n \in F$, such that $\sum_{n=1}^{\infty} u_n y_n$ converges absolutely in F for every $\mathfrak{u} = (u_n) \in \lambda^\times$, which means that $\sum_{n=1}^{\infty} |u_n| q_U(y_n) < \infty$ for every $\mathfrak{u} \in \lambda^\times$ and every $U \in \{U\}$. An equivalent condition is that the sequence $(q_U(y_n))$ is in λ for every $U \in \{U\}$.

The topology $\mathfrak{T}$ on $\lambda\{F\}$ is defined by the semi-norms

$$\pi_{\mathfrak{u},U}(y) = \sum_{n=1}^{\infty} |u_n| q_U(y_n), \qquad y = (y_n) \in \lambda\{F\}, \qquad \mathfrak{u} \in \lambda^\times,\ U \in \{U\}.$$

(1) *The completion of* $\lambda\{F\}$ *is* $\lambda\{\tilde{F}\}$.

The proof is standard: A Cauchy net $y^{(\alpha)} = (y_n^{(\alpha)})$ is a Cauchy net in every coordinate and the sequence consisting of the coordinatewise limits in $\tilde{F}$ is the limit of $y^{(\alpha)}$.

Let φF denote the space of all sequences (y_n), $y_n \in F$, such that $y_n = \mathfrak{o}$ for n greater than some n_0. Obviously, $\lambda\{F\} \supset \varphi F$ and

(2) φF *is dense in* $\lambda\{F\}$.

This is obvious since there exists always a section

$$y^{(n_0)} = (y_1, \ldots, y_{n_0}, \mathfrak{o}, \mathfrak{o}, \ldots)$$

such that $\pi_{\mathfrak{u},U}(y - y^{(n_0)}) = \sum_{n_0+1}^{\infty} |u_n| q_U(y_n) < \varepsilon$. Every $y \in \lambda\{F\}$ is the limit of its sections.

For $\mathfrak{x} = (x_n) \in \lambda$ and $y^{(0)} \in F$ we define the map $(\mathfrak{x}, y^{(0)}) \to (x_n y^{(0)})$ of $\lambda \times F$ in $\lambda\{F\}$. It is bilinear and induces therefore a linear map $J\left(\sum_{i=1}^{k} \mathfrak{x}^{(i)} \otimes y^{(i)}\right) = \left(\sum_{i=1}^{k} x_n^{(i)} y^{(i)}\right)$ of $\lambda \otimes F$ in $\lambda\{F\}$. J is one-one, which becomes obvious if one chooses the $y^{(i)}$ linearly independent in F.

(3) *J is an isomorphism of $\lambda \otimes_\pi F$ onto a dense subspace of $\lambda\{F\}$ such that*

(4) $$\pi_{\mathfrak{u},U}(Jz) = p_{\mathfrak{u}} \otimes q_U(z), \qquad z \in \lambda \otimes_\pi F.$$

Proof. If $z = \sum_{i=1}^{k} \mathfrak{x}^{(i)} \otimes y^{(i)}$, then

$$\pi_{\mathfrak{u},U}(Jz) = \sum_{n=1}^{\infty} |u_n| q_U\left(\sum_{i=1}^{k} x_n^{(i)} y^{(i)}\right) \leq \sum_n \sum_i |u_n| |x_n^{(i)}| q_U(y^{(i)}).$$

Since $p_{\mathfrak{u}} \otimes q_U(z) = \inf \sum_i \left(\sum_{n=1}^{\infty} |u_n| |x_n^{(i)}|\right) q_U(y^{(i)})$ and since $\pi_{\mathfrak{u},U}(Jz)$ is independent of the particular representation of z as a sum, it follows that $\pi_{\mathfrak{u},U}(Jz) \leq p_{\mathfrak{u}} \otimes q_U(z)$.

Next we show that $p_{\mathfrak{u}} \otimes q_U(z) \leq \pi_{\mathfrak{u},U}(Jz)$ for $z \in \varphi \otimes F$ or, equivalently, $Jz \in \varphi F$. For the unit vector $\mathfrak{e}_n \in \lambda$ and $y_n \in F$, $J(\mathfrak{e}_n \otimes y_n)$ is the sequence with only the nth member y_n different from $\mathfrak{o}$. One has

$$p_{\mathfrak{u}} \otimes q_U(\mathfrak{e}_n \otimes y_n) = |u_n| q_U(y_n) = \pi_{\mathfrak{u},U}(J(\mathfrak{e}_n \otimes y_n)).$$

An arbitrary element of φF has the form $J\left(\sum_{n=1}^{N} \mathfrak{e}_n \otimes y_n\right)$ and one has

$$p_{\mathfrak{u}} \otimes q_U\left(\sum_{n=1}^{N} \mathfrak{e}_n \otimes y_n\right) \leqq \sum_{n=1}^{N} |u_n| q_U(y_n) = \pi_{\mathfrak{u},U}\left(J \sum_{n=1}^{N} \mathfrak{e}_n \otimes y_n\right).$$

Therefore (4) is true for $z \in \varphi \otimes F$. Since $\varphi \otimes F$ is dense in $\lambda \otimes F$ and φF is dense in $\lambda\{F\}$ by (2), (4) follows by continuity for every $z \in \lambda \otimes F$. Finally, $J(\lambda \otimes F) \supset \varphi F$ is dense in $\lambda\{F\}$.

As an immediate consequence of (1) and (3) we obtain

(5) $\lambda \tilde{\otimes}_\pi F$ *can be identified with* $\lambda\{\tilde{F}\}$.

In particular, $l^1 \tilde{\otimes}_\pi F$ for complete F can be identified with the space $l^1\{F\}$ of all absolutely summable sequences $y = (y_n)$, $y_n \in F$. The topology $\mathfrak{T}_\pi$ is given by the semi-norms $\pi_U(y) = \sum_{n=1}^{\infty} q_U(y_n)$, $U \in \{U\}$, a neighbourhood base of F.

It is interesting to note (compare 3.(13)) that

(6) *The canonical mapping* $\bar{\psi}$ *of* $\lambda \tilde{\otimes}_\pi F$, F *complete, into* $\mathfrak{L}_e(\lambda_k^\times, F)$ *is one-one.*

Let $z = \sum_{n=1}^{n_0} \mathfrak{e}_n \otimes y_n \in \varphi \otimes F$. Then $\psi(z)$ is the mapping $\mathfrak{u} \to \sum_{n=1}^{n_0} u_n y_n$, where $\mathfrak{u} = (u_1, u_2, \ldots) \in \lambda^\times$. Since every $y = (y_n) \in \lambda\{F\}$ is the limit of its sections $y^{(n_0)}$ and, since ψ is continuous, we obtain $\bar{\psi}(y)(\mathfrak{u}) = \sum_{n=1}^{\infty} u_n y_n$. Thus $\bar{\psi}(y) = \mathfrak{o}$ if and only if $y = \mathfrak{o}$.

We leave it to the reader to verify the following representation of $\lambda\{F\}'$. It consists of all sequences $v = (v_n)$, $v_n \in F'$, such that there exists an absolutely convex neighbourhood $U \ni \mathfrak{o}$ in F and a $\mathfrak{u} \in \lambda^\times$ with $v_n \in |u_n| U^\circ$ for every n. The duality is given by $\langle v, y\rangle = \sum_{n=1}^{\infty} v_n y_n$. The bilinear functional B corresponding to v is given by $B(\mathfrak{x}, y_0) = \sum_{n=1}^{\infty} x_n(v_n y_0)$.

The reader should also reconsider the example in 5. as a particular case of these results.

b) The same method can be applied to obtain a representation of the spaces $L^1_{X,\mu} \tilde{\otimes}_\pi F$, where F is a real resp. complex locally convex space and $L^1_{X,\mu}$ is the (B)-space of all equivalence classes of absolutely μ-summable real resp. complex functions f on the locally compact space X and μ a positive Radon measure on X. The norm is defined by $\|f\|_1 = \int_X |f(X)|\, d\mu$. We assume some elementary facts on integration theory. We treated a particular case in § 14, 10.

Let S be the subspace of $L^1_{X,\mu}$ consisting of the equivalence classes of all simple functions $\sigma(t)$ on X, $\sigma(t) = \sum_{i=1}^{n} \chi_i(t)\sigma_i$, where the σ_i are real resp. complex numbers and the χ_i are the characteristic functions of pairwise disjoint measurable sets I_i. As in the case of § 14, 10., it easily follows that S is dense in $L^1_{X,\mu}$.

Next we consider the functions of the form $s(t) = \sum_{i=1}^{n} \chi_i(t)y_i$, $t \in X$, $y_i \in F$, where the χ_i again belong to pairwise disjoint measurable sets and are determined only almost everywhere. It is easy to see that the set of all these F-valued "functions" is a vector space; we equip it with the topology $\mathfrak{T}$ defined by the semi-norms

$$\pi_U(s) = \sum_{i=1}^{n} \mu(I_i)q_U(y_i) = \int_X q_U(s(t))\, d\mu, \tag{7}$$

where q_U is a semi-norm on F corresponding to a neighbourhood U of a neighbourhood base $\{U\}$ of $\circ$ in F, and $\mu(I_i)$ is the measure of I_i. It follows immediately from (7) that $\mathfrak{T}$ is Hausdorff.

We denote this locally convex space by $S\{F\}$ and its completion by $L^1_{X,\mu}\{\tilde{F}\}$. This space is called the space of absolutely μ-summable F-valued functions. A justification for this terminology will be given below.

Our aim is to prove that $L^1_{X,\mu} \tilde{\otimes}_\pi F$ is isomorphic to $L^1_{X,\mu}\{\tilde{F}\}$. S is a normed space as a subspace of $L^1_{X,\mu}$. The mapping $(\sigma, y) \to \sigma y$ of $S \times F$ in $S\{F\}$ is bilinear and therefore generates a linear map J of $S \otimes F$ in $S\{F\}$. J is one-one because for linearly independent $y_i \in F$, $J(\sum \sigma_i \otimes y_i) = \sum \sigma_i y_i = \circ$ if and only if all $\sigma_i = \circ$ in S.

We prove now that J is an isomorphism of $S \otimes_\pi F$ into $S\{F\}$. Let $z = \sum \sigma_i \otimes y_i$. Then from

$$\pi_U(Jz) = \int q_U\Big(\sum_i \sigma_i(t) y_i\Big)\, d\mu \leqq \sum_i \|\sigma_i\|_1 q_U(y_i)$$

follows $\pi_U(Jz) \leqq \inf \sum \|\sigma_i\|_1 q_U(y_i) = p \otimes q_U(z)$, where p is the norm in $L^1_{X,\mu}$.

Conversely, every $Jz = \sum_{i=1}^{n} \sigma_i y_i$ can be written as $\sum_k \chi_k y'_k = J\Big(\sum_k \chi_k \otimes y'_k\Big)$, where the χ_k belong to pairwise disjoint measurable sets I'_k and we have

$$p \otimes q_U(z) \leqq \sum \mu(I'_k) q_U(y'_k) = \int q_U(Jz)\, d\mu = \pi_U(Jz).$$

Thus $\pi_U(Jz) = p \otimes q_U(z)$ and J is a topological isomorphism. Since S is dense in $L^1_{X,\mu}$, $S \otimes_\pi F$ is dense in $L^1 \otimes_\pi F$ and, by the definition of $L^1_{X,\mu}\{\tilde{F}\}$, we obtain

(8) *For every locally convex space F we have the isomorphism* $L^1_{X,\mu} \tilde{\otimes}_\pi F \cong L^1_{X,\mu}\{\tilde{F}\}$.

If F is a (B)-*space, this is a norm isomorphism.*

This elegant proof was given by SCHAEFER [1′], III, 6. 4. We introduced L^1 in § 14,10. as the completion of its subspace of continuous functions. We could have introduced L^1 also as the completion of S. We then gave a concrete representation of the elements of L^1 as classes of measurable functions f for which $\int |f(t)|\, dt < \infty$. If F is a (B)-space, one can do exactly the same and we find a representation of the elements of $L^1_{X,\mu}\{F\}$ as classes of F-valued functions f for which now $\pi(f) = \int_X \|f(t)\|\, d\mu$ has to be $< \infty$ (for detailed information see BOURBAKI [7], Chap. IV). Hence, in the case of (B)-spaces F, the terminology introduced above is completely justified.

For arbitrary locally convex F the situation is more complex. There are cases in which not all elements of $L^1_{X,\mu}\{F\}$, F complete, are representable as classes of F-valued functions.

§ 42. Compact and nuclear mappings

1. Compact linear mappings. Let E and F be locally convex spaces. A continuous linear mapping A of E in F is called precompact resp. compact if there exists a neighbourhood U of $\mathfrak{o}$ in E such that $A(U)$ is precompact resp. relatively compact in F.

Every $A \in \mathfrak{L}(E, F)$ of finite rank is obviously compact.

If F is quasi-complete, then every precompact A is compact.

The identity I on E is precompact if and only if E is finite dimensional (§ 15, 7.(1)).

We denote by $\mathfrak{C}_p(E, F)$ resp. $\mathfrak{C}(E, F)$ the set of all precompact resp. compact $A \in \mathfrak{L}(E, F)$.

(1) $\mathfrak{C}_p(E, F)$ *resp.* $\mathfrak{C}(E, F)$ *is a subspace of* $\mathfrak{L}(E, F)$. *If* $A \in \mathfrak{L}(E, F)$, $B \in \mathfrak{L}(F, G)$, *and if A or B is precompact resp. compact, then BA is precompact resp. compact.*

Proof. Let $A_1, A_2 \in \mathfrak{C}_p(E, F)$ resp. $\mathfrak{C}(E, F)$. There exist neighbourhoods U_1, U_2 of $\mathfrak{o}$ such that $A_1(U_1)$ and $A_2(U_2)$ are precompact resp. relatively compact. Then $(\alpha_1 A_1 + \alpha_2 A_2)(U_1 \cap U_2) \subset \alpha_1 A_1(U_1) + \alpha_2 A_2(U_2)$ and this, by § 15, 6.(8), is again a precompact resp. relatively compact set.

The second statement follows from the fact that the continuous linear image of a precompact resp. compact set is again a precompact resp. compact set.

(2) *If E is normed and F locally convex, then $\mathfrak{C}_p(E, F)$ is a closed subspace of $\mathfrak{L}_b(E, F)$.*
If F is, moreover, quasi-complete, then $\mathfrak{C}(E, F)$ is closed in $\mathfrak{L}_b(E, F)$.

It is sufficient to prove the first statement. This is a special case of the following lemma:

(3) *Let $\mathfrak{M}$ be a class of bounded subsets of E defining a locally convex topology $\mathfrak{T}_{\mathfrak{M}}$ on $\mathfrak{L}(E, F)$, E, F locally convex. The set H of all $A \in \mathfrak{L}(E, F)$ such that $A(M)$ is precompact for every $M \in \mathfrak{M}$ is closed in $\mathfrak{L}_{\mathfrak{M}}(E, F)$.*

Let A_0 be an adherent point of H in $\mathfrak{L}_{\mathfrak{M}}(E, F)$, $M \in \mathfrak{M}$. If U is an absolutely convex neighbourhood of $\mathfrak{o}$ in F, there exists $A \in H$ such that $A_0 z \in Az + U$ for all $z \in M$. Since $A(M)$ is precompact, there exist $x_1, \ldots, x_n$ in E such that $A(M) \subset \bigcup_{i=1}^{n} (Ax_i + U)$. It follows that

$$A_0 z \in Az + U \subset (Ax_i + U) + U \subset Ax_i + 2U$$

for some i; thus $A_0(M) \subset \bigcup_{i=1}^{n} (Ax_i + 2U)$, which implies that $A_0(M)$ is totally bounded for every $M \in \mathfrak{M}$ or that $A_0 \in H$.

(4) *If E is a* (B)*-space, then $\mathfrak{C}(E, E) = \mathfrak{C}(E)$ is a two-sided closed ideal in the Banach algebra $\mathfrak{L}_b(E)$.*

This follows from (1) and (2).

In general, $\mathfrak{C}(E)$ is not closed: Take $E = \omega$, where ω is the (F)-space endowed with $\mathfrak{T}_b(\varphi)$. Let I_n be the mapping $I_n \mathfrak{x} = \mathfrak{x}_n$, $\mathfrak{x} \in \omega$ and $\mathfrak{x}_n$ the nth section of $\mathfrak{x}$. Then I_n is of finite rank and therefore compact. It is easy to see that I_n converges to the identity I in $\mathfrak{L}_b(\omega)$ and I is not compact.

We study now the duality properties of precompact and compact mappings. Recall that on E', $\mathfrak{T}_c$ denotes the topology of uniform convergence on the precompact subsets of E.

(5) *Let E, F be locally convex. If $A \in \mathfrak{L}(E, F)$ is precompact, then $A' \in \mathfrak{L}(F_c', E_c')$ is compact.*

Proof. Let U be an absolutely convex neighbourhood of $\mathfrak{o}$ such that $C = A(U)$ is precompact. Then C° is a neighbourhood of $\mathfrak{o}$ in F'. We have $C^\circ = A'^{(-1)}(U^\circ)$; thus $A'(C^\circ) \subset U^\circ$ and U° is $\mathfrak{T}_c$-compact by § 21, 6.(3).

An (M)-space E is quasi-complete and reflexive, E_b' is again an (M)-space, and every bounded subset of an (M)-space is relatively compact. From this and (5) follows immediately

(6) *Let E and F be* (M)-*spaces.* $A \in \mathfrak{L}(E, F)$ *is compact if and only if* $A' \in \mathfrak{L}(F_b', E_b')$ *is compact.*

This is nearly obvious. A deeper result is the first duality theorem for compact mappings, the theorem of SCHAUDER:

(7) *Let E, F be normed spaces.* $A \in \mathfrak{L}(E, F)$ *is precompact if and only if* $A' \in \mathfrak{L}(F_b', E_b')$ *is compact.*

We obtain it as a corollary of the following theorem of GROTHENDIECK which generalizes § 21, 7.(1):

(8) *Let $\langle E', E\rangle$ and $\langle F', F\rangle$ be dual pairs, $\mathfrak{M}$ a saturated collection of weakly bounded subsets M of E which cover E, $\mathfrak{N}$ a similar collection of subsets N of F'; finally, let A be a weakly continuous linear mapping of E in F. Then the following statements are equivalent:*

a) *$A(M)$ is $\mathfrak{T}_{\mathfrak{N}}$-precompact for all $M \in \mathfrak{M}$;*

b) *$A'(N)$ is $\mathfrak{T}_{\mathfrak{M}}$-precompact for all $N \in \mathfrak{N}$;*

c) *the restriction of A to every $M \in \mathfrak{M}$ is uniformly continuous for the topologies $\mathfrak{T}_s(E')$ on E and $\mathfrak{T}_{\mathfrak{N}}$ on F';*

d) *the restriction of A' to every $N \in \mathfrak{N}$ is uniformly continuous for the topologies $\mathfrak{T}_s(F)$ on F' and $\mathfrak{T}_{\mathfrak{M}}$ on E'.*

Proof. It follows from § 32, 2.(1) that a) is equivalent to the statement that A' is a (uniformly) continuous mapping of $F'[\mathfrak{T}_c(F[\mathfrak{T}_{\mathfrak{N}}])]$ in $E'[\mathfrak{T}_{\mathfrak{M}}]$, where $\mathfrak{T}_c(F[\mathfrak{T}_{\mathfrak{N}}])$ is the topology of uniform convergence on the $\mathfrak{T}_{\mathfrak{N}}$-precompact subsets of F. Every $N \in \mathfrak{N}$ is $\mathfrak{T}_{\mathfrak{N}}$-equicontinuous in $F' \subset F[\mathfrak{T}_{\mathfrak{N}}]'$ and by § 21, 6.(2) the topologies $\mathfrak{T}_s(F)$ and $\mathfrak{T}_c(F[\mathfrak{T}_{\mathfrak{N}}])$ coincide on N. Therefore the restriction of A' to N is uniformly continuous for the topologies $\mathfrak{T}_s(F)$ on F' and $\mathfrak{T}_{\mathfrak{M}}$ on E'. Thus a) implies d).

We assume now d). Every $N \in \mathfrak{N}$ is weakly precompact as an equicontinuous subset of $F' \subset F[\mathfrak{T}_{\mathfrak{N}}]'$. It follows from d) that $A'(N)$ is $\mathfrak{T}_{\mathfrak{M}}$-precompact. Thus d) implies b).

By symmetry, b) implies c) and c) implies a) and this concludes the proof.

If one takes for $\mathfrak{M}$ and $\mathfrak{N}$ the classes of strongly bounded subsets in the normed space E resp. F', (7) becomes a special case of (8).

For (F)-spaces we obtain from (8)

(9) *Let E, F be* (F)-*spaces.* $A \in \mathfrak{L}(E, F)$ *maps all bounded sets into relatively compact sets if and only if* $A' \in \mathfrak{L}(F_b', E_b')$ *has the same property.*

This is true even if E and F are both only barrelled and F, moreover, quasi-complete (§ 23, 1.(3)).

Thus (8) specializes to the theorem of SCHAUDER for normed spaces, but even in the case of (F)-spaces it gives a theorem which says nothing on precompact mappings. A slightly different approach will give us more information.

(10) *Let $\langle E', E\rangle$ and $\langle F', F\rangle$ be dual pairs, M and N bounded weakly closed absolutely convex subsets of E' and F, respectively, and E'_M and F_N the associated normed spaces.*

Let A be in $\mathfrak{L}(E_s, F_s)$. Then $A(M^\circ)$ is precompact in F_N if and only if $A'(N^\circ)$ is precompact in E'_M.

Proof. We assume that $A(M^\circ)$ is precompact in F_N. Then $A(M^\circ) \subset \rho N$ for some $\rho > 0$ and it follows by polarity that $A'(N^\circ) \subset \rho M^{\circ\circ} = \rho M \subset E'_M$.

Let $\varepsilon > 0$ be given. Since $A(M^\circ)$ is precompact in F_N, there exist $x_1, \ldots, x_n$ in M° such that

$$\sup_{u\in N^\circ} |uA(x - x_i)| \leqq \frac{\varepsilon}{3} \tag{11}$$

for all $x \in M^\circ$ and a suitable x_i depending on x.

The set of all vectors $(uAx_1, \ldots, uAx_n)$, $u \in N^\circ$, is bounded in K^n and therefore precompact in l_n^∞. Hence there exist $u_1, \ldots, u_m$ such that

$$\sup_i |(u - u_k)(Ax_i)| = \sup_i |(A'(u - u_k))x_i| \leqq \frac{\varepsilon}{3}$$

for all $u \in N^\circ$ and u_k depending on u. From this inequality and (11) follows for $x \in M^\circ$ and $u \in N^\circ$

$$|(A'(u - u_k))x| \leqq |(A'(u - u_k))(x - x_i)| + |(A'(u - u_k))x_i| \leqq \frac{2\varepsilon}{3} + \frac{\varepsilon}{3} = \varepsilon.$$

Thus $\sup_{x\in M^\circ} |(A'u - A'u_k)x| \leqq \varepsilon$ and $A'(N^\circ)$ is precompact in E'_M.

The converse statement follows by symmetry.

An equivalent result was proved in KÖTHE [8'] using SCHAUDER'S theorem. The proof above allows an even more general statement (compare pp. 200, 443 of [1] by GARNIR, DE WILDE, and SCHMETS.

We note that (7) is also a special case of (10), so we have two different proofs for the theorem of SCHAUDER.

For metrizable spaces we obtain

(12) *Let E be locally convex, F metrizable locally convex. If $A \in \mathfrak{L}(E, F)$ is precompact, then $A' \in \mathfrak{L}(F'_b, E'_b)$ is precompact.*

For the proof we need the following lemma:

(13) *If C is a precompact subset of the metrizable locally convex space F, there exists an absolutely convex closed precompact subset $C_1 \supset C$ such that C is precompact in the normed space F_{C_1}.*

Proof. By § 41, 4.(3) C is contained in the closed absolutely convex cover of a sequence $x_n \in F$ converging to $\mathfrak{o}$. By § 28, 3.(1) there exist $\rho_n > 0$, $\rho_n \to \infty$ such that $y_n = \rho_n x_n$ converges to $\mathfrak{o}$ too. The closed absolutely convex cover C_1 of the y_n is precompact in F and C is precompact in F_{C_1}.

Proof of (12). Let U be an absolutely convex neighbourhood of $\mathfrak{o}$ such that $A(U) = C$ is precompact in F. By (13) there exists $C_1 \supset C$, C_1 precompact in F, such that $A(U)$ is precompact in F_{C_1}. It follows from (10) that $A'(C_1^\circ)$ is precompact in E'_{U°. Since U° is strongly bounded in E' (§ 21, 5.(1)), the norm topology on E'_{U° is finer than the topology induced from $\mathfrak{T}_b(E)$ and thus $A'(C_1^\circ)$ is precompact in E'_b. This implies the statement.

The converse of (12) is false, as is shown by the following example. We recall the situation of § 31, 5., where a linear continuous mapping A of the (FM)-space λ onto l^1 was defined, which is a homomorphism. It was proved in § 31, 5. that the mapping A' of l^∞ in $\lambda'_b = \lambda^\times_b$ is compact. But A is not compact: Let $\hat{A}$ be the isomorphism of $\lambda/N[A]$ onto l^1; then if $\hat{A}$ would be compact, $\hat{A}\hat{A}^{-1}$ the identity on l^1 too would be compact, which is not the case.

2. Weakly compact linear mappings. Let E, F be locally convex and $A \in \mathfrak{L}(E, F)$. A will be called weakly compact if there exists a neighbourhood U of $\mathfrak{o}$ in E such that $A(U)$ is relatively weakly compact in F, that is, if A is a compact mapping from E in $F[\mathfrak{T}_s(\mathfrak{T}')]$ in the sense of 1. For (B)-spaces we introduced this notion before, in § 40, 3. If A is compact, then A is weakly compact.

Since the weakly bounded and the weakly precompact subsets of F coincide (§ 20, 9.(3)), $A \in \mathfrak{L}(E, F)$ will be called bounded (instead of weakly precompact) if, for some neighbourhood U of $\mathfrak{o}$, $A(U)$ is a bounded subset of F.

If F is weakly quasi-complete, E locally convex, then weakly compact and bounded $A \in \mathfrak{L}(E, F)$ coincide.

One has the following basic result of Grothendieck ([7], [11]):

(1) *Let E, F be locally convex, $A \in \mathfrak{L}(E, F)$. The following two conditions are equivalent:*

(i) *A maps every bounded subset of E in a relatively weakly compact subset of F;*

(ii) *A'' maps E'' in F.*

(i) *or* (ii) *implies*

(iii) *A' maps the equicontinuous subsets of F' in relatively $\mathfrak{T}_s(E'')$-compact subsets of E'.*

If F is quasi-complete, (iii) *is equivalent to* (i) *and* (ii).

Proof. (i) $\curvearrowright$ (ii). A'' is a weakly continuous mapping of E'' into F'' and an extension of A. The space E'' is the union of all sets $\bar{B}$, B bounded in E and $\bar{B}$ the weak closure taken in E'' (§ 23, 2.(1)). Hence $A''(E'')$ is contained in the union of all sets $A''(\bar{B}) \subset \overline{A(B)}$, the weak closure being taken in F''. But since by assumption $A(B)$ is relatively weakly compact in F, $\overline{A(B)} \subset F$.

(ii) $\curvearrowright$ (i). Assume $A''(E'') \subset F$. If B is bounded in E, then its weak closure $\bar{B}$ is weakly compact. Thus $A''(\bar{B})$ is weakly compact in F and $A(B)$ is relatively weakly compact.

(ii) $\curvearrowright$ (iii). $A''(E'') \subset F$ implies that A'' is continuous from $E''[\mathfrak{T}_s(E')]$ in $F[\mathfrak{T}_s(F')]$. Its adjoint A' is therefore continuous from $F'[\mathfrak{T}_s(F)]$ in $E'[\mathfrak{T}_s(E'')]$. A' therefore maps equicontinuous subsets of F', which are relatively weakly compact, in relatively $\mathfrak{T}_s(E'')$-compact subsets of E'.

Finally, we suppose that F is quasi-complete and that A' satisfies (iii). We will prove that (ii) holds. Let $M \subset F'$ be equicontinuous. It follows from (iii) that $\mathfrak{T}_s(E'')$ and $\mathfrak{T}_s(E)$ coincide on $A'(M)$. Thus A' restricted to M is continuous for $\mathfrak{T}_s(F)$ and $\mathfrak{T}_s(E'')$. This implies that the linear form on F', defined by $\langle z_0, A'v\rangle = \langle A''z_0, v\rangle$, $z_0 \in E''$, $v \in F'$, has a $\mathfrak{T}_s(F)$-continuous restriction to M. This is true for every equicontinuous M and it follows from § 21, 9.(2) that $A''z_0 \in \tilde{F}$, the completion of F.

z_0 is in the weak closure of an absolutely convex bounded subset B of E; thus $A''z_0$ is contained in the closure of $A(B)$ in $\tilde{F}$ (§ 20, 7.(6)). Since F is assumed to be quasi-complete, $A''z_0$ lies even in F and (ii) is satisfied.

We obtain as a special case the theorem of GANTMACHER-NAKAMURA:

(2) *Let E be a normed space, F a Banach space, $A \in \mathfrak{L}(E, F)$. The following conditions are equivalent:*

(i) *A is weakly compact;*

(ii) *$A''(E'') \subset F$;*

(iii) *A' is weakly compact as a mapping of F'_b in E'_b.*

We note the following easy consequence of the definitions:

(3) *Let E, F be locally convex. If E or F is a reflexive* (B)-*space, then every $A \in \mathfrak{L}(E, F)$ is weakly compact.*

By analogy to 1.(1) one has

(4) *Let E, F be locally convex, $\mathfrak{W}(E, F)$ the set of all weakly compact mappings of $\mathfrak{L}(E, F)$. Then $\mathfrak{W}(E, F)$ is a subspace of $\mathfrak{L}(E, F)$. If $A \in \mathfrak{L}(E, F)$, $B \in \mathfrak{L}(F, G)$, and if A or B are weakly compact, then BA is weakly compact.*

The proof of 1.(1) works with minor changes also in this case.

Corresponding to 1.(2) and 1.(4) we have

(5) *If E is normed and F complete, then $\mathfrak{W}(E, F)$ is a closed subspace of $\mathfrak{L}_b(E, F)$.*

In particular, if E is a (B)*-space, $\mathfrak{W}(E) = \mathfrak{W}(E, E)$ is a closed two-sided ideal in the Banach algebra $\mathfrak{L}_b(E)$.*

We prove only the first statement. Let A_0 be an adherent point of $\mathfrak{W}(E, F)$ in $\mathfrak{L}_b(E, F)$. Then there exists a net $A_\alpha \in \mathfrak{W}(E, F)$ converging to A_0 in $\mathfrak{L}_b(E, F)$. By (1) $A_\alpha''(E'') \subset F$ for every α and we have to show that $A_0''(E'') \subset F$.

If V is a closed absolutely convex neighbourhood of $\circ$ in F, there exists α_0 such that $A_\alpha x - A_0 x \in V$ for $\alpha \geqq \alpha_0$ and all $x \in E$, $\|x\| \leqq 1$. Since the closed unit ball in E'' is the weak closure of the unit ball in E and the A_α'' and A_0'' are weakly continuous, it follows that

$$A_\alpha'' z - A_0'' z \in \overline{V} \quad \text{for all } z \in E'', \ \|z\| \leqq 1,$$

where $\overline{V}$ is the closure of V in F''. Hence $A_\alpha'' z$ is a Cauchy net in F since $A_0''(E'') \subset F$. Its limit $A_0'' z$ is in F since F is complete.

3. Completely continuous mappings. Examples. We come back to Theorem 1.(8). An immediate consequence of the equivalence of a) and c) in this theorem is

(1) *If $A \in \mathfrak{L}(E_s, F_s)$ satisfies*

a) *$A(M)$ is $\mathfrak{T}_\mathfrak{N}$-precompact for all $M \in \mathfrak{M}$, then the following condition is satisfied too:*

e) *if $x_n \in M \in \mathfrak{M}$ and x_n converges weakly to $\circ$, then Ax_n $\mathfrak{T}_\mathfrak{N}$-converges to $\circ$.*

We are interested in conditions on E and F such that e) implies a). A first result is

(2) *Let E, F be locally convex and E_b' separable. If $A \in \mathfrak{L}(E_s, F_s)$ maps every sequence $x_n \in E$ which converges weakly to $\circ$ onto a sequence Ax_n which converges to $\circ$ in F, then A maps bounded sets of E on precompact sets of F.*

Proof. Since E_b' is separable, the topology $\mathfrak{T}_s(E')$ on every bounded absolutely convex set $M \subset E$ is metrizable (§ 21, 3.(4)). By assumption the restriction of A to M is sequentially continuous at $\circ$ and therefore continuous at $\circ$ and by § 21, 6.(5) uniformly continuous on M for the weak topology on M and the given topology on F. Thus condition c) of 1.(8) is satisfied.

A second result is

(3) *Let E be a reflexive* (F)-*space or a semi-reflexive strict* (LF)-*space, F metrizable locally convex, $A \in \mathfrak{L}(E_s, F_s)$. Then A maps bounded sets of E onto relatively compact sets of F if the weak convergence of x_n to $\circ$ in E implies $Ax_n \to \circ$ in F.*

Proof. Let M be a bounded subset of E. Since F is metrizable, it is sufficient to show that $A(M)$ is relatively sequentially compact. Let Ax_n, $n = 1, 2, \ldots$, be a sequence in $A(M)$; then the set $\{x_1, x_2, \ldots\} \subset M$ is relatively weakly compact and by § 24, 1.(3) and (4) there exists a subsequence x_{n_j} which converges weakly to an element $x_0 \in E$. The sequence $x_{n_j} - x_0$ converges weakly to $\circ$; thus by assumption $Ax_{n_j} \to Ax_0$ in F and $A(M)$ is relatively sequentially compact.

We consider now the case where E and F are (B)-spaces. $A \in \mathfrak{L}(E, F)$ is called completely continuous if A maps weakly convergent sequences into norm convergent sequences. A is completely continuous if A maps every sequence which converges weakly to $\circ$ into a sequence which converges to $\circ$ in the norm.

It follows from (1) that every compact A is completely continuous.

As an immediate consequence of (2) and (3) we obtain

(4) *Let E, F be* (B)-*spaces. If E_b' is separable or if E is reflexive, then completely continuous and compact $A \in \mathfrak{L}(E, F)$ coincide.*

The injection J of l^1 into l^2 is completely continuous but not compact: J is continuous by § 14, 8.(9) and weak and norm convergence of sequences in l^1 coincide (§ 22, 4.(2)); thus J is completely continuous. The set of all e_i, $i = 1, 2, \ldots$, is contained in the image of the unit ball of l^1 and this set is not relatively compact in l^2.

The same argument shows that for any cardinal d the injection J of l_d^1 into l_d^2 is completely continuous. That J is not compact follows for $d > \aleph_0$ immediately from

(5) *A precompact $A \in \mathfrak{L}(E, F)$, E and F locally convex, F metrizable, has a separable range.*

$A(E) = \bigcup_{n=1}^{\infty} nA(U)$, where $A(U)$ is precompact in F and $A(U)$ is relatively compact in the completion $\tilde{F}$. By § 4, 5.(2) $A(U)$ is separable.

Another consequence of 1.(8) is

(6) *Let E, F be* (B)-*spaces, F separable. $A \in \mathfrak{L}(E, F)$ is compact if and only if the $\mathfrak{T}_s(F)$-convergence of a sequence $v_n \in F'$ implies always the strong convergence of $A'v_n$ in E'.*

Proof. By 1.(8) A is compact if and only if the restriction of A' to the unit ball M of F' is uniformly continuous for the topologies $\mathfrak{T}_s(F)$ on M and $\mathfrak{T}_b(E)$ on E'. From the separability of F it follows that M is metrizable for $\mathfrak{T}_s(F)$ and, as in the proof of (2), the sequential continuity of A' on M implies uniform continuity.

In l_d^1 weak and norm convergent sequences coincide. It follows immediately that for arbitrary (B)-spaces E, F every $A \in \mathfrak{L}(E, l_d^1)$ and every $A \in \mathfrak{L}(l_d^1, F)$ is completely continuous.

We see from (4) that if E is reflexive or if E_b' is separable, then every $A \in \mathfrak{L}(E, l_d^1)$ is compact.

We note some results for c_0.

(7) *For any* (B)-*space* F *a weakly compact* $A \in \mathfrak{L}(c_0, F)$ *is always compact.*

Proof. A' is weakly compact in $\mathfrak{L}(F_b', l^1)$ by 2.(2). In l^1 weakly compact sets are compact (§ 22, 4.(3)); thus A' is compact. Finally, A is compact by SCHAUDER's theorem 1.(7).

If F is reflexive, then every $A \in \mathfrak{L}(c_0, F)$ is weakly compact since the bounded sets in F are relatively weakly compact. Thus (7) implies: *Every continuous linear mapping of* c_0 *in a reflexive* (B)-*space is compact.*

This result can be slightly improved.

(8) *Let* E, F *be* (B)-*spaces,* E_b' *separable,* F *weakly sequentially complete. Then every* $A \in \mathfrak{L}(E_s, F_s)$ *is weakly compact.*

Proof. A bounded set $M \subset E$ is metrizable for $\mathfrak{T}_s(E')$ and $\mathfrak{T}_s(E')$-precompact. Therefore every sequence $x_n \in M$ contains a weak Cauchy sequence x_{n_j}. Since F is weakly sequentially complete, the weak Cauchy sequence Ax_{n_j} has a limit in F and thus $A(M)$ is relatively weakly sequentially compact. By the theorems of SMULIAN and EBERLEIN (§ 24, 3.(8)) $A(M)$ is relatively weakly compact.

Using (7) we have the special case

(9) *Every continuous linear mapping of* c_0 *in a weakly sequentially complete* (B)-*space* F *is compact.*

The following interesting result is due to PITT [1'].

(10) *Let* $1 \leqq p < r < \infty$; *then every* $A \in \mathfrak{L}(l^r, l^p)$ *is compact.*

Proof. We denote by P_n (resp. Q_n) the projection of l^r (resp. l^p) which maps every element $x = (x_1, x_2, \ldots)$ onto its nth section $(x_1, \ldots, x_n, 0, 0, \ldots)$. Setting $A_{n,m} = (I - Q_n)A(I - P_m)$ we have the decomposition

$$(11) \quad A = A_{n,m} + Q_nA(I - P_m) + (I - Q_n)AP_m + Q_nAP_m.$$

We need the following fact:

$$(12) \qquad \lim_{m\to\infty} \|Q_n A(I - P_m)\| = 0 \quad \text{for every } n = 1, 2, \ldots.$$

This is easy to prove: Q_nA is of finite rank and has therefore a representation $Q_nAx = \sum_1^s \langle u_i, x\rangle y_i$, $y_i \in l^p$, $u_i \in (l^r)' = l^{r'}$, $1/r + 1/r' = 1$. Hence $Q_nA(I - P_m)x = \sum_1^s \langle (I - P'_m)u_i, x\rangle y_i$. From $\lim_{m\to\infty} \|(I - P'_m)u_i\| = 0$ for every i follows (12).

In (11) the three last mappings are of finite rank. If $\|A_{n_i,m_i}\| \to 0$ for two sequences n_i, m_i of integers, then A will be compact as the limit of a sequence of mappings of finite rank.

We assume that A is not compact. Then there exists $\delta > 0$ such that $\|A_{n,m}\| > \delta$ for every n, m.

Let $\alpha = (\alpha_1, \alpha_2, \ldots)$, $\alpha_i > 0$, be an element of l^r, $\|\alpha\|_r = 1$, which does not lie in l^p (compare § 14, 8. for the construction of such an element), and let ε_i, $i = 1, 2, \ldots$ be a sequence of positive numbers such that $\varepsilon_i < \delta/2$ and $\sum_{i=1}^{\infty} \alpha_i\varepsilon_i = c < \infty$.

There exists in l^r an element $x^{(1)} = (x_1^{(1)}, \ldots, x_{m_1}^{(1)}, 0, 0, \ldots)$, $\|x^{(1)}\|_r = 1$, such that $\|Ax^{(1)}\|_p > \delta$ and there exists n_1 such that the n_1th section $y^{(1)}$ of $Ax^{(1)}$ satisfies $\|y^{(1)}\|_p > \delta$ and $z^{(1)} = Ax^{(1)} - y^{(1)}$ satisfies $\|z^{(1)}\|_p < \varepsilon_1$.

It follows from (12) that there exists $m'_2 > m_1$ such that $\|Q_{n_1}A(I - P_{m'_2})\| < \varepsilon_2$. There exists in l^r an element $x^{(2)} = (0, \ldots, 0, x^{(2)}_{m'_2+1}, \ldots, x^{(2)}_{m_2}, 0, 0, \ldots)$, $\|x^{(2)}\|_r = 1$, such that $\|A_{n_1,m'_2}x^{(2)}\|_p > \delta$. Since $P_{m'_2}x^{(2)} = o$, it follows from (11) that

$$Ax^{(2)} = A_{n_1,m'_2}x^{(2)} + Q_{n_1}A(I - P_{m'_2})x^{(2)} = A_{n_1,m'_2}x^{(2)} + s^{(2)}, \qquad \|s^{(2)}\|_p < \varepsilon_2.$$

We choose now $n_2 > n_1$ such that the n_2th section

$$y^{(2)} = (0, \ldots, 0, y^{(2)}_{n_1+1}, \ldots, y^{(2)}_{n_2}, 0, 0, \ldots)$$

of $A_{n_1,m'_2}x^{(2)}$ satisfies $\|y^{(2)}\|_p > \delta$ and $z^{(2)} = A_{n_1,m'_2}x^{(2)} - y^{(2)}$ satisfies $\|z^{(2)}\|_p < \varepsilon_2$. Then $\|Ax^{(2)}\|_p > \|y^{(2)}\|_p - \|s^{(2)}\|_p - \|z^{(2)}\|_p > \delta - 2\varepsilon_2$.

We continue this construction by induction and obtain a sequence $x^{(i)}$, $\|x^{(i)}\|_r = 1$, of elements of l^r with nonoverlapping nonzero coordinates and this is true also for the corresponding sequence $y^{(i)}$, $\|y^{(i)}\|_p > \delta$, in l^p. The sequence $t_n = \alpha_1 x^{(1)} + \cdots + \alpha_n x^{(n)}$ is bounded in l^r, since

$$\|t_n\|_r = \left(\sum_1^n \alpha_i^r \sum_{m'_i+1}^{m_i} |x_k^{(i)}|^r\right)^{1/r} = \left(\sum_1^n \alpha_i^r\right)^{1/r} \leqq \|\alpha\|_r.$$

For the corresponding sequence $At_n = \sum_1^n \alpha_i(y^{(i)} + s^{(i)} + z^{(i)})$ one obtains $\|At_n\|_p \geqq \delta\left(\sum_1^n \alpha_i^p\right)^{1/p} - 2\sum_1^\infty \alpha_i\varepsilon_i$; thus At_n is unbounded by our assumption on α. But this is a contradiction since A is continuous.

In contrast to (10), the canonical injection of l^p in l^r, $1 \leqq p < r < \infty$, is not compact. This follows as in the example after (4) from the fact that the set of all e_i, $i = 1, 2, \ldots$, is not relatively compact in l^r.

For compact mappings between L^p-spaces see ROSENTHAL [1′].

KATO [1′] introduced the following notion: Let E, F be (B)-spaces, $A \in \mathfrak{L}(E, F)$. Then A is called strictly singular if A has no bounded inverse on any infinite dimensional subspace of its range.

Every compact A is strictly singular, but the converse does not hold. The strictly singular endomorphisms of a (B)-space E constitute a two-sided closed ideal in $\mathfrak{L}_b(E)$. For the theory of strictly singular mappings we refer the reader to GOLDBERG [1′], LACEY and WHITLEY [1′], and PEŁCZYNSKI [2′].

There exist (B)-spaces E with the property that every weakly compact $A \in \mathfrak{L}(E, F)$, F any (B)-space, maps every weakly compact set in a compact set of F. Such an E has the Dunford–Pettis property. Examples for these spaces are the spaces $C(K)$, K compact, and $L_1(\mu)$, μ any measure. There exists a rather deep theory of these and related spaces and their weakly compact mappings. We refer the reader to GROTHENDIECK [7], EDWARDS [1′], Chap. 9, and, for further references, to BATT [1′].

4. Compact mappings in Hilbert space. We assume that the reader is familiar with the elements of Hilbert space theory. We denote the scalar product of two elements x, y of a (real or complex) Hilbert space by (x, y). The following representation of compact mappings of Hilbert spaces is a consequence of the spectral theory of compact symmetric operators.

(1) *Let H_1, H_2 be Hilbert spaces, $A \in \mathfrak{L}(H_1, H_2)$ compact and not of finite rank. Then there exist orthonormal systems $\{e_n\}$, $n = 1, 2, \ldots$, in H_1 and $\{f_n\}$, $n = 1, 2, \ldots$, in H_2 such that*

$$Ax = \sum_{n=1}^{\infty} \lambda_n(x, e_n)f_n, \qquad x \in H_1, \tag{2}$$

where $\lambda_n > 0$ and $\lambda_n \to 0$.

Proof. Since A is compact, A^*A is compact too and positive, where A^* denotes the adjoint in the sense of the scalar product. It follows from spectral theory that there exists an orthonormal sequence of eigenvectors e_n, $n = 1, 2, \ldots$, and eigenvalues $\lambda_n^2 > 0$, $\lambda_n^2 \to 0$ such that

$$A^*Ax = \sum_{n-1}^{\infty} \lambda_n^2(x, e_n)e_n.$$

A^*A is zero on the orthogonal complement H of the closed subspace spanned by all the e_n. But then A is zero too on H: Take $y \in H$ and suppose $Ay \neq o$. Then $(Ay, Ay) = (y, A^*Ay) \neq 0$. But this would imply $A^*Ay \neq o$.

Therefore we have a representation $Ax = \sum_{n=1}^{\infty} (x, e_n)Ae_n$. Define now $f_n = (1/\lambda_n)Ae_n$. Then $Ax = \sum_{n=1}^{\infty} \lambda_n(x, e_n)f_n$ and our proposition will be proved if we show that $\{f_n\}$ is an orthonormal system. But

$(f_i, f_k) = (\lambda_i^{-1}Ae_i, \lambda_k^{-1}Ae_k) = \lambda_i^{-1}\lambda_k^{-1}(A^*Ae_i, e_k) = \lambda_i^{-1}\lambda_k^{-1}(\lambda_i^2 e_i, e_k) = \delta_{ik}.$

(3) *Conversely, every mapping $A \in \mathfrak{L}(H_1, H_2)$ which has a representation* (2) *with $\lambda_n > 0$, $\lambda_n \to 0$ is compact.*

Let A_k be $\sum_{n=1}^{k} \lambda_n(x, e_n)f_n$; then $\|(A - A_n)x\|^2 \leqq \sum_{n=k+1}^{\infty} \lambda_n^2|(x, e_n)|^2 \leqq \varepsilon^2\|x\|^2$ if $|\lambda_n| \leqq \varepsilon$ for $n > k(\varepsilon)$. Thus A is compact as the limit of the A_n in $\mathfrak{L}_b(H_1, H_2)$.

From this proof and (1) follows immediately

(4) *Let H_1, H_2 be Hilbert spaces. Then every compact $A \in \mathfrak{L}_b(H_1, H_2)$ is the limit of a sequence of mappings of finite rank.*

The λ_n of (2) are called the singular values of A and the non-increasing sequence of all singular values of A is uniquely determined by A.

The representation (2) can be written in a different way using linear forms instead of scalar products for the coefficients of the f_n.

The scalar product (x, y) in Hilbert space H is linear in x for y fixed; thus it defines a linear functional $\langle \bar{y}, x\rangle = (x, y)$, where $\bar{y}$ is uniquely determined. One calls $\bar{y}$ the conjugate element to y. There exists an orthonormal basis $\{e_\alpha\}$, $\alpha \in \mathrm{A}$, of H such that for $x = \sum_\alpha \xi_\alpha e_\alpha$, $y = \sum_\alpha \eta_\alpha e_\alpha$

$$(x, y) = \sum_\alpha \xi_\alpha \bar{\eta}_\alpha = \langle \bar{y}, x\rangle.$$

Since this is true for all $x \in H$, it follows that $\bar{y} = \sum_\alpha \bar{\eta}_\alpha e_\alpha$; the coefficients of $\bar{y}$ are the conjugates of the coefficients of y.

The following properties of the mapping $y \to \bar{y}$ are immediate consequences:

$$\overline{(\alpha_1 y_1 + \alpha_2 y_2)} = \bar{\alpha}_1\bar{y}_1 + \bar{\alpha}_2\bar{y}_2, \quad \bar{\bar{y}} = y, \quad (\bar{x}, \bar{y}) = \overline{(x, y)}, \quad \|\bar{y}\| = \|y\|.$$

Hence the conjugate system $\{\bar{v}_\beta\}$ of an orthonormal system $\{v_\beta\}$ is again an orthonormal system. For the basis $\{e_\alpha\}$ one has, obviously, $\bar{e}_\alpha = e_\alpha$ and $(x, e_\alpha) = \langle x, e_\alpha\rangle$.

It follows from these considerations that we can replace the representation (2) by the representation

$$(2') \qquad Ax = \sum_{n=1}^{\infty} \lambda_n \langle \bar{e}_n, x \rangle f_n, \qquad x \in H_1,$$

where $\bar{e}_n$ and f_n are orthonormal sequences in H_1 and H_2, respectively, $\lambda_n > 0$, $\lambda_n \to 0$.

We consider a subclass of the class of all compact linear mappings of Hilbert spaces. Let H_1, H_2 be Hilbert spaces. $A \in \mathfrak{L}(H_1, H_2)$ is called a Hilbert–Schmidt mapping if, using orthonormal bases $\{e_\alpha\}$ in H_1 and $\{f_\beta\}$ in H_2, we have

$$\|A\|_h^2 = \sum_{\alpha,\beta} |(Ae_\alpha, f_\beta)|^2 < \infty.$$

$\|A\|_h$ is the Hilbert–Schmidt norm of A and we have $\|A\| \leqq \|A\|_h$ for every Hilbert–Schmidt mapping:

$$\|Ax\| = \left\| A\left(\sum_\alpha \xi_\alpha e_\alpha\right)\right\| \leqq \sum_\alpha |\xi_\alpha| \, \|Ae_\alpha\| \leqq \left(\sum_\alpha |\xi_\alpha|^2\right)^{1/2} \left(\sum_\alpha \|Ae_\alpha\|^2\right)^{1/2}$$
$$= \|A\|_h \|x\|.$$

The norm $\|A\|_h$ is independent of the special choice of the orthonormal bases, since

$$(5) \qquad \sum_{\alpha,\beta} |(Ae_\alpha, f_\beta)|^2 = \sum_\alpha \|Ae_\alpha\|^2 = \sum_\beta \|A^* f_\beta\|^2.$$

(5) implies that A is Hilbert–Schmidt if and only if A^* is Hilbert–Schmidt and we have $\|A^*\|_h = \|A\|_h$.

From $\left(\sum_\alpha \|(A + B)e_\alpha\|^2\right)^{1/2} \leqq \left(\sum_\alpha \|Ae_\alpha\|^2\right)^{1/2} + \left(\sum_\alpha \|Be_\alpha\|^2\right)^{1/2}$ follows $\|A + B\|_h \leqq \|A\|_h + \|B\|_h$; thus the sum of two Hilbert–Schmidt mappings is again a Hilbert–Schmidt mapping. Together with $\|\alpha A\|_h = |\alpha| \, \|A\|_h$ for α complex, this shows that the class $\mathfrak{H}(H_1, H_2)$ of all Hilbert–Schmidt mappings $A \in \mathfrak{L}(H_1, H_2)$ is a normed space.

It is even a (B)-space: A Cauchy sequence A_r is a Cauchy sequence in $\mathfrak{L}(H_1, H_2)$ and has therefore a strong limit A_0 and it follows from $\sum_\alpha \|(A_r - A_s)e_\alpha\| \leqq \varepsilon$ for $r, s \geqq r_0$ that $\sum_\alpha \|(A_r - A_0)e_\alpha\| \leqq \varepsilon$; thus $A_0 \in \mathfrak{H}(H_1, H_2)$ and $\|A_0 - A_r\|_h \to 0$.

If we define $(A, B) = \sum_\alpha (Ae_\alpha, Be_\alpha)$ for $A, B \in \mathfrak{H}(H_1, H_2)$, then (A, B) is a scalar product such that $\|A\|_h = (A, A)^{1/2}$. The proof is straightforward.

We collect these facts in the following proposition:

(6) *Let H_1, H_2 be two Hilbert spaces, $\{e_\alpha\}$ an orthonormal basis of H_1.*

If we introduce in $\mathfrak{H}(H_1, H_2)$ *the scalar product* $(A, B) = \sum_\alpha (Ae_\alpha, Be_\alpha)$, *then* $\mathfrak{H}(H_1, H_2)$ *is a Hilbert space. The Hilbert–Schmidt norm* $\|A\|_h = (A, A)^{1/2}$ *is stronger than the norm* $\|A\|$ *of* A *in* $\mathfrak{L}_b(H_1, H_2)$.

We remark that (A, B) is independent of the choice of the basis $\{e_\alpha\}$, since the Hilbert–Schmidt norm is independent and the scalar product can be expressed as a linear combination of norms.

Every continuous linear mapping A of finite rank is a Hilbert–Schmidt mapping, since there exists an orthonormal basis $\{e_\alpha\}$ such that only a finite number of the Ae_α are $\neq \mathfrak{o}$.

If A is Hilbert–Schmidt, then it follows from $\sum_\alpha \|Ae_\alpha\|^2 < \infty$ that only for countably many α can Ae_α be different from $\mathfrak{o}$. Thus we may suppose that A is of the form $Ax = \sum_{n=1}^{\infty} \langle x, e_n\rangle Ae_n$. Then A_p defined by $A_px = \sum_{n=1}^{p} \langle x, e_n\rangle Ae_n$ is of finite rank and A is the limit of A_p in $\mathfrak{H}(H_1, H_2)$. This implies in particular that A is compact.

(7) $\mathfrak{H}(H_1, H_2)$ *is the completion of the space of linear continuous mappings of finite rank for the Hilbert–Schmidt norm.* $\mathfrak{H}(H_1, H_2)$ *consists of compact mappings.* $\mathfrak{H}(H)$ *is a two-sided ideal in* $\mathfrak{L}(H)$.

We have only to prove the last statement. If $A \in \mathfrak{H}(H_1, H_2)$, $B \in \mathfrak{L}(H_2, H_3)$, then $\|BA\|_h^2 = \sum_\alpha \|BAe_\alpha\|^2 \leqq \|B\|^2 \sum_\alpha \|Ae_\alpha\|^2 = \|B\|^2\|A\|_h^2$; thus $\|BA\|_h \leqq \|B\|\,\|A\|_h$. If $A \in \mathfrak{H}(H_2, H_3)$, $B \in \mathfrak{L}(H_1, H_2)$, then by (5)

$$\|AB\|_h^2 = \sum_\beta \|B^*A^*f_\beta\|^2 \leqq \|B\|^2\|A\|_h^2$$

and $\|AB\|_h \leqq \|A\|_h\|B\|$. The assertion is a special case of these results.

(8) *A compact mapping* $A \in \mathfrak{L}(H_1, H_2)$ *is Hilbert–Schmidt if and only if* $\sum_i \lambda_i^2 < \infty$, *where the* λ_i *are the singular values of* A. *Moreover,* $\|A\|_h = \sqrt{\sum_i \lambda_i^2}$.

If A is compact, it has a representation (2), so that $Ae_i = \lambda_i f_i$ and $\|A\|_h^2 = \sum_{i=1}^{\infty} \lambda_i^2$ by (5), which implies the statement.

5. Nuclear mappings. We establish the connection with the results of § 41. We showed in § 41, 3.(13) that for complete locally convex E, F there exists a canonical mapping ψ of $E \otimes_\pi F$ in $\mathfrak{L}_e(E'_k, F)$ which has a continuous extension $\bar{\psi}$ to $E \tilde{\otimes}_\pi F$. If $\acute{A} = \sum_i x_i \otimes y_i \in E \otimes F$, then $\psi(\acute{A}) = \tilde{A}$ is given by $\tilde{A}u = \sum_i (ux_i)y_i$. These $\tilde{A}$ are of finite rank. If $\acute{A} \in E \tilde{\otimes}_\pi F$, then

$\tilde{A}$ is the limit of a net $\tilde{A}_\alpha$, $\tilde{A}_\alpha$ of finite rank, in the topology of uniform convergence on the equicontinuous subsets of E'. It follows from 1.(3) that $\tilde{A}$ maps every equicontinuous subset of E' onto a relatively compact subset of F. Hence

(1) *Let E, F be complete locally convex spaces. Every $\dot{A} \in E \tilde{\otimes}_\pi F$ defines an $\tilde{A} \in \mathfrak{L}_e(E'_k, F)$ which maps equicontinuous sets of E' in relatively compact subsets of F and $\tilde{A}$ is the $\mathfrak{T}_e$-limit of mappings of finite rank.*

By § 41, 3.(11) we have as a special case

(2) *Let E, F be* (B)*-spaces. Every $\dot{A} \in E \tilde{\otimes}_\pi F$ defines a compact mapping $\tilde{A} \in \mathfrak{L}_b(E'_b, F)$ which is the $\mathfrak{T}_b$-limit of mappings of finite rank.*

It will be interesting to study more closely the class of these compact mappings. So far we have defined them only for the special case where the first space is a dual space. It is easy to deal with the general case.

Let E, F be (B)-spaces. An element $\dot{A} = \sum_i u_i \otimes y_i$ of $E'_b \otimes F$ generates a mapping $\psi(\dot{A}) = \tilde{A} \in \mathfrak{L}(E, F)$ by defining $\tilde{A}x = \sum_i (u_i x) y_i$ for $x \in E$, so that ψ is an algebraic isomorphism of $E'_b \otimes F$ with the subspace of all maps of finite rank in $\mathfrak{L}(E, F)$.

The injection ψ of $E'_b \otimes_\pi F$ in $\mathfrak{L}_b(E, F)$ is continuous since one has $\|\psi(\dot{A})\| = \|\tilde{A}\| \leqq \|\dot{A}\|_\pi$, as in the proof of § 41, 3.(11). Therefore

(3) *Let E, F be* (B)*-spaces. The canonical continuous injection ψ of $E'_b \otimes_\pi F$ in $\mathfrak{L}_b(E, F)$ has a uniquely determined extension to a continuous linear mapping $\bar{\psi}$ of $E'_b \tilde{\otimes}_\pi F$ in $\mathfrak{L}_b(E, F)$.*

As before we obtain

(4) *Let E, F be* (B)*-spaces. If $\dot{A} \in E'_b \tilde{\otimes}_\pi F$, then the corresponding mapping $\bar{\psi}(\dot{A}) = \tilde{A} \in \mathfrak{L}(E, F)$ is compact as the $\mathfrak{T}_b$-limit of mappings of finite rank.*

The subspace $\bar{\psi}(E'_b \tilde{\otimes}_\pi F)$ of $\mathfrak{L}(E, F)$ is called the space $\mathfrak{N}(E, F)$ of all nuclear mappings of the (B)-space E into the (B)-space F. The nuclear mappings were introduced by GROTHENDIECK.

We have the following characterization:

(5) *Let E, F be* (B)*-spaces. $A \in \mathfrak{L}(E, F)$ is nuclear if and only if A has a representation*

$$(6) \quad Ax = \sum_{n=1}^{\infty} \lambda_n (u_n x) y_n, \qquad u_n \in E', \ \|u_n\| \leqq 1, \ y_n \in F,$$

$$\|y_n\| \leqq 1, \ \sum_{n=1}^{\infty} |\lambda_n| < \infty.$$

Proof. If $A = \bar{\psi}(z)$, $z \in E_b' \tilde{\otimes}_\pi F$, then z has a representation $z = \sum_{n=1}^{\infty} \lambda_n(u_n \otimes y_n)$ by § 41, 4.(1), and then $\bar{\psi}(z)$ has the representation (6) by continuity of $\bar{\psi}$. Conversely, from (6) it follows that $z = \sum_1^\infty \lambda_n u_n \otimes y_n$ is an element of $E_b' \tilde{\otimes}_\pi F$ by the remark following § 41, 4.(1).

We introduce the nuclear norm $\|A\|_\nu$ as the infimum of the sums $\sum_{n=1}^{\infty} |\lambda_n|$ taken over all representations (6) of A. If $\bar{\psi}$ is one-one, then by § 41, 4.(1) $\|A\|_\nu = \|z\|_\pi$, where $A = \bar{\psi}(z)$ and $\mathfrak{N}(E, F)$ is norm isomorphic to $E_b' \tilde{\otimes}_\pi F$. If $\bar{\psi}$ is not one-one, then the nuclear norm is the quotient norm of $(E_b' \tilde{\otimes}_\pi F)/N$, where N is the kernel of $\bar{\psi}$, and $\mathfrak{N}(E, F)$ is norm isomorphic to $(E_b' \tilde{\otimes}_\pi F)/N$.

There is a second characterization of nuclear mappings. Assume that $A \in \mathfrak{L}(E, F)$ has a representation

$$(7) \quad Ax = \sum_{n=1}^{\infty} (\bar{u}_n x)\bar{y}_n, \qquad \bar{u}_n \in E', \bar{y}_n \in F, \sum_{n=1}^{\infty} \|\bar{u}_n\| \|\bar{y}_n\| < \infty.$$

If one defines u_n and y_n by $u_n = \bar{u}_n/\|\bar{u}_n\|$, $y_n = \bar{y}_n/\|\bar{y}_n\|$, then from (7) follows a representation (6) with $\lambda_n = \|\bar{u}_n\| \|\bar{y}_n\|$, so that $\sum |\lambda_n| = \sum \|\bar{u}_n\| \|\bar{y}_n\|$. Conversely, (6) may be written as (7) by setting $\bar{u}_n = \lambda_n u_n$ and $\bar{y}_n = y_n$; then $\sum \|\bar{u}_n\| \|\bar{y}_n\| \leqq \sum_{n=1}^{\infty} |\lambda_n| < \infty$, since $\|u_n\| \leqq 1$, $\|y_n\| \leqq 1$.

Thus the nuclear norm can also be defined by

$$(8) \qquad \|A\|_\nu = \inf \sum \|\bar{u}_n\| \|\bar{y}_n\|,$$

where the infimum is taken over all representations (7).

(9) a) *If* $A \in \mathfrak{N}(E, F)$, $B \in \mathfrak{L}(F, G)$, *then* $BA \in \mathfrak{N}(E, G)$ *and* $\|BA\|_\nu \leqq \|B\| \|A\|_\nu$.

b) *If* $B \in \mathfrak{L}(E, F)$, $A \in \mathfrak{N}(F, G)$, *then* $AB \in \mathfrak{N}(E, G)$ *and* $\|AB\|_\nu \leqq \|A\|_\nu \|B\|$.

Proof. a) If A is represented by $Ax = \sum_{n=1}^{\infty} (u_n x) y_n$ and $\sum_{n=1}^{\infty} \|u_n\| \|y_n\| \leqq \|A\|_\nu + \varepsilon$, then BA is represented by $BAx = \sum_{n=1}^{\infty} (u_n x) B y_n$ and $\sum \|u_n\| \|By_n\| \leqq \|B\|(\|A\|_\nu + \varepsilon)$.

b) $ABx = \sum (u_n(Bx)) y_n = \sum ((B'u_n)x) y_n$; hence

$$\|AB\|_\nu \leqq \sum \|B'u_n\| \|y_n\| \leqq \|B\|(\|A\|_\nu + \varepsilon).$$

(10) *Let* E, F *be* (B)*-spaces. Then* $\mathfrak{N}(E, F)$ *is the completion of the space of linear continuous mappings of finite rank for the nuclear norm.*

$\mathfrak{N}(E) = \mathfrak{N}(E, E)$ *is a two-sided ideal in* $\mathfrak{L}(E)$.

This is an easy consequence of (4), the norm isomorphism $\mathfrak{N}(E, F) \cong (E_b' \tilde{\otimes}_\pi F)/N$, and (9).

It will be important to generalize the notion of nuclear mapping to the case where E and F are normed spaces. We use now (6) as the definition in this more general situation. It is obvious that the set $\mathfrak{N}(E, F)$ of all nuclear mappings is again a vector space and that it is a normed space for the nuclear norm $\|A\|_\nu$ defined as above, and one always has $\|A\| \leqq \|A\|_\nu$. It follows from (7) that every nuclear A is the ν-limit of a sequence of mappings of finite rank and therefore precompact. The range $A(E)$ is always separable (3.(5)).

Since (9) is true also for normed E and F, $\mathfrak{N}(E)$ is again a two-sided ideal in $\mathfrak{L}(E)$.

If F is a (B)-space, then $\mathfrak{N}(E, F)$ and $\mathfrak{N}(\tilde{E}, F)$ can be identified, since every $A \in \mathfrak{L}(E, F)$ has a uniquely determined extension $\tilde{A}$ to $\tilde{E}$ and $\|A\|_\nu = \|\tilde{A}\|_\nu$. In this case $\mathfrak{N}(E, F)$ is a (B)-space for the nuclear norm.

If F is normed and not complete, then $\mathfrak{N}(E, F)$ is obviously a subset of $\mathfrak{N}(E, \tilde{F})$. That this injection is a norm isomorphism was proved by PIETSCH.

(11) *Let E, F, G be normed spaces and F dense in G. If $A \in \mathfrak{L}(E, F)$ is in $\mathfrak{N}(E, G)$, then $A \in \mathfrak{N}(E, F)$ and the nuclear norms coincide; $\|A\|_\nu^F = \|A\|_\nu^G$.*

Proof. a) For every $z \in G$ and $\varepsilon > 0$ there exists a sequence $y_n \in F$ such that $z = \sum_{n=1}^\infty y_n$ and $\sum_{n=1}^\infty \|y_n\| \leqq (1 + \varepsilon)\|z\|$.

To prove this we choose $y_n' \in F$ with $\|z - y_n'\| \leqq (1/2^{n+1})\varepsilon\|z\|$ and set $y_1 = y_1'$, $y_n = y_n' - y_{n-1}'$ for $n > 1$. Clearly, $z = \lim y_n' = \sum_{n=1}^\infty y_n$. Also

$$\|y_1\| \leqq \left(1 + \frac{\varepsilon}{4}\right)\|z\| \quad \text{and} \quad \|y_n\| \leqq \left(\frac{1}{2^{n+1}} + \frac{1}{2^n}\right)\varepsilon\|z\|.$$

It follows that

$$\sum_{n=1}^\infty \|y_n\| \leqq \left[1 + \frac{\varepsilon}{4} + \varepsilon\left(\sum_{n=2}^\infty \frac{1}{2^{n+1}} + \sum_{n=2}^\infty \frac{1}{2^n}\right)\right]\|z\| \leqq (1 + \varepsilon)\|z\|.$$

b) Suppose now $A \in \mathfrak{N}(E, G)$. So A has a representation $Ax = \sum_{n=1}^\infty (u_n x) z_n$, where $u_n \in E'$, $z_n \in G$, and $\sum \|u_n\|\|z_n\| \leqq \|A\|_\nu^G + \varepsilon$. Using a), we can write $z_n = \sum_{m=1}^\infty y_{mn}$, where $y_{mn} \in F$ and $\sum_m \|y_{mn}\| \leqq (1 + \varepsilon)\|z_n\|$.

Hence

$$Ax = \sum_{n=1}^\infty (u_n x)\left(\sum_{m=1}^\infty y_{mn}\right) = \sum_{n=1}^\infty \sum_{m=1}^\infty (u_{mn} x) y_{mn}, \qquad u_{mn} = u_n.$$

Now

$$\|A\|_\nu^F \leqq \sum_n \sum_m \|u_{mn}\| \|y_{mn}\| = \sum_n \|u_n\| \sum_m \|y_{mn}\|$$
$$\leqq \sum_n \|u_n\|(1 + \varepsilon)\|z_n\| \leqq (1 + \varepsilon)(\|A\|_\nu^G + \varepsilon).$$

Since $\varepsilon > 0$ is arbitrary, $\|A\|_\nu^F \leqq \|A\|_\nu^G$. The inverse inequality is trivial since the infimum defining $\|A\|_\nu^G$ is taken over more representations than the infimum defining $\|A\|_\nu^F$. Thus $\|A\|_\nu^F = \|A\|_\nu^G$.

We close this section with the following result.

(12) a) *Let E, F be normed spaces. If $A \in \mathfrak{L}(E, F)$ is nuclear, then the adjoint A' is nuclear and $\|A'\|_\nu \leqq \|A\|_\nu$.*

b) *If F is a reflexive* (B)*-space and if A' is nuclear, then A is nuclear and $\|A'\|_\nu = \|A\|_\nu$.*

Proof. a) Let $A \in \mathfrak{N}(E, F)$, $Ax = \sum (u_n x) y_n$, $\sum \|u_n\| \|y_n\| \leqq \|A\|_\nu + \varepsilon$. If $v \in F'$, then $v(Ax) = \sum_n (u_n x)(v y_n) = (A'v)x = \left[\sum_n (v y_n) u_n\right] x$. Therefore

$$A'v = \sum (y_n v) u_n \quad \text{and} \quad \sum \|y_n\| \|u_n\| \leqq \|A\|_\nu + \varepsilon.$$

But this means that $A' \in \mathfrak{N}(F', E')$ and $\|A'\|_\nu \leqq \|A\|_\nu$.

b) If $A' \in \mathfrak{N}(F', E')$, then $A'' \in \mathfrak{L}(E'', F)$ is nuclear by a) and so is the restriction A of A'' to E, since $A = A''J$, where J is the canonical injection of E into E''. Using a) and (9) b), we obtain $\|A\|_\nu \leqq \|A''\|_\nu \|J\| = \|A''\|_\nu \leqq \|A'\|_\nu$ and, since $\|A'\|_\nu \leqq \|A\|_\nu$, the statement follows.

We remark that we proved in (12) a) that

(13) *If $A = \bar{\psi}\left(\sum_{n=1}^\infty u_n \otimes y_n\right)$, $\sum u_n \otimes y_n \in E_b' \tilde{\otimes}_\pi F$, then*

$$A' = \bar{\psi}\left(\sum_{n=1}^\infty y_n \otimes u_n\right), \qquad \sum_{n=1}^\infty y_n \otimes u_n \in F \tilde{\otimes}_\pi E_b'.$$

For a deeper result on the adjoint of a nuclear mapping see 7.(8).

6. Examples of nuclear mappings. We study nuclear mappings A between Hilbert spaces H_1, H_2. Since A is compact, A has a canonical representation of the form 4.(2) with positive singular values λ_n.

The nuclear mappings have the following characterization:

(1) *Let H_1, H_2 be Hilbert spaces and $A \in \mathfrak{L}(H_1, H_2)$ compact with a representation*

$$Ax = \sum_{n=1}^\infty \lambda_n (x, e_n) f_n, \tag{2}$$

where $\{e_n\}$, $\{f_n\}$ *are orthonormal sets in* H_1 *and* H_2, *respectively, and* $|\lambda_n| \to 0$, λ_n *complex.*

Then A *is nuclear if and only if* $\sum_{n=1}^{\infty} |\lambda_n| < \infty$, *and* $\|A\|_\nu = \sum_{n=1}^{\infty} |\lambda_n|$.

Proof. If a compact A has a representation (2) such that $\sum |\lambda_n| < \infty$, then A is nuclear and $\|A\|_\nu \leqq \sum_{n=1}^{\infty} |\lambda_n|$. We have only to recall that (x, e_n) can be replaced by $\langle \bar{e}_n, x \rangle$.

Conversely, let A be nuclear; then A has a representation

$$(3) \qquad Ax = \sum_{n=1}^{\infty} \mu_n(x, g_n)h_n,$$

$\|g_n\| \leqq 1$, $\|h_n\| \leqq 1$, $\sum |\mu_n| < \infty$. A always has a representation of the form (2), at least its canonical representation. We will show that $\sum_{n=1}^{\infty} |\lambda_n| \leqq \sum_{n=1}^{\infty} |\mu_n|$; then, by the definition of the nuclear norm, $\|A\|_\nu = \sum_{n=1}^{\infty} |\lambda_n|$ and (1) follows.

The following proof is due to S. SIMONS.

From (2) we have $Ae_p = \lambda_p f_p$, which by (3) is equal to $\sum_n \mu_n(e_p, g_n)h_n$; therefore

$$(4) \qquad \lambda_p = \sum_{n=1}^{\infty} \mu_n(e_p, g_n)(h_n, f_p).$$

From this follows

$$\begin{aligned}
(5) \quad \sum_{p=1}^{\infty} |\lambda_p| &\leqq \sum_p \sum_n |\mu_n| |(e_p, g_n)| |(h_n, f_p)| \\
&= \sum_p \sum_n (|\mu_n|^{1/2} |(e_p, g_n)|)(|\mu_n|^{1/2} |(h_n, f_p)|) \\
&\leqq \Big(\sum_p \sum_n |\mu_n| |(e_p, g_n)|^2\Big)^{1/2} \Big(\sum_p \sum_n |\mu_n| |(h_n, f_p)|^2\Big)^{1/2} \\
&= \Big(\sum_n |\mu_n| \sum_p |(e_p, g_n)|^2\Big)^{1/2} \Big(\sum_n |\mu_n| \sum_p |(h_n, f_p)|^2\Big)^{1/2} \\
&\leqq \Big(\sum_n |\mu_n| \|g_n\|^2\Big)^{1/2} \Big(\sum_n |\mu_n| \|h_n\|^2\Big)^{1/2} \leqq \sum |\mu_n|
\end{aligned}$$

since $\|g_n\| \leqq 1$, $\|h_n\| \leqq 1$.

We note that if a compact $A \in \mathfrak{L}(H)$ has a spectral decomposition $Ax = \sum_{n=1}^{\infty} \mu_n(x, u_n)u_n$, where $\{u_n\}$ is an orthonormal system, then by (1) A

is nuclear if and only if $\sum_n |\mu_n| < \infty$, and $\|A\|_\nu = \sum |\mu_n| = \sum \lambda_n$, where λ_n, $n = 1, 2, \ldots$, are the singular values of A.

From (1) and 4.(8) it follows that every nuclear mapping A between Hilbert spaces is Hilbert–Schmidt and that $\|A\| \leqq \|A\|_h \leqq \|A\|_\nu$.

The connection between nuclear and Hilbert–Schmidt mappings is very close, as the following two propositions show.

(6) *The product of two Hilbert–Schmidt mappings A and B is nuclear and* $\|AB\|_\nu \leqq \|A\|_h \|B\|_h$.

Let H_1, H_2, H_3 be Hilbert spaces, $A \in \mathfrak{H}(H_2, H_3)$, $B \in \mathfrak{H}(H_1, H_2)$. Since B is compact, it has a separable range, and if $\{f_n\}$ is an orthonormal basis of $\overline{B(H_1)}$, then $Bx = \sum_{n=1}^{\infty} (Bx, f_n) f_n$ and $ABx = \sum_{n=1}^{\infty} (Bx, f_n) Af_n = \sum_n (x, B^* f_n) Af_n$. Therefore

$$\|AB\|_\nu \leqq \sum_n \|B^* f_n\| \|Af_n\| \leqq \left(\sum \|B^* f_n\|^2\right)^{1/2} \left(\sum \|Af_n\|^2\right)^{1/2} \leqq \|B\|_h \|A\|_h.$$

Conversely,

(7) *Every $A \in \mathfrak{N}(H_1, H_2)$ is the product of two Hilbert–Schmidt mappings.*

Let (2) be the canonical representation of A, $\|A\|_\nu = \sum_n \lambda_n < \infty$. Define $A_1 \in \mathfrak{L}(H_1, H_2)$ by $A_1 x = \sum_{n=1}^{\infty} \lambda_n^{1/2} (x, e_n) f_n$ for $x \in H_1$ and $A_2 \in \mathfrak{L}(H_2)$ by $A_2 y = \sum_{n=1}^{\infty} \lambda_n^{1/2} (y, f_n) f_n$ for $y \in H_2$; then $A = A_2 A_1$ and $\|A_1\|_h = \sqrt{\sum \lambda_n} = \|A_2\|_h$, so that $\|A\|_\nu = \|A_2\|_h \|A_1\|_h$.

A similar factorization is possible for general nuclear maps.

(8) *Let A be nuclear from the normed space E into the* (B)*-space F. Let $1 < p < \infty$, $1/p + 1/q = 1$, $\varepsilon > 0$. Then $A = CB$, where $B \in \mathfrak{L}(E, l^p)$, $C \in \mathfrak{L}(l^p, F)$, and* $\|B\| \leqq (\|A\|_\nu + \varepsilon)^{1/p}$, $\|C\| \leqq (\|A\|_\nu + \varepsilon)^{1/q}$.

Proof. A has a representation $Ax = \sum_{n=1}^{\infty} \lambda_n (u_n x) y_n$, $\|u_n\| \leqq 1$, $\|y_n\| \leqq 1$, $\sum_{n=1}^{\infty} |\lambda_n| \leqq \|A\|_\nu + \varepsilon$. We define B by

$$Bx = (|\lambda_n|^{1/p} (u_n x))_{n=1,2,3,\ldots}.$$

From $\|Bx\| \leqq \|x\| (\sum |\lambda_n|)^{1/p} \leqq (\|A\|_\nu + \varepsilon)^{1/p} \|x\|$ follows the statement for B. Let C be defined by

$$C(\xi_n) = \sum_{n=1}^{\infty} \xi_n |\lambda_n|^{1/q} y_n, \qquad (\xi_n) \in l^p.$$

Since F is complete, this has a meaning and

$$\|C(\xi_n)\| \leqq \sum |\xi_n||\lambda_n|^{1/q} \leqq \left(\sum |\xi_n|^p\right)^{1/p}\left(\sum |\lambda_n|\right)^{1/q} \leqq (\|A\|_\nu + \varepsilon)^{1/q}\|(\xi_n)\|.$$

Hence $C \in \mathfrak{L}(l^p, F)$ and $CBx = Ax$.

As our second example we study nuclear mappings between spaces l^1. Let E be a (B)-space of all $x = \sum_{n=1}^{\infty} x_n e_n$, $\|x\|_1 = \sum |x_n| < \infty$; similarly, let F be the (B)-space of all $y = \sum_{m=1}^{\infty} y_m f_m$, $\|y\|_1 = \sum_{m=1}^{\infty} |y_m| < \infty$, $x_n, y_m \in K$. An $A \in \mathfrak{L}(E, F)$ can be represented by an infinite matrix $\mathfrak{A} = (a_{mn})$, where the a_{mn} are defined by $Ae_n = \sum_{m=1}^{\infty} a_{mn} f_m$. Since the image of the bounded set of all e_n is bounded, continuity of A is equivalent to $\sum_{m=1}^{\infty} |a_{mn}| \leqq M$ for some $M < \infty$ and all n.

(9) *$A \in \mathfrak{L}(E, F)$ represented by the matrix $\mathfrak{A} = (a_{mn})$ is compact if and only if* $\lim_{m\to\infty} \sup_n \sum_{i=m}^{\infty} |a_{in}| = 0$.

Proof. A is compact if and only if the set of all the $Ae_n = \sum_{m=1}^{\infty} a_{mn} f_m$ is relatively compact. The statement follows now from § 22, 4.(3).

(10) *$A \in \mathfrak{L}(E, F)$ represented by the matrix $\mathfrak{A} = (a_{mn})$ is nuclear if and only if $\sum_{m=1}^{\infty} \sup_n |a_{mn}| < \infty$ and this expression is the nuclear norm of A.*

Proof. a) Suppose A nuclear; then A has a representation

$$Ax = \sum_{k=1}^{\infty} (u^{(k)}x)y^{(k)}, \qquad u^{(k)} \in E', y^{(k)} \in F, \sum \|u^{(k)}\|_\infty \|y^{(k)}\|_1 \leqq \|A\|_\nu + \varepsilon.$$

We denote by e_i' the elements of E' defined by $e_i'e_k = \delta_{ik}$, $i, k = 1, 2, \ldots$; the definition of $f_i' \in F'$ is similar. If $u^{(k)} = \sum_{n=1}^{\infty} u_n^{(k)} e_n'$, $y^{(k)} = \sum_{n=1}^{\infty} y_n^{(k)} f_n$, then A is represented by the matrix (a_{mn}), where

$$a_{mn} = f_m'(Ae_n) = \sum_{k=1}^{\infty} (u^{(k)}e_n)(f_m' y^{(k)}) = \sum_{k=1}^{\infty} u_n^{(k)} y_m^{(k)}.$$

Now

$$\sup_n |a_{mn}| = \sup_n \left|\sum_k u_n^{(k)} y_m^{(k)}\right| \leqq \sum_k \|u^{(k)}\|_\infty |y_m^{(k)}|$$

and

$$\sum_m \sup_n |a_{mn}| \leqq \sum_{k=1}^{\infty} \|u^{(k)}\|_\infty \|y^{(k)}\|_1 \leqq \|A\|_\nu + \varepsilon.$$

Therefore $\sum_m \sup |a_{mn}| \leqq \|A\|_\nu < \infty$.

b) Conversely, assume $\sum_m \sup_n |a_{mn}| < \infty$ for $A = (a_{mn}) \in \mathfrak{L}(E, F)$. Denote by a_m the element $\sum_{n=1}^\infty a_{mn}e'_n$ of E'. Then $Ax = \sum_{m=1}^\infty (a_m x) f_m$, where $a_m x = \sum_{n=1}^\infty a_{mn}x_n$. From this representation of A follows $\|A\|_\nu \leqq \sum_m \|a_m\|_\infty \|f_m\|_1 = \sum_{m=1}^\infty \sup_n |a_{mn}| < \infty$. Thus A is nuclear and $\|A\|_\nu = \sum_{m=1}^\infty \sup_n |a_{mn}|$.

7. The trace. Let E be a (B)-space. By § 41, 3.(6) the dual of $E'_b \tilde{\otimes}_\pi E$ can be identified with the space $\mathscr{B}(E'_b \times E)$ of all continuous bilinear forms on $E'_b \times E$. If we take specifically the canonical bilinear form $(u, x) = \langle u, x \rangle = ux$, $u \in E'$, $x \in E$, then we obtain a continuous linear form on $E'_b \tilde{\otimes}_\pi E$ which is called the trace $\operatorname{tr} z$ of the element $z \in E'_b \tilde{\otimes}_\pi E$.

By definition $\operatorname{tr}(u \otimes x) = ux$. If z has a representation

$$(1) \qquad z = \sum_{n=1}^\infty \lambda_n u_n \otimes x_n, \qquad \|u_n\| \leqq 1, \|x_n\| \leqq 1, \sum |\lambda_n| \leqq \|z\|_\pi + \varepsilon,$$

then it follows by continuity that

$$(2) \qquad \operatorname{tr} z = \sum_{n=1}^\infty \lambda_n (u_n x_n),$$

the convergence being absolute, and obviously $|\operatorname{tr} z| \leqq \|z\|_\pi$.

The trace does not depend on the special representation (1) of z. If the canonical mapping $\bar{\psi}$ of $E'_b \tilde{\otimes}_\pi E$ in $\mathfrak{L}(E)$ is one-one, then we define the trace of a nuclear mapping $Ax = \sum \lambda_n (u_n x) x_n$ by

$$\operatorname{tr} A = \operatorname{tr}\left(\sum_{n=1}^\infty \lambda_n u_n \otimes x_n\right) = \sum_{n=1}^\infty \lambda_n (u_n x_n).$$

If $\bar{\psi}$ is not one-one, the trace of a nuclear mapping may not be uniquely defined.

In any case the mapping of $E'_b \otimes_\pi E$ in $\mathfrak{L}(E)$ is one-one, so that the A of finite rank always have a uniquely determined trace.

(3) *If the trace is uniquely defined for the nuclear mappings of* $\mathfrak{L}(E)$, E *a* (B)-*space, then* $\operatorname{tr}(AB) = \operatorname{tr}(BA)$ *for* A, B *in* $\mathfrak{N}(E)$.

Assume

$$Ax = \sum_{n=1}^\infty (u_n x) x_n, \sum \|u_n\| \|x_n\| < \infty, \qquad Bx = \sum_{n=1}^\infty (v_n x) y_n, \sum \|v_n\| \|y_n\| < \infty;$$

then an easy calculation shows that

$$\operatorname{tr}(BA) = \sum_{m=1}^\infty \sum_{n=1}^\infty (v_m x_n)(u_n y_m) = \operatorname{tr}(AB).$$

We consider again the case of Hilbert spaces. Anticipating the general discussion in § 43, we show that the trace is uniquely defined in Hilbert space.

A (B)-space E has the approximation property if $\psi(E' \otimes E)$, the space of all continuous linear mappings of finite rank, is dense in $\mathfrak{L}_c(E)$.

It is sufficient to prove that the identity I is in the closure $\overline{\psi(E' \otimes E)}$ in $\mathfrak{L}_c(E)$: Let $B \in \mathfrak{L}(E)$ be given and assume that the net A_α converges to I, where $A_\alpha \in \psi(E' \otimes E)$. Now every BA_α is of finite rank, so it will be sufficient to prove $BA_\alpha \to B$. Given the neighbourhood $U(M, V)$, M compact in E, there exists a neighbourhood W such that $B(W) \subset V$. Then $(BA_\alpha - B)(M) \subset B(W) \subset V$ if $(A_\alpha - I)(M) \subset W$.

(4) *Every Hilbert space H has the approximation property.*

Proof. Let K be a compact subset of H and U the closed unit ball with radius $\varepsilon > 0$. There exists a finite set $\{x_1, \ldots, x_m\}$ such that $K \subset \bigcup_{i=1}^{m} (x_i + U)$. Let G be the linear span $[x_1, \ldots, x_m]$ and P the orthogonal projection of H onto G. If $x \in K$, then $x = x_i + y_i$, $y_i \in U$, for some i, $Px = x_i + Py_i$, and therefore $\|x - Px\| \leqq \varepsilon$ for all $x \in K$. This means that $I \in \overline{\psi(H' \otimes H)} \subset \mathfrak{L}_c(H)$.

(5) *For a Hilbert space E the mapping $\bar{\psi}$ of $E'_b \tilde{\otimes}_\pi E$ on $\mathfrak{N}(E) \subset \mathfrak{L}(E)$ is one-one. The trace is uniquely defined for all nuclear mappings in $\mathfrak{L}(E)$.*

Proof. We remark that E, E', E'' can be identified. By § 41, 3.(14) it is sufficient to show that $\psi(E \otimes E)$ is $\mathfrak{T}_s(E \tilde{\otimes}_\pi E)$-dense in $\mathscr{B}(E \times E)$.

An element z of $E \tilde{\otimes}_\pi E$ has a representation $z = \sum_{n=1}^{\infty} \lambda_n x_n \otimes y_n$, $\|x_n\| \to 0$, $\|y_n\| \to 0$, and $\sum |\lambda_n| \leqq 1$ (§ 41, 4.(6)). Let K be the compact set consisting of $\mathfrak{o}$ and all x_n, $C = \{y_1, y_2, \ldots\}$ and $U = \varepsilon C^\circ$. Let $B \in \mathscr{B}(E \times E)$ and $\tilde{B}$ the corresponding mapping in $\mathfrak{L}(E)$; then by (4) there exists $A \in E \otimes E$ such that $(\tilde{B} - \tilde{A})(K) \subset \varepsilon C^\circ$. It follows that

$$|\langle B - A, z\rangle| = \left|\sum_{k=1}^{\infty} \lambda_k (B - A)(x_k, y_k)\right|$$

$$= \left|\sum_{k=1}^{\infty} \lambda_k ((\tilde{B} - \tilde{A})x_k) y_k\right| \leqq \sum_{k=1}^{\infty} |\lambda_k| \varepsilon \leqq \varepsilon.$$

Hence $E \otimes E$ is $\mathfrak{T}_s(E \tilde{\otimes}_\pi E)$-dense in $\mathscr{B}(E \times E)$.

We give a second proof of the uniqueness of tr A on $\mathfrak{N}(E)$, E a Hilbert space, without using (5) but, instead, elementary results given in 6.

Let A be nuclear and $Ax = \sum_{n=1}^{\infty} \lambda_n(x, e_n)f_n$ its canonical representation with the singular values $\lambda_n > 0$ and $\|A\|_\nu = \sum_{n=1}^{\infty} \lambda_n$. We define $\operatorname{tr}_0 A = \sum_{n=1}^{\infty} \lambda_n(f_n, e_n) = \sum_{n=1}^{\infty} (Ae_n, e_n)$. If $A = \bar{\psi}(z)$, $z = \sum_{n=1}^{\infty} x^{(n)} \otimes \bar{y}^{(n)}$, $\sum_n \|x^{(n)}\| \times \|\bar{y}^{(n)}\| < \infty$, the uniqueness of the trace will follow from

$$(6) \quad \operatorname{tr}_0 A = \sum_{n=1}^{\infty} (Ae_n, e_n) = \sum_{n=1}^{\infty} (x^{(n)}, y^{(n)}) = \sum_{n=1}^{\infty} \langle x^{(n)}, \bar{y}^{(n)} \rangle = \operatorname{tr} z.$$

Proof of (6). We remark first that $\operatorname{tr}_0 A = \sum_\alpha (Av_\alpha, v_\alpha)$, where $\{v_\alpha\}$ is any orthonormal basis of E: It is obvious that we can enlarge the orthogonal system e_n to an orthonormal basis u_α such that $\operatorname{tr}_0 A = \sum_\alpha (Au_\alpha, u_\alpha)$. By 6.(7) we write A as the product A_2A_1 of two Hilbert–Schmidt mappings; then $\sum_\alpha (Au_\alpha, u_\alpha) = \sum_\alpha (A_1u_\alpha, A_2^*u_\alpha) = (A_1, A_2^*)$, the scalar product defined in 4.(6). But this is independent of the choice of the orthonormal basis.

We take now as basis the basis $\{e_\alpha\}$ introduced in 4. for which $\bar{e}_\alpha = e_\alpha$ and $(x, e_\alpha) = \langle x, e_\alpha \rangle$. We write $x_\alpha = (x, e_\alpha)$ for every $x \in E$. The double sum in

$$\operatorname{tr} z = \sum_{n=1}^{\infty} \langle x^{(n)}, \bar{y}^{(n)} \rangle = \sum_{n=1}^{\infty} \sum_\alpha x_\alpha^{(n)} \bar{y}_\alpha^{(n)}$$

converges absolutely, since

$$\sum_\alpha |x_\alpha^{(n)} \bar{y}_\alpha^{(n)}| \leqq \|x^{(n)}\| \, \|\bar{y}^{(n)}\| \quad \text{and} \quad \sum_{n=1}^{\infty} \|x^{(n)}\| \, \|\bar{y}^{(n)}\| < \infty.$$

On the other hand, $(Ae_\alpha, e_\alpha) = \left(\sum_{n=1}^{\infty} \langle x^{(n)}, e_\alpha \rangle \bar{y}^{(n)}, e_\alpha \right) = \sum_{n=1}^{\infty} x_\alpha^{(n)} \bar{y}_\alpha^{(n)}$; therefore $\operatorname{tr}_0 A = \sum_\alpha (Ae_\alpha, e_\alpha) = \sum_\alpha \sum_{n=1}^{\infty} x_\alpha^{(n)} \bar{y}_\alpha^{(n)}$. Since this double sum converges absolutely it equals $\operatorname{tr} z$.

(6) implies that $\operatorname{tr} A = \operatorname{tr}_0 A$ is a linear functional on $\mathfrak{N}(E)$ and from the definition of $\operatorname{tr} A$ and 6.(1) follows $|\operatorname{tr} A| \leqq \|A\|_\nu$, the continuity of the trace.

Let E, F now be arbitrary (B)-spaces, $\mathscr{B}(E, F)$ the dual of $E \mathbin{\tilde{\otimes}_\pi} F$. We show that the bilinear form $\langle B, z \rangle$, $B \in \mathscr{B}(E, F)$, $z \in E \mathbin{\tilde{\otimes}_\pi} F$, can be expressed by using the trace.

As before, we denote by $\tilde{B}$ the element of $\mathfrak{L}(E, F_b')$ corresponding to B (§ 41, 3.(6)). Let I be the identity on F. Then by § 41, 5.(1) $\tilde{B} \mathbin{\tilde{\otimes}_\pi} I$ is a continuous mapping of $E \tilde{\otimes}_\pi F$ in $F'_b \tilde{\otimes}_\pi F$ determined by the mapping $\tilde{B} \otimes I$ of $E \otimes_\pi F$ in $F'_b \otimes F$. We have

$$(\tilde{B} \otimes I)(x \otimes y) = (\tilde{B}x) \otimes y \quad \text{for } x \in E,\, y \in F;$$

hence

$$\operatorname{tr}((\tilde{B} \otimes I)(x \otimes y)) = (\tilde{B}x)y = B(x, y) = \langle B, x \otimes y \rangle.$$

By continuity this implies the formula

$$(7) \quad \langle B, z \rangle = \operatorname{tr}((\tilde{B} \tilde{\otimes}_\pi I)z), \qquad B \in \mathscr{B}(E \times F),\, z \in E \tilde{\otimes}_\pi F,$$

which is valid for (B)-spaces E and F.

GROTHENDIECK [13] used (7) to obtain the following result, closely related to 5.(12), on nuclear maps:

(8) *Let E, F be* (B)*-spaces and suppose that the canonical mapping $\bar{\psi}$ of $E'' \tilde{\otimes}_\pi E'$ in $\mathfrak{L}(E')$ is one-one. Then $A \in \mathfrak{L}(E, F)$ is nuclear if and only if $A' \in \mathfrak{L}(F', E')$ is nuclear and $\|A\|_\nu = \|A'\|_\nu$.*

We remark that E', E'', F' are always equipped with the strong topology and considered as (B)-spaces.

Proof. a) If A is nuclear, then A' is nuclear by 5.(12) a). We assume now that $A' \in \mathfrak{L}(F', E')$ is nuclear. By 5.(3) $A' = \bar{\psi}(z)$, $z \in F'' \tilde{\otimes}_\pi E'$. We will prove that the nuclearity of A' implies $z \in F \tilde{\otimes}_\pi E'$, $z = \sum_{n=1}^{\infty} y_n \otimes u_n$. Then

$$x(A'v) = \sum (u_n x)(v y_n) = v(Ax) \quad \text{for all } x \in E,\, v \in F'.$$

This implies $A = \bar{\psi}(\sum u_n \otimes y_n)$, $\sum u_n \otimes y_n \in E' \tilde{\otimes}_\pi F$; hence A is nuclear.

b) To prove that $z \in F \tilde{\otimes}_\pi E'$ we remark that $F \tilde{\otimes}_\pi E'$ is a closed subspace of $F'' \tilde{\otimes}_\pi E'$, so it will be sufficient to show that $\langle B, F \tilde{\otimes}_\pi E' \rangle = 0$ implies $\langle B, z \rangle = 0$ for every $B \in \mathscr{B}(F'' \times E')$.

For the corresponding $\tilde{B} \in \mathfrak{L}(F'', E'')$ we have $\tilde{B}(F) = \circ$ and by (7)

$$\langle B, z \rangle = \operatorname{tr} t, \qquad t = (\tilde{B} \tilde{\otimes}_\pi I)z \in E'' \tilde{\otimes}_\pi E'.$$

It will be sufficient to show that $t = \circ$ or, by the assumption on $E'' \tilde{\otimes}_\pi E'$, that $T = \bar{\psi}(t) \in \mathfrak{L}(E')$ is $\circ$.

z has a representation $z = \sum_{n=1}^{\infty} z^{(n)} \otimes u^{(n)} \in F'' \tilde{\otimes}_\pi E'$; hence $t = \sum (\tilde{B}z^{(n)}) \otimes u^{(n)}$ and by 5.(13) we have for all $x'' \in E''$

$$T'x'' = \sum (u^{(n)}x'')(\tilde{B}z^{(n)}).$$

Since $A' = \bar{\psi}(z) = \bar{\psi}(\sum z^{(n)} \otimes u^{(n)})$, $A'' = \bar{\psi}(\sum u^{(n)} \otimes z^{(n)})$ and $\tilde{B}A''x'' = \sum (u^{(n)}x'')(\tilde{B}z^{(n)})$. Hence $T' = \tilde{B}A''$.

Now A is compact and by 2.(1) we have $A''(E'') \subset F$. Since $\tilde{B}(F) = \mathfrak{o}$, it follows that $T' = \mathfrak{o}$ and so $T = \mathfrak{o}$.

c) By 5.(12) a) $\|A'\|_\nu \leqq \|A\|_\nu$. Since $A''(E'') \subset F$, A is the restriction of A'' to E and $\|A\|_\nu \leqq \|A'\|_\nu$ follows as in 5.(12) b).

We will prove in § 43, 2.(7) that the assumption in (8) is equivalent to the assumption that E' has the approximation property.

8. Factorization of compact mappings. TERZIOGLU [3'] gave the following characterization of precompact mappings between normed spaces.

(1) *Let E, F be normed spaces. $A \in \mathfrak{L}(E, F)$ is precompact if and only if A satisfies an inequality of the form*

(2) $\|Ax\| \leqq \sup_n |u_n x|$ *for all* $x \in E$, *where* $u_n \in E'$, $\lim_n \|u_n\| = 0$. *Moreover,* $\|A\| = \inf \sup_n \|u_n\|$, *where the infimum is taken over all sequences* $u_n \in E'$ *satisfying* (2).

Proof. a) Assume A precompact; then A' is compact by 1.(7). Let V be the closed unit ball of F; then $A'(V^\circ)$ is relatively compact in E' and contains only elements u with $\|u\| \leqq \|A'\| = \|A\|$. By § 41, 4.(3) and the remark following this proposition there exist elements $u_n \in E'$, $\|u_n\| \leqq \|A\| + \varepsilon$, $\lim_n \|u_n\| = 0$ such that every $u \in A'(V^\circ)$ has a representation $u = \sum_{n=1}^{\infty} \xi_n u_n$, $\sum |\xi_n| \leqq 1$. Therefore for $v \in F'$

$$\|Ax\| = \sup_{\|v\| \leqq 1} |(A'v)x| \leqq \sup_n |u_n x|$$

and (2) is proved.

It follows that $\|A\| \leqq \sup_n \|u_n\| \leqq \|A\| + \varepsilon$, which implies the second statement.

b) We assume now that A satisfies (2). We define $B \in \mathfrak{L}(E, c_0)$ by $Bx = (u_1 x, u_2 x, \ldots)$. Obviously, $\|B\| = \sup_n \|u_n\|$. If U is the closed unit ball in E, then $B(U)$ is contained in the set K of all $\eta = (\eta_n) \in c_0$ such that $|\eta_n| \leqq \|u_n\|$. Since $\|u_n\| \to 0$, it is easy to see that K is totally bounded in c_0, so that B is compact. If $H = B(E)$, then B is precompact as a mapping of E onto H.

We define now on H a mapping $C \in \mathfrak{L}(H, F)$ by $C(Bx) = Ax$ for all $Bx \in H$. Since $\|C(Bx)\| = \|Ax\| \leqq \sup_n |u_n x| = \|Bx\|$, C is well defined on H and $\|C\| \leqq 1$. Thus we have a factorization $A = CB$, where B is precompact. It follows from 1.(1) that A is precompact and (1) is proved.

We note the following facts:

(3) a) *A subset M of c_0 is relatively compact if and only if it is contained in the normal cover of an element $\xi = (\xi_n) \in c_0$.*

b) *If E is a* (B)*-space, then a compact $A \in \mathfrak{L}(E, c_0)$ has a representation $Ax = (u_1x, u_2x, \ldots)$, where $u_n \in E'$ and $\|u_n\| \to 0$.*

We leave the proof of a) to the reader. That $\|u_n\| \to 0$ in b) follows from a) by contradiction.

If A is precompact, then from the proof of (1) follows the existence of a factorization $A = CB$, $\|B\| \leqq \|A\| + \varepsilon$, $\|C\| \leqq 1$, where B is precompact. By a slight modification we will obtain a factorization in two precompact mappings.

Since $\|u_n\| \to 0$, there exist $\rho_n \geqq 1$, $\rho_n \to \infty$ such that for $u_n' = \rho_n u_n$ we have $\|u_n'\| \to 0$ and $\sup \|u_n'\| = \sup \|u_n\|$. If we define A_1 by $A_1x = (u_1'x, u_2'x, \ldots) \in c_0$ for $x \in E$, then $B = B_1A_1$, where B_1 is the diagonal transformation $B_1\xi = (d_1\xi_1, d_2\xi_2, \ldots)$ in $\mathfrak{L}(c_0)$ and where $d_n = 1/\rho_n \to 0$, $d_n \leqq 1$. By (3) b) B_1 is compact in $\mathfrak{L}(c_0)$ and precompact as a mapping of $A_1(E)$ onto $B(E)$. A is now the product $A_2A_1 = (CB_1)A_1$ of two precompact mappings and we obtain the following factorization theorem (Randtke [2′], Terzioglu [3′]):

(4) *Let E, F be normed spaces, A precompact in $\mathfrak{L}(E, F)$. Let $\varepsilon > 0$ be given. Then there exists a linear subspace H of c_0 such that $A = A_2A_1$, $A_1 \in \mathfrak{L}(E, H)$, $A_2 \in \mathfrak{L}(H, F)$, A_1 and A_2 precompact, $\|A_1\| \leqq \|A\| + \varepsilon$, $\|A_2\| \leqq 1$.*

In this general case we have $H = A_1(E)$. If F is a (B)-space, we replace $H = A_1(E)$ by its closure $\overline{H}$ in c_0; then A_1 is compact in $\mathfrak{L}(E, \overline{H})$, A_2 has a continuous extension A_2 to $\overline{H}$, and we obtain in this way a factorization $A = A_2A_1$ in two compact mappings through a closed subspace of c_0.

It is natural to ask in what cases a precompact $A \in \mathfrak{L}(E, F)$ has a precompact factorization through c_0, not only through a subspace of c_0.

A sequence y_n in a normed space F is called weakly summable in F if $\sup_{\|v\| \leqq 1} \sum_{n=1}^{\infty} |vy_n| < \infty$, where $v \in F'$. A mapping $A \in \mathfrak{L}(E, F)$, E and F normed spaces, is called infinite-nuclear if it has a representation

(5) $Ax = \sum_{n=1}^{\infty} (u_nx)y_n$, $u_n \in E'$, $\lim_n \|u_n\| = 0$, *and the sequence y_n is weakly summable in F. From*

$$\|Ax\| = \sup_{\|v\| \leqq 1} \left| v\left(\sum_n (u_nx)y_n\right) \right| \leqq \left(\sup_n |u_nx|\right) \sup_{\|v\| \leqq 1} \sum_n |vy_n|$$

follows

$$\|A\| \leqq \left(\sup_n \|u_n\|\right) \sup_{\|v\|\leqq 1} \sum_n |vy_n|. \tag{6}$$

This implies that every linear mapping of finite rank is infinite-nuclear and that every infinite-nuclear mapping is precompact as the $\mathfrak{T}_b$-limit of mappings of finite rank.

(7) *If A is infinite-nuclear from E in F, E and F normed spaces, and if E is a subspace of the normed space X, then A has an infinite-nuclear extension $\tilde{A} \in \mathfrak{L}(X, F)$.*

Proof. By HAHN–BANACH there exist continuous extensions $\tilde{u}_n$ of u_n to X such that $\|\tilde{u}_n\| = \|u_n\|$. Then $\tilde{A}z = \sum_n (\tilde{u}_n z) y_n$, $z \in X$, answers the question.

If one defines the infinite-nuclear norm $\|A\|_\nu^\infty$ by

$$\inf \left(\sup_n \|u_n\|\right) \sup_{\|v\|\leqq 1} \sum_n |vy_n|,$$

where the infimum is taken over all representations (5), then $\|\tilde{A}\|_\nu^\infty = \|A\|_\nu^\infty$.

The problem raised above has the following solution:

(8) *A precompact linear mapping $A \in \mathfrak{L}(E, F)$, E and F normed spaces, has a precompact factorization through c_0 if and only if A is infinite-nuclear.*

Proof. a) Assume that $A = A_2 A_1$, $A_1 \in \mathfrak{L}(E, c_0)$ and precompact, $A_2 \in \mathfrak{L}(c_0, F)$. By (3) A_1 has a representation $A_1 x = (u_1 x, u_2 x, \ldots)$, $u_n \in E'$, $\lim_n \|u_n\| = 0$. Let $e_1, e_2, \ldots$ be the unit vectors in c_0; then $Ax = A_2\left(\sum_{n=1}^\infty (u_n x) e_n\right) = \sum_{n=1}^\infty (u_n x) y_n$, $y_n = A_2 e_n$. To see that A is infinite-nuclear we have only to prove that the sequence y_n is weakly summable in F. Now $A_2' v \in l^1$ for every $v \in F'$, so

$$\sup_{\|v\|\leqq 1} \sum_n |vy_n| = \sup_{\|v\|\leqq 1} \sum |(A_2' v) e_n| \leqq \sup_{\|v\|\leqq 1} \|A_2' v\|_1 = \|A_2\| < \infty.$$

We remark that we did not use that A_2 is precompact.

b) Conversely, if A has a representation (5), A can be factored in two precompact mappings $A_1 \in \mathfrak{L}(E, c_0)$, $A_2 \in \mathfrak{L}(c_0, F)$. This follows from writing A as $Ax = \sum ((\rho_n u_n) x)(1/\rho_n) y_n$, where the $\rho_n \geqq 1$, $\rho_n \to \infty$ are chosen such that $\|\rho_n u_n\| \to 0$ and $\|(1/\rho_n) y_n\| \to 0$.

These results can be used to characterize an interesting class of (B)-spaces F. One says that F has the compact extension property if a compact linear mapping from a subspace H of a (B)-space X into F has a compact linear extension mapping X into F. Similarly, F has the c_0-extension property if every compact $A \in \mathfrak{L}(H, F)$, $H \subset c_0$, has an extension $\tilde{A} \in \mathfrak{L}(c_0, F)$ ($\tilde{A}$ does not have to be compact).

(9) *Let F be a* (B)*-space. The following properties of F are equivalent:*

a) *F has the c_0-extension property;*

b) *every compact $A \in \mathfrak{L}(X, F)$, X a normed space, has a compact factorization through c_0;*

c) *compact and infinite-nuclear linear continuous mappings from X into F coincide for every normed space X;*

d) *F has the compact extension property.*

Proof. Suppose a) and consider a compact $A \in \mathfrak{L}(X, F)$. By (4) A has a compact factorization $A = A_2A_1$, $A_1 \in \mathfrak{L}(X, H)$, $A_2 \in \mathfrak{L}(H, F)$, where $H \subset c_0$. Now A_2 has an extension $\tilde{A}_2 \in \mathfrak{L}(c_0, F)$ so that $A = \tilde{A}_2A_1$. By the first part of the proof of (8) it follows that A is infinite-nuclear; hence a) implies c). By (8) we see that b) and c) are equivalent. c) implies d) because of (7). Trivially, d) implies a).

For this theorem and further results compare RANDTKE [1′], [2′], [3′] and TERZIOGLU [4′], [5′].

It is clear from § 38, 3.(5) that every P_λ-space, $\lambda \geqq 1$, satisfies condition a) and therefore has the compact extension property. LINDENSTRAUSS proved in [1′] that a (B)-space X has the compact extension property if and only if the strong bidual X'' is a P_λ-space for some $\lambda \geqq 1$.

Another very satisfactory characterization of our class of (B)-spaces was given later also by LINDENSTRAUSS. We need some definitions.

For two isomorphic (B)-spaces E and F the distance coefficient $d(E, F)$ is defined as $\inf(\|T\| \|T^{-1}\|)$, where the infimum is taken over all isomorphisms T of E onto F.

A (B)-space is called an $\mathscr{L}_{p,\lambda}$-space for some $\lambda \geqq 1$, $1 \leqq p \leqq \infty$, if for every finite dimensional subspace F of E there is a finite dimensional subspace $G \supset F$ such that $d(G, l_n^p) \leqq \lambda$, where n is the dimension of G.

E is called an $\mathscr{L}_p$-space if it is an $\mathscr{L}_{p,\lambda}$-space for some λ.

These spaces were introduced by LINDENSTRAUSS and PEŁCZYNSKI [1′] and their theory has been developed very rapidly during the last years (LINDENSTRAUSS–TZAFRIRI [1′]).

We formulate now LINDENSTRAUSS' theorem:

(10) *A* (B)*-space has the compact extension property if and only if it is an $\mathscr{L}_\infty$-space.*

For the proof we refer the reader to LINDENSTRAUSS and ROSENTHAL [1′]. We state also a dual result of the same authors.

A (B)-space E has the compact lifting property if for any quotient X/Z of (B)-spaces every compact $A \in \mathfrak{L}(E, X/Z)$ has a compact lifting $\hat{A} \in \mathfrak{L}(E, X)$, i.e., $A = K\hat{A}$, where K is the canonical homomorphism $X \to X/Z$.

(11) *A (B)-space has the compact lifting property if and only if it is an $\mathscr{L}_1$-space.*

9. Fixed points and invariant subspaces. Fixed point theorems are the main tool in nonlinear functional analysis. We will present here only one of these theorems, the SCHAUDER–TYCHONOFF theorem, which we will apply immediately.

We need a simple fact on finite dimensional convex sets.

(1) *Let $K \neq \circ$ be a compact subset of P^n; then the closed convex cover $\overline{\subset(K)}$ consists of all convex combinations $\sum_{i=1}^{n+1} \rho_i x_i$, $\sum \rho_i = 1$, $\rho_i \geqq 0$ of $n + 1$ arbitrary points of K. In particular, $\overline{\subset(K)} = \subset(K)$.*

We proceed by induction. (1) is true for a compact set K on the line, because $\overline{\subset(K)}$ is then a closed interval whose endpoints are in K. We assume (1) to be true for all dimensions $\leqq n - 1$. Let K be compact in P^n. If $\overline{\subset(K)}$ is contained in a hyperplane, the statement is true. If $\overline{\subset(K)}$ is not contained in a hyperplane, it is a convex body (see the remark preceding Example 1 in § 16, 2.), so every point of the boundary of $\overline{\subset(K)}$ lies in a supporting hyperplane (§ 17, 5.(1)).

Let x be a point of $\overline{\subset(K)}$ different from the point $p \in K$ and let q be a boundary point of $\overline{\subset(K)}$ which lies on the half-line from p through x but not between p and x. The point q belongs to a supporting hyperplane H of $\overline{\subset(K)}$ and lies therefore in $\overline{\subset(K \cap H)}$. By assumption q is a convex combination of n points $p_1, \ldots, p_n$ of $K \cap H$, so x is a convex combination of $p, p_1, \ldots, p_n$.

We borrow from topology BROUWER's fixed point theorem:

(2) *Let K be a nonvoid convex compact subset of K^n and φ a continuous mapping of K into K. Then φ has a fixed point $x_0 \in K$, i.e., $\varphi(x_0) = x_0$.*

We apply (2) to locally convex spaces.

(3) *Let A be a convex subset of the locally convex space E and let f be a continuous map of A into a compact subset K of A which is contained in a finite dimensional subspace H of E. Then f has a fixed point.*

Proof. The convex cover $\subset(K)$ of K is compact by (1) and contained in A. The restriction of f to $\subset(K)$ has a fixed point by (2).

(4) *Let A be a convex subset of E, K relatively compact in E and $K \subset A$. For every absolutely convex closed neighbourhood U of $\mathfrak{o}$ in E there exists a continuous mapping f of K in a finite dimensional compact subset of A such that $f(x) - x \in U$ for all $x \in K$.*

Proof. There exist $x_1, \ldots, x_m$ in K such that $K \subset \bigcup_{i=1}^{m} (x_i + \frac{1}{2}U)$. Let p be the semi-norm corresponding to U. The function $\alpha_i(x) = \max(0, 1 - p(x - x_i))$ is continuous on E and for every $x \in K$ at least one $\alpha_i(x)$ is $\neq 0$. If we write $\beta_i(x) = \alpha_i(x) \Big/ \sum_{k=1}^{m} \alpha_k(x)$, then the function $f(x) = \sum_{i=1}^{m} \beta_i(x) x_i$ is defined and continuous on K. Every $f(x)$, $x \in K$, is a convex combination of $x_1, \ldots, x_m$, so $f(K)$ is contained in the convex cover of $\{x_1, \ldots, x_m\}$, which is a compact subset of A.

Consider finally $f(x) - x = \sum_{i=1}^{m} \beta_i(x)(x_i - x)$ for $x \in K$. If $\beta_i(x) \neq 0$, then $\alpha_i(x) \neq 0$ and $p(x - x_i) < 1$ or $x_i - x \in U$. This implies $f(x) - x \in U$.

We prove now the fixed point theorem of Schauder–Tychonoff (compare Landsberg [1'] to our exposition).

(5) *Let E be locally convex and A a convex subset of E. Then every continuous map φ of A into a compact subset of A has a fixed point.*

Proof. Let U be an absolutely convex closed neighbourhood of $\mathfrak{o}$ in E. Since $K = \varphi(A)$ is relatively compact in A, there exists a continuous mapping f of K in A with the properties stated in (4). In particular, $f(\varphi(x)) - \varphi(x) \in U$ for all $x \in A$. We write f_U for the mapping $f \circ \varphi$ of A into A and obtain $f_U(x) - \varphi(x) \in U$ for all $x \in A$. The map f_U satisfies the assumptions of (3), so there exists $x_U \in A$ such that $f_U(x_U) = x_U$.

If we write $U_1 \leqq U_2$ for $U_1 \supset U$, then $\{\varphi(x_U)\}$ is a net in K which has an adherent point z in A, since K is relatively compact in A. This means that for every absolutely convex neighbourhood V of $\mathfrak{o}$ in E there exists a cofinal subset $\varphi(x_{U'})$ such that $\varphi(x_{U'}) - z \in V$ for all U'. For every V we choose $U' = U(V)$ such that $U(V) \subset V$. If we set $U(V_1) \leqq U(V_2)$ for $V_1 \supset V_2$, then $\{\varphi(x_{U(V)})\}$ is a net converging to z.

Now since $U(V) \subset V$, we have $\varphi(x) - f_{U(V)}(x) \in V$ for all $x \in A$; in particular, $\varphi(x_{U(V)}) - x_{U(V)} \in V$. Together with $z - \varphi(x_{U(V)}) \in V$, this implies $z - x_{U(V)} \in 2V$ or $\lim_V x_{U(V)} = z$. Since φ is continuous, it follows that $\varphi(z) = z$.

Let E be locally convex, $A \in \mathfrak{L}(E)$. A closed subspace H of E, different from $\mathfrak{o}$ and E, is a nontrivial invariant subspace of A if $A(H) \subset H$.

Aronszajn and Smith [1'] proved in 1954 that every compact $A \in \mathfrak{L}(E)$,

E a complex infinite dimensional (B)-space, has a nontrivial invariant subspace. This result has been generalized in many ways, but even in the case of $E = l^2$ it is not known whether every $A \in \mathfrak{L}(l^2)$ has a nontrivial invariant subspace. Recently LOMONOSOV [1'] proved a very strong result. We reproduce it here in an even more general version due to LINDENSTRAUSS which was communicated to us by DUGUNDJI.

(6) *Let E be a complex infinite dimensional locally convex space, let A, B in $\mathfrak{L}(E)$, $A \neq \circ$ and compact, $B \neq \lambda I$ for every complex λ and B commuting with A. Then the set R of all $C \in \mathfrak{L}(E)$ commuting with B has a common nontrivial invariant subspace.*

Proof. a) There exists $x_0 \in E$ with $Ax_0 \neq \circ$. Let V be an absolutely convex and closed neighbourhood of $\circ$ in E such that $Ax_0 \notin 2V$. Let U be an absolutely convex and closed neighbourhood of $\circ$ such that $A(U) \subset V$ and $A(U)$ is relatively compact in E.

For $S = x_0 + U$ we have $\overline{A(S)} = Ax_0 + \overline{A(U)} \subset Ax_0 + V$ and it follows from $Ax_0 \notin 2V$ that $\overline{A(S)} \cap V$ is void. Thus $\circ \notin \overline{A(S)}$ and $\circ \notin S$.

b) We assume now that R has no common nontrivial invariant subspace. Then for any $y_0 \in \overline{A(S)}$ the set $F(y_0) = \{Cy_0; C \in R\}$ is dense in E, since $F(y_0)$ is invariant for R and $y_0 \neq \circ$. Therefore $x_0 \in \overline{F(y_0)}$ and for every $y_0 \in \overline{A(S)}$ there exists $C_0 \in R$ such that $C_0 y_0 - x_0 \in \frac{1}{2}U$, U a given absolutely convex neighbourhood of $\circ$ in E. Let p be the semi-norm corresponding to U. The set $M_0 = \{y \in E; p(C_0 y - x_0) < 1\}$ is open and contains $y_0 \in \overline{A(S)}$. Since $\overline{A(S)}$ is compact, there exist finitely many sets $M_i = \{y, p(C_i y - x_0) < 1\}$, $i = 1, \ldots, m$, covering $\overline{A(S)}$.

c) We proceed now by analogy to the proof of (4). The function $\alpha_i(y) = \max(0, 1 - p(C_i y - x_0))$ is continuous on $\overline{A(S)}$; $\sum_{k=1}^{m} \alpha_k(y) \neq 0$ for every $y \in \overline{A(S)}$ since $p(C_i y - x_0) < 1$ for at least one i. If we write $\beta_i(y) = \alpha_i(y) \Big/ \sum_{k=1}^{m} \alpha_k(y)$ and $g(y) = \sum_{i=1}^{m} \beta_i(y) C_i y$, then g is a continuous function from $\overline{A(S)}$ in E.

$g(y)$ is a convex combination of the $C_i y$ which have coefficients $\beta_i(y) \neq 0$ only if $C_i y$ is in $x_0 + U = S$, so that $g(\overline{A(S)}) \subset S$. Since $\overline{A(S)}$ is compact, $g(\overline{A(S)})$ is compact and $g \circ A$ is a continuous mapping of S into a compact subset of S. It follows from (5) that there exists $z_0 \in S$ such that $g(Az_0) = z_0$ and $z_0 \neq \circ$ by a).

d) Consider now the continuous linear mapping $A_0 x = \sum_{i=1}^{m} \beta_i(Az_0) C_i A x$, $x \in E$, which is compact and satisfies $A_0 z_0 = g(Az_0) = z_0$. Let $H = \{z; A_0 z = z\}$. This is a closed linear subspace, different from $\circ$ and E since

$A_0 \neq I$. Moreover, H is finite dimensional since $H \subset A_0(E)$ and A_0 is compact.

Since B commutes with every C_i and with A, we have $B(H) \subset H$, because $Bz = BA_0z = A_0Bz$ for every $z \in H$.

Therefore B has an eigenvalue λ with a closed eigenspace $H_\lambda \neq E$. For every $x \in H_\lambda$ and every $C \in R$ we have

$$\lambda Cx = C\lambda x = CBx = BCx.$$

Thus $C(H_\lambda) \subset H_\lambda$ for every $C \in R$. This is a contradiction to the assumption in b).

§ 43. The approximation property

1. Some basic results. Let E and F be locally convex. It will be convenient to write $\mathfrak{F}(E, F)$ for the space of all continuous linear mappings of finite rank of E in F. We recall that $\mathfrak{F}(E, F) = \psi(E' \otimes F)$, where the canonical map ψ is defined by

$$Ax = \psi(\dot{A})x = \sum_{i=1}^{n} (u_i x) y_i \quad \text{for } \dot{A} = \sum_{i=1}^{n} u_i \otimes y_i,\ u_i \in E',\ y_i \in F,\ x \in E.$$

If no difficulties arise we will identify A and $\dot{A}$ and $\mathfrak{F}(E, F)$ and $E' \otimes F$.

In accordance with § 42, 7. we say that a locally convex space has the approximation property if $\mathfrak{F}(E)$ is dense in $\mathfrak{L}_c(E)$, where $\mathfrak{T}_c$ is the topology of uniform convergence on all precompact subsets of E. This is GROTHENDIECK's definition. L. SCHWARTZ and HOGBE-NLEND use a slightly different notion: Let $\mathfrak{T}_{co}$ be the topology of uniform convergence on all convex compact subsets of E. SCHWARTZ defines the approximation property of E by requiring that $\mathfrak{F}(E)$ is dense in $\mathfrak{L}_{co}(E)$. We call this the weak approximation property. If E is quasi-complete, then $\mathfrak{T}_c$ and $\mathfrak{T}_{co}$ coincide on $\mathfrak{L}(E)$. Thus one obtains the same notion for quasi-complete spaces, but a weaker notion for the general case.

To prove the approximation property for E it is sufficient to show that the identity I is a $\mathfrak{T}_c$-adherent point of $\mathfrak{F}(E)$. The proof given in § 42, 7. for (B)-spaces covers the general case.

(1) *Let E, F be locally convex. If E has the approximation property, then $\mathfrak{F}(E, F)$ is dense in $\mathfrak{L}_c(E, F)$ and $\mathfrak{F}(F, E)$ is dense in $\mathfrak{L}_c(F, E)$.*

Proof. a) Assume $A \in \mathfrak{L}(E, F)$, K precompact in E, V a circled neighbourhood of o in F. There exists a neighbourhood U of o in E such that $A(U) \subset V$. Since E has the approximation property, there exists

$B \in \mathfrak{F}(E)$ such that $x - Bx \in U$ for all $x \in K$. It follows that $Ax - ABx \in V$ and this means that $AB \in \mathfrak{F}(E, F)$ is in the neighbourhood $A + W(K, V)$ of A, so that A is in the closure of $\mathfrak{F}(E, F)$ in $\mathfrak{L}_c(E, F)$.

b) Assume $A \in \mathfrak{L}(F, E)$, K precompact in F, U a neighbourhood of $\circ$ in E. Then $A(K)$ is precompact in E and there exists $B \in \mathfrak{F}(E)$ such that $x - Bx \in U$ for all $x \in A(K)$, or $Ay - BAy \in U$ for all $y \in K$. From $BA \in \mathfrak{F}(F, E)$ the statement follows.

(2) *Let H be a dense subspace of a locally convex space E. If E has the approximation property, then H has it also.*

In particular, E has the approximation property if its completion $\tilde{E}$ has it.

Proof. Let K be a precompact subset of H, U an absolutely convex closed neighbourhood of $\circ$ in H, $\overline{U}$ the closure of U in E. By assumption there exists $w = \sum_{i=1}^{n} u_i \otimes x_i \in E' \otimes E$ such that $(w - I)(K) \subset (1/2)\overline{U}$. Let $M > 0$ be such that $|u_i x| \leqq M < \infty$ for all $i = 1, \ldots, n$ and all $x \in K$. Choose $z_i \in H$ such that $z_i - x_i \in (1/2nM)U$ and define $t = \sum_{i=1}^{n} u_i \otimes z_i \in E' \otimes H = H' \otimes H$. One has

$$(t - I)x = (t - w)x + (w - I)x \quad \text{for } x \in K.$$

Since

$$(t - w)x = \sum_i (u_i x)(z_i - x_i) \in nM \frac{1}{2nM} \overline{U} \subset \frac{1}{2} \overline{U} \quad \text{and} \quad (w - I)x \in \frac{1}{2} \overline{U},$$

it follows that $(t - I)(K) \subset U$.

The proof of the approximation property for a locally convex space can be reduced to the case of (B)-spaces in the following way.

(3) *The locally convex space E has the approximation property if E has a fundamental system of absolutely convex neighbourhoods U of $\circ$ such that all the* (B)*-spaces $\tilde{E}_U$ have the approximation property.*

Proof. By (2) it is sufficient to assume that all E_U have the approximation property. Recall that $E_U = E/N(U)$, $N(U) = p^{(-1)}(\circ)$, where p is the semi-norm corresponding to U. If we take U to be open, then $K(U)$ is the open unit ball in E_U, K the canonical map of E onto E_U. One has $U = K^{(-1)}(K(U))$.

Let C be precompact in E; hence $K(C)$ is precompact in E_U. Now by assumption there exists $B = \sum_{i=1}^{n} u_i \otimes Kx_i \in (E_U)' \otimes E_U$ such that $BKx - Kx \in K(U)$ for all $x \in C$. Obviously, u_i can be identified with an element of

E' and $u_i Kx = u_i x$. Thus $BKx - Kx = \sum (u_i x) Kx_i - Kx \in K(U)$. Applying $K^{(-1)}$ to both sides we obtain $Ax - x \in U$ for all $x \in C$, where $A = \sum u_i \otimes x_i \in \mathfrak{F}(E)$.

We proved in § 42, 7.(4) that every Hilbert space has the approximation property. It follows from (3) that

(4) *A locally convex space E has the approximation property if it has a fundamental system of absolutely convex neighbourhoods U such that every $\tilde{E}_U$ is a Hilbert space.*

(3) implies that it will be important to study first the approximation property of (B)-spaces. The following two propositions were obtained by Grothendieck [13].

We proved in § 42, 4.(4) that every compact mapping between Hilbert spaces is the $\mathfrak{T}_b$-limit of a sequence of mappings of finite rank. This is a special case of

(5) *The following statements are equivalent:*

a) *the* (B)-*space E has the approximation property;*

b) *let F be any* (B)-*space. Then every compact $A \in \mathfrak{L}(F, E)$ is the $\mathfrak{T}_b$-limit of a sequence of continuous mappings of finite rank.*

Proof. i) Assume a) and let $A \in \mathfrak{L}(F, E)$ be compact and let V be the closed unit ball in F. Then $A(V)$ is relatively compact in E. By assumption there exists $B \in \mathfrak{F}(E)$ such that $\|Bx - x\| \leqq \varepsilon$ for all $x \in A(V)$. This means that $\|BA - A\| \leqq \varepsilon$, $BA \in \mathfrak{F}(F, E)$; hence $\overline{\mathfrak{F}(F, E)} = \mathfrak{C}(F, E)$.

ii) We assume b). Let C be a compact subset of E. By § 42, 1.(13) there exists an absolutely convex compact subset D of E such that $C \subset D$ and C is compact in the (B)-space E_D. Let K be the canonical map of E_D in E. It is compact and one-one, so $K'(E')$ is weakly dense in $(E_D)'$ and therefore also $\mathfrak{T}_c(E_D)$-dense. It follows that $K'(E)' \otimes E$ is dense in $(E_D)' \otimes E$ in the sense of the topology of $\mathfrak{L}_c(E_D, E)$. By assumption there exists $B \in (E_D)' \otimes E$ such that $\|Bx - Kx\| \leqq \varepsilon/2$ for all $x \in D \subset E_D$. If we determine $A_0 \in K'(E') \otimes E$ such that $\|A_0 x - Bx\| \leqq \varepsilon/2$ for all $x \in C \subset D$, then $\|A_0 x - Kx\| \leqq \varepsilon$ for all $x \in C$.

Let A_0 be $\sum_i (K'u_i) \otimes x_i$ and $A = \sum_i u_i \otimes x_i$; then by identifying Kx and x we obtain $\|Ax - x\| \leqq \varepsilon$ for all $x \in C$, where $A \in \mathfrak{F}(E)$.

(6) *The following statements are equivalent:*

a) *the strong dual E' of the* (B)-*space E has the approximation property;*

b) *let F be any* (B)-*space. Then every compact $A \in \mathfrak{L}(E, F)$ is the $\mathfrak{T}_b$-limit of a sequence of mappings of finite rank.*

Proof. i) Assume a). Let $A \in \mathfrak{L}(E, F)$ be compact. Then A' is compact in $\mathfrak{L}(F', E')$ and $A'(W)$ is precompact in E', where W is the closed unit ball in F'. By assumption there exists $\sum_i w_i \otimes u_i \in E'' \otimes E'$ such that $\|\sum (w_i(A'v))u_i - A'v\| \leqq \varepsilon$ for all $v \in W$. Now $w_i(A'v) = (A''w_i)v$ and it follows from $A''(E'') \subset F$ that $A''w_i = y_i \in F$ (§ 42, 2.(1)). Let B be $\sum u_i \otimes y_i \in E' \otimes F$; then $B' = \sum y_i \otimes u_i$ and we have $\|B' - A'\| \leqq \varepsilon$, which implies $\|B - A\| \leqq \varepsilon$.

ii) We assume b). By (5) it will be sufficient to show that every compact $A \in \mathfrak{L}(F, E')$ is the $\mathfrak{T}_b$-limit of a sequence of elements of $F' \otimes E'$. The adjoint $A' \in \mathfrak{L}(E'', F')$ is compact and so is its restriction $A'_0 \in \mathfrak{L}(E, F')$. For a given $\varepsilon > 0$ there exists by b) an element $\sum u_i \otimes v_i \in E' \otimes F'$ such that $\|\sum (u_i x)v_i - A'_0 x\| \leqq \varepsilon$ for all $x \in E$, $\|x\| \leqq 1$.

Since A is continuous, A' is continuous for the topologies $\mathfrak{T}_s(E')$, $\mathfrak{T}_s(F)$ on E'' resp. F'. It follows that $\|\sum (u_i z)v_i - A'z\| \leqq \varepsilon$ for all $z \in E''$, $\|z\| \leqq 1$, and this implies $\|\sum v_i \otimes u_i - A\| \leqq \varepsilon$.

We note the following corollary to (5) and (6):

(7) *Let E, F be* (B)-*spaces. If E' or F has the approximation property, then every compact $A \in \mathfrak{L}(E, F)$ is the $\mathfrak{T}_b$-limit of a sequence of continuous mappings of finite rank.*

The problem whether every compact mapping between (B)-spaces is the $\mathfrak{T}_b$-limit of mappings of finite rank was raised by Banach. By (5) this is equivalent to the question whether every (B)-space has the approximation property. For a long time a positive answer was expected. Grothendieck made in [13] a deep analysis of this problem. He found many equivalent formulations and consequences but no solution. He conjectured a negative answer.

Only recently Enflo [1'] succeeded in constructing counterexamples. His ingenious but highly complicated methods were simplified to some degree by Davie [1']. We state their results without proofs.

(8) *Every l^p, $2 < p < \infty$, has a closed subspace which is a separable reflexive* (B)-*space not having the approximation property.*

Also, c_0 has a closed subspace without the approximation property.

We note that Grothendieck proved in [13] that if there exists a (B)-space without the approximation property, then there exists a closed subspace of c_0 without the approximation property.

We remark that recently Szankowski [1'] proved (8) also for l^p, $1 \leqq p < 2$.

The construction of the examples in (8) is very involved and there is no simple definition of these spaces. But we will give in 9. an example of a (B)-space without the approximation property which has a nice definition.

2. The canonical map of $E \tilde{\otimes}_\pi F$ in $\mathfrak{B}(E'_s \times F'_s)$. We recall the problem raised in § 41, 3.: Let E, F be complete locally convex spaces. There exists a continuous injection ψ of $E \otimes_\pi F$ in $\mathfrak{B}_e(E'_s \times F'_s)$. It has a continuous extension $\bar{\psi}$ to $E \tilde{\otimes}_\pi F$. When is this canonical map $\bar{\psi}$ one-one?

The key to this problem is the approximation property. We treat first the case of (B)-spaces. We need some auxiliary results.

In accordance with § 41, 7., we denote by $c_0\{F\}$, F a (B)-space, the set of all sequences $y = (y_1, y_2, \ldots)$, $y_i \in F$, $\|y_i\| \to 0$. We introduce the norm $\|y\| = \sup_n \|y_n\|$. Using the elementary methods of § 14, 7., one sees easily that $c_0\{F\}$ is a (B)-space. It is also straightforward to show (§ 14, 7.(11)) that its strong dual can be identified with $l^1\{F'\}$, the space of all $v = (v_1, v_2, \ldots)$, $v_i \in F'$, $\sum_{i=1}^{\infty} \|v_i\| < \infty$, equipped with the norm $\|v\| = \sum_{i=1}^{\infty} \| v_i\|$. The duality $\langle l^1\{F'\}, c_0\{F\}\rangle$ is given by the bilinear form $\langle v, y\rangle = \sum_{n=1}^{\infty} v_n y_n$. We note (§ 20, 9.(5))

(1) *The closed unit ball* $K = \{v; \|v\| = \sum \|v_n\| \leqq 1\}$ *of* $l^1\{F'\}$ *is* $\mathfrak{T}_s(c_0\{F\})$*-compact.*

We use this fact in the proof of

(2) *Let* E, F *be* (B)*-spaces. The dual of* $\mathfrak{L}_c(E, F)$ *can be identified with a quotient of* $E \tilde{\otimes}_\pi F'$.

Proof. a) We show first that every $z \in E \tilde{\otimes}_\pi F'$ defines a $\mathfrak{T}_c$-continuous linear functional on $\mathfrak{L}(E, F)$. By § 41, 4.(6) z has a representation $z = \sum_{i=1}^{\infty} \lambda_i x_i \otimes v_i$, $x_i \in E$, $\|x_i\| \to 0$, $v_i \in F'$, $\|v_i\| \leqq 1$, $\sum_{i=1}^{\infty} |\lambda_i| = 1$.

We recall (§ 41, 3.(6)) that $(E \tilde{\otimes}_\pi F')'$ can be identified with $\mathscr{B}(E \times F')$ and $\mathfrak{L}(E, F'')$, so that we have the dual system $\langle \mathfrak{L}(E, F''), E \tilde{\otimes}_\pi F'\rangle$ and, since $\mathfrak{L}(E, F)$ can be identified with a subspace of $\mathfrak{L}(E, F'')$, every z defines uniquely a linear functional $\langle A, z\rangle = \sum_{i=1}^{\infty} \lambda_i v_i(Ax_i)$ on $\mathfrak{L}(E, F)$.

Let C be the closed absolutely convex cover of the x_i, $i = 1, 2, \ldots$, and V the closed unit ball in F. We remark that C is compact in E. Let W be the neighbourhood of $\circ$ in $\mathfrak{L}_c(E, F)$ consisting of all A such that $A(C) \subset V$. Then one has for all $A \in W$

$$|\langle A, z\rangle| \leqq \sum_{i=1}^{\infty} |\lambda_i||v_i(Ax_i)| \leqq \sum_{i=1}^{\infty} |\lambda_i| = 1$$

and thus z is $\mathfrak{T}_c$-continuous on $\mathfrak{L}(E, F)$.

Since $\mathfrak{L}(E, F)$ is a subspace of $\mathfrak{L}(E, F'')$, the polar $\mathfrak{L}(E, F)^\circ$ in $E \tilde{\otimes}_\pi F'$ may be different from zero, so that not $E \tilde{\otimes}_\pi F'$ itself but the quotient $H = (E \tilde{\otimes}_\pi F')/\mathfrak{L}(E, F)^\circ$ is a subspace of $\mathfrak{L}_c(E, F)'$.

b) We prove that, conversely, every $\mathfrak{T}_c$-continuous linear functional w on $\mathfrak{L}(E, F)$ is given by an element of H.

One has $|\langle w, A\rangle| \leqq 1$ for all A of a $\mathfrak{T}_c$-neighbourhood $W = \{A; A(C) \subset V\}$, where C is absolutely convex and compact in E and V is the closed unit ball in F. We recall that $\mathfrak{L}_s(E, F)' = E \otimes F'$ (§ 39, 7.(2)), so that $E \otimes F'$ is a subspace of H. Since one has $\langle A, x \otimes v\rangle = v(Ax)$ for $x \otimes v \in C \otimes V^\circ$, it follows that W can be written also as $(C \otimes V^\circ)^\circ$, the last polar being taken in $\mathfrak{L}(E, F)$.

We have $E \otimes F' \subset H \subset \mathfrak{L}(E, F)^*$, the algebraic dual of $\mathfrak{L}(E, F)$. Now $w \in W^\circ = (C \otimes V^\circ)^{\circ\circ} = \overline{\Gamma(C \otimes V^\circ)}$, where the last polars in the first two expressions and the $\mathfrak{T}_s(\mathfrak{L}(E, F))$-closure are taken in $\mathfrak{L}(E, F)^*$. Our statement will be proved if we show that $\overline{\Gamma(C \otimes V^\circ)}$ is contained in H.

Since E is a (B)-space, we can assume that C is the closed absolutely convex cover of a sequence $x_1, x_2, \ldots$, $\|x_i\| \to 0$. All $x_i \otimes v_i$, $v_i \in V^\circ$, are in $C \otimes V^\circ \subset H$ and if K_1 denotes the set of all $\hat{z} \in H$, where $z = \sum_{i=1}^\infty \lambda_i x_i \otimes v_i$, $v_i \in V^\circ$, $\sum |\lambda_i| \leqq 1$, then we have $\Gamma(C \otimes V^\circ) \subset K_1 \subset \overline{\Gamma(C \otimes V^\circ)}$. If we show that K_1 is $\mathfrak{T}_s(\mathfrak{L}(E, F))$-compact, then $K_1 = \overline{\Gamma(C \otimes V^\circ)}$ and this set is contained in H.

We define a mapping J of $l^1\{F'\}$ into H by $Jv = J(v_1, v_2, \ldots) = \hat{z}$, where $\hat{z}$ is the residue class in H of $z = \sum_{i=1}^\infty x_i \otimes v_i$ in $E \tilde{\otimes}_\pi F'$ and x_i is the sequence defining C. It is easily verified that $J(K)$, K the closed unit ball in $l^1\{F'\}$, is K_1, so that K_1 will be $\mathfrak{T}_s(\mathfrak{L}(E, F))$-compact by (1) if J is continuous for the topologies $\mathfrak{T}_s(c_0\{F\})$ on $l^1\{F'\}$ and $\mathfrak{T}_s(\mathfrak{L}(E, F))$ on H.

Suppose $A_k \in \mathfrak{L}(E, F)$. Since $\|x_i\| \to 0$, the sequence $y^{(k)} = (A_k x_1, A_k x_2, \ldots)$ lies in $c_0\{F\}$ and one has

$$\langle y^{(k)}, v\rangle = \sum_{i=1}^\infty v_i(A_k x_i) = A_k \sum_{i=1}^\infty x_i \otimes v_i = \langle A_k, Jv\rangle.$$

Let V_1 be the $\mathfrak{T}_s(\mathfrak{L}(E, F))$-neighbourhood $\{\hat{z}; |\langle A_k, \hat{z}\rangle| \leqq \varepsilon, k = 1, \ldots, m\}$ in H and U_1 the $\mathfrak{T}_s(c_0(F))$-neighbourhood $\{v; |\langle y^{(k)}, v\rangle| \leqq \varepsilon, k = 1, \ldots, m\}$ in $l^1\{F'\}$; then $J(U_1) \subset V_1$. Thus J is weakly continuous and (2) is proved.

The proof of (2) yields the following particular case:

(3) *If E, F are* (B)*-spaces, F reflexive, then $\mathfrak{L}_c(E, F)'$ can be identified with $E \tilde{\otimes}_\pi F'$.*

We recall from § 42, 7. the problem of the existence of the trace of a nuclear mapping. It has the following solution:

(4) *Let E be a* (B)*-space. The trace of every nuclear mapping of $\mathfrak{L}(E)$ is uniquely defined if and only if E has the approximation property.*

Proof. Let $\bar{\psi}$ be the canonical map of $E' \tilde{\otimes}_\pi E$ in $\mathfrak{L}(E)$. Then the uniqueness of the trace is obviously equivalent to the statement

(*) $\bar{\psi}(z) = \circ$ *implies* $\operatorname{tr} z = 0$ *for every* $z \in E' \tilde{\otimes}_\pi E$.

The approximation property of E is equivalent to: Every $u \in \mathfrak{L}_c(E)'$ which vanishes on $E' \otimes E \subset \mathfrak{L}(E)$ vanishes on the identity $I \in \mathfrak{L}(E)$. Using the representation of $\mathfrak{L}_c(E)'$ determined in (2), we obtain the following version of the approximation property of E:

(**) *If* $z \in E \tilde{\otimes}_\pi E'$ *vanishes on* $E' \otimes E$, z *vanishes on* $I \in \mathfrak{L}(E)$.

We note

(5) $\operatorname{tr} z = \langle I, z \rangle$ *for every* $z \in E \tilde{\otimes}_\pi E'$,

since for $z = \sum_{n=1}^{\infty} x_n \otimes u_n$, $x_n \in E$, $u_n \in E'$, we have $\operatorname{tr} z = \sum_{n=1}^{\infty} u_n x_n = \sum u_n(Ix_n) = \langle I, z \rangle$.

Furthermore, one has

(6) $\bar{\psi}(z) = \circ$ *if and only if* $z \in (E' \otimes E)^\circ \subset E \tilde{\otimes}_\pi E'$.

To see this recall that $\bar{\psi}(z)x = \sum_{n=1}^{\infty} (u_n x)x_n$; thus $\langle u \otimes x, z \rangle = \sum (ux_n)(u_n x) = \langle u, \bar{\psi}(z)x \rangle$, $u \in E'$, $x \in E$. Thus $\bar{\psi}(z) = \circ$ means $z \in (E' \otimes E)^\circ$.

Now we assume (*). Let z be in $(E' \otimes E)^\circ$. Then $\bar{\psi}(z) = \circ$ by (6) and $\operatorname{tr} z = \circ$ by (*). (5) implies $\langle I, z \rangle = 0$, so (**) is satisfied.

Conversely, suppose (**) and let $\bar{\psi}(z) = \circ$ for $z \in E \tilde{\otimes}_\pi E'$. Then $z \in (E' \otimes E)^\circ$ by (6) and $\langle I, z \rangle = 0$ by (**). (5) implies $\operatorname{tr} z = 0$ and (*) is proved.

We will now answer the question raised at the beginning of this section for the case of (B)-spaces.

(7) *Let E be a* (B)*-space. Each of the following properties is equivalent to the approximation property:*

a) *the canonical map $\bar{\psi}$ of $E' \tilde{\otimes}_\pi E$ into $\mathfrak{L}(E)$ is one-one;*

b) *for every* (B)*-space F the map $\bar{\psi}$ of $F' \tilde{\otimes}_\pi E$ into $\mathfrak{L}(F, E)$ is one-one;*

c) *for every* (B)*-space F the map $\bar{\psi}$ of $F \tilde{\otimes}_\pi E$ into $\mathfrak{L}(F', E)$ is one-one.*

Proof. We show that c) implies b): It follows from c) that the map $\bar{\psi}$ of $F' \tilde{\otimes}_\pi E$ into $\mathfrak{L}(F'', E)$ is one-one. Now $\mathfrak{L}(F'', E)$ can be identified with a subspace of $\mathfrak{L}(F, E)$ and then $\bar{\psi}$ coincides with the canonical map of $F' \tilde{\otimes}_\pi E$ into $\mathfrak{L}(F, E)$ which is therefore one-one.

Obviously, b) implies a) and a) implies (*), so that it follows from (4) that each of a), b), and c) implies the approximation property.

It remains to prove that c) is a consequence of the approximation property. Let $z = \sum_{n=1}^{\infty} y_n \otimes x_n$, $y_n \in F$, $x_n \in E$, be an element of $F \tilde{\otimes}_\pi E$ such that $\bar{\psi}(z) = \mathfrak{o}$, where $\bar{\psi}(z)$ is the corresponding nuclear mapping in $\mathfrak{L}(F', E)$. We have to prove that $z = \mathfrak{o}$ or that $\langle B, z\rangle = 0$ for every $B \in \mathscr{B}(F \times E) = (F \tilde{\otimes}_\pi E)'$.

Let $\tilde{B}$ be the mapping in $\mathfrak{L}(F, E')$ corresponding to B. According to § 42, 7.(7), $\langle B, z\rangle = \operatorname{tr}(\tilde{B} \tilde{\otimes}_\pi I)z$, where $t = (\tilde{B} \tilde{\otimes}_\pi I)z \in E' \tilde{\otimes}_\pi E$. Let $\bar{\psi}(t)$ be the corresponding nuclear mapping in $\mathfrak{L}(E)$. Then it follows from

$$\bar{\psi}(t)x = \sum_{n=1}^{\infty} ((\tilde{B}y_n)x)x_n = \sum (y_n(\tilde{B}'x))x_n = \bar{\psi}(z)(\tilde{B}'x), \qquad x \in E,$$

that $\bar{\psi}(t) = \bar{\psi}(z)\tilde{B}'$. Thus $\bar{\psi}(t) = \mathfrak{o}$ since $\bar{\psi}(z) = \mathfrak{o}$. Now E has the approximation property and by (4)(*) we conclude that $\operatorname{tr} t = \langle B, z\rangle = 0$.

$\mathfrak{L}_b(F', E)$ can be identified with a subspace of $\mathscr{B}_b(F' \times E')$ by defining $B(v, u) = u(\tilde{B}v)$ for $\tilde{B} \in \mathfrak{L}(F', E)$, $v \in F'$, $u \in E'$. Thus we have also a canonical map $\bar{\psi}$ of $E \tilde{\otimes}_\pi F$ into $\mathscr{B}(F' \times E')$.

Let us remark further that to $z = \sum_{n=1}^{\infty} x_n \otimes y_n \in E \tilde{\otimes}_\pi F$ corresponds the mapping $\tilde{B}v = \sum_{n=1}^{\infty} (vy_n)x_n$ in $\mathfrak{L}(F', E)$ and the mapping $\tilde{\tilde{B}}u = \sum_{n=1}^{\infty} (ux_n)y_n$ in $\mathfrak{L}(E', F)$; thus $u(\tilde{B}v) = v(\tilde{\tilde{B}}u) = B(v, u)$, $\tilde{B}$ and $\tilde{\tilde{B}}$ are adjoint to each other, and $z \mapsto \tilde{B}$ is one-one if and only if $z \mapsto \tilde{\tilde{B}}$ is one-one. From (7) c) and these remarks follows

(8) *If one of the* (B)*-spaces E, F has the approximation property, then the canonical maps $\bar{\psi}$ of $E \tilde{\otimes}_\pi F$ into $\mathscr{B}(F' \times E')$ or $\mathfrak{L}(F', E)$ or $\mathfrak{L}(E', F)$ are one-one.*

Similarly, one has

(9) *If one of the* (B)*-spaces F', E has the approximation property, then the canonical map of $F' \tilde{\otimes}_\pi E$ in $\mathfrak{L}(F, E)$ is one-one.*

Proof. By (8) the canonical map of $F' \tilde{\otimes}_\pi E$ in $\mathfrak{L}(F'', E)$ is one-one and $\mathfrak{L}(F'', E)$ is a subspace of $\mathfrak{L}(F, E)$ (compare the proof of (7)).

We note the following improvement of (2):

(10) *If one of the* (B)*-spaces F', E has the approximation property, then the dual of $\mathfrak{L}_c(E, F)$ can be identified with $E \tilde{\otimes}_\pi F'$.*

Proof. Let $\bar{\psi}$ be the canonical map of $F' \tilde{\otimes}_\pi E$ in $\mathfrak{L}(F, E)$ with $\bar{\psi}(z)y = \bar{\psi}\left(\sum_{n=1}^{\infty} v_n \otimes x_n\right)y = \sum_{n=1}^{\infty} (v_n y)x_n$. As in (6), $\bar{\psi}(z) = \mathfrak{o}$ if and only if

$z \in (E' \otimes F)^\circ$, $E' \otimes F \subset \mathfrak{L}(E, F)$. Since $(E' \otimes F)^\circ \supset \mathfrak{L}(E, F)^\circ$, the polars being taken in $F' \tilde{\otimes}_\pi E$, it follows from (9) that $\mathfrak{L}(E, F)^\circ = \mathrm{o}$, which is the statement with $F' \tilde{\otimes}_\pi E$ instead of $E \tilde{\otimes}_\pi F'$; but these are isomorphic.

We come back to the general problem raised at the beginning of 2. We need the following useful remark:

(11) *Let U be an absolutely convex neighbourhood of* o *in the locally convex space E and let K be the canonical mapping of E onto the normed space E_U. Then $J = K'$ is the injection of $(E_U)'$ in E' onto E'_{U° and J is a norm isomorphism of $(E_U)'$ and E'_{U°.*

This follows from $(KU)^\circ = J^{-1}(U^\circ)$.

(12) *Let E be a complete locally convex space with a fundamental system of absolutely convex neighbourhoods U of* o *such that every $\tilde{E}_U$ has the approximation property.*

Then for every complete locally convex F the canonical map $\bar{\psi}$ of $E \tilde{\otimes}_\pi F$ in $\mathfrak{B}(E'_s \times F'_s)$ is one-one.

Proof. By § 41, 3.(14) it is sufficient to show that $E' \otimes F'$ is $\mathfrak{T}_s(E \tilde{\otimes}_\pi F)$-dense in $\mathscr{B}(E \times F) = (E \tilde{\otimes}_\pi F)'$.

Let B be in $\mathscr{B}(E \times F)$. Then $|B(U, V)| \leqq 1$ for some U of the fundamental system of neighbourhoods of E and some absolutely convex neighbourhood V of o in F. Let K_1, K_2 be the canonical mappings of E onto E_U and F onto F_V, respectively. Then $B(x, y)$ depends only on the residue classes $K_1 x$ and $K_2 y$, so that $B_1(K_1 x, K_2 y) = B(x, y)$ defines a $B_1 \in \mathscr{B}(E_U \times F_V)$. By § 40, 3.(2) B_1 has a uniquely determined continuous extension $\tilde{B}_1 \in \mathscr{B}(\tilde{E}_U \times \tilde{F}_V)$.

The mapping $K = K_1 \otimes K_2$ of $E \otimes_\pi F$ into $\tilde{E}_U \otimes_\pi \tilde{F}_V$ has by § 41, 5.(1) a continuous extension $\tilde{K} = K_1 \tilde{\otimes}_\pi K_2$ which maps $E \tilde{\otimes}_\pi F$ in $\tilde{E}_U \tilde{\otimes}_\pi \tilde{F}_V$. From the definition $B_1(K_1 x, K_2 y) = \langle B_1, K(x \otimes y)\rangle = \langle B, x \otimes y\rangle$ follows immediately $\langle B_1, Kz\rangle = \langle B, z\rangle$ for every $z \in E \otimes F$, and the continuity of $\tilde{B}_1, B, \tilde{K}$ implies

$$\text{(13)} \qquad \langle \tilde{B}_1, \tilde{K}z\rangle = \langle B, z\rangle \quad \text{for every } z \in E \tilde{\otimes}_\pi F.$$

Since $\tilde{E}_U$ has the approximation property, it follows from (8) that the canonical map of $\tilde{E}_U \tilde{\otimes}_\pi \tilde{F}_V$ in $\mathscr{B}((\tilde{E}_U)' \times (\tilde{F}_V)')$ is one-one and by § 41, 3.(14) $(\tilde{E}_U)' \otimes (\tilde{F}_V)'$ is $\mathfrak{T}_s(\tilde{E}_U \tilde{\otimes}_\pi \tilde{F}_V)$-dense in $\mathscr{B}(\tilde{E}_U \times \tilde{F}_V)$.

Therefore, if $\varepsilon > 0$ and $z_1, \dots, z_k \in E \tilde{\otimes}_\pi F$ are given, there exists $w \in (\tilde{E}_U)' \otimes (\tilde{F}_V)'$ such that

$$\text{(14)} \qquad |\langle w - \tilde{B}_1, \tilde{K}z_i\rangle| \leqq \varepsilon, \qquad i = 1, \dots, k.$$

Let w be, in particular, $u \otimes v$, $u \in (E_U)'$, $v \in (F_V)'$. Then (11) implies

$$\begin{aligned}\langle u \otimes v, K(x \otimes y)\rangle &= \langle u, K_1 x\rangle\langle v, K_2 y\rangle = \langle J_1 u, x\rangle\langle J_2 v, y\rangle \\ &= \langle J(u \otimes v), x \otimes y\rangle,\end{aligned}$$

where $J = J_1 \otimes J_2$ maps $(E_U)' \otimes (F_V)'$ into $E'_{U^\circ} \otimes F'_{V^\circ} \subset E' \otimes F'$. From this follows by continuity for every $w \in (E_U)' \otimes (F_V)'$ and every $z \in E \tilde{\otimes}_\pi F$ the relation $\langle w, \tilde{K}z\rangle = \langle Jw, z\rangle$, $Jw \in E' \otimes F'$. By this relation and (13) we rewrite the inequalities (14) in the form $|\langle Jw - B, z_i\rangle| \leqq \varepsilon$, $i = 1, \ldots, k$. Thus B is in the $\mathfrak{T}_s(E \tilde{\otimes}_\pi F)$-closure of $E' \otimes F'$.

3. Another interpretation of the approximation property. The tensor product $E \otimes F$ of two locally convex spaces E and F is algebraically a subspace of $\mathfrak{B}_e(E'_s \times F'_s)$, which is isomorphic to $\mathfrak{L}_e(E'_k, F)$ and $\mathfrak{L}_e(F'_k, E)$ by § 40, 4.(5). Instead of introducing the π-topology on $E \otimes F$, as we did in 2., we will consider $E \otimes F$ as equipped with the topology $\mathfrak{T}_e$ of the bi-equicontinuous topology and we will try to determine the closure $\overline{E \otimes F}$ of $E \otimes F$ in $\mathfrak{B}_e(E'_s \times F'_s)$, $\mathfrak{L}_e(E'_k, F)$, and $\mathfrak{L}_e(F'_k, E)$. This will lead us to a new interpretation of the approximation property of E.

If $A \in E \otimes F \subset \mathfrak{B}(E'_s \times F'_s)$, then the corresponding mapping $\tilde{A} \in \mathfrak{L}(E'_k, F)$ is of finite rank and maps the equicontinuous subsets M of E' in relatively compact sets in F. Similarly, $\tilde{\tilde{A}} \in \mathfrak{L}(F'_k, E)$ maps the equicontinuous sets N of F' in relatively compact sets in E. Moreover, these maps are weakly continuous, where in 3. the weak topology on a dual E' will always mean $\mathfrak{T}_s(E)$.

We note

(1) *If a weakly continuous linear mapping A of E' in F maps every equicontinuous set M of E' in a precompact set $A(M)$ in F, then $A(M)$ is always relatively compact.*

M is contained in a weakly compact set M_1; $A(M_1)$ is weakly compact and therefore complete. Hence $A(M)$ is relatively compact.

The following proposition gives different characterizations of the mappings considered in (1).

(2) *Let A be a weakly continuous linear mapping of E' in F, and A' the adjoint weakly continuous mapping of F' in E. Then the following properties are equivalent:*

a) *A maps the equicontinuous subsets M of E' in relatively compact sets in F;*

b) *A' maps the equicontinuous subsets N of F' in relatively compact sets in E;*

c) *$A \in \mathfrak{L}(E'_{co}, F)$, where $\mathfrak{T}_{co}$ is the topology of uniform convergence on the convex compact subsets of E;*

d) $A' \in \mathfrak{L}(F'_{co}, E)$;

e) *the bilinear form* $\mathring{A}(u, v) = \langle Au, v \rangle = \langle u, A'v \rangle$ *is* $(\mathfrak{E}, \mathfrak{E})$*-hypocontinuous on* $E'_{co} \times F'_{co}$, *where* $\mathfrak{E}$ *is the class of equicontinuous subsets.*

Proof. The equivalence of a) and b) is a consequence of § 42, 1.(8) if we use (1).

Suppose a) for A. If $N \subset F'$ is given, N equicontinuous, there exists by b) a convex compact $C \subset E$ such that $A'(N) \subset C$. Hence $A^{(-1)}(N^\circ) \supset C^\circ$, $A(C^\circ) \subset N^\circ$ and this means $A \in \mathfrak{L}(E'_{co}, F)$. Conversely, assume $A \in \mathfrak{L}(E'_{co}, F)$. Then A is weakly continuous. Furthermore, every weakly closed equicontinuous M is $\mathfrak{T}_c$-compact (§ 21, 6.(3)); hence $\mathfrak{T}_{co}$-compact, and therefore $A(M)$ is compact, so that A satisfies a).

The equivalence of d) with a) and b) follows by symmetry.

We prove now that a), b), c), d) imply e). If $A \in \mathfrak{L}(E'_{co}, F)$ and $A' \in \mathfrak{L}(F'_{co}, E)$, then $\mathring{A}$ is separately continuous for the weak topologies and for the topologies $\mathfrak{T}_{co}$ since A and A' are the mappings corresponding to $\mathring{A}$ (§ 40, 1.(2')). Furthermore, if the equicontinuous set $M \subset E'$ is given, there exists an absolutely convex and compact set C such that $A(M) \subset C$ or $|\mathring{A}(M, C^\circ)| \leqq 1$, and this is the hypocontinuity of $\mathring{A}$ with respect to the class of all M. By symmetry it follows that $\mathring{A}$ is $(\mathfrak{E}, \mathfrak{E})$-hypocontinuous.

Conversely, assume that B is $(\mathfrak{E}, \mathfrak{E})$-hypocontinuous on $E'_{co} \times F'_{co}$. Since B is separately continuous, the corresponding mappings $\tilde{B}$ from E'_{co} in $(F'_{co})' = F$ and $\tilde{\tilde{B}}$ from F'_{co} in $(E'_{co})' = E$ are weakly continuous (§ 40, 1.(2')). By assumption there exists for every equicontinuous set $M \subset E'$ an absolutely convex compact set C in F such that $|B(M, C^\circ)| = |\langle \tilde{B}(M), C^\circ \rangle| \leqq 1$; thus $\tilde{B}(M) \subset C$ and $\tilde{B}$ satisfies a). Similarly, $\tilde{\tilde{B}}$ satisfies b).

We determined in (2) the subspace $\mathfrak{X}^{(\mathfrak{E},\mathfrak{E})}(E'_{co} \times F'_{co})$ of $\mathfrak{B}(E'_s \times F'_s)$ and showed that the corresponding spaces of linear mappings are $\mathfrak{L}(E'_{co}, F) \subset \mathfrak{L}(E'_k, F)$ and $\mathfrak{L}(F'_{co}, E) \subset \mathfrak{L}(F'_k, E)$. Using the notations of § 40, 4., the correspondences $B \leftrightarrow \tilde{B} \leftrightarrow \tilde{\tilde{B}}$ generate the topological isomorphisms

$$\mathfrak{X}_e^{(\mathfrak{E},\mathfrak{E})}(E'_{co} \times F'_{co}) \cong \mathfrak{L}_e(E'_{co}, F) \cong \mathfrak{L}_e(F'_{co}, E). \tag{3}$$

In the notation introduced by SCHWARTZ in [3'] these isomorphisms take the form

$$\varepsilon(E, F) \cong E\varepsilon F \cong F\varepsilon E. \tag{3'}$$

$E\varepsilon F = \mathfrak{L}_e(E'_{co}, F)$ and $F\varepsilon E = \mathfrak{L}_e(F'_{co}, E)$ are called the ε-products of the spaces E and F and $\varepsilon(E, F)$ is the space $\mathfrak{X}_e^{(\mathfrak{E},\mathfrak{E})}(E'_{co} \times F'_{co})$, whose elements B are called the ε-hypocontinuous bilinear forms on $E'_{co} \times F'_{co}$.

We also introduce the notation $E \otimes_\varepsilon F$ for $E \otimes F$ equipped with the topology induced by the topology $\mathfrak{T}_e$ of $\mathfrak{B}_e(E'_s \times F'_s)$ and call $E \otimes_\varepsilon F$ the ε-tensor product of E and F. The completion of $E \otimes_\varepsilon F$ will be denoted by $E \tilde{\otimes}_\varepsilon F$. This notion will be studied in detail in § 44.

(4) *$E\varepsilon F$ is a closed subspace of $\mathfrak{L}_e(E'_k, F)$. The closure of $E \otimes_\varepsilon F$ in $\mathfrak{L}_e(E'_k, F)$ is a subspace of $E\varepsilon F$.*

Since $E \otimes F \subset \mathfrak{L}(E'_{co}, F)$, by (2) we have only to prove the first statement. But this is an immediate consequence of (1) and § 42, 1.(3) applied to $\mathfrak{L}_e(E'_k, F)$.

We recall that $\mathfrak{L}_e(E'_k, F)$ is complete if and only if E and F are complete (§ 40, 4.(5)). This implies the following particular case of (4):

(5) *If E and F are complete locally convex spaces, then $E \tilde{\otimes}_\varepsilon F$ is a closed subspace of the complete space $E\varepsilon F$.*

We are now able to formulate and prove the following result of Grothendieck and Schwartz:

(6) a) *Let E be locally convex. E has the weak approximation property if and only if $F \otimes E$ is dense in $F\varepsilon E$ for every locally convex F.*

b) *If E is quasi-complete, then E has the approximation property if and only if $F \otimes E$ is dense in $F\varepsilon E$ for every locally convex F.*

We have to prove only a) (see the remarks at the beginning of 1.).

i) Sufficiency. Take $F = E'_{co}$. Then $(E'_{co})'_{co} = E_\gamma$, where $\mathfrak{T}_\gamma$ is the topology of uniform convergence on the convex relatively $\mathfrak{T}_{co}$-compact subsets of E'. Every weakly closed equicontinuous subset of E' is ($\mathfrak{T}_c$- and therefore) $\mathfrak{T}_{co}$-compact; hence $\mathfrak{T}_\gamma = \mathfrak{T}_{co}(E'_{co})$ is finer than the original topology $\mathfrak{T}$ on E. Thus $\mathfrak{L}(E, E) \subset \mathfrak{L}(E_\gamma, E)$. A fundamental system of the equicontinuous subsets of $(E'_{co})' = E$ is given by the convex relatively compact subsets of E; hence $\mathfrak{L}_e(E_\gamma, E) = \mathfrak{L}_e((E'_{co})'_{co}, E) = E'_{co}\varepsilon E$ induces on $\mathfrak{L}(E, E)$ the topology $\mathfrak{T}_{co}$. Thus $\mathfrak{L}_{co}(E)$ is a subspace of $E'_{co}\varepsilon E$. By assumption $E' \otimes E$ is dense in $E'_{co}\varepsilon E$; hence $E' \otimes E$ is dense in $\mathfrak{L}_{co}(E)$ and E has the weak approximation property.

ii) Necessity. By assumption there exist $A_\alpha \in E' \otimes E$ such that $A_\alpha \to I$ in $\mathfrak{L}_{co}(E)$. Let B be an element of $F\varepsilon E = \mathfrak{L}_e(F'_{co}, E)$. Let N be an equicontinuous subset of F'. Then $B(N)$ is relatively compact in E by (2) and from $(A_\alpha B - B)(N) = (A_\alpha - I)(B(N))$ it follows that $A_\alpha B \in F \otimes E$ converges to B in $F\varepsilon E$. Thus $F \otimes E$ is dense in $F\varepsilon E$.

(5) and (6) imply the corollary

(7) *Let E and F be locally convex and complete. If E or F has the approximation property, then $E \tilde{\otimes}_\varepsilon F = E\varepsilon F$, which means that $E \tilde{\otimes}_\varepsilon F$*

consists of all weakly continuous linear mappings of E' in F which map equicontinuous subsets of E' in relatively compact sets in F.

For (B)-spaces E, F one obtains a sharper result. We remark first that in this case $\mathfrak{L}_e(F'_k, E)$ is the space of all weakly continuous mappings A of F' in E equipped with the strong topology $\mathfrak{T}_b$ generated by the norm $\|A\|$. Moreover, $F\varepsilon E = \mathfrak{L}_b(F'_c, E)$ is the subspace of $\mathfrak{L}_b(F'_k, E)$ consisting of all weakly continuous and compact mappings.

(8) *A* (B)*-space E has the approximation property if and only if $F \otimes E$ is dense in $\mathfrak{L}_b(F'_k, E) = F\varepsilon E$ or $F \tilde{\otimes}_\varepsilon E = F\varepsilon E$ for every* (B)*-space F, which means that every weakly continuous compact mapping of F' in E is the $\mathfrak{T}_b$-limit of weakly continuous mappings of finite rank.*

Proof. Because of (6) we have only to prove sufficiency, and by 1.(5) this will be done if we show that every compact $B \in \mathfrak{L}(F, E)$ is the $\mathfrak{T}_b$-limit of mappings of finite rank.

We take $B'' \in \mathfrak{L}(F'', E'')$ which is a weakly continuous extension of B and satisfies $B''(F'') \subset E$ by § 42, 2.(1). B is compact and weakly continuous in the sense of $\mathfrak{T}_s(F')$, $\mathfrak{T}_s(E')$. Hence $B'' \in \mathfrak{L}((F'')_c, E) = F'\varepsilon E$, and therefore by our assumption applied to F' and E it follows that B'' and also its restriction B is the $\mathfrak{T}_b$-limit of mappings of $F' \otimes E$.

We note that Bierstedt and Meise proved in [1'] that also in (6) b) it is sufficient that $F \otimes E$ be dense in $F\varepsilon E$ for every (B)-space F.

4. Hereditary properties. Since the discovery of Enflo the interest in the hereditary properties of the approximation property has increased. We recall that 1.(2) is a first result of this kind.

The examples of Enflo and Davie (1.(8)) show that there exist even separable reflexive (B)-spaces (the spaces l^p, $2 < p < \infty$) which have the approximation property but have a closed subspace without this property (for the approximation property of l^p see 7.).

A positive result in this direction is

(1) *If the locally convex space E has the approximation property, so has every complemented closed subspace.*

Proof. Let $E = H \oplus H'$ be the direct topological decomposition, P the projection of E onto H with kernel H'. The restriction of P to H is the identity on H. By assumption there exists an $A \in \mathfrak{L}(E)$ of finite rank such that for a given precompact subset K of H one has $(A - P)(K) \subset U \oplus U'$, where U and U' are given absolutely convex neighbourhoods of $\circ$ in H and H', respectively. Since $Px = x$ for every $x \in K$, one has $(AP - P)(K) \subset U \oplus U'$ and $P(AP - P)(K) = (PAP - P)(K) \subset U$, which is the statement, since PAP is of finite rank on H and P is the identity on H.

(2) *The locally convex direct sum* $E = \bigoplus_{\alpha} E_\alpha$ *of locally convex spaces* E_α *has the approximation property if and only if all* E_α *have this property.*

The necessity follows immediately from (1). We prove sufficiency. It follows from § 18, 5.(4) that a precompact subset K of E is contained in a set $K_{\alpha_1} \oplus \cdots \oplus K_{\alpha_n}$, K_{α_k} precompact in E_{α_k}. Let U be a given neighbourhood of $\mathfrak{o}$ in E. It contains a sum $U_1 \oplus \cdots \oplus U_n$, where U_k is a neighbourhood of $\mathfrak{o}$ in E_{α_k}. By assumption there exists $z^{(k)} = \sum_i u_i^{(k)} \otimes x_i^{(k)}$, $u_i^{(k)} \in E'_{\alpha_k}$, $x_i^{(k)} \in E_{\alpha_k}$, such that $(z^{(k)} - I_k)x^{(k)} \in U_k$ for all $x^{(k)} \in K_{\alpha_k}$, where I_k is the identity on E_{α_k}. Identifying the $u_i^{(k)}$ in the obvious way with elements of E', one obtains $\left(\sum_{k=1}^{n} z^{(k)} - I\right)x \in U_1 \oplus \cdots \oplus U_n$ for all $x \in K$, which implies the statement.

(3) *The topological product* $E = \prod_{\alpha} E_\alpha$ *of locally convex spaces* E_α *has the approximation property if and only if all* E_α *have this property.*

Necessity follows from (1). Conversely, let K be a precompact subset of E. It is contained in a set $\prod_{\alpha} K_\alpha$, K_α precompact in E_α. Let U be a neighbourhood of $\mathfrak{o}$ in E. We may suppose U to be of the form $U = \left(\prod_{\alpha_i \in \Phi} U_{\alpha_i}\right) \times \prod_{\alpha \notin \Phi} E_\alpha$, Φ a finite set of indices. We put $\prod_{\alpha_i \in \Phi} E_{\alpha_i} = E_\Phi$, $\prod_{\alpha_i \in \Phi} U_{\alpha_i} = U_\Phi$, $\prod_{\alpha_i \in \Phi} x_{\alpha_i} = x_\Phi$. Applying (2) to E_Φ we find $A_\Phi \in \mathfrak{F}(E_\Phi)$ such that $(A_\Phi - I_\Phi)x_\Phi \in U_\Phi$ for all $x_\Phi \in \prod_{\alpha_i \in \Phi} K_{\alpha_i}$. We extend A_Φ to an $A \in \mathfrak{F}(E)$ be defining $Ax_\alpha = \mathfrak{o}$ for all $x_\alpha \in E_\alpha$, $\alpha \notin \Phi$, and we obtain $(A - I)x \in U$ for all $x \in K$.

It follows from (2) and (3) that all ω_d, φ_d and all spaces of countable degree (§ 23, 5.) have the approximation property.

We proved in § 42, 7.(4) that every Hilbert space has the approximation property. By (3) every topological product of Hilbert spaces also has the approximation property. In this case one can say more:

(4) *Every subspace* E *of a topological product* $F = \prod_{\alpha} H_\alpha$ *of Hilbert spaces* H_α *has the approximation property.*

Since a finite product of Hilbert spaces is isomorphic to a Hilbert space, F has a fundamental system of neighbourhoods U of $\mathfrak{o}$ such that every $F_U = \tilde{F}_U$ is a Hilbert space. Then $\{V\} = \{E \cap U\}$ is a fundamental system of neighbourhoods of $\mathfrak{o}$ in E. Let p be the semi-norm on F corresponding to U. For $y \in E$ the mapping $y + N(V) \mapsto y + N(U)$ of E_V in F_U is one-one and even a norm isomorphism since $\hat{p}(y + N(V)) = \hat{p}(y + N(U)) = p(y)$. It follows that $\tilde{E}_V$ is a Hilbert space and 1.(3) implies the statement.

(5) *Let E be the strict inductive limit $\varinjlim E_n$ of a sequence $E_1 \subset E_2 \subset \cdots$ of locally convex spaces such that every E_n is a proper closed subspace of E_{n+1}. If all E_n have the approximation property, then E has it.*

Proof. By § 19, 4.(4) every precompact subset K of E lies in some E_k and is precompact in E_k. Let U be an open neighbourhood of $\circ$ in E; then $U_k = U \cap E_k$ is a neighbourhood of $\circ$ in E_k. By assumption there exist $u_i \in E_k'$, $x_i \in E_k$ such that $\sum_{i=1}^{n} (u_i x) x_i - x \in U_k$ for all $x \in K$. Let $\tilde{u}_i$ be HAHN–BANACH extensions of the u_i to E; then $(\sum \tilde{u}_i \otimes x_i - I)x \in U$ for all $x \in K$.

(5) is closely related to the following proposition (HOGBE-NLEND [2'], BIERSTEDT and MEISE [1']):

(6) *Let E be the locally convex hull $\sum_\alpha I_\alpha(E_\alpha)$ of locally convex spaces E_α, I_α the injection of E_α in E. Suppose further that every absolutely convex and compact subset of E is contained in some E_α and is compact in E_α.*

Then E has the weak approximation property if all E_α have the weak approximation property. If E is, moreover, quasi-complete, E has the approximation property.

Proof. Let U be an absolutely convex neighbourhood of $\circ$ in E, K an absolutely convex compact subset of E. Then by assumption K lies in some E_α and has there the same properties. $U_\alpha = U \cap E_\alpha$ is a $\mathfrak{T}_\alpha$-neighbourhood of $\circ$ in E_α and by assumption there exist $u_i \in E_\alpha'$, $x_i \in E_\alpha$, such that $\sum_{i=1}^{n} (u_i x) x_i - x \in U_\alpha$ for all $x \in K$. Now I_α is an injection; therefore $I_\alpha'(E')$ is weakly dense and even $\mathfrak{T}_{co}$-dense in E_α'. It follows that there exist $w_i \in E'$ such that $|(w_i - u_i)x| \leqq \varepsilon$ for all $x \in K$ and $i = 1, \ldots, n$. We choose $\varepsilon > 0$ such that $\sum_{i=1}^{n} \alpha_i x_i \in U_\alpha$ for all α_i, $|\alpha_i| \leqq \varepsilon$. Then we obtain for $\sum w_i \otimes x_i \in E' \otimes E$

$$\sum (w_i \otimes x_i)x - x = \sum (u_i \otimes x_i)x - x + \sum ((w_i - u_i)x)x_i \in U_\alpha + U_\alpha \subset 2U.$$

Let E be a locally convex kernel $\mathsf{K}_\alpha A_\alpha^{(-1)}(E_\alpha)$, where the α form a directed set A of indices and there exist for $\alpha < \alpha'$ linear mappings $A_{\alpha\alpha'} \in \mathfrak{L}(E_{\alpha'}, E_\alpha)$ such that

$$A_\alpha = A_{\alpha\alpha'} A_{\alpha'}, \qquad A_{\alpha\alpha'} A_{\alpha'\alpha''} = A_{\alpha\alpha''} \quad \text{for } \alpha < \alpha' < \alpha''.$$

The E_α may be arbitrary locally convex spaces. By § 19, 8.(1) a neighbourhood base of $\circ$ of the kernel topology on E is given by the sets $A_\alpha^{(-1)}(U_\alpha)$, where U_α is a neighbourhood of $\circ$ in E_α.

We will further suppose that E is reduced, that is, that $A_\alpha(E)$ is dense in E_α for every α. A special case of such a locally convex kernel is any reduced projective limit $\varprojlim A_\alpha^{(-1)}(E_\alpha)$ of locally convex spaces E_α.

(7) *A reduced projective limit* $\varprojlim A_\alpha^{(-1)}(E_\alpha)$ *or, more generally, a reduced locally convex kernel* $E = \underset{\alpha}{\mathsf{K}}\, A_\alpha^{(-1)}(E_\alpha)$ *with the properties stated above has the approximation property if all* E_α *have this property.*

Proof. Let K be a precompact subset of E and $A_\alpha^{(-1)}(U_\alpha)$ a neighbourhood of $\circ$ in E. Since $A_\alpha \in \mathfrak{L}(E, E_\alpha)$, the set $A_\alpha(K)$ is precompact in $A_\alpha(E)$, which is dense in E_α. Using 1.(2), we find $v_i \in E'_\alpha$, $A_\alpha x_i \in A_\alpha(E)$, such that $\sum_i (v_i(A_\alpha x))A_\alpha x_i - A_\alpha x \in U_\alpha$ for every $x \in K$. Now $v_i(A_\alpha x) = (A'_\alpha v_i)x$, where $A'_\alpha \in \mathfrak{L}(E'_{\alpha s}, E'_s)$; hence $A'_\alpha v_i = u_i \in E'$ and we obtain

$$\sum_i (u_i x)A_\alpha x_i - A_\alpha x \in U_\alpha \quad \text{for all } x \in K.$$

Therefore

$$\sum_i (u_i x)x_i - x \in A_\alpha^{(-1)}(U_\alpha) \quad \text{for all } x \in K.$$

There are some results on dual spaces.

(8) *Let E be a* (B)*-space. If the strong dual E' has the approximation property, then E has the approximation property.*

In particular, a reflexive (B)*-space E has the approximation property if and only if E' has it.*

Proof. If E' has the approximation property, then by 2.(9) the canonical mapping of $E' \tilde{\otimes}_\pi E$ in $\mathfrak{L}(E)$ is one-one. By 2.(7) this implies the approximation property of E.

By 1.(8) l^p, $2 < p < \infty$, has a closed subspace which does not have the approximation property. It follows from (8) by duality that l^q, $1/p + 1/q = 1$, has a quotient which does not have the approximation property. Hence a quotient of a separable reflexive (B)-space with the approximation property does not always have this property. Not every quotient of l^1 has the approximation property (§ 22, 4.(1)).

We remark that it follows also from Enflo's counterexample and a theorem of Pełczynski [3′] that there exists a separable (B)-space with a basis which therefore has the approximation property (see 5.) such that its strong dual is separable and does not have the approximation property.

(9) *Let E be quasi-complete locally convex. If E'_c is also quasi-complete and has the approximation property, then E has the approximation property.*

Proof. Let C be absolutely convex and compact in E, K absolutely convex, and $\mathfrak{T}_c$-compact in E'. There exists $A \in \mathfrak{L}(E'_c)$ of finite rank such that $(A - I)(K) \subset C^\circ$. By duality we have $(A' - I)(C) \subset K^\circ$, where $A' \in \mathfrak{F}(E)$ and K° is a $\mathfrak{T}_\gamma$-neighbourhood of o in E (see proof of 3.(6)). Since $\mathfrak{T}_\gamma$ is finer than the topology $\mathfrak{T}$ of E, this implies the approximation property of E.

(10) *Let $E[\mathfrak{T}]$ be quasi-complete, $\mathfrak{T}$ the Mackey topology. If E'_c is quasi-complete and E has the approximation property, then E'_c has the approximation property.*

If $\mathfrak{T}$ is the Mackey topology, then $\mathfrak{T} = \mathfrak{T}_\gamma$; hence $E[\mathfrak{T}] = (E'_c)'_c$ and the statement follows from (9).

We recall that $\mathfrak{T} = \mathfrak{T}_\gamma$ is the same as $\mathfrak{T} = \mathfrak{T}^{\circ\circ}$ or that $E[\mathfrak{T}]$ is polar reflexive (§ 23, 9.).

Combining (9) and (10) we obtain

(11) *Let $E[\mathfrak{T}]$ be locally convex and quasi-complete, where $\mathfrak{T}$ is the Mackey topology, and let E'_c be quasi-complete. Then E has the approximation property if and only if E'_c has this property.*

As a particular case we have by § 23, 5.(3)

(12) *A reflexive locally convex space E has the approximation property if and only if E'_c has this property.*

By § 27, 2.(1) every (M)-space E is reflexive and in this case E'_c is the strong dual E'_b.

Using ENFLO's counterexample, HOGBE-NLEND [1′] gave an example of an (M)-space which does not have the approximation property.

5. Bases, Schauder bases, weak bases. Let $E[\mathfrak{T}]$ be locally convex. A sequence (x_n) of elements of E is called a basis of E if every $x \in E$ has a unique representation of the form $x = \sum_{n=1}^{\infty} a_n(x)x_n$, $a_n(x) \in \mathrm{K}$. The convergence of this sum means that the partial sums $S_k x = \sum_{n=1}^{k} a_n(x)x_n$ converge to x in the sense of $\mathfrak{T}$.

(x_n) is a weak basis of E if it is a basis for the weak topology $\mathfrak{T}_s(E')$.

The (weak) basis (x_n) is called a (weak) Schauder basis if $a_n(x)$ is a continuous linear functional on E for every n.

A basis (x_n) is called equicontinuous if the set of corresponding projections S_k, $k = 1, 2, \ldots$, is equicontinuous in $\mathfrak{L}(E)$.

(1) *Let E be a locally convex space with an equicontinuous basis (x_n). Then E has the approximation property.*

Proof. The set H consisting of I and all S_k is also equicontinuous. The sequence $S_k x$ converges to $Ix = x$ for every $x \in E$. But on $H \subset \mathfrak{L}(E)$ the topologies $\mathfrak{T}_s$ and $\mathfrak{T}_c$ coincide (§ 39, 4.(2)) and thus S_k converges to I in $\mathfrak{L}_c(E)$.

As a consequence of the BANACH–STEINHAUS theorem we have, in particular,

(2) *Let E be countably barrelled (which includes barrelled). If E has a Schauder basis, then E has the approximation property.*

The set $\{S_k\}$ is simply bounded and therefore equicontinuous in $\mathfrak{L}(E)$ (§ 39, 5.); hence the statement follows from (1).

The following result is a little stronger.

(3) *If a countably barrelled space $E[\mathfrak{T}]$ has a weak Schauder basis, this basis is a Schauder basis and E has the approximation property.*

Again the set $H = \{S_k\} \cup \{I\}$ is an equicontinuous subset of $\mathfrak{L}(E)$. The sequence $S_k x$ converges to x in the sense of $\mathfrak{T}$ for all x which are finite linear combinations of the x_n. The space N of all these x is dense in E. This means that S_k converges to I in $\mathfrak{L}(E)$ in the sense of $\mathfrak{T}_s(N)$. By § 39, 4.(1) $\mathfrak{T}_s(N)$ and $\mathfrak{T}_s(E)$ coincide on H; therefore $S_k x \to x$ in the sense of $\mathfrak{T}$ for every x and the basis is a Schauder basis.

BANACH proved that a basis of a (B)-space is always a Schauder basis. This result was generalized to (F)-spaces by NEWNS [1′]. BANACH proved, further, that a weak basis of a (B)-space is always a basis. BESSAGA and PEŁCZYNSKI (see EDWARDS [1′], p. 453) generalized this result to (F)-spaces. We will prove here the result of BANACH and NEWNS and a recent rather general weak basis theorem of DE WILDE [3′] which contains the BESSAGA–PEŁCZYNSKI result as a particular case.

(4) (BANACH–NEWNS) *A basis (x_n) in an (F)-space E is always a Schauder basis.*

Proof. Let $p_1(x) \leqq p_2(x) \leqq \cdots$ be a sequence of semi-norms defining the topology $\mathfrak{T}$ of E. We define $p_n^*(x) = \sup_k p_n(S_k x) = \sup_k p_n\left(\sum_1^k a_j(x) x_j\right)$. Since $S_k x$ converges to x for every x, one has always $p_n^*(x) < \infty$ and $p_n(x) \leqq p_n^*(x)$. It is trivial to check that p_n^* is again a semi-norm on E and the topology $\mathfrak{T}^*$ defined by the sequence $p_1^*(x) \leqq p_2^*(x) \leqq \cdots$ is metrizable and finer than $\mathfrak{T}$.

It follows from

$$|a_n(x)|\, p_k(x_n) = p_k(a_n(x) x_n) = p_k(S_n x - S_{n-1} x) \leqq 2 p_k^*(x)$$

that $|a_n(x)| \leqq C p_k^*(x)$ for a k for which $p_k(x_n) \neq 0$. Thus every linear functional a_n is $\mathfrak{T}^*$-continuous. If $E[\mathfrak{T}^*]$ is complete, then $E[\mathfrak{T}^*]$ and $E[\mathfrak{T}]$

are isomorphic by the BANACH–SCHAUDER theorem; hence a_n is then also continuous on $E[\mathfrak{T}]$, which is our statement.

So we have to show that $E[\mathfrak{T}^*]$ is complete. Let y_n be a $\mathfrak{T}^*$-Cauchy sequence. Since a_n is $\mathfrak{T}^*$-continuous, $a_n(y_1), a_n(y_2), \ldots$ is a Cauchy sequence in K and therefore has a limit t_n. We will prove that $\sum_{n=1}^{k} t_n x_n$ $\mathfrak{T}$-converges to an element $y \in E$ and that y is the $\mathfrak{T}^*$-limit of y_n.

Let p be one of the p_k and p^* the corresponding $\mathfrak{T}^*$-semi-norm. Since y_n is $\mathfrak{T}^*$-Cauchy, there exists r_0 for a given $\varepsilon > 0$ such that

$$p\left(\sum_{k=m}^{n} [a_k(y_r) - a_k(y_s)]x_k\right) \leqq \varepsilon \quad \text{for all } n > m \text{ and all } s > r \geqq r_0.$$

Taking the limit $s \to \infty$ we obtain

$$(5) \qquad p\left(\sum_{k=m}^{n} a_k(y_r)x_k - \sum_{m}^{n} t_k x_k\right) \leqq \varepsilon \quad \text{for all } n > m \text{ and all } r \geqq r_0.$$

For a fixed $r \geqq r_0$ one has $p\left(\sum_{m}^{n} a_k(y_r)x_k\right) \leqq \varepsilon$ for all $n > m \geqq m_0$, m_0 sufficiently large, and it follows from (5) that $p\left(\sum_{m}^{n} t_k x_k\right) \leqq 2\varepsilon$ for $n > m \geqq m_0$. Since such an inequality is true for every p_l, it follows that $\sum_{1}^{n} t_k x_k$ is $\mathfrak{T}$-Cauchy; thus it has a $\mathfrak{T}$-limit with the basis representation $y = \sum_{n=1}^{\infty} t_n x_n = \sum_{n=1}^{\infty} a_n(y)x_n$ in E.

Now it follows from (5) for $m = 1$ and all n that

$$p^*(y_r - y) = \sup_n p\left(\sum_{1}^{n} a_k(y_r)x_k - \sum_{1}^{n} t_k x_k\right) \leqq \varepsilon \quad \text{for } r \geqq r_0,$$

or $y = \mathfrak{T}^*\text{-}\lim y_r$.

Using his closed-graph theorem, DE WILDE [3′] was able to prove the following weak basis theorem:

(6) *Let E be bornological, sequentially complete, and strictly webbed. If E has a weak basis (x_n), then it is a Schauder basis and E has the approximation property.*

a) The first part of the proof is similar to the first part of the proof of (4). Let $\{p_\alpha\}$ be the set of semi-norms corresponding to a neighbourhood base $\{U_\alpha\}$ of $\mathfrak{o}$ of the topology $\mathfrak{T}$ of E. We define $p_\alpha^*(x) = \sup_k p_\alpha(S_k x) = \sup_k p_\alpha\left(\sum_{1}^{k} a_n(x)x_n\right)$ as before; then $p_\alpha(x) \leq p_\alpha^*(x)$ and one verifies again that $\{p_\alpha^*\}$ is a system of semi-norms defining a neighbourhood base $\{U_\alpha^*\}$ of $\mathfrak{o}$ of a topology $\mathfrak{T}^* \supset \mathfrak{T}$ on E.

We will show that $E[\mathfrak{T}^*]$ is a webbed space. The identity mapping I of $E[\mathfrak{T}]$ onto $E[\mathfrak{T}^*]$ is closed since $\mathfrak{T}^* \supset \mathfrak{T}$. Now $E[\mathfrak{T}]$ is ultrabornological and $E[\mathfrak{T}^*]$ is webbed; thus it follows from § 35, 2.(2) that I is an isomorphism and $\mathfrak{T}^* = \mathfrak{T}$. But $a_n(x)$ is $\mathfrak{T}^*$-continuous as in the proof of (4); hence $a_n(x)$ is $\mathfrak{T}$-continuous. The last statement follows then from (2).

b) It remains to prove that $E[\mathfrak{T}^*]$ is webbed. Let $\mathscr{W} = \{C_{n_1,\ldots,n_k}\}$ be a strict web on $E[\mathfrak{T}]$. For every $x \in E$ we introduce the set $B(x) = \overline{\Gamma\{x, S_1x, S_2x, \ldots\}}$, which is absolutely convex, closed, and bounded in E. Since E is sequentially complete, it is clear that $B(x)$ is also sequentially complete.

It is proved in § 35, 6.(2) b) that for every $B(x)$ there exists a sequence n_k of integers and a sequence of positive numbers α_k, $k = 1, 2, \ldots$, such that $B(x) \subset \alpha_k C_{n_1,\ldots,n_k}$ for all k.

For the following it will be convenient to replace $\mathscr{W}$ by another strict web, $\mathscr{W}' = \{C'_{n_1,\ldots,n_k}\}$, on E which has the property that we can suppose $\alpha_k = 1$ for all k. This can be done in the following way.

We define

$$\begin{aligned} C'_{n'_1} &= m_1 C_{n_1}, & \text{where } n'_1 &= (m_1, n_1),\\ C'_{n'_1,n'_2} &= C'_{n'_1} \cap m_2 C_{n_1,n_2}, & n'_2 &= (m_2, n_2), \ldots,\\ C'_{n'_1,\ldots,n'_k} &= C'_{n'_1,\ldots,n'_{k-1}} \cap m_k C_{n_1,\ldots,n_k}, & n'_k &= (m_k, n_k), \end{aligned}$$

and so on. These sets are absolutely convex. The defining relations for a web,

$$\text{(w)} \qquad E = \bigcup_{n'_1=1}^{\infty} C'_{n'_1}, \ldots, C_{n'_1,\ldots,n'_{k-1}} = \bigcup_{n'_k=1}^{\infty} C'_{n'_1,\ldots,n'_k},$$

follow easily from the corresponding relations for $\mathscr{W}$. It remains to show that $\mathscr{W}' = \{C'_{n'_1,\ldots,n'_k}\}$ is strict.

Now $\mathscr{W}$ is strict. That means that for every fixed sequence n_k there exists a sequence $\rho_k > 0$ such that for all λ_k, $0 \leqq \lambda_k \leqq \rho_k$, and all $z_k \in C_{n_1,\ldots,n_k}$ the series $\sum_1^\infty \lambda_k z_k$ converges in E and $\sum_{k_0}^\infty \lambda_k z_k$ is contained in $C_{n_1,\ldots,n_{k_0}}$ for every k_0.

Consider a fixed sequence $n'_k = (m_k, n_k)$ and define $\rho'_k = \rho_k/m_k$, where the ρ_k correspond to the n_k in the web $\mathscr{W}$. Suppose $0 \leqq \lambda'_k \leqq \rho'_k$ and $z'_k \in C'_{n'_1,\ldots,n'_k}$. Then $0 \leqq \lambda_k \leqq \rho_k$ for $\lambda_k = \lambda'_k m_k$ and $z_k = z'_k/m_k \in C_{n_1,\ldots n_k}$, since $C'_{n'_1,\ldots,n'_k} \subset m_k C_{n_1,\ldots,n_k}$. It follows that $\sum_1^\infty \lambda'_k z'_k = \sum_1^\infty \lambda_k z_k$ converges in E. Moreover, $\sum_{k_0}^\infty \lambda'_k z'_k = \sum_{k_0}^\infty \lambda_k z_k \in C_{n_1,\ldots,n_j} \subset m_j C_{n_1,\ldots,n_j}$ for all $j \leqq k_0$, since $\mathscr{W}$ is strict. This implies $\sum_{k_0}^\infty \lambda'_k z'_k \in C'_{n'_1,\ldots,n'_{k_0}}$ by the definition of $C'_{n'_1,\ldots n'_{k_0}}$. Hence $\mathscr{W}'$ is a strict web on E.

Finally, if $B(x) \subset \alpha_k C_{n_1,\dots n_k}$ and if $\alpha_k \leqq m_k$, m_k an integer, then $B(x) \subset \bigcap_{j=1}^{k} m_j C_{n_1,\dots,n_j} \subset C'_{n'_1,\dots,n'_k}$, where $n'_j = (m_j, n_j)$.

Therefore we may suppose that there exists a strict web $\mathscr{W} = \{C_{n_1,\dots,n_k}\}$ on E such that for every $x \in E$ there exists a sequence n_k of integers such that $B(x) \subset C_{n_1,\dots,n_k}$.

c) Let $C^*_{n_1,\dots,n_k}$ be the set of all $x \in E$ such that $B(x) \subset C_{n_1,\dots,n_k}$. It is trivial to check condition (w), so $\mathscr{W}^* = \{C^*_{n_1,\dots,n_k}\}$ is a web on E. It remains to prove that $\mathscr{W}^*$ is of type $\mathscr{C}$ in $E[\mathfrak{T}^*]$.

Let n_k be a fixed sequence and ρ_k the sequence of corresponding numbers for the web $\mathscr{W}$. We suppose the ρ_k decreasing and < 1. It will be sufficient to prove that for $\lambda_k \in [0, \rho_k^2]$ and $z_k \in C^*_{n_1,\dots,n_k}$ the series $\sum_1^\infty \lambda_k z_k$ converges in $E[\mathfrak{T}^*]$.

We remark first that by the definition of $B(x)$ the elements z_k, $S_1 z_k$, $S_2 z_k, \dots$ are all contained in $C_{n_1,\dots,n_k}$; hence $\sum_1^\infty \lambda_k z_k$ and $\sum_1^\infty \lambda_k S_m z_k$ converge in $E[\mathfrak{T}]$ to elements y resp. y_m for all $m = 1, 2, \dots$.

Our aim is now to prove that $y_m = S_m y$ and that y_m converges weakly to y. By definition $\sum_{k=1}^N \lambda_k S_1 z_k = \sum_1^N \lambda_k a_1(z_k) x_1$ converges to y_1; hence $y_1 = \beta_1 x_1$, where $\beta_1 = \sum_1^\infty \lambda_k a_1(z_k)$. Similarly, $\sum_1^N \lambda_k a_m(z_k) x_m = \sum_1^N \lambda_k S_m z_k - \sum_1^N \lambda_k S_{m-1} z_k$ converges to $y_m - y_{m-1}$; thus $y_m - y_{m-1} = \beta_m x_m$ with $\beta_m = \sum_1^\infty \lambda_k a_m(z_k)$. It follows that $y_m = \sum_1^m \beta_i x_i$. Because (x_n) is a weak basis it will be sufficient to show that y_m converges weakly to y, since then it follows from the uniqueness of the basis representation that $y_m = S_m y$.

Let u be an element of E'. We write

$$|u(y - y_m)| \leqq \left| u\left(y - \sum_1^N \lambda_k z_k\right)\right| + \left| u\left(\sum_1^N \lambda_k z_k - \sum_1^N \lambda_k S_m z_k\right)\right|$$

$$= \quad A \quad + \quad B$$

$$+ \sup_m \left| u\left(\sum_1^N \lambda_k S_m z_k - y_m\right)\right|$$

$$+ \quad C.$$

We show first that for a given $\varepsilon > 0$ we can choose N_0 such that A and C are $\leqq \varepsilon/3$ for $N \geqq N_0$. We recall that the sequence $\rho_n < 1$ is decreasing, that $0 \leqq \lambda_k \leqq \rho_k^2$, and that $z_k \in C^*_{n_1,\dots,n_k}$ or $B(z_k) \subset C_{n_1,\dots,n_k}$. Hence

$$y - \sum_1^N \lambda_k z_k = \sum_{N+1}^\infty \lambda_k z_k = \rho_{N+1} \sum_{N+1}^\infty \lambda'_k z_k, \qquad \lambda'_k \leqq \rho_k \quad \text{for } k \geqq N + 1.$$

Since $\mathscr{W}$ is strict, we have

$$\sum_{N+1}^{\infty} \lambda'_k z_k \in C_{n_1,\ldots,n_{N+1}} \quad \text{and} \quad y - \sum_{1}^{N} \lambda_k z_k \in \rho_{N+1} C_{n_1,\ldots,n_{N+1}}.$$

Similarly, $y_m - \sum_{1}^{N} \lambda_k S_m z_k \in \rho_{N+1} C_{n_1,\ldots,n_{N+1}}$.

Now let U be a neighbourhood of $\circ$ in $E[\mathfrak{T}]$ such that $|ux| \leqq \varepsilon/3$ for $x \in U$. By § 35, 1.(3) there exists N_0 such that $\rho_{N+1} C_{n_1,\ldots,n_{N+1}} \subset U$ for all $N \geqq N_0$. For such an N obviously $A \leqq \varepsilon/3$ and $C \leqq \varepsilon/3$.

We fix $N \geqq N_0$ and observe that $\sum_{1}^{N} \lambda_k S_m z_k = S_m\left(\sum_{1}^{N} \lambda_k z_k\right)$ converges in m weakly to $\sum_{1}^{N} \lambda_k z_k$; hence for m sufficiently large one has $B \leqq \varepsilon/3$. It follows that y_m converges weakly to y.

d) We come to the last step of De Wilde's proof. We showed that $y_m = S_m y$ for every m. We use this to prove the $\mathfrak{T}^*$-convergence of $\sum_{1}^{N} \lambda_k z_k$ to y in the following way. One has again

$$y - \sum_{1}^{N} \lambda_k z_k \in \rho_{N+1} C_{n_1,\ldots,n_{N+1}}, \qquad S_m y - S_m\left(\sum_{1}^{N} \lambda_k z_k\right) \in \rho_{N+1} C_{n_1,\ldots,n_{N+1}},$$
$$N = 1, 2, \ldots.$$

Let p be a continuous semi-norm on E and let N be such that $p(x) \leqq \varepsilon$ for $x \in \rho_{N+1} C_{n_1,\ldots,n_{N+1}}$. Then it follows that

$$p^*\left(y - \sum_{1}^{N} \lambda_k z_k\right) = \sup_m p\left(S_m y - S_m\left(\sum_{1}^{N} \lambda_k z_k\right)\right) \leqq \varepsilon.$$

This shows that $\sum_{1}^{\infty} \lambda_k z_k$ converges in $E[\mathfrak{T}^*]$.

In § 35, 4. one can find classes of spaces which satisfy (6). For example, it follows from § 35, 4.(8) that a weak basis of a sequentially complete (LF)-space is always a Schauder basis.

6. The basis problem. If the locally convex space E has a basis (x_n), then the finite rational resp. complex rational linear combinations $\sum_{n=1}^{N} \alpha_n x_n$ are dense in E; therefore

(1) *A locally convex space with a basis is separable.*

The basis problem, "Does every separable (B)-space possess a basis?," was raised by Banach in his book [3] and was solved in the negative by Enflo [1′]. He constructed separable (B)-spaces which do not have the approximation property. By 5.(2) and 5.(4) such a space has no basis.

During the forty years between the statement of the basis problem and its negative solution, bases of (B)-spaces and their properties have been studied intensively and the results of these investigations are of great importance for the finer structure of (B)-spaces. Detailed expositions are given in LINDENSTRAUSS–TZAFRIRI [1′], [2′], MCARTHUR [1′], MARTI [1′], and SINGER [1′].

Our interest is at the moment limited to the fact that it follows from the existence of a basis in a (B)-space that the space has the approximation property.

It is trivial to check that the unit vectors define a basis in c_0 and in l^p, $1 \leqq p < \infty$, so these spaces have the approximation property.

The space c of convergent sequences (§ 14, 7.) has a basis consisting of the unit vectors and the vector $\mathfrak{e} = (1, 1, \ldots)$. The space l^∞ is not separable and therefore has no basis. But l^∞ has the approximation property, as we will see in 7.

Bases in the spaces $C[0, 1]$ and $L^p[0, 1]$, $1 \leqq p < \infty$, have been constructed by SCHAUDER (cf. SINGER [1′], I § 2); hence these spaces have the approximation property.

The (F)-space ω has the unit vectors as a basis; the (F)-space $H(\mathfrak{G})$, where $\mathfrak{G}$ is the open unit disc in the complex plane, has $1, z, z^2, \ldots$ as a basis (cf. § 27, 3. for the definition of $H(\mathfrak{G})$).

It seems to be unknown whether the (B)-space $HB(\overline{\mathfrak{G}})$ of all functions analytic on the open unit disc and continuous on the closed unit disc has a basis. But it has the approximation property (see 7.).

A large class of sequence spaces with a basis is given in

(2) *Let $\lambda[\mathfrak{T}]$ be a perfect sequence space, where $\mathfrak{T}$ is the normal topology $\mathfrak{T}_n$ or the Mackey topology $\mathfrak{T}_k(\lambda^\times)$. Then the sequence $\mathfrak{e}_1, \mathfrak{e}_2, \ldots$ of unit vectors is an equicontinuous basis of $\lambda[\mathfrak{T}]$ and $\lambda[\mathfrak{T}]$ has the approximation property.*

Proof. Every $\mathfrak{x} = (x_1, x_2, \ldots) \in \lambda$ is the $\mathfrak{T}$-limit of its sections $\mathfrak{x}_n = x_1\mathfrak{e}_1 + \cdots + x_n\mathfrak{e}_n$ by § 30, 5.(8) and § 30, 5.(10), and $x_n = \mathfrak{e}_n\mathfrak{x}$, where $\mathfrak{e}_n \in \lambda^\times$; hence $\mathfrak{e}_1, \mathfrak{e}_2, \ldots$ is a Schauder basis of $\lambda[\mathfrak{T}]$.

A neighbourhood base of $\mathfrak{o}$ is given by the set of all normal closed neighbourhoods U of $\mathfrak{o}$. This is trivial for the normal topology and follows from § 30, 6.(2) for the topology $\mathfrak{T}_k(\lambda^\times)$. Therefore U contains with $\mathfrak{x}$ its sections $\mathfrak{x}_n = S_n\mathfrak{x}$ and $S_n(U) \subset U$ for $n = 1, 2, \ldots$ means that the basis $(\mathfrak{e}_n)$ is equicontinuous. The last statement in (2) follows from 5.(1).

We proved in § 30, 5.(11) that every perfect $\lambda[\mathfrak{T}]$, where $\mathfrak{T}$ is the normal or the Mackey topology, is sequentially separable, i.e., every element is the limit of a sequence of elements belonging to a fixed countable subset of λ.

By a similar argument one proves the following sharpened form of (1):

(3) *A locally convex space E with a basis is sequentially separable.*

Let (x_n) be the basis and x a fixed element in E, $x = \sum_{n=1}^{\infty} a_n(x)x_n$. Then the $S_n x$ and the $a_n(x)x_n = S_n x - S_{n-1}x$ are contained in an absolutely convex bounded subset B of E. Determine the (complex) rational numbers $\rho_i^{(n)}$, $i = 1, 2, \ldots, n$, such that $|\sigma_i^{(n)}| = |a_i(x) - \rho_i^{(n)}| \leqq (1/n^2)|a_i(x)|$. One has then

$$x - \sum_1^n \rho_i^{(n)} x_i = (x - S_n x) + \sum_1^n \sigma_i^{(n)} x_i.$$

Let U be an absolutely convex neighbourhood of $\mathfrak{o}$ in E. Then there exists n_0 such that for $n \geqq n_0$, $x - S_n x \in U/2$ and also $\sum_1^n \sigma_i^{(n)} x_i \in U/2$, since $\sum_1^n \sigma_i^{(n)} x_i \in B/n$. Hence the countable sets of all $\sum_{i=1}^n \rho_i x_i$, ρ_i rational, $n = 1, 2, \ldots$, is sequentially dense in E.

The importance of sequential separability is demonstrated by the following result of KALTON [2′]:

(4) *The barrelled space ω_d, $d = 2^{\aleph_0}$, is separable but not sequentially separable. ω_d has the approximation property but no basis.*

A sequentially separable space contains at most $2^{\aleph_0}$ elements, but ω_d contains 2^d elements (§ 9, 5.). Hence ω_d is not sequentially separable and has no basis by (3). It has the approximation property by 4.(3). A proof of the curious fact that ω_d is separable can be found in HENRIQUES [1′], where other closely related facts are given.

Bases of barrelled spaces were investigated for the first time by DIEUDONNÉ [1′].

7. Some function spaces with the approximation property. We treat first the case $E = C(K)$, where K is any compact topological space.

(1) *Let R be a normal topological space and $U_1, \ldots, U_n$ open sets such that $R = \bigcup_{i=1}^n U_i$. Then there exist n continuous functions $\varphi_1, \ldots, \varphi_n$ on R with values in* [0, 1] *such that*

(2) $\sum_{i=1}^n \varphi_i(x) = 1$ *for* $x \in R$ *and* $\varphi_i(x) = 0$ *for* $x \in R \sim U_i$.

Such a system $\{\varphi_1, \ldots, \varphi_n\}$ is called a partition of unity on R.

Proof. a) We show first that there exist open sets $O_1, \ldots, O_n$ such that $O_i \subset \bar{O}_i \subset U_i$ and $\bigcup_{i=1}^n O_i = R$.

The set $R \sim \bigcup_{2}^{n} U_i$ is a closed subset of U_1. By § 3, 7.(N') there exists an open set O_1 such that $R \sim \bigcup_{2}^{n} U_i \subset O_1 \subset \bar{O}_1 \subset U_1$. Again $O_1 \cup U_2 \cup \cdots \cup U_n = R$. Repetition of this procedure proves the existence of $O_1, \ldots, O_n$.

b) By URYSOHN's lemma (§ 6, 4.(1)) there exists a continuous function ψ_i on R with values in $[0, 1]$ such that $\psi_i(x) = 1$ on $\bar{O}_i$ and $\psi_i(x) = 0$ on $R \sim U_i$. It is obvious that the functions $\varphi_i = \psi_i/\psi$, where $\psi = \sum_{i=1}^{n} \psi_i$ satisfy (2) and have values in $[0, 1]$.

We note that the support of φ_i, $\operatorname{supp} \varphi_i$, is contained in U_i ($\operatorname{supp} f$ is the closure of the set $\{x \in R; f(x) \neq 0\}$).

(1) is true in particular for compact spaces R (§ 3, 7.(2)).

(3) *$C(K)$ has the approximation property.*

Proof. Let $\{\varphi_i\}$ be a partition of unity on K and $x_i \in \operatorname{supp} \varphi_i$, $i = 1, \ldots, n$. We define the corresponding mapping $A \in \mathfrak{F}(C(K))$ by $Af = \sum_{i=1}^{n} f(x_i)\varphi_i$. Obviously, $\|A\| = \sup_{\|f\| \leqq 1} \|Af\| \leqq 1$; hence the set H of all these mappings is equicontinuous in $\mathfrak{L}(C(K))$. If I is the identity of $\mathfrak{L}(C(K))$, then $H \cup \{I\}$ is equicontinuous and $\mathfrak{T}_s$ and $\mathfrak{T}_c$ coincide on this set. Therefore it will be sufficient to prove that I is a $\mathfrak{T}_s$-adherent point of H.

Let $\varepsilon > 0$ and $f_1, \ldots, f_k \in C(K)$ be given. Since K is compact, there exists a finite covering $K = \bigcup_{j=1}^{m} U_j$, U_j open, such that every f_i has an oscillation $\leqq \varepsilon$ on every U_j. Let $\{\varphi_j\}$ be a corresponding partition of unity and A the corresponding mapping. Then by (2)

$$\begin{aligned} \|f_i - Af_i\| &= \sup_{x \in K} \left| f_i(x) - \sum_{j=1}^{n} f_i(x_j)\varphi_j(x) \right| \\ &\leqq \sup \sum \varphi_j(x) |f_i(x) - f_i(x_j)| \leqq \varepsilon, \qquad i = 1, \ldots, k, \end{aligned}$$

so that I is a $\mathfrak{T}_s$-adherent point of H.

Let R be a locally compact space, $C(R)$ the vector space of all continuous functions on R. Let $\{K_\alpha\}$, $\alpha \in \mathsf{A}$, be a fundamental system of compact subsets of R. The topology of compact convergence on $C(R)$ is then defined by the system of semi-norms $p_\alpha(f) = \sup_{x \in K_\alpha} |f(x)|$.

Let $J_\alpha f$ be the restriction of $f \in C(R)$ to K_α. Then J_α maps $C(R)$ onto $C(K_\alpha)$, as follows easily from § 6, 4.(5). One checks immediately that $C(R)$ is the reduced projective limit $\varprojlim J_\alpha^{(-1)}(C(K_\alpha))$. Thus (3) and 4.(7) imply

(4) *$C(R)$, R locally compact, has the approximation property.*

Let R be a locally compact space which is not compact but countable at infinity (§ 3, 6.) and let $\mathscr{K}(R)$ be the space of all continuous functions on R with compact support. If K is a compact subset of R we denote by $\mathscr{K}(K)$ the (B)-space of all $f \in \mathscr{K}(R)$ with $\operatorname{supp} f \subset K$ and $\|f\|_K = \sup_{x \in K} |f(x)|$. One equips $\mathscr{K}(R)$ with the hull topology of $\sum_K \mathscr{K}(K)$. Since R is countable at infinity there exists a fundamental sequence $K_1 \subset K_2 \subset \cdots$ of compact sets such that K_{n+1} is a neighbourhood of K_n for every n, $R = \bigcup_{n=1}^{\infty} K_n$, and $\mathscr{K}(R) = \varinjlim \mathscr{K}(K_n)$. This inductive limit is strict by § 19, 4.(1) and complete by § 19, 5.(3).

(5) *Let R be locally compact, noncompact, and countable at infinity. Then $\mathscr{K}(R)$ has the approximation property.*

Proof. Let M be a compact subset of $\mathscr{K}(R)$. By § 19, 4.(4) M lies in some $\mathscr{K}(K_n)$ and is therefore compact in $\mathscr{K}(K_n)$. It will be sufficient to define for a given $\varepsilon > 0$ an A of finite rank which maps $\mathscr{K}(R)$ into $\mathscr{K}(K_{n+1})$ such that $|(Af)(x) - f(x)| \leqq \varepsilon$ for all $x \in K_{n+1}$ and all $f \in M$.

Let Jf be the restriction of $f \in \mathscr{K}(R)$ to K_{n+1}. Obviously, J maps $\mathscr{K}(R)$ continuously in $C(K_{n+1})$ and $J(M)$ is compact in $C(K_{n+1})$. By (3) there exists $B \in \mathfrak{F}(C(K_{n+1}))$ such that

(6) $|(B(Jf))(x) - (Jf)(x)| \leqq \varepsilon$ for all $x \in K_{n+1}$ and $f \in M$.

Let $\alpha(x)$ be a continuous function on R with values in $[0, 1]$, identically 1 on K_n and identically 0 on $R \sim K_{n+1}$. We define $A = \alpha(x)BJ$, then $A \in \mathfrak{F}(\mathscr{K}(R))$, and it follows from (6) that $|(Af)(x) - f(x)| \leqq \varepsilon$ for all $x \in K_{n+1}$ and $f \in M$, since $\operatorname{supp} f \in K_n$.

We remark that by a refinement of the method of proof of (3) it is also possible to prove the approximation property for $\mathscr{K}(K)$, K a compact subset of R (see Bierstedt [1′]). Then (5) follows from this result and 4.(6).

We give another application of (3). Let S be a completely regular space and $CB(S)$ the space of all continuous and bounded functions f on S equipped with the norm $\|f\| = \sup_{x \in S} |f(x)|$. Clearly, $CB(S)$ is a (B)-space. We denote $CB(S)'$ by $\mathfrak{M}(S)$ as in the case of a compact S (§ 24, 5.).

We define the mapping $\Phi(x) = \delta_x$ (where $\delta_x(f) = f(x)$) of S into the unit ball of $\mathfrak{M}(S)$. Φ is one-one and one has $\|\delta_x\| = 1$ for any $x \in S$ as a consequence of § 6, 6.(V).

(7) *$\Phi(S)$ equipped with $\mathfrak{T}_s(CB(S))$ is homeomorphic to S.*

Proof. A weak neighbourhood of δ_{x_0} consists of all δ_x such that $|(\delta_x - \delta_{x_0})f_i| < \varepsilon$ or $|f_i(x) - f_i(x_0)| < \varepsilon$, $i = 1, \ldots, n$. Since the f_i are continuous, this is true for some neighbourhood of x_0. Hence Φ is continuous.

But Φ is also open: Let $\{f_\alpha\}$, $\alpha \in \mathsf{A}$, be the set of all continuous functions on S with values in $[0, 1]$. The set of all $[f_\alpha < 1]$ is a base of open sets in S (§ 6, 6.). The set of all $u \in \mathfrak{M}(S)$ such that $\langle f_\alpha, u \rangle = u(f_\alpha) = 1$ is a closed hyperplane in $\mathfrak{M}(S)$. This hyperplane cuts $\Phi(S)$ in $\{\delta_x; \delta_x(f_\alpha) = 1\}$, which is the complement of $\Phi([f_\alpha < 1]) = \{\delta_x; \delta_x(f_\alpha) < 1\}$ in $\Phi(S)$. Hence $\Phi([f_\alpha < 1])$ is open.

Let βS be the weak closure of $\Phi(S)$ in $\mathfrak{M}(S)$. Since βS is contained in the weakly compact unit ball of $\mathfrak{M}(S)$, βS is a compact space. It follows from (7) that βS can be considered as a compact extension of S and S is dense in βS.

One calls βS the STONE–ČECH compactification of S.

We use this construction in the following proposition:

(8) *Let S be completely regular. Then $CB(S)$ is norm isomorphic to $C(\beta S)$ and has the approximation property.*

Every $f \in CB(S)$ has a continuous extension to βS which has the same norm. Conversely, every continuous function on βS is bounded and has a restriction to S with the same norm. The last statement follows from (3).

(9) *l_d^∞ has the approximation property for every cardinal d.*

Proof. The elements of l_d^∞ are of the form $\mathfrak{x} = (\xi_\alpha)$, where α runs through an index set A with cardinality d. We consider A as a discrete topological space and A is therefore completely regular. Hence l_d^∞ is the space $CB(\mathsf{A})$ and (8) implies the statement.

We indicate a direct proof: Let $\mathsf{A} = \mathsf{A}_1 \cup \cdots \cup \mathsf{A}_n$ be a partition of A into n disjoint subsets and α_i a fixed element of A_i. Let $e(\mathsf{A}_i)$ be the characteristic function of A_i which is an element of l_d^∞. Then A defined by $A\mathfrak{x} = \sum_{i=1}^{n} \xi_{\alpha_i} e(\mathsf{A}_i)$ is in $\mathfrak{F}(l_d^\infty)$ and $\|A\| = 1$; hence the set H of all these A is equicontinuous. It is easy to determine A in such a way that $\|A\mathfrak{x}_j - \mathfrak{x}_j\| \leqq \varepsilon$ for a finite set of $\mathfrak{x}_j \in l_d^\infty$. Hence I is a $\mathfrak{T}_s$-adherent point of H and the statement follows as in the proof of (3).

The same method, which goes back to PHILLIPS [1], will also settle the case of L^p-spaces. We will consider these spaces in greater generality than in § 14, 10. and refer the reader to BOURBAKI [7] for detailed information.

Let R be a locally compact space and μ a positive Radon measure on R. Then $L^p(R, \mu)$ is the (B)-space of equivalence classes of functions on R which are μ-integrable in the pth power with the norm $\|f\|_p = \left(\int |f|^p \, d\mu\right)^{1/p}$. We note that the subspace $\mathscr{K}(R)$ of all continuous functions with compact support is dense in $L^p(R, \mu)$.

(10) $L^p(R, \mu)$, $1 \leqq p < \infty$, *has the approximation property.*

Proof. Let $K = K_1 \cup \cdots \cup K_n$ be a decomposition of the compact subset K of R in disjoint relatively compact subsets, $m = \mu(K)$, $m_i = \mu(K_i)$, and let χ_i be the characteristic function of K_i. Then we define the mapping $Af = \sum_{i=1}^{n} \left(\int f\chi_i \, d\mu\right)(\chi_i/m_i)$, which obviously lies in $\mathfrak{F}(L^p)$. From Hölder's inequality it follows with $f_i = f\chi_i$ that

$$\left|\int f\chi_i \, d\mu\right| = \left|\int f_i\chi_i \, d\mu\right| \leqq \|f_i\|_p \|\chi_i\|_q = \|f_i\|_p m_i^{1/q}.$$

Hence

$$|Af| \leqq \sum_i \|f_i\|_p m_i^{-1/p}\chi_i \quad \text{and} \quad \int |Af|^p \, d\mu \leqq \int \sum \|f_i\|_p^p m_i^{-1}\chi_i \, d\mu$$
$$= \sum \|f_i\|_p^p \leqq \|f\|_p^p.$$

Therefore $\|A\| \leqq 1$ and the set of all A is equicontinuous.

Let now $f^{(1)}, \ldots, f^{(m)}$ be given functions in $\mathscr{K}(R)$ and let K be a compact set containing the support of all these $f^{(k)}$. We decompose K in disjoint relatively compact subsets, $K = K_1 \cup \cdots \cup K_n$, such that the oscillation of all $f^{(k)}$ on every K_j is $\leqq \varepsilon/m^{1/p}$. Let A be the mapping corresponding to this decomposition of K.

One checks easily that for every $x \in K$ one has $|Af^{(k)}(x) - f^{(k)}(x)| \leqq \varepsilon/m^{1/p}$; hence $\|Af^{(k)} - f^{(k)}\|_p \leqq \varepsilon$ for $k = 1, \ldots, m$. This implies that the identity I is an adherent point of the equicontinuous set of all mappings A for the topology of simple convergence on $\mathscr{K}(R)$ and therefore I is adherent point for $\mathfrak{T}_c$ also (§ 39, 4.(1) and (2)).

The space $L^\infty(R, \mu)$, R locally compact, consists of the equivalence classes of all locally measurable functions and locally almost everywhere bounded functions on R and is a (B)-space with the norm $\|f\|_\infty = \inf\{c; |f(x)| \leqq c \text{ locally almost everywhere}\}$.

(11) $L^\infty(R, \mu)$ *has the approximation property.*

This can be proved directly with the method indicated in the second proof for l_d^∞. It is also an immediate consequence of the norm isomorphism of $L^\infty(R, \mu)$ with a space $C(K)$, K compact, which can be obtained as the Gelfand representation of the Banach algebra $L^\infty(R, \mu)$ defined by pointwise multiplication.

(12) *The* (B)-*space* $HB(\overline{\mathfrak{G}})$, $\mathfrak{G}$ *the open unit disc in the complex plane, has the approximation property.*

Proof. To $f \in HB(\overline{\mathfrak{G}})$ one introduces $f_n(z) = f(z/(1 + 1/n))$. One has $\|f_n\| \leqq \|f\|$ and $f_n \to f$ in the norm. Let $T_{kn}f$ be the kth partial sum of the Taylor expansion of f_n at the point $z = 0$. Then T_{kn} is of finite rank, $\|T_{kn}\| \leqq 1$ and $\|T_{kn}f - f\| \leqq \|T_{kn}f - f_n\| + \|f_n - f\| \leqq \varepsilon$ for k, n sufficiently large. This implies the statement.

We remarked in 5. that it is unknown whether $HB(\overline{\mathfrak{G}})$ has a basis.

Most of the examples in 7. can be found in GROTHENDIECK [13], PHILLIPS [1], and SCHWARTZ [1'].

8. The bounded approximation property. In 7. we proved the approximation property for some (B)-spaces. In every case we showed that for a given compact set K and $\varepsilon > 0$ there exists $A \in \mathfrak{F}(E)$ *with* $\|A\| \leqq 1$ such that $\|Ax - x\| \leqq \varepsilon$ for all $x \in K$. This sharper form of the approximation property is called the metric approximation property. If $A \in \mathfrak{F}(E)$ can always be chosen such that $\|A\| \leqq \lambda$, then E is said to have the λ-metric approximation property. E has the bounded approximation property if it has the λ-metric approximation property for some λ.

A (B)-space E with a Schauder basis always has the bounded approximation property, since the set $\{S_n\}$ of 5.(2) is equicontinuous. It follows from the proof of the theorem of BANACH–NEWNS (5.(4)) that it is possible to introduce an equivalent norm on E such that $\|S_n\| \leqq 1$ in this new norm, so that E has the (1-)metric approximation property.

Recently, FIGIEL and JOHNSON [1'] constructed a separable (B)-space E which has the approximation property but not the bounded approximation property and thus has no basis. They use ENFLO's counterexample and PEŁCZYNSKI's result cited before 4.(9). E can be chosen to have a separable conjugate E' and in this case the authors show that there exists a non-nuclear mapping $A \in \mathfrak{L}(E)$ whose adjoint A' is nuclear in $\mathfrak{L}(E')$.

It is not known whether there exists a separable (B)-space without a basis but with the metric approximation property.

If one follows the reasoning of the proof of 1.(1), one obtains easily

(1) *Let E, F be* (B)-*spaces. If E has the metric approximation property, then the unit ball of $\mathfrak{F}(E, F)$ is $\mathfrak{T}_c$-dense in the unit ball of $\mathfrak{L}_b(E, F)$ and the unit ball of $\mathfrak{F}(F, E)$ is $\mathfrak{T}_c$-dense in the unit ball of $\mathfrak{L}_b(F, E)$.*

GROTHENDIECK proved in [13] some deep results on the metric approximation property that are based on some facts on bilinear integral forms, which will be considered in § 45.

One of his results was proved in a more elementary way by JOHNSON [1']. We reproduce his proof.

Let E, F be (B)-spaces. We recall from 3. that $\mathfrak{L}_b(E'_k, F)$ is the space of all weakly continuous mappings A of E' in F and the topology is given by the norm $\|A\|$. If F is finite dimensional, then $\mathfrak{L}_b(E'_k, F)$ contains only mappings of finite rank and we have in this case $\mathfrak{L}_b(E'_k, F) = E \otimes_\varepsilon F$. We need the following lemma:

(2) *Let E, F be* (B)*-spaces, F finite dimensional. Then* $(E \otimes_\varepsilon F)'' = E'' \otimes_\varepsilon F$.

We remark that X' resp. X'' always means the strong dual resp. strong bidual of X. The proof of (2) will be given in § 45, 1.(11).

(3) *Let E, F be* (B)*-spaces, F finite dimensional, and H a finite dimensional subspace of E'. Suppose $A \in \mathfrak{L}_b(E', F) = E'' \otimes_\varepsilon F$, $\delta > 0$. Then there exists $B \in \mathfrak{L}_b(E'_k, F) = E \otimes_\varepsilon F$ such that B coincides with A on H and $\|B\| \leqq \|A\| + \delta$.*

Proof. $H \otimes F'$ is a finite dimensional subspace of $(E \otimes_\varepsilon F)'$. By (2) A is an element of $(E \otimes_\varepsilon F)''$ and it defines on $H \otimes F'$ a linear functional with norm $\leqq \|A\|$. We apply HELLY's theorem (§ 38, 1.(11)) to this situation and find an element B of $E \otimes_\varepsilon F = \mathfrak{L}_b(E'_k, F)$ which coincides on H with A and has norm $\|B\| \leqq \|A\| + \delta$.

We are now able to prove

(4) *Let E be a* (B)*-space. If E' has the λ-metric approximation property, so has E.*

A reflexive (B)*-space E has the λ-metric approximation property if and only if E' has this property.*

Proof. We assume that E' has the λ-metric approximation property. If $u_1, \dots, u_m \in E'$ and $\varepsilon > 0$ are given, there exists then $A \in \mathfrak{F}(E')$ such that $\|A\| \leqq \lambda$ and $\|Au_k - u_k\| \leqq \varepsilon/2$, $k = 1, \dots, m$. By (3) there exists $B \in \mathfrak{F}(E'_s) = E \otimes E'$ such that B coincides on $H = [u_1, \dots, u_m]$ with A and $\|B\| \leqq \|A\| + \delta \leqq \lambda + \delta$ for a given $\delta > 0$. We put $C = [\lambda/(\lambda + \delta)]B$; then $\|C\| \leqq \lambda$. For a suitable δ one has

$$(5) \qquad \|Cu_k - u_k\| \leqq \left\| \left(B - \frac{\delta B}{\lambda + \delta}\right) u_k - u_k \right\| \leqq \|Au_k - u_k\| + \delta \|u_k\|$$

$$\leqq \frac{\varepsilon}{2} + \frac{\varepsilon}{2} = \varepsilon.$$

Thus I is $\mathfrak{T}_s$-adherent point of the convex set of all $C \in E \otimes E'$ with bound λ. The adjoints $C' \in E' \otimes E \in \mathfrak{F}(E)$ determine also a convex set

M with bound λ. (5) implies $|u_k(C'x - x)| \leqq \varepsilon$, $k = 1, \ldots, m$, for a fixed x, $\|x\| \leqq 1$; hence x is a weak adherent point of the set $\{C'x\}$ of all $C'x$, $C' \in M$. Since M is convex, it follows from § 20, 7.(6) that x is also a strong adherent point of $\{C'x\}$. This means that there exists $C'_0 \in M$ such that $\|C'_0x - x\| \leqq \varepsilon_1$, $\varepsilon_1 > 0$ given. Thus I is $\mathfrak{T}_s$-adherent point of M. Since M is equicontinuous, the statement now follows from § 39, 4.(2).

9. Johnson's universal space. Shortly before ENFLO discovered his counterexample, JOHNSON constructed in [2′] a (B)-space C'_1 with the property that if C'_1 has the approximation property, then every separable (B)-space has the approximation property. Since this is not the case, C'_1 is another example of a (B)-space which fails to have the approximation property. C'_1 has also another interesting property, so we present this example in detail.

We recall the notion of the distance coefficient $d(E, F) = \inf(\|T\|\,\|T^{-1}\|)$ of two isomorphic (B)-spaces from § 42, 8.

(1) *There exists a sequence* G_n, $n = 1, 2, \ldots$, *of finite dimensional* (B)-*spaces with the following property: For every finite dimensional* (B)-*space* F *and every* $\varepsilon > 0$ *there exists* n_0 *such that* $\dim G_{n_0} = \dim F$ *and* $d(F, G_{n_0}) < 1 + \varepsilon$.

It is sufficient to construct such a sequence for all F of a fixed dimension $N > 0$. The norm $p(x)$ of F defines a continuous function on the N-dimensional Euclidean unit ball $K = \{x; \|x\|_2 \leqq 1\}$. Since $C(K)$ is separable, the subset of all norms $p(x)$ is also separable (§ 4, 5.(1)), so there exists a sequence of norms $p_1, p_2, \ldots$ such that $|p(x) - p_k(x)| \leqq \varepsilon_1$ for all $x \in K$ and some k depending on p and ε_1.

Let $G_1, G_2, \ldots$ be the sequence of N-dimensional (B)-spaces with the norms $p_1, p_2, \ldots$ and let I be the identity map of F onto G_k. The closed unit ball U of F is contained in ρK for some $\rho > 0$. For $y \in U$ we have therefore $p(y) = \rho p(y/\rho)$, where $y/\rho \in K$; hence $p_k(y) \leqq \rho(p(y/\rho) + \varepsilon_1) \leqq 1 + \rho\varepsilon_1$. This implies $\|I\| \leqq 1 + \rho\varepsilon_1$.

The closed unit ball V of G_k is contained in $2\rho K$: Otherwise there would exist an x with $\|x\|_2 = 2\rho$, $p_k(x) \leqq 1$, and $p(x) \geqq 2$, which contradicts $|p(x) - p_k(x)| \leqq 2\rho\varepsilon_1$ for ε_1 small enough. The same reasoning as before shows that $\|I^{-1}\| \leqq 1 + 2\rho\varepsilon_1$. For ε_1 small enough $\|I\|\,\|I^{-1}\|$ will be $< 1 + \varepsilon$, which implies the statement for a fixed dimension N.

Let C_1 be the space $l^1(G_n)$ of all sequences $x = (x_n)$, $x_n \in G_n$, $\|x\| = \sum_{n=1}^{\infty} \|x_n\| < \infty$. Its strong dual is the space $C'_1 = l^\infty(G'_n)$ of all $u = (u_n)$, $u_n \in G'_n$, $\|u\| = \sup_n \|u_n\| < \infty$.

We need the following lemma of JOHNSON:

(2) *Let E, F be* (B)*-spaces and $A \in \mathfrak{L}(E, F)$. Let F_α, $\alpha \in \mathsf{A}$, be a net of subspaces of F, directed by inclusion, such that $\bigcup_\alpha F_\alpha = F_0$ is dense in F. Assume, further, that for every α there exists $B_\alpha \in \mathfrak{L}(F_\alpha, E)$ such that $AB_\alpha = I_{F_\alpha}$ and $\limsup_\alpha \|B_\alpha\| = \lambda < \infty$.*

Then A' is an isomorphism of F' into E' with inverse S, $\|S\| \leqq \lambda$, and there exists a projection P of E' onto $A'(F')$ such that $\|P\| \leqq \lambda\|A\|$.

The method of proof is rather interesting; it uses a compactness argument going back to LINDENSTRAUSS.

We extend B_α to $\tilde{B}_\alpha$ defined on F_0 by setting $\tilde{B}_\alpha y = \mathsf{o}$ for $y \in F_0 \sim F_\alpha$. Then $\tilde{B}_\alpha$ is a noncontinuous and even nonlinear map of F_0 in E.

Let $\overline{\mathsf{K}}$ be the one-point compactification of the scalar field K. We define S_α by $(S_\alpha u)y = u(\tilde{B}_\alpha y)$ for every $u \in E'$ and every $y \in F_0$. S_α is a mapping of E' into $\overline{\mathsf{K}}^{F_0}$. The net S_α, $\alpha \in \mathsf{A}$, is contained in the compact space $(\overline{\mathsf{K}}^{F_0})^{E'}$ and has therefore an adherent point $\tilde{S}$. Thus for every $\alpha \in \mathsf{A}$ and every neighbourhood U of $\tilde{S}$ there exists an $\alpha' = \alpha'(\alpha, U) \geqq \alpha$ in A such that $S_{\alpha'}$ is contained in U. The set B of all $\beta = (\alpha, U)$ is directed by setting $(\alpha_1, U_1) \leqq (\alpha_2, U_2)$ if $\alpha_1 \leqq \alpha_2$ and $U_1 \supset U_2$. Hence all $S_{\alpha'(\beta)}$, $\beta \in \mathsf{B}$, form a net over B which converges to $\tilde{S}$. This implies

$$(\tilde{S}u)y = \lim_\beta (S_{\alpha'(\beta)}u)y = \lim_\beta u(\tilde{B}_{\alpha'(\beta)}y) \quad \text{for every } y \in F_0,\ u \in E',$$

where the limit is taken in $\overline{\mathsf{K}}$.

Now every y lies in some F_α and so $\tilde{B}_{\alpha'(\beta)}y = B_{\alpha'(\beta)}y$ for $\beta \geqq \beta_0$, where $\beta_0(y)$ is (α, U) for some U. It follows that

$$(\tilde{S}u)y = \lim_{\beta \geqq \beta_0(y)} u(B_{\alpha'(\beta)}y).$$

Recalling $\limsup_\alpha \|B_\alpha\| = \lambda < \infty$ we see that the limit is always finite and it follows also that $(\tilde{S}u)y$ is a bilinear form. More precisely,

$$|(\tilde{S}u)y| \leqq \lim_\beta |u(B_{\alpha'(\beta)}y)| \leqq \limsup_\alpha \|B_\alpha\|\,\|u\|\,\|y\|,$$

so $\tilde{S} \in \mathfrak{L}(E', F_0')$ and $\|\tilde{S}\| \leqq \lambda$. Extending every $\tilde{S}u$ from F_0 to F we obtain $S \in \mathfrak{L}(E', F')$ such that $\|S\| \leqq \lambda$.

Furthermore, $(\tilde{S}A'v)y = \lim_\beta (A'v)(B_{\alpha'(\beta)}y) = \lim v(AB_{\alpha'(\beta)}y) = vy$ for every $y \in F_0$, $v \in F'$, which implies $SA' = I_{F'}$.

Finally, $P = A'S$ is the projection of E' onto $A'(F')$, $\|P\| \leqq \lambda\|A\|$.

Now we prove the following universal property of C_1':

(3) *Let F be a separable* (B)*-space of infinite dimension. Then the strong dual F' is norm isomorphic to a complemented subspace H of C_1' and H is the range of a norm one projection of C_1'.*

Proof. Let $F_1 \subset F_2 \subset \cdots$ be a sequence of subspaces of F such that $\dim F_n = n$ and $\bigcup_{n=1}^{\infty} F_n = F_0$ is dense in F. By (1) there exists an n-dimensional space $G_{k(n)}$ and an isomorphism T_n of $G_{k(n)}$ onto F_n such that $\|T_n\| = 1$ and $\|T_n^{-1}\| \leqq 1 + 1/n$. Let $A \in \mathfrak{L}(C_1, F)$ be defined by $Ax = A(x_1, x_2, \ldots) = \sum_{n=1}^{\infty} T_n x_{k(n)}$, $x_n \in G_n$. Then $\|Ax\| \leqq \sum_{n=1}^{\infty} \|x_{k(n)}\| \leqq \|x\|$ implies $\|A\| \leqq 1$. We now define $B_n = T_n^{-1} \in \mathfrak{L}(F_n, C_1)$ and have $\lim_n \|B_n\| = 1$ and $AB_n = I_{F_n}$.

It follows from (2) that A' is an isomorphism of F' into C_1' with inverse S and that $\|A'\| = \|S\| = 1$, so A' is even a norm isomorphism of F' onto a subspace H of C_1'. The projection $P = A'S$ of C_1' onto H has norm one.

(4) *The* (B)*-space* C_1' *does not have the approximation property.*

We assume that C_1' has the approximation property. Then by (3) and 4.(1) every strong dual of a separable (B)-space has this property and by 4.(8) so does every separable (B)-space. This contradicts Enflo's result.

§ 44. The injective tensor product and the ε-product

1. Compatible topologies on $E \otimes F$. We introduced in § 41, 2.(4) the π-topology on the tensor product $E \otimes F$ of two locally convex spaces E and F. This was done in a rather natural way and we studied the properties of $E \otimes_\pi F$ and its completion $E \tilde{\otimes}_\pi F$. In § 43, 3. we were led to introduce the ε-tensor product $E \otimes_\varepsilon F$ and so we obtained a second topology on $E \otimes F$. Thus the problem arises of finding a nice class of topologies on the tensor products $E \otimes F$ of locally convex spaces which will contain the π- and the ε-topology as particular cases.

Following Grothendieck [13] we will say that a locally convex topology $\mathfrak{T}_\tau$ on $E \otimes F$ is compatible with the tensor product if it satisfies:

a) the canonical map χ of $E \times F$ into $(E \otimes F)[\mathfrak{T}_\tau] = E \otimes_\tau F$ is separately continuous;

b) every $u \otimes v$, $u \in E'$, $v \in F'$, is in $(E \otimes_\tau F)'$;

c) if $G_1 \subset E'$ is equicontinuous on E and $G_2 \subset F'$ is equicontinuous on F, then $G_1 \otimes G_2 \subset E' \otimes F'$ is equicontinuous on $E \otimes_\tau F$.

The meaning of a) is clear from

(1) *Condition* a) *implies* $(E \otimes_\tau F)' \subset \mathfrak{B}(E \times F)$, *the space of separately continuous bilinear forms.*

Proof. We assume a). Let x_0 be an element of E and W an absolutely convex neighbourhood of $\mathfrak{o}$ in $E \otimes_\tau F$. Then there exists a neighbourhood $V \ni \mathfrak{o}$ in F such that $\chi(x_0, V) = x_0 \otimes V \subset W$. Hence, if $\dot{B} \in (E \otimes_\tau F)'$ and $|\dot{B}(W)| \leqq \varepsilon$, then, using the notations of § 41, 1.(1), we have

$$|B(x_0, y)| = |\dot{B}\chi(x_0, y)| = |\dot{B}(x_0 \otimes y)| \leqq \varepsilon \quad \text{for all } y \in V$$

and $B(x, y)$ is continuous at $\mathfrak{o}$ in the second variable. Using the same argument for the first variable, it follows that B is a separately continuous bilinear form.

We note that conditions a) and b) together imply

$$(2) \qquad E' \otimes F' \subset (E \otimes_\tau F)' \subset \mathfrak{B}(E \times F).$$

Another definition of the compatible topologies on $E \otimes F$ is contained in

(3) *A locally convex topology $\mathfrak{T}_\tau$ on $E \otimes F$ is compatible if and only if it is a topology $\mathfrak{T}_{\mathfrak{M}}$ of uniform convergence on a class $\mathfrak{M}$ of subsets M of $\mathfrak{B}(E \times F)$ satisfying the following two conditions:*

α) every $M \in \mathfrak{M}$ is separately equicontinuous, i.e., for every $x_0 \in E$ the set $\tilde{M}(x_0)$ is equicontinuous in F' and for every $y_0 \in F$ the set $\tilde{\tilde{M}}(y_0)$ is equicontinuous in E';

β) $\mathfrak{M}$ contains all sets $G_1 \otimes G_2$, where G_1 and G_2 are equicontinuous subsets of E' and F', respectively.

Proof. i) We suppose that $\mathfrak{T}_\tau$ is compatible. A class $\mathfrak{M}$ defining $\mathfrak{T}_\tau$ consists of all equicontinuous subsets of $(E \otimes_\tau F)'$ and it follows from condition c) that condition β) is satisfied.

Let $\dot{M}$ be an equicontinuous subset of $(E \otimes_\tau F)'$; then $|\dot{M}(W)| \leqq 1$ for some neighbourhood W of $\mathfrak{o}$ in $E \otimes_\tau F$. It follows now from condition a) that we can choose an absolutely convex neighbourhood V of $\mathfrak{o}$ in F such that $\chi(x_0, V) \subset W$. Then $|\dot{M}(x_0 \otimes V)| \leqq 1$ or equivalently $|\tilde{M}(x_0)(V)| \leqq 1$, $\tilde{M}(x_0) \subset V^\circ$; hence $\tilde{M}(x_0)$ is equicontinuous in F'. This is condition α).

ii) Conversely, assume α) and β) to be satisfied. Obviously, β) implies b) and c) and $\mathfrak{T}_{\mathfrak{M}}$ is Hausdorff. It remains to prove a). It is sufficient to show that for a given $x_0 \in E$ and a given neighbourhood W of $\mathfrak{o}$ in $E \otimes_\tau F$ there exists a suitable neighbourhood V of $\mathfrak{o}$ in F such that $x_0 \otimes V \subset W$. We can assume $W = M^\circ$, where M is absolutely convex and separately equicontinuous by α). It follows that $\tilde{M}(x_0)$ is equicontinuous in F'; hence there exists a V such that $|\tilde{M}(x_0)V| \leqq 1$, $|M(x_0 \otimes V)| \leqq 1$, so that $x_0 \otimes V \subset W$.

The compatible topologies on $E \otimes F$ have the following important property:

(4) *Every subspace $x_0 \otimes F$, $x_0 \neq \mathfrak{o}$, of $E \otimes_\tau F$ is isomorphic to F; every subspace $E \otimes y_0$, $y_0 \neq \mathfrak{o}$, of $E \otimes_\tau F$ is isomorphic to E.*

Proof. We denote the topology of F by $\mathfrak{T}$. The subset $F_1 = \{(x_0, y); y \in F\}$ of $E \times F$ is obviously homeomorphic to $F[\mathfrak{T}]$. The map χ of F_1 onto $x_0 \otimes F$ is one-one and continuous by property a); hence $\mathfrak{T}_\tau \subset \mathfrak{T}$ on $x_0 \otimes F$. Conversely, let V be an absolutely convex closed neighbourhood of $\mathfrak{o}$ in F, $u_0 \in E'$, $u_0 x_0 = 1$; then $u_0 \otimes V^\circ$ is a $\mathfrak{T}_\tau$-equicontinuous subset of $E' \otimes F'$ and $(u_0 \otimes V^\circ)^\circ \cap (x_0 \otimes F) = x_0 \otimes V$ is a $\mathfrak{T}_\tau$-neighbourhood of $\mathfrak{o}$ in $x_0 \otimes F$. Thus $\mathfrak{T}_\tau \supset \mathfrak{T}$ on $x_0 \otimes F$ and $\mathfrak{T}_\tau = \mathfrak{T}$ implies the statement.

Obviously, there is a finest compatible topology on the tensor product $E \otimes F$ of two locally convex spaces, the topology of uniform convergence on all separately equicontinuous subsets of $\mathfrak{B}(E \times F)$. This topology is called the inductive tensor product topology $\mathfrak{T}_{in}$ and $E \otimes_{in} F$ is the inductive tensor product of E and F.

An immediate consequence of (1) and (3) is

(5) *$\mathfrak{T}_{in}$ is the finest locally convex topology $\mathfrak{T}$ on $E \otimes F$ such that the canonical map χ of $E \times F$ in $(E \otimes F)[\mathfrak{T}]$ is separately continuous. The dual $(E \otimes_{in} F)'$ can be identified with $\mathfrak{B}(E \times F)$.*

There is a close connection between the inductive and the projective tensor product. We recall (§ 41, 2.(4)) that $\mathfrak{T}_\pi$ is the finest locally convex topology $\mathfrak{T}_\tau$ on $E \otimes F$ such that χ is a continuous bilinear mapping of $E \times F$ into $E \otimes_\tau F$. This and the fact that every set $G_1 \otimes G_2$ is an equicontinuous set of linear forms on $E \otimes_\pi F$ show that $\mathfrak{T}_\pi$ is compatible with the tensor product and obviously weaker than $\mathfrak{T}_{in}$. Both topologies coincide in the following cases:

(6) *Let E and F be locally convex. The inductive and the projective tensor products $E \otimes_{in} F$ and $E \otimes_\pi F$ coincide if* a) *E and F are both barrelled and metrizable, or* b) *if E and F are both barrelled* (DF)*-spaces.*

Proof. The continuity theorems § 40, 2.(2) and § 40, 2.(11) assure that $\mathfrak{B}(E \times F) = \mathscr{B}(E \times F)$ in the cases a) and b) and that separately equicontinuous sets and equicontinuous sets of $\mathfrak{B}(E \times F)$ coincide. Recalling § 41, 3.(4), we see that this implies the identity of $\mathfrak{T}_\pi$ and $\mathfrak{T}_{in}$.

So far the inductive tensor product did not have many applications in analysis. For further details we refer the reader to GROTHENDIECK's thesis [13], p. 73.

2. The injective tensor product. Let $E \otimes F$ be the tensor product of two locally convex spaces. It follows from the definition that there exists a weakest compatible topology on $E \otimes F$, the topology $\mathfrak{T}_\varepsilon$ of uniform convergence on the class of all sets $G_1 \otimes G_2 \subset E' \otimes F'$, where G_1 and G_2 are equicontinuous subsets of E' and F', respectively. $\mathfrak{T}_\varepsilon$ is Hausdorff and is called the injective or ε-topology on $E \otimes F$ and $E \otimes_\varepsilon F$ is the injective tensor product or ε-tensor product of E and F. We will see in a moment that this notion coincides with the ε-tensor product we introduced in § 43, 3. Again $E \tilde{\otimes}_\varepsilon F$ will denote the completion of $E \otimes_\varepsilon F$.

Evidently we have

(1) *Let E and F be locally convex spaces. The topology of $E \otimes_\varepsilon F$ is determined by the system of semi-norms*

$$(2)\quad \varepsilon_{G_1,G_2}(z) = \sup_{u \otimes v \in G_1 \otimes G_2} |(u \otimes v)z| = \sup_{(u,v) \in G_1 \times G_2} \left| \sum_{i=1}^{n} (ux_i)(vy_i) \right|,$$

where $z = \sum_{i=1}^{n} x_i \otimes y_i \in E \otimes F$ and G_1, G_2 are equicontinuous subsets of E' and F', respectively.

If E and F are normed spaces, then $E \otimes_\varepsilon F$ has a natural norm, the ε-norm, defined by

$$(3)\quad \varepsilon(z) = \|z\|_\varepsilon = \sup_{\|u\| \leqq 1, \|v\| \leqq 1} |(u \otimes v)z| = \sup_{\|u\| \leqq 1, \|v\| \leqq 1} \left| \sum_{i=1}^{n} (ux_i)(vy_i) \right|.$$

If U, V are the closed unit balls of E and F, respectively, then $W = (U^\circ \otimes V^\circ)^\circ$ is the closed unit ball $\{z; \varepsilon(z) \leqq 1\}$ of $E \otimes_\varepsilon F$.

Let the topologies on E and F be given by the systems of semi-norms $\{p\}$ and $\{q\}$, respectively, and let G_1 be the polar of $U = \{x \in E; p(x) \leqq 1\}$ and $G_2 = V^\circ$, $V = \{y \in F; q(y) \leqq 1\}$; then we will also write

$$\varepsilon_{G_1,G_2}(z) = p \otimes_\varepsilon q(z).$$

We note further that for bases $\{U\}$, $\{V\}$ of absolutely convex neighbourhoods of $\circ$ in E and F, respectively, $\{W\}$ with $W = (U^\circ \otimes V^\circ)^\circ$ is a base of absolutely convex neighbourhoods of $\circ$ in $E \otimes_\varepsilon F$.

From (2) follows immediately

$$(4)\qquad p \otimes_\varepsilon q(x \otimes y) = p(x)q(y) \quad \text{for } x \in E,\ y \in F$$

(see the corresponding relation § 41, 2.(8) a) for the projective tensor norm).

(5) *If E and F are metrizable locally convex spaces with defining semi-norms $p_1 \leqq p_2 \leqq \cdots$ and $q_1 \leqq q_2 \leqq \cdots$, respectively, then $E \otimes_\varepsilon F$ is metrizable with defining semi-norms $p_1 \otimes_\varepsilon q_1 \leqq p_2 \otimes_\varepsilon q_2 \leqq \cdots$ and $E \tilde{\otimes}_\varepsilon F$ is an* (F)*-space.*

This is trivial and corresponds to § 41, 2.(7).

Let us remark that it is unknown whether $E \otimes_\varepsilon F$ or $E \tilde{\otimes}_\varepsilon F$ are always (DF)-spaces if E and F are (DF)-spaces, contrary to the situation for the π-tensor product (§ 41, 4.(7)).

Let us now establish the connection with § 43, 3. We know that $E \otimes F$ can be algebraically imbedded in $\mathfrak{B}(E'_s \times F'_s)$. The element B of $\mathfrak{B}(E'_s \times F'_s)$ corresponding to $\dot{B} = \sum x_i \otimes y_i$ is defined by $B(u, v) = \sum (ux_i)(vy_i)$. The topology $\mathfrak{T}_\varepsilon$ on $E \otimes F$ is given by the neighbourhoods

$$\{\dot{B} \in E \otimes F; |(G_1 \otimes G_2)\dot{B}| \leqq 1\}.$$

The bi-equicontinuous topology $\mathfrak{T}_e$ on $\mathfrak{B}(E'_s \times F'_s)$ is given by the neighbourhoods $\{B \in \mathfrak{B}(E'_s \times F'_s); |B(G_1, G_2)| \leqq 1\}$. Obviously, $\mathfrak{T}_e$ and $\mathfrak{T}_\varepsilon$ coincide on $E \otimes F$.

This implies that we may define $E \otimes_\varepsilon F$ also as the subspace $E \otimes F$ of $\mathfrak{B}(E'_s \times F'_s)$ equipped with the topology induced by $\mathfrak{T}_e$. This we did in § 43, 3.

We introduced there also the ε-product $E\varepsilon F$ of two locally convex spaces which is closely related to the ε-tensor product. For the convenience of the reader we recall some of the results of § 43, 3. which are fundamental in the study of the properties of the injective tensor product and the ε-product of two locally convex spaces.

$E\varepsilon F$ consists of all weakly continuous mappings of E' in F which map equicontinuous subsets of E' in relatively compact sets in F and the topology on $E\varepsilon F$ is $\mathfrak{T}_e$, the topology of uniform convergence on the equicontinuous subsets of E'.

$E \otimes_\varepsilon F$ is the subspace of $E\varepsilon F$ consisting of all weakly continuous mappings of finite rank. If E and F are complete spaces, then $E\varepsilon F$ is complete and contains $E \tilde{\otimes}_\varepsilon F$ as a subspace and $E \tilde{\otimes}_\varepsilon F = E\varepsilon F$ if E or F has the approximation property (§ 43, 3.(7)).

If E and F are (B)-spaces, then $E\varepsilon F$ can be identified with $\mathfrak{L}_b(E'_c, F)$, the space of all weakly continuous compact mappings of the (B)-space E' in F and $E \tilde{\otimes}_\varepsilon F$ is the closed subspace of $\mathfrak{L}_b(E'_c, F)$ consisting of all mappings which are $\mathfrak{T}_b$-limits of weakly continuous mappings of finite rank. We recall that the topology $\mathfrak{T}_b$ on $E\varepsilon F = \mathfrak{L}_b(E'_c, F)$ is defined by the norm $\|A\|$, $A \in \mathfrak{L}(E', F)$.

If, moreover, E or F has the approximation property, then every weakly continuous compact mapping is in $E \tilde{\otimes}_\varepsilon F$ (§ 43, 3.(7)).

The following result is a useful corollary.

(6) *Let E and F be* (B)-*spaces. Then $E'\varepsilon F$ is norm isomorphic to the subspace $\mathfrak{C}_b(E, F)$ of $\mathfrak{L}_b(E, F)$ consisting of all compact mappings and $E' \tilde{\otimes}_\varepsilon F$ is the space of all compact mappings which are $\mathfrak{T}_b$-limits of mappings of finite rank.*

If, moreover, E' or F has the approximation property, then $E' \tilde{\otimes}_\varepsilon F = \mathfrak{C}_b(E, F)$.

Proof. We recall that $E'\varepsilon F$ can be identified with $\mathfrak{L}_b((E'')_c, F)$, the space of all linear continuous compact mappings of E'' in F which are also $\mathfrak{T}_s(E')$–$\mathfrak{T}_s(F')$-continuous. Let A be in $\mathfrak{C}(E, F)$. Then A'' is a continuous and $\mathfrak{T}_s(E')$–$\mathfrak{T}_s(F')$-continuous map of E'' in F'' (§ 32, 2.(6)). From SCHAUDER's theorem (§ 42, 1.(7)) and § 42, 2.(1) it follows that A'' is a compact mapping of E'' in F; hence $A'' \in \mathfrak{L}_b((E'')_c, F)$.

Clearly, the map $J: A \to A''$ of $\mathfrak{C}_b(E, F)$ into $\mathfrak{L}_b((E'')_c, F)$ is one-one and a norm isomorphism because of $\|A''\| = \|A\|$. Finally, the map J is onto since every $A_0 \in \mathfrak{L}_b((E'')_c, F)$, $A_0 \neq o$, has a restriction $A \neq o$ to E which is in $\mathfrak{C}(E, F)$ and $A_0 = A''$ (§ 32, 2.(6)). This proves $\mathfrak{C}_b(E, F) = E'\varepsilon F$.

The remaining statements of (6) are immediate consequences of the remarks preceding (6) and of § 43, 1.(7).

There is another connection with the results of § 43. Since $\mathfrak{T}_\varepsilon$ is the weakest compatible topology on $E \otimes F$, the identity map ψ of $E \otimes_\pi F$ onto $E \otimes_\varepsilon F$ is continuous and its extension $\bar{\psi}$ to a map of $E \tilde{\otimes}_\pi F$ into $E \tilde{\otimes}_\varepsilon F$ is also continuous. If E and F are both complete locally convex spaces, then $E \tilde{\otimes}_\varepsilon F$ is a subspace of $\mathfrak{B}_e(E'_s \times F'_s)$ or $\mathfrak{L}_e(E'_k, F)$. In this case the problem whether $\bar{\psi}$ is one-one is identical with the problem treated in § 43, 2., where we have seen that the solution depends on the approximation property of the spaces involved.

We note that, in general, $E \tilde{\otimes}_\varepsilon F$ will not be contained in $E\varepsilon F$. It follows from 1.(4) that $E \otimes_\varepsilon \tilde{F} \subset E \tilde{\otimes}_\varepsilon F$; therefore in $E \tilde{\otimes}_\varepsilon F$ there will lie mappings with a range which is not contained in F, but $E\varepsilon F$ consists only of mappings of E' into F.

Similar to (5) is

(7) *If E and F are metrizable locally convex spaces, then $E\varepsilon F$ is metrizable; if E and F are* (F)*-spaces, then $E\varepsilon F$ is an* (F)*-space.*

We leave the proof to the reader.

We close with the following useful proposition of SCHWARTZ:

(8) *If E and F are quasi-complete locally convex spaces, then $E\varepsilon F$ is quasi-complete.*

Proof. Let $\tilde{B}_\alpha$, $\alpha \in \mathrm{A}$, be a Cauchy net on a bounded subset N of $\mathfrak{L}_e(E'_{co}, F)$ (§ 43, 3.). Then by § 39, 1.(5) $\tilde{B}_\alpha u$ is a Cauchy net on the bounded subset $N(u)$ of F for every $u \in E'$ and has a limit $\tilde{B}_0 u \in F$ by assumption. Hence $\tilde{B}_0 \in L(E', F)$.

By § 43, 3.(2) it will be sufficient to show that $\tilde{B}_0$ is weakly continuous and that it maps every equicontinuous subset M of E' in a relatively compact subset of F.

By § 43, 3.(3′) $E\varepsilon F$ is isomorphic to $F\varepsilon E$ and this isomorphism takes $\tilde{B}_\alpha$ into its adjoint $\tilde{B}'_\alpha = \tilde{\tilde{B}}_\alpha \in \mathfrak{L}_e(F'_{co}, E)$; hence $\tilde{\tilde{B}}_\alpha$ is a Cauchy net on N' and, as before, $\tilde{\tilde{B}}_\alpha v$ has a limit $\tilde{\tilde{B}}_0 v \in E$ for every $v \in F'$ and $\tilde{\tilde{B}}_0 \in L(F', E)$. From $v(\tilde{B}_\alpha u) = (\tilde{\tilde{B}}_\alpha v)u$ follows $v(\tilde{B}_0 u) = (\tilde{\tilde{B}}_0 v)u$ or $\tilde{\tilde{B}}_0 = \tilde{B}'_0$ and it maps F' into E. Now § 20, 4.(1) implies that $\tilde{B}_0$ is weakly continuous; thus $\tilde{B}_0 \in \mathfrak{L}(E'_s, F_s)$.

$\mathfrak{L}_e(E'_{co}, F)$ is a subspace H of $\mathfrak{L}_e(E'_s, F_s)$. It follows from the equivalence of a) and c) in § 43, 3.(2) that § 42, 1.(3) can be applied and we see that $\tilde{B}_0(M)$ is precompact in F for every equicontinuous M. Since F is quasi-complete, it follows that $\tilde{B}_0(M)$ is relatively compact.

3. Relatively compact subsets of $E\varepsilon F$ and $E \tilde{\otimes}_\varepsilon F$. We start with some simple observations on bounded subsets of $E\varepsilon F$.

We recall from § 43, 3. that $\varepsilon(E, F) \cong E\varepsilon F$ can be written as $\mathfrak{X}_e^{(\mathfrak{E},\mathfrak{E})}(E'_{co} \times F'_{co})$, the space of ε-hypocontinuous bilinear forms on $E'_{co} \times F'_{co}$ with the topology $\mathfrak{T}_e$ of uniform convergence on the sets $M \times N$, where M and N are equicontinuous subsets of E' and F', respectively. One has the isomorphisms § 43, 3.(3) and if $B \in \mathfrak{X}$, then $\tilde{B}$ and $\tilde{\tilde{B}} = \tilde{B}'$ are the corresponding elements in $\mathfrak{L}_e(E'_{co}, F)$ and $\mathfrak{L}_e(F'_{co}, E)$, respectively.

(1) *Let E and F be locally convex and M and N arbitrary absolutely convex equicontinuous subsets of E' and F', respectively. A subset H of $\mathfrak{X}_e^{(\mathfrak{E},\mathfrak{E})}(E'_{co} \times F'_{co}) = \varepsilon(E, F)$ is bounded if and only if one of the following equivalent conditions is satisfied:*

a) $|H(M, N)| = k(M, N) < \infty$ *for every pair M, N;*

b) $\tilde{H}(M)$ *is a bounded subset of F for every M;*

c) $\tilde{\tilde{H}}(N)$ *is a bounded subset of E for every N;*

d) *H is an ε-equihypocontinuous subset of $\mathfrak{X}_e^{(\mathfrak{E},\mathfrak{E})}(E'_b \times F'_b)$.*

Proof. a) is an immediate consequence of the definition of the topology $\mathfrak{T}_e$ on $\varepsilon(E, F)$; b) and c) are obviously equivalent formulations of a) in $E\varepsilon F$ and $F\varepsilon E$, respectively. It remains to prove the equivalence of d) with the boundedness of H.

i) We show first that $\mathfrak{X}^{(\mathfrak{E},\mathfrak{E})}(E'_{co} \times F'_{co}) \subset \mathfrak{X}^{(\mathfrak{E},\mathfrak{E})}(E'_b \times F'_b)$: A bilinear and separately continuous form B on $E'_{co} \times F'_{co}$ is ε-hypocontinuous if for given equicontinuous sets M and N in E' and F', respectively, there exist always $\mathfrak{T}_{co}$-neighbourhoods U, V of $\circ$ in E' and F' such that $|B(M, V)| \leqq 1$ and $|B(U, N)| \leqq 1$.

Since every $\mathfrak{T}_{co}$-neighbourhood of $\circ$ is a $\mathfrak{T}_b$-neighbourhood of $\circ$, it is obvious that B is also separately continuous and ε-hypocontinuous on $E'_b \times F'_b$.

ii) Let $H \subset \varepsilon(E, F)$ be ε-equihypocontinuous in $\mathfrak{X}^{(\mathfrak{E},\mathfrak{E})}(E'_b \times F'_b)$. Then there exist $\mathfrak{T}_b$-neighbourhoods V_1 of $\circ$ in F' such that $|H(M, V_1)| \leqq 1$ for a given equicontinuous M in E'. An arbitrary equicontinuous $N \subset F'$ is strongly bounded; hence there exists $k > 0$ such that $N \subset kV_1$ and $|H(M, N)| \leqq |H(M, kV_1)| \leqq k < \infty$ for every N. This implies a).

iii) Assume, conversely, that H is bounded in $\varepsilon(E, F)$. By b) $\tilde{H}(M)$ is a bounded set B_2 in F. It follows that $|\langle \tilde{H}(M), B_2^\circ \rangle| = |H(M, B_2^\circ)| \leqq 1$. Similarly, $B_1 = \tilde{\tilde{H}}(N)$ is bounded in E and $|H(B_1^\circ, N)| \leqq 1$. Both inequalities together imply the ε-equihypocontinuity of H in $\mathfrak{X}^{(\mathfrak{E},\mathfrak{E})}(E'_b \times F'_b)$.

We note that for (B)-spaces E, F a subset H of $E \tilde{\otimes}_\varepsilon F$ or $E\varepsilon F$ is obviously bounded if and only if the elements of H are uniformly bounded in norm.

We give now a characterization of the relatively compact subsets of $E\varepsilon F$ and $E \tilde{\otimes}_\varepsilon F$, essentially due to SCHWARTZ [3′], p. 22.

(2) *Let E, F be locally convex and quasi-complete and H a subset of $\varepsilon(E, F) = \mathfrak{X}_e^{(\mathfrak{E},\mathfrak{E})}(E'_{co} \times F'_{co})$. Then the following statements are equivalent:*

i) *H is an ε-equihypocontinuous subset of $\varepsilon(E, F)$;*

ii) *$\tilde{H}$ is equicontinuous in $E\varepsilon F = \mathfrak{L}_e(E'_{co}, F)$ and $\tilde{\tilde{H}}$ is equicontinuous in $F\varepsilon E = \mathfrak{L}_e(F'_{co}, E)$;*

iii) *$\tilde{H}(M)$ is relatively compact in F for every equicontinuous subset M of E' and $\tilde{\tilde{H}}(N)$ is relatively compact in E for every equicontinuous subset N of F';*

iv) *$\tilde{H}$ is relatively compact in $E\varepsilon F$.*

Proof. a) i) and iii) are equivalent: i) means that for given absolutely convex, weakly closed equicontinuous subsets M and N of E' and F', respectively, there exist absolutely convex closed neighbourhoods V and W of $\circ$ in E'_{co} and F'_{co}, respectively, such that

$$(3) \qquad |H(M, W)| \leqq 1, \qquad |H(V, N)| \leqq 1.$$

Since $V = C^\circ$, $W = D^\circ$, where C and D are absolutely convex and compact in E and F, respectively, the inequalities (3) are equivalent to $\tilde{H}(M) \subset D$ and $\tilde{\tilde{H}}(N) \subset C$, and this is iii). Conversely, if iii) holds, $D \supset \tilde{H}(M)$ and $C \supset \tilde{\tilde{H}}(N)$ can always be chosen absolutely convex and compact since E and F are quasi-complete; hence (3) follows from iii).

b) ii) and iii) are equivalent: By § 39, 3.(4) $\tilde{H}$ is equicontinuous if and only if for every equicontinuous $N \subset F'$ the set $\tilde{H}'(N) = \tilde{\tilde{H}}(N)$ is equicontinuous in $(E'_{co})' = E$. This is equivalent to $\tilde{\tilde{H}}(N) \subset C$, where C is absolutely convex and compact in E. This is for a quasi-complete E the second condition of iii). Similarly, the equicontinuity of $\tilde{\tilde{H}}$ is equivalent to the first condition of iii).

c) iv) implies iii): Let $\tilde{H}$ be a compact subset of $\mathfrak{L}_e(E'_{co}, F)$ and M an absolutely convex, weakly closed equicontinuous subset of E'. We equip M with the topology $\mathfrak{T}_{co}(E)$. Then M is compact for $\mathfrak{T}_{co}(E)$ since it is $\mathfrak{T}_s(E)$-compact (§ 21, 6.(3)). $\tilde{H}(M)$ will be compact in F if the mapping $J(\tilde{B}, u) = \tilde{B}u$ of $\mathfrak{L}_e(E'_{co}, F) \times M[\mathfrak{T}_{co}(E)]$ in F is continuous.

Let N° be a given neighbourhood of $\mathfrak{o}$ in F and $\tilde{B}_0 \in \mathfrak{L}$, $u_0 \in M$ fixed. We take as the neighbourhood of $\tilde{B}_0$ the set $\tilde{B}_0 + U$, where

$$U = \{\tilde{B}; \tilde{B}(M) \subset \tfrac{1}{2}N^\circ\}.$$

There exists an absolutely convex compact set $C \subset E$ such that $\tilde{B}_0(C^\circ) \subset \frac{1}{2}N^\circ$ and we take $(u_0 + C^\circ) \cap M$ as a $\mathfrak{T}_{co}$-neighbourhood of u_0 in M. The continuity of J in $(\tilde{B}_0, u_0)$ is now a consequence of

$$\begin{aligned} J(\tilde{B}_0 + U, (u_0 + C^\circ) \cap M) &\subset \tilde{B}_0((u_0 + C^\circ) \cap M) + U((u_0 + C^\circ) \cap M) \\ &\subset \tilde{B}_0 u_0 + \tilde{B}_0(C^\circ) + U(M) \subset \tilde{B}_0 u_0 + N^\circ. \end{aligned}$$

Since $E\varepsilon F$ is isomorphic to $F\varepsilon E$, iv) implies that $\tilde{\tilde{H}}$ is relatively compact in $F\varepsilon E$ and it follows similarly that $\tilde{\tilde{H}}(N)$ is relatively compact for every equicontinuous $N \subset F'$, which is iii).

d) iii) implies iv): We recall from § 39, 1. that $\mathfrak{L}_s(E'_{co}, F) \subset L_s(E', F) = F^{\mathsf{A}}$, where $\mathfrak{T}_s$ is the simple topology on $L(E', F)$ and A the index set of a linear basis $\{u_\alpha\}$, $\alpha \in \mathsf{A}$, of E'. Every set $\tilde{H}(u_\alpha)$ is relatively compact in F by iii); hence $\tilde{H}$, as contained in the topological product of these sets in F^{A}, is relatively compact in F^{A}. By ii) $\tilde{H}$ is equicontinuous in $\mathfrak{L}(E'_{co}, F)$ and by § 39, 4.(3) the $\mathfrak{T}_s$-closure $\bar{\tilde{H}}$ of $\tilde{H}$ in F^{A} is contained in $\mathfrak{L}(E'_{co}, F)$; hence $\bar{\tilde{H}}$ is equicontinuous and $\mathfrak{T}_s$-compact in $\mathfrak{L}(E'_{co}, F)$.

But $\bar{\tilde{H}}$ is also $\mathfrak{T}_c$-compact by § 39, 4.(2). We have to prove that $\bar{\tilde{H}}$ is $\mathfrak{T}_e$-compact in $\mathfrak{L}(E'_{co}, F)$ and this will be obvious if we show $\mathfrak{T}_c \supset \mathfrak{T}_e$ on $\mathfrak{L}(E'_{co}, F)$.

Let N be a weakly closed equicontinuous subset of E'. Then N is weakly compact and $\mathfrak{T}_{co}$-compact. Thus the class of all equicontinuous subsets of E' is a subclass of the class of all precompact subsets of E'_{co} and $\mathfrak{T}_c \supset \mathfrak{T}_e$ on $\mathfrak{L}(E'_{co}, F)$.

For complete locally convex spaces E and F the completed ε-tensor product $E \tilde{\otimes}_\varepsilon F$ is a closed subspace of the complete ε-product $E\varepsilon F$ (§ 43, 3.(5)). Hence we have the following corollary to (2):

(4) *Let E and F be complete locally convex spaces. A subset H of $E \tilde{\otimes}_\varepsilon F$ is relatively compact if and only if the following condition is satisfied:*

iii) *$\tilde{H}(M)$ is relatively compact in F for every equicontinuous subset M of E' and $\tilde{\tilde{H}}(N)$ is relatively compact in E for every equicontinuous subset N of F'.*

Note. The following example shows that in (2) in ii) and iii) we need both conditions. Let E be a (B)-space. Following the remarks in the proof of § 43, 4.(10), the spaces $(E'_c)'_c$ and E can be identified. We consider $\mathfrak{L}((E'_c)'_c, E) = \mathfrak{L}(E)$. The set $\tilde{H} = \{\tilde{B} \in \mathfrak{L}(E); \|\tilde{B}\| \leqq 1\}$ is equicontinuous in $\mathfrak{L}(E)$, but for every $x \neq \circ$, $x \in E$, the set $\tilde{H}(x)$ is obviously not relatively compact in E. From b) in the proof of (2) it follows that $\tilde{\tilde{H}}$ is not equicontinuous in $\mathfrak{L}(E'_c, E'_c)$. Similarly, only the second condition of iii) is satisfied.

We have $E \tilde{\otimes}_\varepsilon F \cong F \tilde{\otimes}_\varepsilon E$ and $E \tilde{\otimes}_\varepsilon (F \tilde{\otimes}_\varepsilon G) \cong (E \tilde{\otimes}_\varepsilon F) \tilde{\otimes}_\varepsilon G$ for the completed ε-tensor product. The isomorphism $E\varepsilon F \cong F\varepsilon E$ was stated in § 43, 3.(3′). That the ε-product is also associative will be proved now.

(5) *Let E_1, E_2, E_3 be locally convex. Then $E_1\varepsilon(E_2\varepsilon E_3)$ is isomorphic to $(E_1\varepsilon E_2)\varepsilon E_3$. In the case of* (B)*-spaces there exists even a norm isomorphism.*

Proof. Let $T(u_1, u_2, u_3)$ be a trilinear form on $(E_1)'_{co} \times (E_2)'_{co} \times (E_3)'_{co}$. Such a T is called ε-hypocontinuous if the following conditions are satisfied: To given equicontinuous subsets $M_1 \subset E'_1$, $M_2 \subset E'_2$ there exists a neighbourhood W_3 of $\circ$ in $(E_3)'_{co}$ such that $|T(M_1, M_2, W_3)| \leqq 1$; similarly, there exist W_1, W_2 such that $|T(W_1, M_2, M_3)| \leqq 1$ and $|T(M_1, W_2, M_3)| \leqq 1$ for given M_1, M_2, M_3.

We denote by $\varepsilon(E_1, E_2, E_3)$ the space of all ε-hypocontinuous trilinear forms on $(E_1)'_{co} \times (E_2)'_{co} \times (E_3)'_{co}$ equipped with the topology $\mathfrak{T}_e$ of uniform convergence on all products $M_1 \times M_2 \times M_3$ of equicontinuous sets. A fundamental set of $\mathfrak{T}_e$-neighbourhoods of $\circ$ is given by the sets $\{T; |T(M_1, M_2, M_3)| \leqq 1\}$ with M_1, M_2, M_3 given.

Since $\varepsilon(E_1, E_2, E_3)$ is symmetric in its arguments and since $E\varepsilon F \cong \varepsilon(E, F)$, it will be sufficient to prove that $\varepsilon(E_1, E_2, E_3)$ and $E_1\varepsilon(\varepsilon(E_2, E_3)) = \mathfrak{L}_e((E_1)'_{co}, \varepsilon(E_2, E_3))$ are topologically isomorphic.

To every $T(u_1, u_2, u_3) \in \varepsilon(E_1, E_2, E_3)$ corresponds a $\tilde{T}$ which maps $u_1 \in E'_1$ into a bilinear form $\tilde{T}u_1$ on $E'_2 \times E'_3$ defined by $(\tilde{T}u_1)(u_2, u_3) = T(u_1, u_2, u_3)$. It follows from $|(\tilde{T}u_1)(M_2, W_3)| = |T(u_1, M_2, W_3)| \leqq 1$ and $|(\tilde{T}(u_1))(W_2, M_3)| \leqq 1$ that $\tilde{T}u_1 \in \varepsilon(E_2, E_3)$. Obviously, $\tilde{T}$ is linear on E'_1 and it follows from $|T(W_1, M_2, M_3)| = |(\tilde{T}(W_1))(M_2, M_3)| \leqq 1$ that $\tilde{T}$ maps W_1 in the $\mathfrak{T}_e$-neighbourhood $\{B; |B(M_2, M_3)| \leqq 1\}$ of $\circ$ in $\varepsilon(E_2, E_3)$. Hence $\tilde{T}$ corresponding to T is an element of $\mathfrak{L}((E_1)'_{co}, \varepsilon(E_2, E_3))$.

Conversely, every $\tilde{T} \in \mathfrak{L}((E_1)'_{co}, \varepsilon(E_2, E_3))$ defines a trilinear form $T(u_1, u_2, u_3) = (\tilde{T}u_1)(u_2, u_3)$. We have to show that T is ε-hypocontinuous. It is obvious that for given M_2, M_3 there exists W_1 such that $|T(W_1, M_2, M_3)| \leqq 1$. We prove the two other conditions. By § 43, 3.(2) a) $\tilde{T}$ maps a set M_1 into a relatively compact subset of $\varepsilon(E_2, E_3)$ which is by (2) ε-equihypocontinuous. Hence there exist W_2, W_3 such that $|(\tilde{T}(M_1))(M_2, W_3)| \leqq 1$ and $|(\tilde{T}(M_1))(W_2, M_3)| \leqq 1$, which implies the ε-hypocontinuity of T. Thus every T corresponding to a $\tilde{T}$ is in $\varepsilon(E_1, E_2, E_3)$.

Finally, it follows from the equivalence of $|T(M_1, M_2, M_3)| \leqq 1$ and $|(\tilde{T}(M_1))(M_2, M_3)| \leqq 1$ that the topologies $\mathfrak{T}_e$ on both spaces coincide. If E_1, E_2, E_3 are (B)-spaces, then the spaces are in this way even norm isomorphic.

The bounded subsets of an ε-product were determined in (1). For some classes of locally convex spaces sharper results can be obtained.

Let B_1, B_2 be absolutely convex bounded subsets of the locally convex spaces E and F, respectively, B_1°, B_2° their polars in E' and F', respectively. Using the dual pair $\langle E' \otimes F', \varepsilon(E, F)\rangle$ with the bilinear form $\left\langle \sum_{i=1}^{n} u_i \otimes v_i, B \right\rangle = \sum_{i=1}^{n} B(u_i, v_i)$, one identifies $(B_1^\circ \otimes B_2^\circ)^\circ$ with the set $\{B \in \varepsilon(E, F); \sup_{u \in B_1^\circ, v \in B_2^\circ} |B(u, v)| \leqq 1\} = \{B \in \varepsilon(E, F); |B(B_1^\circ, B_2^\circ)| \leqq 1\}$. Hence $(B_1^\circ \otimes B_2^\circ)^\circ$ is obviously an equicontinuous set of bilinear forms on $E_b' \times E_b'$. Since an equicontinuous set in E' or F' is always strongly bounded (§ 21, 5.(1)), $(B_1^\circ \otimes B_2^\circ)^\circ$ is also ε-equihypocontinuous on $E_b' \times F_b'$, thus bounded in $\varepsilon(E, F)$ by (1). Conversely, a subset of $\varepsilon(E, F)$ which is an equicontinuous set of bilinear forms on $E_b' \times F_b'$ is always contained in some $(B_1^\circ \otimes B_2^\circ)^\circ$. Therefore

(6) *Let E, F be locally convex. The class of all sets $(B_1^\circ \otimes B_2^\circ)^\circ$ and their subsets (where B_1, B_2 are bounded subsets of E and F, respectively) coincides with the class of all bounded subsets of $\varepsilon(E, F)$ which are equicontinuous sets of bilinear forms on $E_b' \times F_b'$.*

The following result is due to R. Hollstein.

(7) *Let E, F be* (F)*-spaces. The bounded subsets of $\varepsilon(E, F)$ or of $E \tilde{\otimes}_\varepsilon F$ coincide with the subsets of the sets $(B_1^\circ \otimes B_2^\circ)^\circ$, where B_1 and B_2 are arbitrary absolutely convex bounded subsets of E and F, respectively, and where the polar of $B_1^\circ \otimes B_2^\circ$ is taken in $\varepsilon(E, F)$ resp. $E \tilde{\otimes}_\varepsilon F$.*

Proof. By (1) a bounded subset H of $\varepsilon(E, F)$ is an ε-equihypocontinuous subset of $\mathfrak{X}^{(\mathfrak{E},\mathfrak{E})}(E_b' \times F_b')$. Since E and F are barrelled, the equicontinuous subsets of E' and F' coincide with the weakly bounded subsets and by § 20, 11.(3) with the bounded subsets of E_b' and F_b'. Hence H is an equihypocontinuous subset of $\mathfrak{X}(E_b' \times F_b')$. Now E_b' and F_b' are (DF)-spaces and by § 40, 2.(10) H is equicontinuous on $E_b' \times F_b'$. The statement follows now from (6).

For (DF)-spaces one has similarly

(8) *Let E, F be complete* (DF)*-spaces, $B_1 \subset B_2 \subset \cdots$ and $C_1 \subset C_2 \subset \cdots$ fundamental sequences of bounded sets in E and F, respectively. Then*

$$(B_1^\circ \otimes C_1^\circ)^\circ \subset (B_2^\circ \otimes C_2^\circ)^\circ \subset \cdots$$

is a fundamental sequence of bounded sets in $\varepsilon(E, F)$ *resp.* $E \tilde{\otimes}_\varepsilon F$ *(the polars are taken as in* (7)).

Consequently, the strong duals of $E\varepsilon F$ *and of* $E \tilde{\otimes}_\varepsilon F$ *are metrizable.*

We prove this only for $\varepsilon(E, F) \cong E\varepsilon F$. A bounded subset H of $\varepsilon(E, F)$ is by (1) an ε-equihypocontinuous subset of $\mathfrak{X}^{(\mathfrak{C},\mathfrak{C})}(E'_b \times F'_b)$, where E'_b and F'_b are (F)-spaces (§ 29, 3.(1)). It follows from the definitions that an ε-equihypocontinuous set of bilinear forms is separately equicontinuous on $E'_b \times F'_b$. Using § 40, 2.(2), we see that H is equicontinuous and the statement follows now from (6).

A similar result holds for relatively compact subsets:

(9) *Let* E, F *be* (F)-*spaces. The relatively compact subsets of* $\varepsilon(E, F)$ *or* $E \tilde{\otimes}_\varepsilon F$ *are the subsets of the sets* $(C_1^\circ \otimes C_2^\circ)^\circ$, *where* C_1, C_2 *are absolutely convex and compact in* E *and* F, *respectively, and where the polar of* $C_1^\circ \otimes C_2^\circ$ *is taken in* $\varepsilon(E, F)$ *or in* $E \tilde{\otimes}_\varepsilon F$.

Proof. A relatively compact subset H of $\varepsilon(E, F)$ is by (2) an ε-equihypocontinuous set of bilinear forms on $E'_c \times F'_c$. Since the equicontinuous subsets of E' and F' are the bounded subsets, H is an equihypocontinuous set of bilinear forms on $E'_c \times F'_c$. By a theorem of HOLLSTEIN (§ 45, 3.(4)) H is equicontinuous on $E'_c \times F'_c$. Hence there exist absolutely convex and compact $C_1 \subset E$, $C_2 \subset F$ such that

$$H \subset (C_1^\circ \otimes C_2^\circ)^\circ = \left\{B \in \varepsilon(E, F);\ \sup_{u \in C_1^\circ, v \in C_2^\circ} |B(u, v)| \leqq 1\right\}.$$

Conversely, every set $K = (C_1^\circ \otimes C_2^\circ)^\circ$ is relatively compact: K is equicontinuous on $E'_c \times F'_c$, hence equihypocontinuous and, finally, relatively compact by (2).

If E and F are complete and if every bounded subset of E and F is relatively compact, then this is true also for $\varepsilon(E, F)$ and $E \tilde{\otimes}_\varepsilon F$, as follows immediately from (1) and (2). In particular,

(10) *If* E *and* F *are* (FM)-*spaces, then* $E\varepsilon F$ *and* $E \tilde{\otimes}_\varepsilon F$ *are* (FM)-*spaces.*

4. Tensor products of mappings. Let E_1, E_2, F_1, F_2 be locally convex, $A_1 \in \mathfrak{L}(E_1, F_1)$, $A_2 \in \mathfrak{L}(E_2, F_2)$. In § 41, 5. we defined the linear map $A_1 \otimes A_2$ of $E_1 \otimes E_2$ into $F_1 \otimes F_2$.

The following proposition corresponds to § 41, 5.(1) and (2).

(1) $A_1 \otimes A_2$ *is a continuous linear mapping of* $E_1 \otimes_\varepsilon E_2$ *into* $F_1 \otimes_\varepsilon F_2$ *and* $A_1 \otimes A_2$ *has a uniquely determined continuous extension* $A_1 \tilde{\otimes}_\varepsilon A$ *in* $\mathfrak{L}(E_1 \tilde{\otimes}_\varepsilon E_2, F_1 \tilde{\otimes}_\varepsilon F_2)$. *If all the spaces are* (B)-*spaces then* $\|A_1 \otimes A_2\| = \|A_1 \tilde{\otimes}_\varepsilon A_2\| = \|A_1\| \cdot \|A_2\|$.

Proof. a) Let G_1, G_2 be absolutely convex equicontinuous subsets of F_1' and F_2', respectively. Then $A_1'(G_1)$ and $A_2'(G_2)$ are equicontinuous subsets of E_1' and E_2', respectively (§ 32, 1.(10)). One has

$$\langle A_1'v_1 \otimes A_2'v_2, \sum_{i=1}^{n} x_i \otimes y_i\rangle = \langle v_1 \otimes v_2, \sum A_1x_i \otimes A_2y_i\rangle$$

for all $x_i \in E_1$, $y_i \in E_2$, $v_1 \in G_1$, $v_2 \in G_2$, and this implies that for every $z = \sum_{i=1}^{n} x_i \otimes y_i \in (A_1'(G_1) \otimes A_2'(G_2))^\circ$ the image $(A_1 \otimes A_2)z$ is contained in $(G_1 \otimes G_2)^\circ \subset F_1 \otimes_\varepsilon F_2$; hence $A_1 \otimes A_2$ is continuous.

b) For (B)-spaces one has

$$\begin{aligned}\|A_1 \otimes A_2\| &= \sup_{\|v_1\|,\|v_2\|\leqq 1,\|z\|_\varepsilon\leqq 1} |\langle v_1 \otimes v_2, (A_1 \otimes A_2)z\rangle| \\ &= \sup |\langle A_1'v_1 \otimes A_2'v_2, z\rangle| \leqq \|A_1\| \cdot \|A_2\|\end{aligned}$$

and, conversely,

$$\|A_1 \otimes A_2\| \geqq \sup_{\|x_1\|,\|x_2\|,\|v_1\|,\|v_2\|\leqq 1} |\langle v_1 \otimes v_2, A_1x_1 \otimes A_2x_2\rangle| = \|A_1\| \cdot \|A_2\|.$$

We give another interpretation of the map $A_1 \otimes A_2$. We recall that $E_1 \otimes_\varepsilon E_2$ can be identified with $\mathfrak{F}_e(E_1', E_2)$, the space of all weakly continuous mappings of finite rank of E_1' into E_2.

If $z = \sum_{i=1}^{n} x_i^{(1)} \otimes x_i^{(2)}$ is an element of $E_1 \otimes_\varepsilon E_2$, then the corresponding $Z \in \mathfrak{F}_e(E_1', E_2)$ is defined by $Zu = \sum_{i=1}^{n} (ux_i^{(1)})x_i^{(2)}$, $u \in E_1'$. To $A_1 \otimes A_2$ corresponds the map

$$\begin{aligned}((A_1 \otimes A_2)Z)v &= \sum_{i=1}^{n} (v(A_1x_i^{(1)}))(A_2x_i^{(2)}) \\ &= \sum_{i=1}^{n} ((A_1'v)x_i^{(1)})(A_2x_i^{(2)}) = A_2ZA_1'v, \qquad v \in F_1'.\end{aligned}$$

Hence, if we consider $A_1 \otimes A_2$ as a map of $\mathfrak{F}_e(E_1', E_2)$ into $\mathfrak{F}_e(F_1', F_2)$, it has the form

$$(2) \qquad (A_1 \otimes A_2)Z = A_2ZA_1', \qquad Z \in \mathfrak{F}_e(E_1', E_2)$$

and is continuous by (1).

Formula (2) suggests that the domain of definition of $A_1 \otimes A_2$ may be enlarged. Indeed one has

(3) *For all $Z \in E_1\varepsilon E_2$ resp. for all $Z \in \mathfrak{L}_e((E_1)_k', E_2)$, (2) defines a continuous linear map of $E_1\varepsilon E_2$ in $F_1\varepsilon F_2$ resp. of $\mathfrak{L}_e((E_1)_k', E_2)$ in $\mathfrak{L}_e((F_1)_k', F_2)$.*

Proof. a) We verify first that $A_1' \in \mathfrak{L}((F_1')_{co}, (E_1')_{co})$ if $A_1 \in \mathfrak{L}(E_1, F_1)$. Let C be absolutely convex and compact in E_1; then $D = A_1(C)$ has the same properties in F_1, and from $A_1(C) \subset D$ follows $A_1'(D^\circ) \subset C^\circ$, which is the statement.

If $Z \in E_1 \varepsilon E_2 = \mathfrak{L}_e((E_1)'_{co}, E_2)$, it follows now that $A_2 Z A_1'$ is the product of three continuous mappings, hence is continuous from $(F_1)'_{co}$ in F_2 or $A_2 Z A_1' \in F_1 \varepsilon F_2$.

Similar arguments with $\mathfrak{T}_k$ instead of $\mathfrak{T}_{co}$ show $A_2 Z A_1' \in \mathfrak{L}_e((F_1)'_k, F_2)$ for $Z \in \mathfrak{L}_e((E_1)'_k, E_2)$.

b) It remains to prove the continuity of $A_1 \otimes A_2$ in both cases. A $\mathfrak{T}_e$-neighbourhood of o in $\mathfrak{L}_e((F_1)'_{co}, F_2)$ is defined by $W = \{Y; Y(M) \subset V\}$, where M is an equicontinuous subset of F_1' and V a o-neighbourhood in F_2. Now $N = A_1'(M)$ is equicontinuous in E_1' and there exists a o-neighbourhood U in E_2 such that $A_2(U) \subset V$. But then the $A_1 \otimes A_2$-image of $W_1 = \{Z; Z(N) \subset U\}$ is contained in W since $A_2 Z A_1'(M) \subset V$.

The same argument settles the second case.

It is natural to introduce the notation $A_1 \varepsilon A_2$ for the map defined by (2) from $E_1 \varepsilon E_2$ into $F_1 \varepsilon F_2$.

The second mapping of (3) has another interpretation. We recall from § 40, 4.(5) that $\mathfrak{L}_e((E_1)'_k, E_2)$ is isomorphic to $\mathfrak{B}_e((E_1)'_s \times (E_2)'_s)$. Let $\tilde{B} \in \mathfrak{L}_e$ and $B \in \mathfrak{B}_e$ be corresponding elements. The bilinear form in $\mathfrak{B}_e((F_1)'_s \times (F_2)'_s)$ corresponding to $(A_1 \otimes A_2)\tilde{B} \in \mathfrak{L}_e((F_1)'_k, F_2)$ is given by

$$v_2((A_2 \tilde{B} A_1')v_1) = (A_2' v_2)(\tilde{B}(A_1' v_1)) = B(A_1' v_1, A_2' v_2)$$

(using (2) and § 40, 1.(1')).

Introducing the notation $A_1 \boxtimes A_2$ for the second case of (3) and the isomorphic situation for bilinear forms, we have

$$\text{(4)} \qquad ((A_1 \boxtimes A_2)B)(v_1, v_2) = B(A_1' v_1, A_2' v_2),$$

where $A_1 \in \mathfrak{L}(E_1, F_1)$, $A_2 \in \mathfrak{L}(E_2, F_2)$, $B \in \mathfrak{B}_e((E_1)'_s \times (E_2)'_s)$, $v_1 \in F_1'$, $v_2 \in F_2'$; hence $(A_1 \boxtimes A_2)B \in \mathfrak{L}_e(F_1)'_s \times (F_2)'_s$.

We also write $A_1 \otimes_\varepsilon A_2$ for $A_1 \otimes A_2$ considered as a map of $E_1 \otimes_\varepsilon E_2$ in $F_1 \otimes_\varepsilon F_2$. With these notations we have

(5) *Let E_1, E_2, F_1, F_2 be locally convex, $A_1 \in \mathfrak{L}(E_1, F_1)$, $A_2 \in \mathfrak{L}(E_2, F_2)$. If A_1 and A_2 are one-one, then $A_1 \otimes_\varepsilon A_2$, $A_1 \varepsilon A_2$, and $A_1 \boxtimes A_2$ are one-one. If E_1, E_2, F_1, F_2 are complete, then also $A_1 \tilde{\otimes}_\varepsilon A_2$ is one-one.*

Proof. By § 43, 3. one has $E_1 \otimes_\varepsilon E_2 \subset E_1 \varepsilon E_2 \subset \mathfrak{L}_e((E_1)'_k, E_2) = \mathfrak{B}_e((E_1)'_s \times (E_2)'_s)$ and $E_1 \tilde{\otimes}_\varepsilon E_2 \subset E_1 \varepsilon E_2$ in the case of complete spaces. Hence all the mappings considered are restrictions of $A_1 \boxtimes A_2$ and it will be sufficient to prove that $A_1 \boxtimes A_2$ is one-one.

It follows from the assumption that $A_1'(F_1')$ and $A_2'(F_2')$ are weakly dense in E_1' and E_2', respectively. Let us assume that $((A_1 \boxtimes A_2)B)(v_1, v_2) = 0$ for all $v_1 \in F_1'$, $v_2 \in F_2'$. Since $B(u_1, u_2)$ is weakly continuous in each variable, (4) implies that $B(u_1, u_2) = 0$ for all $u_1 \in E_1'$, $u_2 \in E_2'$. Hence $A_1 \boxtimes A_2$ is one-one from $\mathfrak{B}_e((E_1)_s' \times (E_2)_s')$ in $\mathfrak{B}_e((F_1)_s' \times (F_2)_s')$.

An analogous important result is

(6) *Let E_1, E_2, F_1, F_2 be locally convex, $A_1 \in \mathfrak{L}(E_1, F_1)$, $A_2 \in \mathfrak{L}(E_2, F_2)$. If A_1 and A_2 are monomorphisms, then $A_1 \otimes_\varepsilon A_2$, $A_1 \tilde{\otimes}_\varepsilon A_2$, $A_1 \varepsilon A_2$, $A_1 \boxtimes A_2$ are also monomorphisms.*

If E_1, E_2, F_1, F_2 are normed spaces and A_1 and A_2 are norm isomorphisms in F_1 and F_2, respectively, then $A_1 \otimes_\varepsilon A_2$, $A_1 \tilde{\otimes}_\varepsilon A_2$, $A_1 \varepsilon A_2$, $A_1 \boxtimes A_2$ are also norm isomorphisms.

In particular, if H_1 and H_2 are subspaces of E_1 and E_2, respectively, then $H_1 \otimes_\varepsilon H_2$ can be identified with the subspace $H_1 \otimes H_2$ of $E_1 \otimes E_2$ equipped with the topology resp. norm induced by $E_1 \otimes_\varepsilon E_2$.

Proof. If $A_1 \otimes_\varepsilon A_2$ is a monomorphism resp. a norm isomorphism, then its continuous extension $A_1 \tilde{\otimes}_\varepsilon A_2$ has these properties too. Using the same argument as in the foregoing proof, we see that we have to consider again only the case $A_1 \boxtimes A_2$.

A_1 is a monomorphism if and only if A_1' maps the class of all equicontinuous subsets of F_1' onto the class of all equicontinuous subsets of E_1' (§ 32, 4.). Thus a $\mathfrak{T}_e$-neighbourhood of $\circ$ in $\mathfrak{B}_e((E_1)_s' \times (E_2)_s')$ can be assumed to be of the form $U = \{B; |B(A_1'(G_1), A_2'(G_2))| \leq 1\}$, where G_1, G_2 are equicontinuous subsets of F_1' and F_2', respectively. Using (4), we see that $B \in U$ if and only if $(A_1 \boxtimes A_2)B \in V = \{C; |C(G_1, G_2)| \leq 1\}$, where $C \in \mathfrak{B}_e((F_1)_s' \times (F_2)_s')$. Hence $A_1 \boxtimes A_2$ is open and a monomorphism by (3).

In the case of normed spaces we take as G_1, G_2 the closed unit balls in F_1', F_2'; then $A_1'(G_1)$ and $A_2'(G_2)$ are the closed unit balls in E_1' and E_2', U is the closed unit ball in $\mathfrak{B}_e((E_1)_s' \times (E_2)_s')$, and $(A_1 \boxtimes A_2)(U)$ is the closed unit ball in the range of $A_1 \boxtimes A_2$; in particular, we have

$$\|(A_1 \boxtimes A_2)B\| = \sup_{\|v_1\|, \|v_2\| \leq 1} |B(A_1'v_1, A_2'v_2)| = \sup_{\|u_1\|, \|u_2\| \leq 1} |B(u_1, u_2)| = \|B\|,$$

so that $A_1 \boxtimes A_2$ is a norm isomorphism.

Speaking of "injections" instead of "monomorphisms," the tensor products $A_1 \otimes_\varepsilon A_2$ and $A_1 \tilde{\otimes}_\varepsilon A_2$ of two injections are again injections and so (6) is the reason for using the term "injective tensor product" for the ε-tensor product.

We will see later in 4. that the product $A_1 \tilde{\otimes}_\varepsilon A_2$ of two homomorphisms onto will in general not be a homomorphism of $E_1 \tilde{\otimes}_\varepsilon E_2$ in $F_1 \tilde{\otimes}_\varepsilon F_2$. Thus we have a kind of dual behaviour of ε- and π-tensor products, since

$A_1 \tilde{\otimes}_\pi A_2$ for homomorphisms A_1 and A_2 onto is a homomorphism, whereas the product $A_1 \tilde{\otimes}_\pi A_2$ of two monomorphisms into is in general not a monomorphism into (see the results of § 41, 5.).

Before studying ε-tensor products of homomorphisms we prove a related but simpler result.

We recall that for an absolutely convex neighbourhood U of $\mathfrak{o}$ in a locally convex space E and the corresponding semi-norm $p(x)$ the space E_U is defined as the quotient $E/N[U]$ considered as a normed space with norm $\hat{p}(\hat{x}) = p(x)$, $\hat{x} = x + N[U]$, $N[U] = p^{(-1)}(0)$.

Consider similarly $F_V = F/N[V]$, the corresponding semi-norm being $q(y)$. As we remarked in 2., the pair U, V determines the neighbourhood $(U^\circ \otimes V^\circ)^\circ = \{z; p \otimes_\varepsilon q(z) \leqq 1\}$ in $E \otimes_\varepsilon F$. We describe the structure of the corresponding normed space in

(7) *$(E \otimes_\varepsilon F)_{(U^\circ \otimes V^\circ)^\circ}$ is norm isomorphic to $E_U \otimes_\varepsilon F_V$; in particular, $N[(U^\circ \otimes V^\circ)^\circ] = N[U] \otimes F + E \otimes N[V]$.*

Proof. We know from § 41, 1. that $E_U \otimes F_V$ is algebraically isomorphic to $(E \otimes F)/D$, $D = N[U] \otimes F + E \otimes N[V]$, in the following manner: Let $z = \sum_{i=1}^{n} x_i \otimes y_i$ be an element of $E \otimes F$, $\check{z} = \sum \hat{x}_i \otimes \hat{y}_i$ the corresponding element in $E_U \otimes F_V$, and $\hat{z}$ the residue class of z in $(E \otimes F)/D$; then $\check{z} \leftrightarrow \hat{z}$ is the algebraic isomorphism. This is even a norm isomorphism if we define $\|\hat{z}\| = \|\check{z}\|_\varepsilon$.

We obtain

$$\|\hat{z}\| = \|\check{z}\|_\varepsilon = \hat{p} \otimes_\varepsilon \hat{q}(\check{z}) = \sup_{u \in U^\circ, u \in V^\circ} \left| \sum (u\hat{x}_i)(v\hat{y}_i) \right|$$

$$= \sup_{u \in U^\circ, v \in V^\circ} \left| \sum (ux_i)(vy_i) \right| = p \otimes_\varepsilon q(z).$$

It follows immediately that $p \otimes_\varepsilon q(z) = 0$ if and only if $z \in D$; hence $D = N[(U^\circ \otimes V^\circ)^\circ]$ and $\|\hat{z}\|$ is the norm on $(E \otimes_\varepsilon F)_{(U^\circ \otimes V^\circ)^\circ}$, which proves the norm isomorphism.

We study now the product $A_1 \tilde{\otimes}_\varepsilon A_2$ of two homomorphisms onto.

Let E, F be (B)-spaces, K the canonical homomorphism of F onto a quotient F/H, I the identity map on E. Then $I \tilde{\otimes}_\varepsilon K$ is a continuous map of $E \tilde{\otimes}_\varepsilon F$ into $E \tilde{\otimes}_\varepsilon (F/H)$. Since $(I \otimes K)(\sum x_i \otimes y_i) = \sum x_i \otimes Ky_i$, $x_i \in E$, $y_i \in F$, the range of $I \tilde{\otimes}_\varepsilon K$ contains $E \otimes (F/H)$ and is therefore dense in $E \tilde{\otimes}_\varepsilon (F/H)$. Hence $I \tilde{\otimes}_\varepsilon K$ is a homomorphism if and only if its range is $E \tilde{\otimes}_\varepsilon (F/H)$.

Let us assume that E', the strong dual of E, has the approximation property. Then $E' \tilde{\otimes}_\varepsilon F$ can be identified with $\mathfrak{C}_b(E, F)$ and $E' \tilde{\otimes}_\varepsilon (F/H)$ with $\mathfrak{C}_b(E, F/H)$, as follows immediately from 2.(6).

If $Z \in \mathfrak{C}_b(E, F)$, I is the identity on E', and K is as before, then $(I \tilde{\otimes}_\varepsilon K)Z = (I\varepsilon K)Z = KZ$ by (3) and $I \tilde{\otimes}_\varepsilon K$ will be a homomorphism if and only if every compact mapping of E in F/H has the form KZ, i.e., has a compact lifting.

We know from § 42, 8.(11) that this is the case for a given (B)-space E for every quotient F/H, F any (B)-space (E has the compact lifting property), if and only if E is an $\mathscr{L}_1$-space. Thus we have

(8) *Let E be a* (B)*-space such that E' has the approximation property. Let I be the identity map on E' and let K be the canonical homomorphism of a* (B)*-space F onto its quotient F/H.*

Then $I \tilde{\otimes}_\varepsilon K$ is a homomorphism of $E' \tilde{\otimes}_\varepsilon F$ in $E' \tilde{\otimes}_\varepsilon (F/H)$ for every quotient F/H if and only if E is an $\mathscr{L}_1$-space.

This shows that a product $A_1 \tilde{\otimes}_\varepsilon A_2$ of two homomorphisms onto is not necessarily a homomorphism.

The same is true also for $A_1 \otimes_\varepsilon A_2$, as the following observations show.

(9) *Let E_1 E_2, F_1, F_2 be normed spaces, $A_1 \in \mathfrak{L}(E_1, F_1)$, $A_2 \in \mathfrak{L}(E_2, F_2)$ homomorphisms onto such that $A_1 \tilde{\otimes}_\varepsilon A_2$ is not a homomorphism of $E_1 \tilde{\otimes}_\varepsilon E_2$ into $F_1 \tilde{\otimes}_\varepsilon F_2$. Then $A_1 \otimes_\varepsilon A_2$ is a continuous map of $E_1 \otimes_\varepsilon E_2$ onto $F_1 \otimes_\varepsilon F_2$ but not a homomorphism.*

Assuming that $A_1 \otimes_\varepsilon A_2$ is a homomorphism, we arrive immediately at a contradiction by using

(10) *Let A be a homomorphism of X into Y, X and Y metrizable spaces. Then the continuous extension $\tilde{A}$ which maps $\tilde{X}$ into $\tilde{Y}$ is again a homomorphism.*

Proof. A and $\tilde{A}$ have the same adjoint $\tilde{A}' = A' \in \mathfrak{L}(Y', X')$ because $\tilde{X}' = X'$ and $\tilde{Y}' = Y'$. Since A is a homomorphism, $A'(F')$ is $\mathfrak{T}_s(E)$-closed in E' by § 32, 3.(2). Then $A'(F')$ is also $\mathfrak{T}_s(\tilde{E})$-closed and $\tilde{A}$ is a homomorphism by § 33, 4.(2).

A systematic investigation of the relations between lifting properties and the ε-tensor product has recently been made by Kaballo [1'].

5. Hereditary properties. As an easy consequence of 4.(6) one obtains

(1) *Let E_1 and E_2 be dense subspaces of the locally convex spaces F_1 and F_2, respectively. Then $E_1 \otimes_\varepsilon E_2$ is dense in $F_1 \otimes_\varepsilon F_2$.*

If follows that $E_1 \tilde{\otimes}_\varepsilon E_2$ is always isomorphic to $\tilde{E}_1 \tilde{\otimes}_\varepsilon \tilde{E}_2$. If E_1 and E_2 are normed spaces, this isomorphism is even a norm isomorphism.

Proof. Let $\sum_{i=1}^{n} y_i^{(1)} \otimes x_i^{(2)}$ be an element of $F_1 \otimes E_2$ and let $p \otimes_\varepsilon q$ be a $\mathfrak{T}_\varepsilon$-semi-norm on $F_1 \otimes F_2$. Since E_1 is dense in F_1, there exist $x_i^{(1)} \in E_1$ such that $p(y_i^{(1)} - x_i^{(1)}) \leqq \varepsilon/n$ for $i = 1, \ldots, n$. Using 2.(4), one has

$$p \otimes_\varepsilon q\left(\sum_{i=1}^{n} (y_i^{(1)} - x_i^{(1)}) \otimes x_i^{(2)}\right) \leqq \sum_{i=1}^{n} p(y_i^{(1)} - x_i^{(1)}) q(x_i^{(2)}) \leqq \varepsilon \sup_i q(x_i^{(2)}),$$

which implies that $E_1 \otimes E_2$ is dense in $F_1 \otimes_\varepsilon E_2$ and this space again is dense in $F_1 \otimes_\varepsilon F_2$. That $E_1 \otimes_\varepsilon E_2$ is also a subspace of $F_1 \otimes_\varepsilon F_2$ in the sense of the topologies is an immediate consequence of 4.(6).

Proposition § 41, 5.(5) is true also for ε-tensor products.

(2) *Let F_1, F_2 be locally convex, P_1, P_2 continuous projections with ranges $P_1(F_1) = E_1$, $P_2(F_2) = E_2$ and kernels N_1, N_2. Then $P_1 \otimes P_2$ is a continuous projection of $F_1 \otimes_\varepsilon F_2$ onto the subspace $E_1 \otimes_\varepsilon E_2$ with kernel $D[N_1, N_2]$ and $P_1 \tilde{\otimes}_\varepsilon P_2$ is a continuous projection of $F_1 \tilde{\otimes}_\varepsilon F_2$ onto the subspace $E_1 \tilde{\otimes}_\varepsilon E_2$ with kernel $\overline{D[N_1, N_2]}$, the closure being taken in $F_1 \tilde{\otimes}_\varepsilon F_2$.*

As in the proof of § 41, 5.(5), it is obvious that $P_1 \otimes P_2$ is a projection of $F_1 \otimes_\varepsilon F_2$ onto $E_1 \otimes E_2$. By 4.(6) the topology induced by $F_1 \otimes_\varepsilon F_2$ is the topology of $E_1 \otimes_\varepsilon E_2$; hence $P_1 \otimes P_2$ is continuous with kernel $D[N_1, N_2]$. The second statement follows immediately, as in the proof of § 41, 5.(5).

In particular,

(3) *Let $E_1, \ldots, E_n, F$ be locally convex. Then one has always*

$$\left(\bigoplus_{i=1}^{n} E_i\right) \otimes_\varepsilon F = \bigoplus_{i=1}^{n} (E_i \otimes_\varepsilon F), \qquad \left(\bigoplus_{i=1}^{n} E_i\right) \tilde{\otimes}_\varepsilon F = \bigoplus_{i=1}^{n} (E_i \tilde{\otimes}_\varepsilon F).$$

We could have written $\prod_{i=1}^{n} E_i$ instead of $\bigoplus_{i=1}^{n} E_i$ and so the problem arises immediately whether these relations remain true for infinite products and infinite direct sums. We will answer these questions even for ε-products by using structure theorems from § 39, 8.

(4) *Let E, F_α, $\alpha \in \mathsf{A}$, be locally convex and F the locally convex kernel $\mathsf{K}_\alpha A_\alpha^{(-1)}(F_\alpha)$. Then $E\varepsilon\left(\mathsf{K}_\alpha A_\alpha^{(-1)}(F_\alpha)\right) \cong \mathsf{K}_\alpha (I\varepsilon A_\alpha)^{(-1)}(E\varepsilon F_\alpha)$.*

Furthermore, $E\varepsilon \prod_\alpha F_\alpha \cong \prod_\alpha (E\varepsilon F_\alpha)$.

By definition $E\varepsilon G \cong \mathfrak{L}_e(E'_{co}, G)$ and (4) is a particular case of § 39, 8.(10):

$$E\varepsilon F \cong \mathfrak{L}_e(E'_{co}, F) \cong \mathsf{K}_\alpha A_\alpha^{(-1)}\mathfrak{L}_e(E'_{co}, F_\alpha) \cong \mathsf{K}_\alpha (I\varepsilon A_\alpha)^{(-1)}(E\varepsilon F_\alpha),$$

where the last equality is a consequence of 4.(3).

For topological products and projective limits we obtain

(5) a) *Let* $E, F_\alpha, \alpha \in \mathsf{A}$, *be locally convex. Then* $E \tilde{\otimes}_\varepsilon \prod_\alpha F_\alpha \cong \prod_\alpha (E \tilde{\otimes}_\varepsilon F_\alpha)$.

b) *Let* $E = \varprojlim A_{\alpha\alpha'}(E_{\alpha'})$ *and* $F = \varprojlim B_{\beta\beta'}(F_{\beta'})$ *be reduced projective limits of locally convex spaces* E_α, $\alpha \in \mathsf{A}$, *and* F_β, $\beta \in \mathsf{B}$, *respectively. Then* $E \tilde{\otimes}_\varepsilon F$ *is isomorphic to the reduced projective limit*

$$X = \varprojlim (A_{\alpha\alpha'} \tilde{\otimes}_\varepsilon B_{\beta\beta'})(E_{\alpha'} \tilde{\otimes}_\varepsilon F_{\beta'}).$$

a) $E \otimes \left(\bigoplus_\alpha F_\alpha\right)$ is dense in $E \tilde{\otimes}_\varepsilon \prod_\alpha F_\alpha$ and $\bigoplus_\alpha (E \otimes F_\alpha)$ is dense in $\prod_\alpha (E \tilde{\otimes}_\varepsilon F_\alpha)$ by (1). It follows from (4) that on $H = E \otimes \left(\bigoplus_\alpha F_\alpha\right) = \bigoplus_\alpha (E \otimes F_\alpha)$ the topologies induced by $E \otimes_\varepsilon \prod_\alpha F_\alpha$ and by $\prod_\alpha (E \otimes_\varepsilon F_\alpha)$ coincide, which implies the statement.

The second formula of (3) is a particular case of (5) a). We note that $E \otimes_\varepsilon \prod_{n=1}^\infty F_n$ is in general not isomorphic to $\prod_{n=1}^\infty (E \otimes_\varepsilon F_n)$. To see this one takes $E = \varphi$ and $F_n = \mathsf{K}$ such that $E \otimes_\varepsilon \prod_n F_n = \varphi \otimes_\varepsilon \omega$, which is incomplete, whereas $\prod_{n=1}^\infty (\varphi \otimes_\varepsilon \mathsf{K}) \cong \omega\varphi$ is complete.

Proof of b). E and F are locally convex kernels, $E = \mathop{\mathsf{K}}_\alpha A_\alpha^{(-1)}(E_\alpha)$, $F = \mathop{\mathsf{K}}_\beta B_\beta^{(-1)}(F_\beta)$, and one has for E the relations

$$(*) \qquad A_\alpha = A_{\alpha\alpha'} A_{\alpha'}, \quad \alpha < \alpha', \qquad A_{\alpha\alpha'} A_{\alpha'\alpha''} = A_{\alpha\alpha''}, \quad \alpha < \alpha' < \alpha'';$$

similarly for F.

Every $x \in E$ can be written as an element of $\prod_\alpha E_\alpha$, $x = (x_\alpha)$, $x_\alpha = A_\alpha x \in E_\alpha$; similarly, $y \in F$ as $y = (y_\beta) \in \prod_\beta F_\beta$, $y_\beta = B_\beta y \in F_\beta$. Then $E \otimes F$ is the linear span of the elements $x \otimes y = (x_\alpha) \otimes (y_\beta) \in \left(\prod_\alpha E_\alpha\right) \otimes \left(\prod_\beta F_\beta\right)$. Obviously, $E \otimes F$ can algebraically be represented as a kernel

$$\mathop{\mathsf{K}}_{\alpha,\beta} (A_\alpha \otimes B_\beta)^{(-1)}(E_\alpha \otimes F_\beta);$$

in particular, $x \otimes y$ is represented as a vector $(x_\alpha \otimes y_\beta) \in \prod_{\alpha,\beta} (E_\alpha \otimes F_\beta)$, where $x_\alpha \otimes y_\beta = (A_\alpha \otimes B_\beta)(x \otimes y)$. Using a), we see that this correspondence $(x_\alpha) \otimes (y_\beta) \to (x_\alpha \otimes y_\beta)$ is generated by the topological isomorphism $\left(\prod_\alpha E_\alpha\right) \tilde{\otimes}_\varepsilon \left(\prod_\beta F_\beta\right) \cong \prod_{\alpha,\beta} (E_\alpha \tilde{\otimes}_\varepsilon F_\beta)$. It follows from 4.(6) that $E \otimes_\varepsilon F$ is isomorphic to $Z = \mathop{\mathsf{K}}_{\alpha,\beta} (A_\alpha \otimes B_\beta)^{(-1)}(E_\alpha \otimes_\varepsilon F_\beta)$ (equipped with the kernel topology). Consequently, $E \tilde{\otimes}_\varepsilon F$ is isomorphic to the completion $\tilde{Z}$. We have to show that $\tilde{Z} = X$.

We introduce $Y = \varprojlim (A_{\alpha\alpha'} \otimes B_{\beta\beta'})(E_{\alpha'} \otimes_\varepsilon F_{\beta'})$. This projective limit exists since the relations of type $(*)$ for the mappings $A_\alpha \otimes B_\beta$ and $A_{\alpha\alpha'} \otimes B_{\beta\beta'}$ are immediate consequences of these relations for E and F. It follows, as in part a) of the proof of § 41, 6.(3), that Z is contained and dense in Y and that Y is reduced. Finally, § 41, 6.(4) implies that $\tilde{Z} = \tilde{Y} = X$ and that X is reduced.

In contrast to these positive results, $E\varepsilon\left(\bigoplus_{n=1}^{\infty} F_n\right)$ resp. $E \mathbin{\tilde{\otimes}_\varepsilon} \left(\bigoplus_{n=1}^{\infty} F_n\right)$ is in general different from $\bigoplus_{n=1}^{\infty} (E\varepsilon F_n)$ resp. $\bigoplus_{n=1}^{\infty} (E \mathbin{\tilde{\otimes}_\varepsilon} F_n)$.

We use the known structure of φ, ω and § 39, 8.(14) and verify that

$$\omega \mathbin{\tilde{\otimes}_\varepsilon} \varphi = \omega\varepsilon\varphi = \mathfrak{L}_b(\varphi, \varphi) \cong \omega\varphi$$

and that for $\varphi = \bigoplus_{n=1}^{\infty} \mathsf{K}_n$, $\mathsf{K}_n = \mathsf{K}$, and $\omega_n = \omega$

$$\bigoplus_{n=1}^{\infty} (\omega \mathbin{\tilde{\otimes}_\varepsilon} \mathsf{K}_n) = \bigoplus_{n=1}^{\infty} (\omega_n \mathbin{\tilde{\otimes}_\varepsilon} \mathsf{K}_n) = \bigoplus_{n=1}^{\infty} \mathfrak{L}_b(\varphi, \mathsf{K}_n) = \bigoplus_{n=1}^{\infty} \omega_n = \varphi\omega.$$

One has $\mathfrak{L}(\varphi, \varphi) \supset \bigoplus_{n=1}^{\infty} \mathfrak{L}(\varphi, \mathsf{K}_n)$ by § 39, 8.(11) and in our case the sign $\supset$ is strict, as was shown immediately after § 39, 8.(11). Thus $\omega \mathbin{\tilde{\otimes}_\varepsilon} \left(\bigoplus_{n=1}^{\infty} \mathsf{K}_n\right)$ is different from $\bigoplus_{n=1}^{\infty} (\omega \mathbin{\tilde{\otimes}_\varepsilon} \mathsf{K}_n)$.

ω is an (F)-space and, in view of our counterexample, the following positive result is quite interesting.

(6) *Let E be a* (DF)-*space, $F_1, F_2, \ldots$ locally convex spaces. Then one has the following isomorphisms:* $E\varepsilon\left(\bigoplus_{n=1}^{\infty} F_n\right) \cong \bigoplus_{n=1}^{\infty} (E\varepsilon F_n)$, $E \mathbin{\tilde{\otimes}_\varepsilon} \left(\bigoplus_{n=1}^{\infty} F_n\right) \cong \bigoplus_{n=1}^{\infty} (E \mathbin{\tilde{\otimes}_\varepsilon} F_n)$, *and* $E \otimes_\varepsilon \left(\bigoplus_{n=1}^{\infty} F_n\right) \cong \bigoplus_{n=1}^{\infty} (E \otimes_\varepsilon F_n)$.

Proof. The first statement reads $\mathfrak{L}_e\left(E'_{co}, \bigoplus_{n=1}^{\infty} F_n\right) \cong \bigoplus_{n=1}^{\infty} \mathfrak{L}_e(E'_{co}, F_n)$ and looks similar to § 39, 8.(12).

We note the following fact on (DF)-spaces E, which is an immediate consequence of § 39, 8.(7):

$(*)$ Let $M_1, M_2, \ldots$ be a sequence of equicontinuous subsets of E'; then there exist $\rho_i > 0$ such that $M = \bigcup_{i=1}^{\infty} \rho_i M_i$ is again equicontinuous in E'.

Using § 39, 8.(11) and (*), one shows, as in the proof of § 39, 8.(12), that $\mathfrak{L}\left(E'_{co}, \bigoplus_{n=1}^{\infty} F_n\right) = \bigoplus_{n=1}^{\infty} \mathfrak{L}(E'_{co}, F_n)$. Again by using (*) for the class $\mathfrak{M}$ of equicontinuous subsets of E', one deduces the identity of the topologies on both spaces, as in the second part of the proof of § 39, 8.(12).

The second statement of (6) is a simple consequence of the third statement.

We remarked in the proof of (5) a) that the two spaces in the last statement of (6) are algebraically identical, which implies that they are also topologically identical.

L. SCHWARTZ [3′] proved the following result:

(7) *Let E and F be locally convex.*

a) *If E and F have the weak approximation property, then $E\varepsilon F$ has this property.*

b) *If E and F are quasi-complete and have the approximation property, then $E\varepsilon F$ has the approximation property.*

c) *If E and F are complete and have the approximation property, then $E \tilde{\otimes}_\varepsilon F$ has the approximation property.*

We prove first a) and use § 43, 3.(6) a). We have to show that $G \otimes (E\varepsilon F)$ is dense in $G\varepsilon(E\varepsilon F)$ for every locally convex space G. Since E has the weak approximation property, $G \otimes E$ is dense in $G\varepsilon E$; hence $(G \otimes E) \otimes F$ is dense in $(G\varepsilon F) \otimes F$. Since F has the weak approximation property, $(G\varepsilon E) \otimes F$ is dense in $(G\varepsilon E)\varepsilon F$. The associativity of $\otimes$ is trivial and the associativity of ε was proved in 3.(5); hence $G \otimes (E \otimes F)$ is dense in $G\varepsilon(E\varepsilon F)$, which implies that $G \otimes (E\varepsilon F)$ is dense in $G\varepsilon(E\varepsilon F)$. Thus a) is proved.

b) follows from a) and 2.(8), c) follows from b) and $E\varepsilon F = E \tilde{\otimes}_\varepsilon F$ in this case.

We note without proof the following result of DE WILDE [3′], p. 79:

(8) *Let E be an* (F)*-space and F a complete webbed resp. strictly webbed space. Then $E \tilde{\otimes}_\varepsilon F$ is again webbed resp. strictly webbed.*

6. Further results on tensor product mappings. For the π- and the ε-tensor product we investigated quite thoroughly under what conditions the product $A_1 \otimes A_2$ of two homomorphisms is again a homomorphism. Obviously, there are many questions of this type and we will answer a few of them (the easy ones) here.

(1) *Let E_1, E_2, F_1, F_2 be locally convex spaces, $A_1 \in \mathfrak{L}(E_1, F_1), A_2 \in \mathfrak{L}(E_2, F_2)$. If A_1 and A_2 are compact, then $A_1 \tilde{\otimes}_\pi A_2$ is a compact map of $E_1 \tilde{\otimes}_\pi E_2$ into $F_1 \tilde{\otimes}_\pi F_2$.*

Proof. We have $A_1(U_1) \subset C_1$, $A_2(U_2) \subset C_2$ for suitable neighbourhoods U_1, U_2, where C_1, C_2 are compact subsets of F_1, F_2. Since the canonical map χ of $F_1 \times F_2$ in $F_1 \otimes_\pi F_2$ is continuous, the subsets $C_1 \otimes C_2$ and $\overline{\Gamma(C_1 \otimes C_2)}$ of $F_1 \tilde{\otimes}_\pi F_2$ are compact. The statement follows from

$$(A_1 \tilde{\otimes}_\pi A_2)(\overline{\Gamma(U_1 \otimes U_2)}) \subset \overline{\Gamma(A_1(U_1) \otimes A_2(U_2))} \subset \overline{\Gamma(C_1 \otimes C_2)}.$$

(1) and the next proposition were proved for (B)-spaces by Holub [2′].

(2) *Let E_1, E_2, F_1, F_2 be complete locally convex spaces, $A_1 \in \mathfrak{L}(E_1, F_1)$, $A_2 \in \mathfrak{L}(E_2, F_2)$. If A_1 and A_2 are compact, then $A_1 \varepsilon A_2$ and $A_1 \tilde{\otimes}_\varepsilon A_2$ are compact mappings of $E_1 \varepsilon E_2$ in $F_1 \varepsilon F_2$ and $E_1 \tilde{\otimes}_\varepsilon E_2$ in $F_1 \tilde{\otimes}_\varepsilon F_2$, respectively.*

Proof. Since $E \tilde{\otimes}_\varepsilon F$ is a complete subspace of $E \varepsilon F$ for complete spaces, it will be sufficient to prove (2) for the ε-product.

By assumption there exist absolutely convex weakly closed equicontinuous subsets $M_1 \subset E_1'$, $M_2 \subset E_2'$ and compact sets $C_1 \subset F_1$, $C_2 \subset F_2$ such that $A_1(M_1^\circ) \subset C_1$, $A_2(M_2^\circ) \subset C_2$.

Recalling 3.(2), it will be sufficient to determine a o-neighbourhood W in $E_1 \varepsilon E_2 = \mathfrak{L}_e((E_1)'_{co}, E_2)$ such that its image $\tilde{H} = (A_1 \varepsilon A_2)(W)$ has the property that $\tilde{H}(G_1)$ is relatively compact in F_2 for every absolutely convex weakly closed equicontinuous $G_1 \subset F_1'$ (this is half of condition iii) in 3.(2)), and then to show also that $\tilde{\tilde{H}}(G_2)$ is relatively compact in F_1 for every absolutely convex weakly closed equicontinuous $G_2 \subset F_2'$, where $\tilde{\tilde{H}} = \tilde{H}' \subset \mathfrak{L}_e((F_2)'_{co}, F_1)$.

We define $W = \{Z \in E_1 \varepsilon E_2; Z(M_1) \subset M_2^\circ\}$; then $\tilde{H} = \{A_2 Z A_1', Z \in W\}$. Let us determine $A_2 Z A_1'(G_1)$. Since $A_1(M_1^\circ) \subset C_1$, one has $A_1'(C_1^\circ) \subset M_1^{\circ\circ} = M_1$. For some $\rho > 0$ one has $G_1 \subset \rho C_1^\circ$; hence $A_1'(G_1) \subset \rho M_1$. Therefore for every $Z \in W$ we have $A_2 Z A_1'(G_1) \subset \rho C_2$ or $\tilde{H}(G_1) \subset \rho C_2$; $\tilde{H}(G_1)$ is relatively compact in F_2.

By transposition we get

$$W' = \{Z'; Z'(M_2) \subset M_1^\circ\} \quad \text{and} \quad \tilde{H}' = \{A_1 Z' A_2'; Z' \in W'\}$$

and the same argument proves that $\tilde{H}'(G_2)$ is relatively compact in F_1.

We recall from § 42, 5. the definition of a nuclear mapping and of the nuclear norm. The following proposition is due to Holub [1′].

(3) *Let E_1, E_2, F_1, F_2 be normed spaces and $A_1 \in \mathfrak{L}(E_1, F_1)$, $A_2 \in \mathfrak{L}(E_2, F_2)$. If A_1 and A_2 are nuclear mappings, then $A_1 \tilde{\otimes}_\pi A_2$ and $A_1 \tilde{\otimes}_\varepsilon A_2$ are also nuclear mappings and one has*

$$\|A_1 \tilde{\otimes}_\pi A_2\|_\nu \leqq \|A_1\|_\nu \|A_2\|_\nu, \qquad \|A_1 \tilde{\otimes}_\varepsilon A_2\|_\nu \leqq \|A_1\|_\nu \|A_2\|_\nu.$$

Proof. Recalling § 42, 5.(7) and (8), we see that there exist representations of A_1 and A_2 of the form

$$A_1x^1 = \sum_{n=1}^{\infty} (u_n^1x^1)y_n^1, \quad x^1 \in E_1, u_n^1 \in E_1', y_n^1 \in F_1, \quad \sum \|u_n^1\| \|y_n^1\| \leqq \|A_1\|_\nu + \varepsilon,$$

$$A_2x^2 = \sum_{n=1}^{\infty} (u_n^2x^2)y_n^2, \quad x^2 \in E_2, u_n^2 \in E_2', y_n^2 \in F_2, \quad \sum \|u_n^2\| \|y_n^2\| \leqq \|A_2\|_\nu + \varepsilon.$$

Consequently, for $\sum_{i=1}^{m} x_i^1 \otimes x_i^2 \in E_1 \otimes E_2$ one has

$$(A_1 \otimes A_2)\left(\sum_{i=1}^{m} x_i^1 \otimes x_i^2\right) = \sum_{i=1}^{m} \left(\sum_{n=1}^{\infty} (u_n^1x_i^1)y_n^1\right) \otimes \left(\sum_{k=1}^{\infty} (u_k^2x_k^2)y_k^2\right)$$

$$= \sum_{n,k=1}^{\infty} \left((u_n^1 \otimes u_k^2) \sum_{i=1}^{m} (x_i^1 \otimes x_i^2)\right)(y_n^1 \otimes y_k^2)$$

and this double sum is absolutely convergent in the π- and the ε-norm since

$$\|u_n^1 \otimes u_k^1\|_\pi \|y_n^1 \otimes y_k^2\|_\pi = \|u_n^1\| \|u_k^2\| \|y_n^1\| \|y_k^2\| = \|u_n^1 \otimes u_k^2\|_\varepsilon \|y_n^1 \otimes y_k^2\|_\varepsilon$$

and

$$\sum_{n,k=1}^{\infty} \|u_n^1\| \|u_k^2\| \|y_n^1\| \|y_k^2\| = \left(\sum_{n=1}^{\infty} \|u_n^1\| \|y_n^2\|\right)\left(\sum_{k=1}^{\infty} \|u_k^1\| \|y_k^2\|\right)$$

$$\leqq (\|A_1\|_\nu + \varepsilon)(\|A_2\|_\nu + \varepsilon).$$

It follows that $A_1 \otimes A_2$ is nuclear on $E_1 \otimes E_2$ for both norms and so are the continuous extensions $A_1 \tilde{\otimes}_\pi A_2$ and $A_1 \tilde{\otimes}_\varepsilon A_2$. The inequalities for the nuclear norms are obvious.

More and deeper results on tensor product mappings $A_1 \otimes A_2$ of (B)-spaces are contained in Holub [1'], [2'], [3']. Even the cases where A_1 and A_2 are unbounded have been thoroughly investigated in connection with spectral theory. We refer to the work of Ichinose [1'], [2'].

7. Vector valued continuous functions. We study an important class of examples for the ε-tensor product.

Let X be a locally compact Hausdorff space, E a locally convex space. We denote by $C(X, E)$ the space of all continuous functions on X with values in E. Its natural topology is the topology of compact convergence on X defined by the semi-norms $p_K(f) = \sup_{t \in K} p(f(t))$, where K is a compact subset of X and p is a semi-norm of a system of semi-norms defining the topology of E.

(1) *$C(X, E)$ is complete if and only if E is complete.*

Proof. If E is complete and if f_α is a Cauchy net in $C(X, E)$, then $f_\alpha(t)$ is a Cauchy net in E for every fixed $t \in X$ and it has a limit point $f_0(t)$. Since the convergence $f_\alpha \to f_0$ is uniform on compact subsets, it follows that f_0 is continuous and $C(X, E)$ is complete.

Conversely, if $C(X, E)$ is complete and x_α is a Cauchy net in E and $\varphi(t) \equiv 1$ on X, then $\varphi(t)x_\alpha = x_\alpha$ is a Cauchy net in $C(X, E)$ which has a limit in $C(X, E)$ and also in E, so E is complete.

If E is an (F)-space and X is countable at infinity, then $C(X, E)$ is an (F)-space. If E is a normed space and X compact, then $C(X, E)$ is normed with norm $\|f\| = \sup_{t \in X} \|f(t)\|$.

We consider now $C(X) \otimes E$. The mapping $(\varphi, x) \to f(t) = \varphi(t)x$ of $C(X) \times E$ into $C(X, E)$ is bilinear; therefore it generates an algebraic homomorphism ψ of $C(X) \otimes E$ into $C(X, E)$, $\psi\left(\sum_{i=1}^{n} \varphi_i \otimes x_i\right) = \sum_{i=1}^{n} \varphi_i(t)x_i$.
By assuming the x_i linearly independent one verifies that ψ is an algebraic isomorphism. In this way $C(X) \otimes E$ can be identified with the subspace of $C(X, E)$ consisting of all functions on X having their range in a finite dimensional subspace of E.

One can say much more:

(2) a) *ψ is an isomorphism of $C(X) \otimes_\varepsilon E$ on a dense subspace of $C(X, E)$.*

b) *If E is complete, then $C(X) \tilde{\otimes}_\varepsilon E$ is isomorphic to $C(X, E)$.*

c) *If E is a* (B)*-space and X is compact, then $C(X) \tilde{\otimes}_\varepsilon E$ is norm isomorphic to $C(X, E)$.*

Proof. i) We show first that $C(X) \otimes E$ is dense in $C(X, E)$. Let $f \in C(X, E)$ be given, K a compact subset of X, p a continuous semi-norm on E. We note that there exists a compact subset K_1 of X which contains an open neighbourhood of K. Let $O_1, \ldots, O_m$ be a finite open cover of K_1 such that

$$\sup_{t_1, t_2 \in O_j} p(f(t_1) - f(t_2)) \leqq \varepsilon, \qquad j = 1, \ldots, m,$$

and let $\tilde{\varphi}_1, \ldots, \tilde{\varphi}_m$ be a corresponding partition of unity on K_1 (§ 43, 7.(1)) and $\alpha \in C(X)$, $0 \leqq \alpha(t) \leqq 1$, where $\alpha \equiv 1$ on K and $\alpha \equiv 0$ outside K_1. Then $\varphi_i = \tilde{\varphi}_i\alpha \in C(X)$, $\sum_{i=1}^{m} \varphi_i = \alpha$. We choose a fixed $t_k \in O_k$ for every $k = 1, \ldots, m$. Then $g(t) = \sum_{k=1}^{m} \varphi_k \otimes f(t_k) \in C(X) \otimes E$ and for every $t \in K$

we have

$$p(f(t) - g(t)) = p\left(\sum_{k=1}^{m} \varphi_k(t)(f(t) - f(t_k))\right)$$
$$\leqq \sum_{k=1}^{m} \varphi_k(t) \sup_{s \in O_k} p(f(s) - f(t_k)) \leqq \alpha\varepsilon = \varepsilon$$

or $p_K(f - g) \leqq \varepsilon$.

ii) Next we show that ψ is a topological isomorphism. Let K be a compact subset of X. Consider the subset P_K of $C(X)'$ consisting of all point measures δ_t, $t \in K$. The absolute polar of P_K is $U_K = \{f; \sup_{t \in K} |f(t)| \leqq 1\}$ and $G_K = \overline{\Gamma(P_K)} = U_K^\circ$ is an absolutely convex, weakly closed, equicontinuous subset of $C(X)'$. If K varies over all compact subsets of X we obtain a class $\{G_K\}$ defining the topology of $C(X)$.

Let p be a continuous semi-norm on E and G_p the polar of $\{x \in E; p(x) \leqq 1\}$ in E'. For the semi-norms defining the ε-tensor product topology on $C(X) \otimes E$ we now obtain

$$\varepsilon_{G_K, G_p}\left(\sum \varphi_i \otimes x_i\right) = \sup_{\mu \in G_K, u \in G_p} \left|\sum \mu(\varphi_i)\langle u, x_i\rangle\right| = \sup_{\mu \in \Gamma(P_K), u \in G_p} |\cdots|$$
$$= \sup_{t_k \in K, u \in G_p, \sum|\lambda_k| \leq 1} \left|\sum_i \left\langle \sum_k \lambda_k \delta_{t_k}, \varphi_i \right\rangle \langle u, x_i\rangle\right|$$
$$= \sup_{t \in K, u \in G_p} \left|\sum \langle \delta_t, \varphi_i\rangle\langle u, x_i\rangle\right|$$
$$= \varepsilon_{G_K, G_p}\left(\sum \varphi_i \otimes x_i\right).$$

Consequently, $\varepsilon_{G_K, G_p}(\sum \varphi_i \otimes x_i) = \sup_{t \in K, u \in G_p} |\sum \langle \alpha_t, \varphi_i\rangle\langle u, x_i\rangle| = \sup_{t \in K} p(f(t))$, where $f(t) = \sum \varphi_i(t) x_i = \psi(\sum \varphi_i \otimes x_i)$. This proves a).

iii) b) is an immediate consequence of (1) and a). In the case c) our proof shows that the isomorphism of $C(X, E)$ is even a norm isomorphism, so c) is true also.

We consider the particular case $E = C(Y)$, Y locally compact. Then (2) shows that $C(X, C(Y))$ is isomorphic to $C(X) \tilde{\otimes}_\varepsilon C(Y)$. There is a better result:

(3) *Let X, Y be locally compact Hausdorff spaces. Then $C(X) \tilde{\otimes}_\varepsilon C(Y)$ is isomorphic to $C(X \times Y)$. If X and Y are compact, this isomorphism is even a norm isomorphism.*

We have to prove the (norm) isomorphism $C(X \times Y) \cong C(X, C(Y))$. Let $\varphi = \varphi(s, t)$ be in $C(X \times Y)$. We define $\varphi_s \in C(Y)$ by $\varphi_s(t) = \varphi(s, t)$ and $\tilde{\varphi}$ by $\tilde{\varphi}(s) = \varphi_s$. We will show that $\tilde{\varphi} \in C(X, C(Y))$. This means that for $\varepsilon > 0$ and a compact subset K_2 of Y there exists a compact neighbourhood K_1 of $s_0 \in X$ such that $p_{K_2}(\varphi_s - \varphi_{s_0}) = \sup_{t \in K_2} |\varphi_s(t) - \varphi_{s_0}(t)| \leqq \varepsilon$ for all $s \in K_1$. But if K_1' is a compact neighbourhood of s_0 in X, then $\varphi(s, t)$ is continuous on $K_1' \times K_2$ and the existence of K_1 follows from the compactness of $K_1' \times K_2$.

Conversely, every element of $C(X, C(Y))$ is the image of an element of $C(X \times Y)$. Hence $\varphi \to \tilde{\varphi}$ is an algebraic isomorphism of $C(X \times Y)$ onto $C(X, C(Y))$. That it is also a topological resp. norm isomorphism follows from

$$\sup_{(s,t) \in K_1 \times K_2} |\varphi(s, t)| = \sup_{s \in K_1} \left(\sup_{t \in K_2} |\varphi_s(t)| \right) = \sup_{s \in K_1} p_{K_2}(\tilde{\varphi}(s)).$$

The next statement follows immediately from (2) b), § 43, 7.(4) and 5.(7) c):

(4) *Let X be locally compact, E complete locally convex. Then $C(X, E)$ has the approximation property if E has this property.*

Extensive use of the ε-product and the ε-tensor product has been made by SCHWARTZ in his theory of vector valued distributions and recently by BIERSTEDT [1′] and BIERSTEDT and MEISE [2′] in their theory of vector valued functions.

8. ε-tensor product with a sequence space. In § 41, 7. we found a concrete representation of $\lambda \tilde{\otimes}_\pi E$ for arbitrary perfect sequence spaces λ and an arbitrary locally convex space E. Following again PIETSCH [1′], [2′], we will now obtain a similar representation of $\lambda \tilde{\otimes}_\varepsilon E$.

Let A be a set of indices. The class $\Phi = \{\varphi\}$ of all finite subsets φ of A, partially ordered by inclusion, forms a directed set. Let E be a locally convex space and $x = (x_\alpha)_{\alpha \in \mathsf{A}}$, $x_\alpha \in E$; then the vector x is called s u m m a b l e if the net s_φ, $\varphi \in \Phi$, $s_\varphi = \sum_{\alpha \in \varphi} x_\alpha$, is a Cauchy net in E. It has a limit $s = \lim_\varphi s_\varphi$ in $\tilde{E}$ (not necessarily in E); s is called the s u m $\sum_{\alpha \in \mathsf{A}} x_\alpha$ of the vector $x = (x_\alpha)$.

It is easy to see that for $\mathsf{A} = N$, the set of natural numbers, the summability of a sequence $x = (x_1, x_2, \ldots)$, $x_n \in E$, is equivalent to the unconditional convergence of $\sum_{n=1}^{\infty} x_n$.

If $E = \mathsf{K}$, the real or complex field, one has in generalization of a classical result of RIEMANN:

(1) *A vector $x = (x_\alpha)_{\alpha \in \mathsf{A}}$ of real resp. complex numbers is summable if and only if it is absolutely summable, i.e., $x \in l^1_{\mathsf{A}}$.*

If $\sup_{\varphi \in \Phi} |s_\varphi| = M$, *then* $\|x\|_1 = \sum_\alpha |x_\alpha| \leqq 4M$.

Proof. a) Let $x = (x_\alpha)$ be summable with sum s. Then there exists $\varphi_0 \in \Phi$ such that $|s - s_{\varphi'}| \leqq 1$ for all $\varphi' \supseteq \varphi_0$. Now let φ be arbitrary in Φ. Then

$$|s_\varphi| = \left|\sum_{\alpha \in \varphi} x_\alpha\right| = \left|\sum_{\varphi \cup \varphi_0} x_\alpha - \sum_{\varphi_0 \sim \varphi} x_\alpha\right| \leqq |s_{\varphi \cup \varphi_0} - s| + |s - s_{\varphi_0 \sim \varphi}|$$

$$\leqq 1 + |s| + |s_{\varphi_0 \sim \varphi}| \leqq 1 + |s| + \sum_{\alpha \in \varphi_0} |x_\alpha| = \gamma < \infty.$$

Hence $\sup_{\varphi \in \Phi} |s_\varphi| = M < \infty$.

If the x_α are real numbers, then

$$\sum_{\alpha \in \varphi} |x_\alpha| = \left|\sum_{x_\alpha > 0, \alpha \in \varphi} x_\alpha\right| + \left|\sum_{x_\alpha < 0, \alpha \in \varphi} x_\alpha\right| \leqq 2M.$$

For complex x_α one has

$$\sum_{\alpha \in \varphi} |x_\alpha| \leqq \sum_{\alpha \in \varphi} |\Re(x_\alpha)| + \sum_{\alpha \in \varphi} |\Im(x_\alpha)| \leqq 4M,$$

and this implies $\sum_{\alpha \in \mathsf{A}} |x_\alpha| \leqq 4M < \infty$.

b) Conversely, if $\sum_{\alpha \in \mathsf{A}} |x_\alpha| < \infty$, then $x_\alpha \neq o$ only for countably many $\alpha \in \mathsf{A}$. Thus one has only to consider the case of an absolutely summable sequence $(x_1, x_2, \ldots)$. It is easy to see that such a sequence is summable.

Thus l^1_{A} is the space of all summable scalar vectors $x = (x_\alpha)_{\alpha \in \mathsf{A}}$, and this space has a natural topology defined by the norm $\|x\|_1 = \sum_{\alpha \in \mathsf{A}} |x_\alpha|$.

Let now E be a locally convex space. We denote by $l^1_{\mathsf{A}}(E)$ the space of all summable vectors $x = (x_\alpha)_{\alpha \in \mathsf{A}}$, $x_\alpha \in E$. This space has also a natural topology, as we will see now.

Let $x^\circ = (x^\circ_\alpha)$ be an element of $l^1_{\mathsf{A}}(E)$. Consider the set $B \subset E$ of all $\sum_{\alpha \in \varphi} \gamma_\alpha x^\circ_\alpha$, $|\gamma_\alpha| \leqq 1$, $\varphi \in \Phi$. For every $u \in E'$ the vector $(ux^\circ_\alpha)_{\alpha \in \mathsf{A}}$ is summable in K and therefore absolutely summable by (1). Then

$$\left|\left\langle u, \sum_{\alpha \in \varphi} \gamma_\alpha x^\circ_\alpha \right\rangle\right| \leqq \sum_{\alpha \in \varphi} |ux^\circ_\alpha| \leqq \sum_{\alpha \in \mathsf{A}} |ux^\circ_\alpha| < \infty.$$

Hence B is weakly bounded and therefore bounded in E.

Let U be an absolutely convex neighbourhood of o in E. Then $|uy| \leqq \rho < \infty$ for $u \in U^\circ$, $y \in B$. For every $\varphi \in \Phi$ there exist γ_α, $\alpha \in \varphi$, such that $\left\langle u, \sum_{\alpha \in \varphi} \gamma_\alpha x^\circ_\alpha \right\rangle = \sum_{\alpha \in \varphi} |ux^\circ_\alpha| \leqq \rho$. Therefore for every $x^\circ = (x^\circ_\alpha) \in l^1_{\mathsf{A}}(E)$ the expression

$$(2) \qquad \varepsilon_U(x^\circ) = \sup_{u \in U^\circ} \sum_{\alpha \in A} |u x_\alpha^\circ|$$

is finite and is obviously a semi-norm on $l_A^1(E)$.

The topology on $l_A^1(E)$ is now defined by the system of all the semi-norms (2). It is easy to see that $l_A^1(E)$ is locally convex. If E is normed, then $l_A^1(E)$ is normed by (2), where U is the unit ball; if E is metrizable, then $l_A^1(E)$ is metrizable. We note

(3) *If* $x^\circ = (x_\alpha^\circ)$ *is summable, then every* $(\gamma_\alpha x_\alpha^\circ)$, $|\gamma_\alpha| \leqq 1$, *is again summable.*

This follows easily from the finiteness of (2).

Let now λ be a perfect sequence space, F locally convex. We define $\lambda(F)$ as the space of all sequences $y = (y_n)_{n=1,2,\ldots}$, $y_n \in F$, such that for every $\mathfrak{u} = (u_n) \in \lambda^\times$ the vector $(u_n y_n)$ is summable in F. It follows from the previous remarks that for every $\mathfrak{u} \in \lambda^\times$ and every absolutely convex neighbourhood U the expression

$$(4) \qquad \varepsilon_{\mathfrak{u},U}(y) = \sup_{v \in U^\circ} \sum_{n=1}^{\infty} |u_n(v y_n)|$$

is a semi-norm on $\lambda(F)$ and the system of all semi-norms (4) defines a natural topology on $\lambda(F)$. For $\lambda = l^1$ this new definition of $l^1(E)$ coincides with the previous one because of (3).

If we recall § 41, 7. we see that $\lambda\{F\} \subset \lambda(F)$ and that $\varepsilon_{\mathfrak{u},U}(y) \leqq \pi_{\mathfrak{u},U}(y)$ for every $y \in \lambda\{F\}$. Since $\lambda \otimes F$ can be identified with a subspace of $\lambda\{F\}$, $\lambda \otimes F$ is also a subspace of $\lambda(F)$ and for the induced topology one has

(5) *The topology of* $\lambda(F)$ *induces on* $\lambda \otimes F$ *the* ε*-topology.*

Proof. We recall that the topology of λ is the normal topology; hence, using 2.(2), we see that the ε-topology on $\lambda \otimes F$ is defined by semi-norms of the type

$$\varepsilon_{\{\mathfrak{u}\}^n, U^\circ}\left(\sum_{i=1}^{k} \mathfrak{x}^{(i)} \otimes y^{(i)}\right) = \sup_{|\eta_n| \leqq |u_n|, v \in U^\circ} \left| \sum_{i=1}^{k} \left(\sum_{n=1}^{\infty} \eta_n \xi_n^{(i)} \right) (v y^{(i)}) \right|,$$

$$\mathfrak{u} = (u_n) \in \lambda^\times, \ \mathfrak{x}^{(i)} = (\xi_n^{(i)}) \in \lambda, \ y^{(i)} \in F.$$

To $\sum_{i=1}^{k} \mathfrak{x}^{(i)} \otimes y^{(i)} \in \lambda \otimes F$ corresponds the element $\left(\sum_{i=1}^{k} \xi_n^{(i)} y^{(i)}\right)_{n=1,2,\ldots}$ in

$\lambda(F)$ (§ 41, 7.) and for the semi-norm on $\lambda(F)$ corresponding to $\mathfrak{u}$ and U one obtains from (4)

$$\varepsilon_{\mathfrak{u},U}\left(\left(\sum_{i=1}^{k} \xi_n^{(i)} y^{(i)}\right)\right) = \sup_{v\in U^\circ} \sum_{n=1}^{\infty} \left| u_n \sum_{1}^{k} \xi_n^{(i)}(vy^{(i)})\right|$$

$$= \sup_{|\eta_n|\leqq |u_n|, v\in U^\circ} \left|\sum_{n=1}^{\infty} \eta_n \sum_{i=1}^{k} \xi_n^{(i)}(vy^{(i)})\right|$$

$$= \varepsilon_{\{\mathfrak{u}\}^n, U^\circ}\left(\sum_{i=1}^{k} \mathfrak{x}^{(i)} \otimes y^{(i)}\right).$$

This implies the statement.

Similarly, one proves

(6) *The topology of $l_{\mathsf{A}}^1(F)$ induces on $l_{\mathsf{A}}^1 \otimes F$ the ε-topology.*

We note that for a normed F the norm on $l_{\mathsf{A}}^1(F)$ and the norm on $l_{\mathsf{A}}^1 \otimes_\varepsilon F$ coincide with $\varepsilon\left(\sum_{i=1}^{k} \mathfrak{x}^{(i)} \otimes y^{(i)}\right) = \sup_{\|v\|\leqq 1} \sum_{\alpha} \left|\sum_{i=1}^{k} \xi_\alpha^{(i)}(vy^{(i)})\right|$.

(5) and (6) indicate that we have a situation corresponding to that in § 41, 7. and our aim is now to prove in analogy to § 41, 7.(5) that $\lambda \tilde{\otimes}_\varepsilon F$ is isomorphic to $\lambda(\tilde{F})$ and that $l_{\mathsf{A}}^1 \tilde{\otimes}_\varepsilon F$ is isomorphic to $l_{\mathsf{A}}^1(\tilde{F})$.

We investigate first the case l_{A}^1. A vector $x = (x_\alpha)_{\alpha\in\mathsf{A}}$, $x_\alpha \in E$, is called **weakly summable** if the vector (ux_α) is summable for every $u \in E'$. Following PIETSCH we denote this space of all weakly summable vectors over E by $l_{\mathsf{A}}^1[E]$. The arguments leading from the definition of $l_{\mathsf{A}}^1(E)$ to the formula (2) are valid also for the elements x of $l_{\mathsf{A}}^1[E]$, so that the semi-norms (2) define also a topology on $l_{\mathsf{A}}^1[E]$ and $l_{\mathsf{A}}^1[E]$ is the space of all vectors $x = (x_\alpha)_{\alpha\in\mathsf{A}}$ with finite $\varepsilon_U(x)$ for all U.

Obviously, $l_{\mathsf{A}}^1(E)$ is a subspace of $l_{\mathsf{A}}^1[E]$ and it can be a strict subspace as is shown by the following example:

Consider $l^1[c_0]$. The vector $x_0 = (e_1, e_2, \ldots)$ of the unit vectors is weakly summable, so $x_0 \in l^1[c_0]$, but is not summable in c_0, so $x_0 \notin l^1(c_0)$.

Let $x = (x_\alpha)_{\alpha\in\mathsf{A}}$, $x_\alpha \in E$, be an element of $l_{\mathsf{A}}^1[E]$ and let φ be a finite subset of A. The **finite section** x^φ of x is the vector with $x_\alpha^\varphi = x_\alpha$ for $\alpha \in \varphi$, $x_\alpha^\varphi = \mathfrak{o}$ for $\alpha \notin \varphi$.

(7) $x = (x_\alpha)$, $x_\alpha \in E$, *lies in* $l_{\mathsf{A}}^1(E)$ *if and only if for every absolutely convex neighbourhood* U *of* $\mathfrak{o}$ *in* E *there exists a finite subset* φ_0 *of* A *such that* $\varepsilon_U(x^\psi) \leqq 1$ *for all finite* $\psi \in \mathsf{A} \sim \varphi_0$.

Proof. a) Assume $x \in l_{\mathsf{A}}^1(E)$; hence $s_\varphi = \sum_{\alpha\in\varphi} x_\alpha$, $\varphi \in \Phi$, is a Cauchy net in E. There exists φ_0 such that $s_\varphi - s_{\varphi_0} \in \frac{1}{4}U$ for $\varphi \supset \varphi_0$ or $s_\psi \in \frac{1}{4}U$ for all $\psi \subset \mathsf{A} \sim \varphi_0$. This means $\sup_{u\in U^\circ} |us_\psi| = \sup \left|\sum_{\alpha\in\psi} ux_\alpha\right| \leqq \frac{1}{4}$. By application of (1) to the numerical vector $(ux_\alpha)_{\alpha \in \mathsf{A} \sim \varphi_0}$ we obtain $\sup_{u \in U^\circ} \sum_{\alpha \in \psi} |ux_\alpha| \leqq 1$ or $\varepsilon_U(x^\psi) \leqq 1$.

b) If the condition is satisfied, then the net s_φ corresponding to x is obviously a Cauchy net in E, so x is summable.

We have the corollary

(8) *$l^1_\mathsf{A} \otimes E$ is dense in $l^1_\mathsf{A}(E)$.*

A finite section x^φ of $x = (x_\alpha) \in l^1_\mathsf{A}(E)$ can be written as the element $\sum_{\alpha \in \varphi} e_\alpha \otimes x_\alpha$ of $l^1_\mathsf{A} \otimes E$ and by (7) there exists φ_0 such that $\varepsilon_U(x - x^\varphi) \leqq 1$ for $\varphi \supset \varphi_0$ and this proves the statement.

(9) *$l^1_\mathsf{A} \tilde{\otimes}_\varepsilon E$ is isomorphic to $l^1_\mathsf{A}(\tilde{E})$, even norm isomorphic in the case of a normed space E.*

Since $l^1_\mathsf{A} \otimes E$ is dense in $l^1_\mathsf{A} \otimes_\varepsilon \tilde{E}$, it follows from (8) that $l^1_\mathsf{A} \otimes E$ is dense in $l^1_\mathsf{A}(\tilde{E})$; hence $l^1_\mathsf{A}(\tilde{E}) \subset l^1_\mathsf{A} \tilde{\otimes}_\varepsilon E$, so we have to prove only that $l^1_\mathsf{A}(E)$ is complete for complete E.

Let $x^{(\beta)}$ be a Cauchy net in $l^1_\mathsf{A}(E)$. For $U \supset \circ$ there exists $\beta_0(U)$ such that $\varepsilon_U(x^{(\beta)} - x^{(\beta')}) = \sup_{v \in U^\circ} \sum_\alpha |v(x^{(\beta)}_\alpha - x^{(\beta')}_\alpha)| \leqq \delta$ for all $\beta, \beta' \geqq \beta_0(U)$. Hence for every $\alpha \in \mathsf{A}$, $x^{(\beta)}_\alpha$ is a Cauchy net in E with a limit $x^{(\mathsf{o})}_\alpha$. Put $x^{(\mathsf{o})} = (x^{(\mathsf{o})}_\alpha)$. It follows easily that $\varepsilon_U(x^{(\beta)} - x^{(\mathsf{o})}) \leqq \delta$ for all $\beta \geqq \beta_0(U)$ and $x^{(\mathsf{o})}$ lies in $l^1_\mathsf{A}[E]$.

By (7) there exists φ_0 such that $\sup_{v \in U^\circ} \sum_{\alpha \in \psi} |vx^{(\beta)}_\alpha| \leqq \delta$ for all finite $\psi \in \mathsf{A} \sim \varphi_0$, where $\beta \geqq \beta_0(U)$ is fixed. Hence

$$\sup_{v \in U^\circ} \sum_{\alpha \in \psi} |vx^{(\mathsf{o})}_\alpha| \leqq \sup_{v \in U^\circ} \sum_{\psi} |v(x^{(\mathsf{o})}_\alpha - x^{(\beta)}_\alpha)| + \sup \sum_{\psi} |vx^{(\beta)}_\alpha| \leqq 2\delta$$

and $x^{(\mathsf{o})} \in l^1_\mathsf{A}(E)$ by (7).

We have the corresponding results for $\lambda(F)$.

(10) *$y = (y_n)_{n=1,2,\ldots}$, $y_n \in F$, lies in $\lambda(F)$ if and only if for every $\mathfrak{u} = (u_1, u_2, \ldots) \in \lambda^\times$ and every absolutely convex neighbourhood U of $\circ$ of F there exists $n_0(U)$ such that $\sup_{v \in U^\circ} \sum_{k=n_0+1}^{\infty} |u_k(vy_k)| \leqq 1$.*

$\lambda \otimes F$ is dense in $\lambda(F)$; $\lambda \tilde{\otimes}_\varepsilon F$ is isomorphic to $\lambda(\tilde{F})$.

The first statement follows easily from (7) applied to the different spaces of summable sequences of which $\lambda(F)$ is the intersection. The details of the proof are left to the reader.

§ 45. Duality of tensor products

1. First results. Let E, F be locally convex spaces. We recall that the dual of $E \otimes_\pi F$ can be identified with $\mathscr{B}(E \times F)$, the space of all continuous bilinear forms on $E \times F$. The π-equicontinuous subsets of $(E \otimes_\pi F)'$ are exactly the equicontinuous subsets of $\mathscr{B}(E \times F)$ (cf. § 41, 3.(3) and (4)).

In this paragraph we are interested in the dual $(E \otimes_\varepsilon F)'$ and its equicontinuous subsets. Since $\mathfrak{T}_\varepsilon$ is weaker on $E \otimes F$ than $\mathfrak{T}_\pi$, every continuous linear functional on $E \otimes_\varepsilon F$ can be represented by a uniquely defined element of $\mathscr{B}(E \times F)$, so that $(E \otimes_\varepsilon F)'$ will become a subspace $\mathfrak{J}(E \times F)$ of $\mathscr{B}(E \times F)$. The elements of $\mathfrak{J}(E \times F)$ are called integral bilinear forms on $E \times F$. The reason for this notation will become clear in 4.

We recall from § 44, 1.(2) the relation

$$E' \otimes F' \subset \mathfrak{J}(E \times F) \subset \mathscr{B}(E \times F). \tag{1}$$

The following statement is rather obvious.

(2) *Let E, F be locally convex: then $\mathfrak{J}(E \times F) = (E \otimes_\varepsilon F)' = (E \tilde{\otimes}_\varepsilon F)'$ is the union of all sets $\overline{\Gamma(G_1 \otimes G_2)}$, where G_1 and G_2 are equicontinuous subsets of E' and F', respectively, and where the closure of $\Gamma(G_1 \otimes G_2)$ is taken in $\mathscr{B}(E \times F)$ for the $\mathfrak{T}_s(E \otimes F)$-topology or the $\mathfrak{T}_s(E \tilde{\otimes}_\varepsilon F)$-topology.*

Every set $\overline{\Gamma(G_1 \otimes G_2)}$ is equicontinuous and every equicontinuous subset of $\mathfrak{J}(E \times F)$ is contained in some $\overline{\Gamma(G_1 \otimes G_2)}$.

Proof. We consider the dual pair $\langle \mathfrak{J}(E \times F), E \otimes F \rangle$. The ε-topology on $E \otimes F$ is defined by the polars $(G_1 \otimes G_2)^\circ$, where $G_1 \otimes G_2 \subset E' \otimes F' \subset \mathfrak{J}(E \times F)$ and where G_1 and G_2 are absolutely convex, weakly closed, equicontinuous subsets of E' and F', respectively. The bipolar $(G_1 \otimes G_2)^{\circ\circ} = \overline{\Gamma(G_1 \otimes G_2)}$ in $\mathfrak{J}(E \times F)$ is equicontinuous and weakly complete (ALAOGLU–BOURBAKI), hence coincides with the $\mathfrak{T}_s(E \otimes F)$- resp. $\mathfrak{T}_s(E \tilde{\otimes}_\varepsilon F)$-closure of $\overline{\Gamma(G_1 \otimes G_2)}$ in $\mathscr{B}(E \times F)$.

Conversely, every equicontinuous set in $\mathfrak{J}(E \times F)$ is contained in the polar of some absolutely convex $\mathfrak{T}_\varepsilon$-neighbourhood of $\mathfrak{o}$, hence in some $\overline{\Gamma(G_1 \otimes G_2)}$.

If E and F are normed spaces, then the norm of $(E \otimes_\varepsilon F)_b'$ is called the integral norm $\| \; \|_I$ on $\mathfrak{J}(E \times F)$, and it is given by

$$\|w\|_I = \sup_{\|z\|_\varepsilon \leq 1} |\langle w, z \rangle|, \qquad w \in \mathfrak{J}(E \times F),\ z \in E \otimes_\varepsilon F. \tag{3}$$

We remark that in the following we will usually write E' for the strong dual space E_b' of a normed space.

We compare the integral norm with the π-norm.

(4) *Let E, F be* (B)*-spaces and $w \in E' \otimes F'$. Then $\|w\|_I \leq \|w\|_\pi$.*

Proof. Let $u \in E'$, $v \in F'$, $z = \sum_{k=1}^{n} x_k \otimes y_k \in E \otimes F$. Then by the definition of the ε-norm we have $|\langle u \otimes v, z\rangle| \leqq \|u\| \|v\| \|z\|_\varepsilon$. If $w = \sum_{i=1}^{m} u_i \otimes v_i$, then $|\langle w, z\rangle| \leqq \left(\sum_i \|u\|_i \|v\|_i\right) \|z\|_\varepsilon$.

Let us suppose that we have a representation of w such that $\sum_i \|u\|_i \|v\|_i < \|w\|_\pi + \varepsilon$; then it follows that $\|w\|_I = \sup_{\|z\|_\varepsilon \leqq 1} |\langle w, z\rangle| < \|w\|_\pi + \varepsilon$. Since there exists such a representation of w for every $\varepsilon > 0$, we conclude that $\|w\|_I \leqq \|w\|_\pi$.

We make the useful observation

(5) *Let E, F be* (B)*-spaces. Then $E' \otimes_\varepsilon F'$, $E' \tilde{\otimes}_\varepsilon F'$, and $E' \varepsilon F'$ can be norm isomorphically embedded in $\mathscr{B}_b(E \times F) = (E \tilde{\otimes}_\pi F)' = \mathfrak{L}_b(E, F')$.*

Proof. By § 41, 3.(6) $\mathscr{B}_b(E \times F)$ and $\mathfrak{L}_b(E, F')$ are norm isomorphic. Furthermore, by § 44, 2.(6) $E' \varepsilon F'$ is norm isomorphic to $\mathscr{C}_b(E, F') \subset \mathfrak{L}_b(E, F')$. The other statements are immediate consequences.

Remark 1. If, moreover, E or F has finite dimension, then $E \tilde{\otimes}_\pi F = E \otimes_\pi F$ and $(E \tilde{\otimes}_\pi F)'_b = \mathfrak{L}_b(E, F') = E' \otimes_\varepsilon F'$ by (5).

Remark 2. If E and F both have finite dimension, then $E \otimes_\pi F$ and $E' \otimes_\varepsilon F'$, similarly $E \otimes_\varepsilon F$ and $E' \otimes_\pi F'$, are the strong duals of each other.

Next we consider the case that E and F are locally convex and E of finite dimension k, $E = \bigoplus_{i=1}^{k} \mathsf{K} e_i$. It follows from § 41, 6.(5) that $E \otimes_\pi F$ is isomorphic to $\bigoplus_{i=1}^{k} (\mathsf{K} e_i \otimes_\pi F)$. Now $\mathsf{K} e_i \otimes_\pi F = e_i \otimes F$ is by § 44, 1.(4) isomorphic to F; hence $E \otimes_\pi F$ is isomorphic to $F^k = \prod_{i=1}^{k} F_i$, $F_i = F$.

Similarly, using § 44, 5.(3), we get $E \otimes_\varepsilon F \cong F^k$, so we have

(6) *Let E, F be locally convex, E k-dimensional. Then one has the isomorphism $E \otimes_\pi F \cong F^k \cong E \otimes_\varepsilon F$.*

By taking strong duals we obtain

(7) $\quad (E \otimes_\varepsilon F)'_b \cong (E \otimes_\pi F)'_b \cong (F^k)'_b \cong (F'_b)^k \cong E \otimes_\pi F'_b \cong E \otimes_\varepsilon F'_b,$

where the central isomorphism can be found in § 22, 5.

We recall the definition of the natural topology $\mathfrak{T}_n$ on the bidual F'' of a locally convex space F (§ 23, 4.). By using § 22, 5.(1) we see that the equicontinuous sets of $(F^k)' = (F'_b)^k$ coincide with the subsets of products of equicontinuous subsets of F'; hence $((F'')_n)^k \cong ((F^k)'')_n$. Taking duals in (7) and using (6), we obtain the isomorphisms

$$(8) \qquad (E \otimes_\varepsilon F)''_n \cong (E \otimes_\pi F)''_n \cong ((F^k)'')_n \cong ((F'')_n)^k \cong E \otimes_\pi F''_n \cong E \otimes_\varepsilon F''_n.$$

These results are nearly trivial. For (B)-spaces the situation is much more delicate because in this case we are interested in norm isomorphisms and not only in isomorphisms as in (6), (7), and (8).

(9) *Let E, F be* (B)*-spaces, E of finite dimension. Then*

$$(E \otimes_\varepsilon F)'_b = \mathfrak{J}_I(E \times F) = E' \otimes_\pi F',$$

where equality means norm isomorphisms.

Since $\mathfrak{L}_b(E, F) = E' \otimes_\varepsilon F$, one has equivalently $\mathfrak{L}_b(E, F)'_b = E \otimes_\pi F'$.

We reproduce a proof for the second statement due to Lotz [2'].

a) Since E is finite dimensional, we have

$$E \otimes F' = E'' \otimes F' \subset \mathfrak{J}(E' \times F) \subset \mathfrak{L}(E', F') = E \otimes F',$$

so that every $w \in \mathfrak{L}_b(E, F)'$ is an element of $E \otimes F'$ and $\|w\|_I \leqq \|w\|_\pi$ by (4).

b) Next we construct for a given $\varepsilon > 0$ a topological isomorphism of $\mathfrak{L}_b(E, F)$ into a space $l^\infty_m(F)$, where m depends on ε.

The unit sphere S of E is compact; therefore for given $\varepsilon > 0$ there exists a set $M_\varepsilon = \{x_1, \ldots, x_m\} \subset S$ such that $S \subset \bigcup_{i=1}^m (x_i + \varepsilon U)$, U being the closed unit ball of E. Then $U = \complement(S) \subset \complement(M_\varepsilon) + \varepsilon U$. If x'_0 is a given element in U, then for $k = 1, 2, \ldots$ there exist $a_k \in \complement(M_\varepsilon)$ and $x'_k \in U$ such that $x'_{k-1} = a_k + \varepsilon x'_k$. This implies $x'_0 = \sum_{k=1}^\infty \varepsilon^{k-1} a_k \in [1/(1-\varepsilon)]\complement(M_\varepsilon)$, since $\complement(M_\varepsilon)$ is closed. Hence $U \subset [1/(1-\varepsilon)]\complement(M_\varepsilon)$.

Now let $l^\infty_m(F)$ be the (B)-space with the elements $(y_1, \ldots, y_m)$, $y_i \in F$, and $\|(y_1, \ldots, y_m)\| = \sup_i \|y_i\|$ (see § 26, 8.). For every $A \in \mathfrak{L}(E, F)$ we define $J(A) = (Ax_1, \ldots, Ax_m) \in l^\infty_m(F)$. Obviously, $\|J(A)\| = \sup_i \|Ax_i\| \leqq \sup_{\|x\| \leqq 1} \|Ax\| = \|A\|$, so J is linear and continuous.

Clearly, we have $(1-\varepsilon)U \subset \complement(M_\varepsilon)$; hence for $A \in \mathfrak{L}(E, F)$

$$(1-\varepsilon)\|A\| = \sup_{x \in (1-\varepsilon)U} \|Ax\| \leqq \sup_{z \in \complement(M_\varepsilon)} \|Az\| = \sup_{x_i \in M_\varepsilon} \|Ax_i\| = \|JA\|$$

holds. Therefore

$$(10) \qquad (1-\varepsilon)\|A\| \leqq \|JA\| \leqq \|A\|,$$

J is an isomorphism of $\mathfrak{L}_b(E, F)$ on a closed subspace of $l^\infty_m(F)$.

c) Take $w \in \mathfrak{L}_b(E, F)' = J_I(E' \times F) = (E \otimes F')_I$ (see a) and Remark 1); $\|w\| = 1$. We define w' on the closed subspace $J(\mathfrak{L}(E, F))$ of $l_m^\infty(F)$ by $\langle w', JA \rangle = \langle w, A \rangle$. The inequality $1 \leqq \|w'\| \leqq 1/(1 - \varepsilon)$ is an easy consequence of (10).

$l_m^1(F')$ is the strong dual of $l_m^\infty(F)$ (this is nearly obvious; see also § 26, 8. for related proofs). Using HAHN–BANACH, we obtain a norm-preserving extension $\hat{w}$ of w' to $l_m^\infty(F)$ and $\hat{w}$ is of the form $\hat{w} = (\hat{v}_1, \ldots, \hat{v}_m)$, $\hat{v}_i \in F'$, $1 \leqq \|\hat{w}\| = \sum_{i=1}^m \|\hat{v}_i\| \leqq 1/(1 - \varepsilon)$. Hence

$$\langle w, A \rangle = \langle \hat{w}, JA \rangle = \langle \hat{w}, (Ax_1, \ldots, Ax_m) \rangle$$
$$= \sum_{i=1}^m \hat{v}_i(Ax_i) = \left\langle \sum_i \hat{v}_i \otimes x_i, A \right\rangle = \langle \tilde{w}, A \rangle.$$

Therefore $w \in E \otimes F'$ is represented by an element $\tilde{w} = \sum_{i=1}^m \hat{v}_i \otimes x_i$ of $E \otimes F'$ such that $\sum_i \|\hat{v}_i\| \|x_i\| \leqq 1/(1 - \varepsilon)$. Therefore $\|w\|_\pi \leqq \sum_i \|\hat{v}_i\| \|x_i\| \leqq 1/(1 - \varepsilon)$. Our arguments are true for every $\varepsilon > 0$; hence we have $\|w\|_\pi \leqq 1 = \|w\|_I$. This, together with a), gives $\|w\|_\pi = \|w\|_I$ and this implies (9).

We note the corollary

(11) *Let E, F be* (B)*-spaces, E of finite dimension. Then*

$$(E \otimes_\varepsilon F)'' = E \otimes_\varepsilon F'' \quad \textit{and} \quad \mathfrak{L}_b(E, F)'' = \mathfrak{L}_b(E, F'').$$

Proof. $(E \otimes_\varepsilon F)'_b = E' \otimes_\pi F'$, $(E' \otimes_\pi F')'_b = \mathscr{B}_b(E' \times F') = E \otimes_\varepsilon F''$ and $\mathfrak{L}_b(E, F) = E' \otimes_\varepsilon F$, $\mathfrak{L}_b(E, F)'' = E' \otimes_\varepsilon F'' = \mathfrak{L}_b(E, F'')$. This is Lemma § 43, 8.(2).

2. A theorem of SCHATTEN. The results of 1. will certainly not be the final answer to our problem of characterizing $\mathfrak{J}(E \times F)$. The first important result in this direction is due to SCHATTEN [2'], who settled the case of Hilbert spaces by using Hilbert space methods. His main result corresponds to 1.(9).

(1) *If E, F are Hilbert spaces, then $\mathfrak{J}_I(E \times F)$ is norm isomorphic to $E' \tilde{\otimes}_\pi F'$.*

Proof. a) We will use a representation of Hilbert space mappings by diagonal mappings going back to KÖTHE [9'].

We recall the polar decomposition of any $A \in \mathfrak{L}(E, F)$. The mapping $C = (A^*A)^{1/2} \in \mathfrak{L}(E)$ is nonnegative, $(Cx, x) \geqq 0$, and one has

$$\|Cx\|^2 = (Cx, Cx) = (C^2x, x) = (A^*Ax, x) = (Ax, Ax) = \|Ax\|^2.$$

From this it follows that the equation $UCx = Ax$ defines unambiguously an isometry U of $C(E)$ into F since

$$(UCx, UCx) = (Ax, Ax) = (Cx, Cx).$$

Extending U to $\overline{C(E)}$ and defining $U = 0$ on $C(E)^{\perp}$, we find a partial isometry $U \in \mathfrak{L}(E, F)$ and $A = UC$ is the polar decomposition of $A \in \mathfrak{L}(E, F)$.

b) We determine now a decomposition of C. Let $C = \int_{0-}^{\infty} \lambda \, dP_\lambda$ be the spectral decomposition of C, where P_λ is continuous in λ from the right. Since $C \geqq 0$, one has $P_\lambda = 0$ for $\lambda < 0$. For $\lambda \geqq 0$ we write $\lambda = \delta(\lambda)\varphi(\lambda)$, where

$$\delta(\lambda) = 2^n \quad \text{for } 2^n < \lambda \leqq 2^{n+1}, \qquad n = 0, \pm 1, \pm 2, \ldots, \delta(0) = 0,$$
$$\varphi(\lambda) = \lambda/2^n \quad \text{for } 2^n < \lambda \leqq 2^{n+1}, \qquad n = 0, \pm 1, \pm 2, \ldots, \varphi(0) = 1.$$

Observe that $1 \leqq \varphi(\lambda) \leqq 2$ and $1/2 \leqq 1/\varphi(\lambda) \leqq 1$ for $\lambda \geqq 0$. We define $P = \int_{0-}^{\infty} \varphi(\lambda) \, dP_\lambda$; then

$$P^{-1} = \int_{0-}^{\infty} \frac{1}{\varphi(\lambda)} \, dP_\lambda, \qquad \|P\| \leqq 2, \qquad \|P^{-1}\| \leqq 1.$$

Finally, we define $D = \int_{0-}^{\infty} \delta(\lambda) \, dP_\lambda$. By the rules of the functional calculus we obtain immediately the decomposition

$$C = PD = DP, \qquad A = UPD,$$

where $D \in \mathfrak{L}(E, F)$ is a diagonal mapping since D can be written as

$$D = \sum_{n=-\infty}^{+\infty} 2^n (P_{2^{n+1}} - P_{2^n}).$$

c) We need the representation $A = UPD$ in a more concrete form. The kernel of D is $P_0(E)$ and its orthogonal complement $(I - P_0)(E) = \overline{\sum_{n=-\infty}^{\infty} (P_{2^{n+1}} - P_{2^n})(E)}$ has an orthonormal base $\{e_\alpha\}$, $\alpha \in \mathrm{A}$, such that $De_\alpha = \lambda_\alpha e_\alpha$ with $\lambda_\alpha = 2^n$ for $e_\alpha \in (P_{2^{n+1}} - P_{2^n})(E)$. We write $UPe_\alpha = f_\alpha$; then $Ae_\alpha = \lambda_\alpha f_\alpha$ and we have

$$Ax = \sum_\alpha \lambda_\alpha (x, e_\alpha) f_\alpha \quad \text{for } x \in E, \ \lambda_\alpha \geqq 0, \tag{2}$$

where only countably many (x, e_α) are $\neq 0$.

We note that $\{f_\alpha\}$ is not necessarily an orthogonal system, but $\|f_\alpha\| = \|UPe_\alpha\| \leqq \|P\| \leqq 2$.

Every e_α is in the range $D(E)$; hence $Pe_\alpha \in PD(E) = C(E)$ and $P^{-1}e_\alpha \in P^{-1}(D(E)) = DP^{-1}(E) \subset D(E) = DPP^{-1}(E) \subset DP(E) = C(E)$ and so $P^{-1}e_\alpha \in C(E)$. We define $g_\alpha = UP^{-1}e_\alpha \in F$ and observe $\|g_\alpha\| \leqq \|P^{-1}\| \leqq 1$. Since U is an isometry on $C(E)$, we obtain the relations

(3) $(g_\beta, f_\alpha) = (UP^{-1}e_\beta, UPe_\alpha) = (P^{-1}e_\beta, Pe_\alpha) = (e_\beta, e_\alpha) = \delta_{\beta\alpha}$.

With these facts from Hilbert space theory the proof of (1) will now be straightforward.

d) Suppose $B \in \mathfrak{J}(E \times F) \subset \mathscr{B}(E \times F)$. The corresponding $\tilde{B} = A \in \mathfrak{L}(E, F')$ has by (2) a representation

$$Ax = \sum_\alpha \lambda_\alpha (x, e_\alpha) f_\alpha = \sum \lambda_\alpha \langle \bar{e}_\alpha, x\rangle f_\alpha, \qquad x \in E,\ \bar{e}_\alpha \in E',\ f_\alpha \in F'.$$

We use the notations of § 42, 4. and recall that $\{\bar{e}_\alpha\}$, $\alpha \in \mathrm{A}$, is an orthonormal system in E'. It follows that for $y \in F$

$$\begin{aligned} B(x, y) = (\tilde{B}x)y &= \sum_\alpha \lambda_\alpha \langle \bar{e}_\alpha, x\rangle\langle f_\alpha, y\rangle \\ &= \sum_\alpha \lambda_\alpha (x, e_\alpha)(y, \bar{f}_\alpha), \qquad \lambda_\alpha \geqq 0,\ \bar{f}_\alpha \in F. \end{aligned}$$

Since $B \in \mathfrak{J}(E \times F)$, we have $|\langle B, z\rangle| \leqq \|B\|_I < \infty$ for all $z \in E \otimes F$, $\|z\|_\varepsilon \leqq 1$.

Consider now an element $z = \sum_{\beta\in\Phi} (e_\beta \otimes \bar{g}_\beta)$, where Φ is a finite subset of A and where the $g_\beta \in F'$ are defined as before. Then one has

$$\begin{aligned} \|z\|_\varepsilon &\leqq \sup_{\|u\|\leqq 1, \|v\|\leqq 1} \sum_\beta |\langle u, e_\beta\rangle\langle v, \bar{g}_\beta\rangle| \\ &\leqq \sup_{\|u\|\leqq 1} \Big(\sum_\beta |\langle u, e_\beta\rangle|^2\Big)^{1/2} \sup_{\|v\|\leqq 1} \Big(\sum_\beta |\langle v, \bar{g}_\beta\rangle|^2\Big)^{1/2}. \end{aligned}$$

Clearly, $\sum_\beta |\langle u, e_\beta\rangle|^2 \leqq 1$. Also

$$\begin{aligned} \sum_\beta |\langle v, \bar{g}_\beta\rangle|^2 &= \sum_\beta |\langle \bar{v}, g_\beta\rangle|^2 = \sum_\beta |\langle \bar{v}, UP^{-1}e_\beta\rangle|^2 \\ &= \sum_\beta |\langle (P^{-1})'U'\bar{v}, e_\beta\rangle|^2 \leqq 1, \end{aligned}$$

since $\|(P^{-1})'U'\bar{v}\| \leqq 1$. It follows that $\|z\|_\varepsilon \leqq 1$ and (3) implies

$$|\langle B, z\rangle| = \Big|\sum_\beta B(e_\beta, \bar{g}_\beta)\Big| = \Big|\sum_\beta \sum_\alpha \lambda_\alpha (e_\beta, e_\alpha)(\bar{g}_\beta, \bar{f}_\alpha)\Big| = \sum_{\beta\in\Phi} \lambda_\beta \leqq \|B\|_I.$$

Since $\sum_{\beta\in\Phi} \lambda_\beta \leqq \|B\|_I$ for every Φ, we have $\sum_\alpha \lambda_\alpha \leqq \|B\|_I < \infty$.

It follows that $A = \tilde{B}$ is nuclear and B can be identified with the element $\sum_\alpha \lambda_\alpha(\bar{e}_\alpha \otimes f_\alpha)$ of $E' \tilde{\otimes}_\pi F'$ since Hilbert space has the approximation property. Thus $\mathfrak{J}(E \times F)$ is algebraically a subspace of $E' \tilde{\otimes}_\pi F'$.

e) Conversely, assume $B_1 \in E' \tilde{\otimes}_\pi F'$, which means that $A_1 = \tilde{B}_1 \in \mathfrak{L}(E, F')$ is nuclear. By § 42, 6.(1) B_1 has a representation $B_1 = \sum_{n=1}^\infty \lambda_n \langle e_n, x\rangle\langle f_n, y\rangle$ with $\lambda_n \geqq 0$ and $\{\bar{e}_n\}$ and $\{f_n\}$ orthonormal systems in E' and F', respectively, such that $\|B_1\|_\pi = \|A_1\|_\nu = \sum_{n=1}^\infty \lambda_n$. We conclude, as in d), that $\|z\|_\varepsilon \leqq 1$ for every $z = \sum e_\beta \otimes \bar{f}_\beta \in E \otimes F$ and one verifies, as in d),

$$\|B_1\|_\pi = \|A_1\|_\nu = \sum_{n=1}^\infty \lambda_n \leqq \|B_1\|_I.$$

Since $\|B_1\|_I \leqq \|B_1\|_\pi$ by 1.(4) for all $B_1 \in E' \otimes F'$, we have $\|B_1\|_I = \|B_1\|_\pi$ on $E' \otimes F'$. But then for every $B_1 \in E' \tilde{\otimes}_\pi F'$ follows

$$\|B_1\|_\pi = \lim_{m\to\infty} \left\| \sum_{n=1}^m \lambda_n(\bar{e}_n \otimes f_n) \right\|_\pi = \lim_{m\to\infty} \sum_{n=1}^m \lambda_n$$

$$= \lim_{m\to\infty} \left\| \sum_{n=1}^m \lambda_n(\bar{e}_n \otimes f_n) \right\|_I = \|B\|_I.$$

This and the result of d) imply the algebraic and norm isomorphism of $\mathfrak{J}_I(E \times F)$ and $E' \tilde{\otimes}_\pi F'$.

From (1) and § 41, 3.(6) we obtain the theorem of SCHATTEN:

(4) *Let E and F be Hilbert spaces. Then $E \tilde{\otimes}_\varepsilon F$ is the subspace of all compact mappings of $\mathfrak{L}_b(E', F)$. Its strong dual $\mathfrak{J}_I(E \times F)$ coincides with $E' \tilde{\otimes}_\pi F'$, the space of all nuclear mappings in $\mathfrak{L}(E, F')$, and the π-norm coincides with the nuclear norm ν.*

The strong dual of $E' \tilde{\otimes}_\pi F'$, which is the strong bidual of $E \tilde{\otimes}_\varepsilon F$, coincides with $\mathscr{B}_b(E' \times F') = \mathfrak{L}_b(E', F)$, the space of all continuous linear mappings of E' in F.

We observe that if E and F are reflexive (B)-spaces, (4) shows that $E \tilde{\otimes}_\varepsilon F$, $E \tilde{\otimes}_\pi F$, and $\mathfrak{L}_b(E, F)$ need not be reflexive spaces.

On the other hand, (4) shows also that in the case of Hilbert space the integral and the π-norm coincide on $\mathfrak{J}(E, F)$. This is not true in general as we will see in 6.

3. BUCHWALTER's results on duality. We give an exposition of recent results of BUCHWALTER [1'] on a duality between ε- and π-tensor products of (F)- and (DF)-spaces. BIERSTEDT and MEISE [2'] have pointed out that by using the ε-product instead of the ε-tensor product it is possible to drop the assumption of the approximation property in one of BUCHWALTER's theorems. We will show that this is possible also in the second theorem.

The first duality theorem of BUCHWALTER says

(1) *Let* E, F *be* (F)-*spaces; then* $(E \tilde{\otimes}_\pi F)'_c = E'_c \varepsilon F'_c$ *and* $(E'_c \varepsilon F'_c)'_c = E \tilde{\otimes}_\pi F$.

Proof. By § 41, 3.(3) we have $(E \tilde{\otimes}_\pi F)' = \mathscr{B}(E \times F)$. On the other hand, it follows from § 43, 3.(3) and from the polar reflexivity of (F)-spaces (§ 23, 9.(5)) that $E'_c \varepsilon F'_c = \mathfrak{X}_e^{(\mathfrak{C},\mathfrak{C})}((E'_c)'_c \times (F'_c)'_c) = \mathfrak{X}_e^{(\mathfrak{C},\mathfrak{C})}(E \times F)$, the space of $(\mathfrak{C}, \mathfrak{C})$-hypocontinuous bilinear forms on $E \times F$ (where $\mathfrak{C}$ is the class of all relatively compact subsets of E resp. F). Since every separately continuous bilinear form on $E \times F$ is continuous (§ 40, 2.(1)), it follows that $\mathfrak{X}^{(\mathfrak{C},\mathfrak{C})}(E \times F) = \mathscr{B}(E \times F)$; thus $(E \tilde{\otimes}_\pi F)'$ and $E'_c \varepsilon F'_c$ coincide algebraically.

Using the remark following § 41, 4.(5), we see that $\mathfrak{T}_c$ on $(E \tilde{\otimes}_\pi F)'$ is the topology of uniform convergence on the sets $C_1 \otimes C_2$, where C_1 and C_2 are relatively compact subsets of E and F, respectively. But this topology coincides with the topology $\mathfrak{T}_e$ of bi-equicontinuous convergence, since the equicontinuous subsets in E and F for E'_c and F'_c, respectively, are the relatively compact subsets. But $\mathfrak{T}_e$ is also the topology on $E'_c \varepsilon F'_c$; hence $E'_c \varepsilon F'_c = \mathscr{B}_e(E \times F) = (E \tilde{\otimes}_\pi F)'_c$. By polar reflexivity we obtain $((E \tilde{\otimes}_\pi F)'_c)'_c = E \tilde{\otimes}_\pi F$ and this implies the second statement in (1).

The proof of the second duality theorem needs some preparation.

ADASCH and ERNST [1'] call a locally convex space $E[\mathfrak{T}]$ locally topological if it has the following property: An absolutely convex set $U \subset E$ is a $\mathfrak{T}$-neighbourhood of $\mathfrak{o}$ in E if $M \cap U$ is a $\mathfrak{T}$-neighbourhood of $\mathfrak{o}$ in M for every bounded subset M of E containing $\mathfrak{o}$.

E is called σ-locally topological if E has, moreover, a fundamental sequence $M_1 \subset M_2 \subset \cdots$ of absolutely convex bounded sets.

By § 29, 3.(2) every (DF)-space is σ-locally topological.

(2) *Let E be σ-locally topological with the fundamental sequence* $M_1 \subset M_2 \subset \cdots$ *of absolutely convex bounded subsets. Let* U_n, $n = 1, 2, \ldots$, *be a sequence of absolutely convex neighbourhoods of* $\mathfrak{o}$ *in E.*

Then $U = \bigcap_{k=1}^{\infty} (U_k + M_k)$ *is a neighbourhood of* $\mathfrak{o}$ *in E which is absorbed by every* U_n, $n = 1, 2, \ldots$.

Proof. One has $U_m + M_m \supset M_n$ for $m \geqq n$; hence $U \cap M_n \supset \left(\bigcap_{k=1}^{n} U_k\right) \cap M_n$ is for every n a neighbourhood of $\mathfrak{o}$ in M_n and U is a neighbourhood in E. From $U \subset U_n + M_n \subset (1 + \rho_n)U_n$ for suitable $\rho_n > 0$ follows the second statement.

HOLLSTEIN [1′] proved the following generalization of § 40, 2.(10):

(3) *Let E, F be σ-locally topological spaces and let H be an equihypocontinuous set of bilinear mappings of $E \times F$ into the locally convex space G. Then H is equicontinuous.*

Proof. Let M_n, N_n, $n = 1, 2, \dots$, be fundamental sequences of bounded subsets of E and F, respectively. Let W be an absolutely convex neighbourhood of $\mathfrak{o}$ in G. Then there exist neighbourhoods U_n, V_n of $\mathfrak{o}$ in E and F, respectively, such that

$$H(U_n, N_n) \subset W, \qquad H(M_n, V_n) \subset W.$$

It follows that $H(U_n + M_n, N_n \cap V_n) \subset 2W$, $H\left(\bigcap_{k=1}^{\infty} (U_k + M_k), N_n \cap V_n\right) \subset 2W$ for all $n = 1, 2, \dots$.

$U = \bigcap_{k=1}^{\infty} (U_k + M_k)$ is a neighbourhood of $\mathfrak{o}$ in E by (2) and we have $H(U, N_n \cap V_n) \subset 2W$ for all n. Since F is σ-locally topological, $V = \Gamma_{n=1}^{\infty} (N_n \cap V_n)$ is a neighbourhood of $\mathfrak{o}$ in F. Hence $H(U, V) \subset 2W$; H is equicontinuous.

If F is an (F)-space, then by the BANACH–DIEUDONNÉ theorem (§ 21, 10.(1)) the topology $\mathfrak{T}_c$ coincides on E' with the topology given by the absolutely convex sets which intersect all bounded sets containing $\mathfrak{o}$ in $\mathfrak{T}_c$-neighbourhoods of $\mathfrak{o}$. Hence E'_c is σ-locally topological and we note as a particular case of (3)

(4) *Let E, F be* (F)*-spaces, H an equihypocontinuous set of bilinear forms on $E'_c \times F'_c$. Then H is equicontinuous.*

We are now able to prove the second duality theorem of BUCHWALTER [1′]:

(5) *Let E, F be* (F)*-spaces. Then $(E\varepsilon F)'_c = E'_c \tilde{\otimes}_\pi F'_c$ and $(E'_c \tilde{\otimes}_\pi F'_c)'_c = E\varepsilon F$.*

Proof. a) $E\varepsilon F = \mathscr{B}_e(E'_c \times F'_c)$. By § 43, 3.(3) $E\varepsilon F = \mathfrak{X}_e^{(\mathfrak{E},\mathfrak{E})}(E'_c \times F'_c)$. The equicontinuous subsets of E'_c and F'_c are the bounded subsets; hence $\mathfrak{X}_e^{(\mathfrak{E},\mathfrak{E})}(E'_c \times F'_c) = \mathfrak{X}_e(E'_c \times F'_c)$, the class of all hypocontinuous bilinear forms on $E'_c \times F'_c$. It follows from (4) that $\mathfrak{X}_e(E'_c \times F'_c) = \mathscr{B}_e(E'_c \times F'_c)$, the class of all continuous bilinear forms on $E'_c \times F'_c$.

b) $E' \otimes F'$ is dense in $(E\varepsilon F)'_c$. A $\mathfrak{T}_e$-neighbourhood of $\mathfrak{o}$ in $\mathscr{B}_e(E'_c \times F'_c)$ is of the form $\{B \in \mathscr{B}_e; |B(M \times N)| \leqq 1\}$, where M and N are bounded subsets of E' and F', respectively. Therefore

$$\{B; |B(u, v)| \leqq 1\} = \{B; |\langle B, u \otimes v\rangle| \leqq 1\}$$

is a $\mathfrak{T}_e$-neighbourhood; hence $u \otimes v$ is, for $u \in E'$, $v \in F'$, a $\mathfrak{T}_e$-continuous linear form on $E\varepsilon F$ by a) and $E' \otimes F' \subset (E\varepsilon F)'$.

$E\varepsilon F$ is an (F)-space by § 44, 2.(7); hence $((E\varepsilon F)'_c)'_c = E\varepsilon F$ by § 23, 9.(5). We consider $H = E' \otimes F'$ as a subspace of $(E\varepsilon F)'_c$. Then $H^\circ = \mathfrak{o}$ in $((E\varepsilon F)'_c)'_c = E\varepsilon F$ and $(E\varepsilon F)'_c = H^{\circ\circ}$ is the $\mathfrak{T}_c$-closure $\overline{H}$ of H in $(E\varepsilon F)'_c$.

c) We determine the topology $\mathfrak{T}_c$ on $(E\varepsilon F)'$. The relatively compact subsets of $E\varepsilon F$ are by § 44, 3.(2) the ε-equihypocontinuous subsets of $\mathfrak{X}^{(\mathfrak{C},\mathfrak{C})}(E'_c \times F'_c)$, which coincide with the equihypocontinuous subsets, and hence by (4) with the equicontinuous subsets of $\mathscr{B}(E'_c \times F'_c)$. Using § 41, 3.(4), we see that $\mathfrak{T}_c$ coincides with the π-topology on $E'_c \otimes F'_c$; hence $E'_c \otimes_\pi F'_c \subset (E\varepsilon F)'_c$. Since $(E\varepsilon F)'_c$ is complete by § 44, 2.(7) and § 21, 6.(4), we have $E'_c \tilde{\otimes}_\pi F'_c \subset (E\varepsilon F)'_c$. Using b), we obtain $(E\varepsilon F)'_c = E'_c \tilde{\otimes}_\pi F'_c$.

The second statement in (5) follows by taking the duals in the first statement and equipping them again with the topology $\mathfrak{T}_c$.

(6) *If E or F has the approximation property, then $E'_c\varepsilon F'_c = E'_c \tilde{\otimes}_\varepsilon F'_c$ in* (1) *and $E\varepsilon F = E \tilde{\otimes}_\varepsilon F$ in* (5).

This is a consequence of § 43, 3.(7) in both cases, recalling that E'_c and F'_c are complete for (F)-spaces E and F.

The theorems resulting from (1) and (5) by using (6) are the original theorems of BUCHWALTER.

The duality established by (1) and (5) can be described in the following way. Denote by $(\mathscr{F})$ the class of all (F)-spaces. $(\mathscr{F})$ contains with two spaces E and F the completed π-tensor product $E \tilde{\otimes}_\pi F$ and the ε-product $E\varepsilon F$ by § 44, 2.(7). Next we construct for every E the $\mathfrak{T}_c$-dual E'_c; then we have a one-one correspondence of $(\mathscr{F})$ and the class $(\mathscr{F}'_c)$ of all $\mathfrak{T}_c$-duals of (F)-spaces. If we take once more the $\mathfrak{T}_c$-dual, then we come back to $E = (E'_c)'_c \in (\mathscr{F})$. It follows immediately from (1) and (5) that $(\mathscr{F}'_c)$ contains with G and H the spaces $G \tilde{\otimes}_\pi H$ and $G\varepsilon H$ and that taking $\mathfrak{T}_c$-duals of spaces interchanges the completed π-tensor product and the ε-product. It seems remarkable that for this duality it is essential to use the ε-product and not the completed ε-tensor product.

Next we consider the class $(\mathscr{F}\mathscr{M})$ of all (FM)-spaces, which is a subclass of $(\mathscr{F})$. The topology $\mathfrak{T}_c$ on E' coincides with the strong topology $\mathfrak{T}_b$ and by § 27, 2.(2) E'_b is a (DFM)-space, i.e., a (DF)-space which is also an (M)-space (we note that a (DFM)-space is always complete by § 29, 5.(3)). Conversely, if F is a (DFM)-space, then $F'_c = F'_b$ is an (FM)-space by § 29, 3.(1) and again § 27, 2.(2).

Thus by taking the strong duals we obtain a one-one correspondence between the classes $(\mathscr{F}\mathscr{M})$ and $(\mathscr{D}\mathscr{F}\mathscr{M})$. The duality theorem (5) takes the following form:

(7) *If E and F are in $(\mathscr{F}\mathscr{M})$, then $E\varepsilon F$ is in $(\mathscr{F}\mathscr{M})$. If E and F are in $(\mathscr{D}\mathscr{F}\mathscr{M})$, then $E \tilde{\otimes}_\pi F$ is in $(\mathscr{D}\mathscr{F}\mathscr{M})$. Furthermore,*
i) $(E\varepsilon F)'_b = E'_b \tilde{\otimes}_\pi F'_b$ *for E, F in $(\mathscr{F}\mathscr{M})$,*
ii) $(E \tilde{\otimes}_\pi F)'_b = E'_b \varepsilon F'_b$ *for E, F in $(\mathscr{D}\mathscr{F}\mathscr{M})$.*

Proof. By § 44, 3.(10) $E\varepsilon F$ is an (FM)-space and i) follows immediately from (5). Reading i) from the right, one sees that the completed π-product of two (DFM)-spaces is again a (DFM)-space and ii) is a consequence of the second equality in (5).

This is only half of the duality we expect to be true for (FM)- and (DFM)-spaces. Unfortunately, we do not know whether $E \tilde{\otimes}_\pi F$ is again an (FM)-space if E and F are (FM)-spaces or, equivalently, whether $E\varepsilon F$ of two (DFM)-spaces is again a (DFM)-space. A positive answer would give the full duality. A partial solution of this problem was recently given by Hollstein [2′].

4. Canonical representations of integral bilinear forms. Let A be an element of $\mathfrak{L}(E, F)$, where E and F are locally convex. Then $B_A(x, v) = \langle Ax, v\rangle$, $x \in E$, $v \in F'$, is a continuous bilinear form on $E \times F'_b$. We say that A is an integral mapping if $B_A(x, v)$ is an integral bilinear form on $E \times F'_b$. We denote by $\mathfrak{L}^I(E, F)$ the vector space of all integral mappings of E in F.

If the spaces involved are normed, one introduces again the integral norm (cf. 1.(3)) on $\mathfrak{L}^I(E, F)$:

$$(1)\quad \|A\|_I = \|B_A\|_I = \sup_{\|z\|_\varepsilon \leq 1} |\langle B_A, z\rangle| = \sup_{\|\sum x_i \oplus v_i\|_\varepsilon \leq 1} \left|\sum_i v_i(Ax_i)\right|.$$

In the case of (B)-spaces the correspondence $A \to B_A$ gives a norm isomorphism of $\mathfrak{L}^I(E, F)$ into $\mathfrak{J}_I(E \times F')$. Since $E' \otimes F'' \subset \mathfrak{J}(E \times F')$ by 1.(1), it is obvious that in general the image of $\mathfrak{L}^I(E, F)$ is a strict subspace of $\mathfrak{J}(E \times F')$.

Results on integral bilinear forms can be translated into results on integral mappings. A first example is

(2) *Let $B(y_1, y_2)$ be an integral bilinear form on $F_1 \times F_2$ and let $A_1 \in \mathfrak{L}(E_1, F_1)$, $A_2 \in \mathfrak{L}(E_2, F_2)$. Then the bilinear form $C(x_1, x_2) = B(A_1x_1, A_2x_2)$ on $E_1 \times E_2$ is integral.*

For normed spaces one has $\|C\|_I \leqq \|A_1\|\,\|A_2\|\,\|B\|_I$.

Proof. B can be considered as a continuous linear form on $F_1 \otimes_\varepsilon F_2$.

Now $A_1 \otimes A_2$ is a continuous mapping from $E_1 \otimes_\varepsilon E_2$ in $F_1 \otimes_\varepsilon F_2$ (§ 44, 4.(1)). This implies the first statement; the second follows from $\|A_1 \otimes A_2\| = \|A\|_1 \|A_2\|$.

The version for integral mappings is

(3) *Let B be an integral mapping from F_1 into F_2, $A_1 \in \mathfrak{L}(E_1, F_1)$, $A_2 \in \mathfrak{L}(F_2, E_2)$. Then the mapping A_2BA_1 from E_1 in E_2 is again integral.*

For normed spaces one has $\|A_2BA_1\|_I \leqq \|A_1\| \|A_2\| \|B\|_I$.

Proof. The bilinear form corresponding to B is $\langle By_1, v_2\rangle$, $y_1 \in F_1$, $v_2 \in F_2'$, and (2) gives the result in the form $C(x_1, u_2) = (BA_1x_1, A_2'u_2)$, where $x_1 \in E_1$, $u_2 \in E_2'$.

The use of the term "integral" is justified by a representation of the elements of $\mathfrak{J}(E \times F)$ due to GROTHENDIECK [13] which we now discuss.

The space $C(K)$ of all continuous functions on a compact topological space K is a (B)-space for the sup norm; its strong dual $\mathfrak{M}(K)$ is the space of all (Radon) measures μ on K. For $f \in C(K)$ one writes $\langle \mu, f\rangle = \mu(f) = \int_K f\, d\mu$.

Now let E, F be locally convex and G_1 and G_2 weakly closed equicontinuous subsets of E' and F', respectively. Then G_1 is $\mathfrak{T}_s(E)$-compact, G_2 is $\mathfrak{T}_s(F)$-compact, and $G_1 \times G_2$ is compact for the product topology $\mathfrak{T}_s(E) \times \mathfrak{T}_s(F)$. The spaces $C(G_1 \times G_2)$ and $\mathfrak{M}(G_1 \times G_2)$ are well defined.

$E \otimes F$ is embedded in $\mathscr{B}(E_s' \times F_s')$ by § 41, 2.(5) and § 41, 3.(3); thus every element z of $E \otimes F$ is a continuous bilinear form on $E_s' \times F_s'$ and has therefore a restriction $\tilde{z}$ to $G_1 \times G_2$ which lies in $C(G_1 \times G_2)$. Since the elements of $E \tilde{\otimes}_\varepsilon F$ are on $G_1 \times G_2$ uniform limits of elements of $E \otimes F$, each $z \in E \tilde{\otimes}_\varepsilon F$ has a restriction $\tilde{z}$ to $G_1 \times G_2$ which lies in $C(G_1 \times G_2)$. This implies the relation (cf. § 44, 2.)

$$(4) \quad \varepsilon_{G_1, G_2}(z) = \sup_{u \otimes v \in G_1 \otimes G_2} |(u \otimes v)z| = \sup_{(u,v) \in G_1 \times G_2} |\tilde{z}(u, v)| = \|\tilde{z}\|_{G_1 \times G_2}, \qquad z \in E \tilde{\otimes}_\varepsilon F,$$

where the last norm is the sup norm in $C(G_1 \times G_2)$.

Now let μ be an element of $\mathfrak{M}(G_1 \times G_2)$. We define a linear functional w on $E \tilde{\otimes}_\varepsilon F$ by

$$(5) \qquad w(z) = \mu(\tilde{z}) = \int_{G_1 \times G_2} \tilde{z}\, d\mu \quad \text{for every } z \in E \tilde{\otimes}_\varepsilon F.$$

It follows from (4) that

$$(5') \qquad |w(z)| \leqq \|\mu\| \|\tilde{z}\| = \|\mu\| \varepsilon_{G_1, G_2}(z);$$

hence $w \in (E \tilde{\otimes}_\varepsilon F)'$ and the bilinear form $B(x, y) \in \mathfrak{J}(E \times F)$ representing w is integral with

$$(6) \qquad B(x, y) = w(x \otimes y) = \int_{G_1 \times G_2} (ux)(vy)\, d\mu, \; x \in E, \; y \in F.$$

We will now prove that, conversely, every $w \in (E \tilde{\otimes}_\varepsilon F)'$ resp. $B \in \mathfrak{J}(E \times F)$ has an integral representation (5) resp. (6).

(7) i) *Let E, F be locally convex, G_1 and G_2 weakly closed equicontinuous subsets of E' and F', respectively. Then the equicontinuous subset $\overline{\Gamma(G_1 \otimes G_2)}$ of $\mathfrak{J}(E \times F)$ is the set of all bilinear forms* (6) *with $\mu \in \mathfrak{M}(G_1 \times G_2)$ and $\|\mu\| \leqq 1$. The corresponding elements of $(E \tilde{\otimes}_\varepsilon F)'$ are given by* (5).

ii) *If, moreover, G_1 and G_2 are circled, then we obtain all elements of $\overline{\Gamma(G_1 \otimes G_2)}$ by using only positive measures μ with $\|\mu\| \leqq 1$.*

Proof. i) If w has the form (5) and $\|\mu\| \leqq 1$, then $|w(z)| \leqq \varepsilon_{G_1 \times G_2}(z)$ by (5'); hence $w \in \Gamma(G_1 \otimes G_2)^{\circ\circ} = \overline{\Gamma(G_1 \otimes G_2)}$ (using the duality $\langle \mathfrak{J}(E \times F), E \tilde{\otimes}_\varepsilon F \rangle$).

Conversely, let w be in $\overline{\Gamma(G_1 \otimes G_2)}$. This is equivalent to $|w(z)| \leqq \varepsilon_{G_1 \times G_2}(z)$. We define $\tilde{w}(\tilde{z}) = w(z)$ for the restriction $\tilde{z}$ of z to $G_1 \times G_2$. It follows from (4) that $|\tilde{w}(\tilde{z})| \leqq \|\tilde{z}\|$ and thus $\tilde{w}(\tilde{z})$ is well defined on the subspace $H = \{\tilde{z}; z \in E \tilde{\otimes}_\varepsilon F\}$ of $C(G_1 \times G_2)$. Using HAHN–BANACH, one extends $\tilde{w}$ to a measure μ on $G_1 \times G_2$ such that $\|\mu\| \leqq 1$ and one has

$$w(z) = \tilde{w}(\tilde{z}) = \mu(\tilde{z}) = \int_{G_1 \times G_2} \tilde{z}\, d\mu \quad \text{for all } z \in E \tilde{\otimes}_\varepsilon F.$$

Thus w has a representation of the form (5) with $\|\mu\| \leqq 1$ and (6) follows.

ii) The point measure $\delta_{(u,v)}$ is a positive measure on $G_1 \times G_2$ of norm 1. Let D be the set of all $\delta_{(u,v)}$, $(u, v) \in G_1 \times G_2$. We denote by $\hat{D}$ the set of all restrictions $\hat{\delta}_{(u,v)}$ of the $\delta_{(u,v)}$ to the subspace H of $C(G_1 \times G_2)$. We have $\hat{D} \subset H'$ and $\Gamma(G_1 \otimes G_2) \subset H'$.

Obviously, $\Gamma(\hat{D})^\circ = \{\tilde{z} \in H; \|\tilde{z}\| \leqq 1\} = \Gamma(G_1 \otimes G_2)^\circ$, the polars being taken in H, so that $\Gamma(\hat{D})^{\circ\circ} = \overline{\Gamma(G_1 \otimes G_2)}$ in H', where the closure is the $\mathfrak{T}_s(H)$-closure in H'. G_1 and G_2 are circled by assumption. It follows from the bilinearity of z that for every complex α, $|\alpha| \leqq 1$, $u \in G_1$, $v \in G_2$,

$$\langle \alpha \hat{\delta}_{(u,v)}, \tilde{z} \rangle = \alpha \tilde{z}(u, v) = \tilde{z}(\alpha u, v) = \langle \hat{\delta}_{(\alpha u, v)}, \tilde{z} \rangle.$$

Since $\alpha u \in G_1$, it follows that $\hat{D}$ is also circled; hence $\Gamma(\hat{D}) = C(\hat{D})$ and $\overline{\Gamma(\hat{D})} = \overline{C(\hat{D})}$ for the $\mathfrak{T}_s(H)$-closure. The subset $\overline{C(D)}$ of $\mathfrak{M}(G_1 \times G_2)$ consists of positive measures of norm $\leqq 1$ and is compact and closed for the topology $\mathfrak{T}_s(C(C_1 \times G_2))$. Let K be the canonical mapping of

$\mathfrak{M}(G_1 \times G_2)$ onto its quotient H'. Since K is weakly continuous, $K(\overline{\subset(D)})$ is $\mathfrak{T}_s(H)$-compact and we have $K(\overline{\subset(D)}) = \overline{\subset(\hat{D})} = \overline{\Gamma(\hat{D})} = \overline{\Gamma(G_1 \otimes G_2)}$. This means that every $\hat{\mu} \in \overline{\Gamma(G_1 \otimes G_2)}$ can be represented by a $\mu \in \overline{\subset(D)}$, $\|\mu\| \leqq 1$, $\mu \geqq 0$. Such a μ is a $\mathfrak{T}_s(C(G_1 \times G_2)$-limit of positive measures and therefore positive.

For normed spaces (7) can be replaced by

(8) *If E and F are normed spaces, then every $w \in \mathfrak{J}(E \times F)$ has a representation*

(8′) $w(z) = \mu(\tilde{z}) = \int_{U_1^\circ \times U_2^\circ} \tilde{z}\, d\mu$, *$\mu$ positive on $U_1^\circ \times U_2^\circ$, $\|w\|_I = \|\mu\|$,*

U_1 and U_2 being the closed unit balls in E and F, respectively.

Proof. It is sufficient to prove this for a w of norm $\|w\|_I = 1$. Then all the statements follow from (7) except $\|\mu\| = 1$. But (7) implies $\|\mu\| \leqq 1$ and it follows from (5′) that $\|w\|_I \leqq \|\mu\|$.

We note an important example. Let K be a compact space, ν a positive Radon measure on K, $\|\nu\| = \nu(K) = 1$. The formula

(9) $$J_0(f, g) = \int_K f(t)g(t)\, d\nu, \quad f \text{ and } g \text{ in } C(K),$$

defines a bilinear form on $C(K) \times C(K)$.

(10) *The bilinear form* (9) *is integral on $C(K) \times C(K)$ and $\|J_0\|_I = \|\nu\| = 1$.*

Proof. The mapping $t \rightarrow \delta_t$ embeds K into the unit ball of $\mathfrak{M}_s(K)$. Let μ be a measure on K. We define the measure $\tilde{\mu}$ on $K \times K$ by

$$\tilde{\mu}(h) = \int_{K \times K} h(t, t')\, d\tilde{\mu} = \int_K h(t, t)\, d\mu, \qquad h \in C(K \times K).$$

(9) can be written in the form

(9′) $$J_0(f, g) = \int_{K \times K} \langle \delta_t, f \rangle \langle \delta_{t'}, g \rangle\, d\tilde{\nu}.$$

This is a special case of (6) and therefore (9′) defines an integral bilinear form on $C(K) \times C(K)$ and a linear functional on $C(K) \tilde{\otimes}_\varepsilon C(K)$. We remark that $\tilde{\nu}$ is again positive, $\|\tilde{\nu}\| = \tilde{\nu}(K \times K) = \nu(K) = 1$, and we note that J_0 depends only on K and ν; finally, $\|J_0\|_I = \|\nu\| = 1$.

We know from § 44, 7.(3) that $C(K \times K) = C(K) \tilde{\otimes}_\varepsilon C(K)$; hence

(11) $$\mathfrak{J}_I(C(K) \times C(K)) = (C(K) \tilde{\otimes}_\varepsilon C(K))'_b = C(K \times K)'_b = \mathfrak{M}(K \times K).$$

We use this example to give a representation of an integral bilinear form $B(x, y) \in \mathfrak{J}(E \times F)$ which is slightly different from (6). Let $B(x, y)$ be defined by (6) for $K = G_1 \times G_2$ and $d\mu = d\nu$, $\nu(K) = 1$ (this is no restriction, since we may replace K by any positive multiple ρK). Then we introduce the mappings $A_1 \in \mathfrak{L}(E, C(K))$ and $A_2 \in \mathfrak{L}(F, C(K))$ defined by

$$A_1x: (u, v) \to (ux), \qquad A_2y: (u, v) \to (vy), \qquad (u, v) \in G_1 \times G_2.$$

Using (6) and (9), we obtain

(12) $$B(x, y) = \int_{G_1 \times G_2} (ux)(vy)\, d\nu = J_0(A_1x, A_2y),$$

where $\nu \in \mathfrak{M}(G_1 \times G_2)$, $\nu(G_1 \times G_2) = 1$.

We note that in the case of normed spaces one has always $\|A_1\| \leqq 1$ and $\|A_2\| \leqq 1$.

If f and g are arbitrary elements of $\mathscr{L}_\nu^\infty(K)$, K compact, the measure ν on K positive and $\|\nu\| = 1$, then

(13) $$J_\infty(f, g) = \int_K f(t)g(t)\, d\nu$$

is a bilinear form on $\mathscr{L}_\nu^\infty(K) \times \mathscr{L}_\nu^\infty(K)$ and, similar to (10), we have

(14) *J_∞ is integral on $\mathscr{L}_\nu^\infty(K) \times \mathscr{L}_\nu^\infty(K)$ and* $\|J_\infty\|_I = 1$.

Proof. We show first that

$$|\langle J_\infty, z\rangle| \leqq \|z\|_\varepsilon \quad \text{for all } z = \sum_i f_i \otimes g_i \in \mathscr{L}_\nu^\infty(K) \otimes_\varepsilon \mathscr{L}_\nu^\infty(K).$$

Since the simple functions on K are dense in $\mathscr{L}_\nu^\infty(K)$, we will assume that the f_i and g_i are simple functions.

A simple function has the form $s = \sum_{i=1}^n \alpha_i \chi_{M_i}$, where the α_i are complex numbers and χ_{M_i} is the characteristic function of M_i and $K = \bigcup_{i=1}^n M_i$ is a disjoint union of sets of positive measure $\nu(M_i)$.

If $s' = \sum_j \beta_j \chi_{N_j}$ is a second simple function on K, then $s + s' = \sum (\alpha_i + \beta_j)\chi_{M_i \cap N_j}$ is again a simple function (we omit sets $M_i \cap N_j$ with measure 0).

Using this remark, we see that it is sufficient to prove $|\langle J_\infty, z\rangle| \leqq \|z\|_\varepsilon$

for elements of the form $z' = \sum_{p,q} \gamma_{pq}(\chi_{M_p} \otimes \chi_{N_q})$, where $K \times K = \bigcup_{p,q} M_p \times N_q$ is a union of disjoint sets of positive measure.

From (13) and $\nu(K) = 1$ follows

$$|\langle J_\infty, z' \rangle| = \left| \sum_{p,q} \gamma_{pq} \int_K \chi_{M_p}(t) \chi_{N_q}(t) \, d\nu \right| = \left| \sum_{p,q} \gamma_{pq} \int \chi_{M_p \cap N_q} \, d\nu \right| \leqq \sup_{p,q} |\gamma_{pq}|.$$

Next we prove

$$(15) \qquad \|z'\|_\varepsilon = \sup_{p,q} |\gamma_{pq}|.$$

The unit ball of $\mathscr{L}_\nu^1(K)$ is weakly dense in the unit ball of $\mathscr{L}_\nu^\infty(K)'$; hence for $u, v \in \mathscr{L}_\nu^1(K)$ one has

$$\begin{aligned} \|z'\|_\varepsilon &= \sup_{\|u\| \leqq 1, \|v\| \leqq 1} \left| \sum_{p,q} \gamma_{pq} \langle u, \chi_{M_p} \rangle \langle v, \chi_{N_q} \rangle \right| \\ &= \sup \left| \int_{K \times K} u(t) v(t') \sum_{p,q} \gamma_{pq} \chi_{M_p}(t) \chi_{N_q}(t') \, d(\nu \times \nu) \right| \\ &\leqq \left\| \sum_{p,q} \gamma_{pq} \chi_{M_p}(t) \chi_{N_q}(t') \right\|_{\mathscr{L}_{\nu \times \nu}^\infty(K \times K)} \leqq \sup |\gamma_{pq}|. \end{aligned}$$

If we choose $u = [1/\nu(M_r)]\chi_{M_r}$ and $v = [1/\nu(N_s)]\chi_{N_s}$, we have $|\langle u \otimes v, z' \rangle| = |\sum \gamma_{pq} \langle u, \chi_{M_p} \rangle \langle v, \chi_{N_q} \rangle| = \gamma_{rs}$, which implies (15). Thus we have proved $\|J_\infty\|_I \leqq 1$.

For $z = 1 \otimes 1$ one has $J_\infty(1, 1) = \nu(K) = 1$ and (14) follows.

We use J_∞ for a representation of integral bilinear forms similar to (12).

(16) *Let E, F be locally convex (normed) spaces. $B \in \mathscr{B}(E \times F)$ is integral (and $\|B\|_I \leqq 1$) if and only if there exists a compact space K, a positive Radon measure ν on K with $\nu(K) = 1$, and mappings $A_1 \in \mathfrak{L}(E, \mathscr{L}_\nu^\infty(K))$, $A_2 \in \mathfrak{L}(F, \mathscr{L}_\nu^\infty(K))$ (with $\|A_1\| \leqq 1$ and $\|A_2\| \leqq 1$) such that*

$$(17) \qquad B(x, y) = J_\infty(A_1 x, A_2 y) = \int_K A_1(x)(t) A_2(y)(t) \, d\nu.$$

Proof. If (17) is true, then B is integral by (2) since J_∞ is integral by (14). If the spaces are normed, then $\|B\|_I \leqq 1$ follows again by (2).

Conversely, if B is integral we have the representation (12), which is for $K = G_1 \times G_2$ a representation (17) since A_1 and A_2 may be considered as mappings into $\mathscr{L}_\nu(K)$ instead of $C(K)$ (with norms $\leqq 1$ in the case of normed spaces E and F).

5. Integral mappings. The results of 4. on integral bilinear forms can be translated into factorization theorems for integral mappings.

We need an extension property of bilinear forms. Let E, F be locally convex spaces. It follows from § 40, 3.(5) that a continuous bilinear form $B \in \mathscr{B}(E \times F)$ is separately weakly continuous on E and F, respectively, and that B has a uniquely determined extension $\hat{B}$ to $E \times F''_n$, $\hat{B} \in \mathscr{B}(E \times F''_n)$ which is again separately continuous on E for $\mathfrak{T}_s(E')$ and on F'' for $\mathfrak{T}_s(F')$.

We recall the definition: $\hat{B}(x, z) = \lim_\alpha B(x, y_\alpha)$, where $y_\alpha \in F$ is a net weakly convergent to $z \in F''$.

(1) *Let E, F be locally convex.*

a) *$B_0 \in \mathscr{B}(E \times F)$ is integral if and only if $\hat{B}_0 \in \mathscr{B}(E \times F''_n)$ is integral, where $\mathfrak{T}_n$ denotes the natural topology of the bidual.*

Moreover, if E and F are normed spaces, then $\|\hat{B}_0\|_I = \|B_0\|_I$, so that $\mathfrak{J}_I(E \times F)$ is norm isomorphically embedded in $\mathfrak{J}_I(E \times F'')$.

b) *The subspace $E \otimes_\varepsilon F$ is $\mathfrak{T}_s(\mathfrak{J}(E \times F))$-dense in $E \otimes_\varepsilon F''_n$.*

Moreover, if E, F are (B)*-spaces, then the closed unit ball U of $E \otimes_\varepsilon F$ is $\mathfrak{T}_s(\mathfrak{J}(E \times F))$-dense in the closed unit ball V of $E \otimes_\varepsilon F''$.*

Proof. i) We assume that $\hat{B}_0$ is integral, i.e., continuous on $E \otimes_\varepsilon F''_n$. By § 44, 4.(6) the restriction B_0 of $\hat{B}_0$ to $E \otimes_\varepsilon F$ is again integral.

ii) We prove b) first for locally convex E and F. Every element of $E \otimes_\varepsilon F''_n$ is of the form $\sum_{i=1}^{g} x_i \otimes z_i$, $x_i \in E$, $z_i \in F''$. It is sufficient to show that every $x \otimes z$ is the $\mathfrak{T}_s(\mathfrak{J}(E \times F))$-limit of a net $x \otimes y_\alpha$, where $y_\alpha \in F$. By § 23, 2.(3) z is the $\mathfrak{T}_s(F')$-limit of a bounded net $y_\alpha \in F$ and for every $B \in \mathfrak{J}(E \times F)$, $B(x, y) = B_x(y)$ is weakly continuous in y; therefore $\lim_\alpha B_x(y_\alpha) = B_x(z) = \hat{B}(x, z)$. Hence $x \otimes y_\alpha$ $\mathfrak{T}_s(\mathfrak{J}(E \times F))$-converges to $x \otimes z$ and b) is true for locally convex E and F.

We have also shown that every $B \in \mathfrak{J}(E \times F)$ has a $\mathfrak{T}_s(\mathfrak{J}(E \times F))$-extension to $E \otimes_\varepsilon F''_n$ which coincides with $\hat{B}$.

iii) Let E, F be (B)-spaces. Then the closed unit ball of $(E \otimes_\varepsilon F)''$ is the $\mathfrak{T}_s(\mathfrak{J}(E \times F))$-closure of U by § 23, 2.; hence U will be $\mathfrak{T}_s(\mathfrak{J}(E \times F))$-dense in V if we prove $E \otimes_\varepsilon F'' \subset (E \otimes_\varepsilon F)''$.

Now $E \otimes_\varepsilon F'' = \bigcup_H H \otimes_\varepsilon F''$, where H is a finite dimensional subspace of E and $H \otimes_\varepsilon F'' = (H \otimes_\varepsilon F)''$ by 1.(11). If $Y \subset X$ for (B)-spaces, then $Y'' \subset X''$ canonically; hence $H \otimes_\varepsilon F'' \subset (E \otimes_\varepsilon F)''$ and $E \otimes_\varepsilon F'' = \bigcup_H H \otimes_\varepsilon F'' \subset (E \otimes_\varepsilon F)''$ and this proves b).

iv) We assume now that B_0 is integral. Then B_0 is $\mathfrak{T}_s(\mathfrak{J}(E \times F))$-continuous on $E \otimes_\varepsilon F$ and there exists an absolutely convex $\mathfrak{T}_\varepsilon$-neighbourhood U of $\mathfrak{o}$ in $E \otimes_\varepsilon F$ such that $\sup_{s \in U} |\langle B_0, s\rangle| = 1$. B_0 has by b) the uniquely defined $\mathfrak{T}_s(\mathfrak{J}(E \times F))$-continuous extension $\hat{B}_0$ to $E \otimes_\varepsilon F''_n$ and one has also $\sup_{t \in \bar{U}} |\langle \hat{B}_0, t\rangle| = 1$ for the $\mathfrak{T}_s(\mathfrak{J}(E \times F))$-closure $\bar{U}$ of U in $E \otimes_\varepsilon F''_n$.

Now § 44, 4.(6) implies that $\overline{U}$ is a $\mathfrak{T}_\varepsilon$-neighbourhood of $\circ$ in $E \otimes_\varepsilon F''_n$ and so $\hat{B}_0$ is integral.

If we take for U a multiple of the closed unit ball in the case of normed spaces, we obtain $\|\hat{B}_0\|_I = \|B_0\|_I$.

We note a simple corollary to (1):

(2) *Let E, F be locally convex, $B \in \mathscr{B}(E \times F)$, and let $\tilde{B}$ be the corresponding mapping in $\mathfrak{L}(E, F'_b)$, where $B_{\tilde{B}}$ is the bilinear form in $\mathscr{B}(E \times F''_n)$ corresponding to $\tilde{B}$. Then, if one of these three objects is integral, all three are integral.*

Proof. We recall that for $x \in E$, $y \in F$, $B(x, y) = \langle \tilde{B}(z), y \rangle$, $\tilde{B}(x) \in F'$, and that for $z \in F''_n$ and a net $y_\alpha \in F$ weakly converging to z, $B_{\tilde{B}}(x, z) = \langle \tilde{B}(x), z \rangle = \lim_\alpha \langle \tilde{B}(x), y_\alpha \rangle = \lim B(x, y_\alpha) = \hat{B}(x, z)$. Hence B and $B_{\tilde{B}}$ are exactly in the situation of B_0 and $\hat{B}_0$ in (1) a).

Hence, if B is integral, then $B_{\tilde{B}}$ is integral by (1) a); hence $\tilde{B}$ is integral by definition of an integral mapping. If $\tilde{B}$ is integral, then $B_{\tilde{B}}$ is integral by definition and B by (1). If $B_{\tilde{B}}$ is integral, then B is integral by (1).

Let A be an element of $\mathfrak{L}(E, F)$, E and F locally convex. The corresponding bilinear form is $B_A(x, v) = \langle Ax, v \rangle$, $x \in E$, $v \in F'$. We denote by N the canonical injection of F into F'' and by $J_{1,\infty}$ the canonical injection of $\mathscr{L}^\infty_\nu(K)$ into $\mathscr{L}^1_\nu(K)$, K compact, $\nu \geqq 0$, $\|\nu\| = 1$. The corresponding bilinear form J_∞ on $\mathscr{L}^\infty_\nu(K) \times \mathscr{L}^\infty_\nu(K)$ is integral by 4.(14); hence $J_{1,\infty}$ is also integral and $\|J_{1,\infty}\|_I = \|J_\infty\|_I = 1$ by 4.(14).

We give now the factorization theorem corresponding to 4.(16):

(3) *Let E, F be locally convex resp. normed. A mapping $A \in \mathfrak{L}(E, F)$ is integral (and $\|A\|_I \leqq 1$) if and only if there exist a compact space K, a positive measure ν on K with $\nu(K) = 1$, and mappings $C_1 \in \mathfrak{L}(E, \mathscr{L}^\infty_\nu(K))$ and $C_2 \in \mathfrak{L}(\mathscr{L}^\infty_\nu(K), F''_n)$ (with $\|C_1\| \leqq 1$ and $\|C_2\| \leqq 1$) such that NA has the factorization $NA = C_2 J_{1,\infty} C_1$.*

Instead of the last equality one uses also the equivalent statement that the diagram

(4)

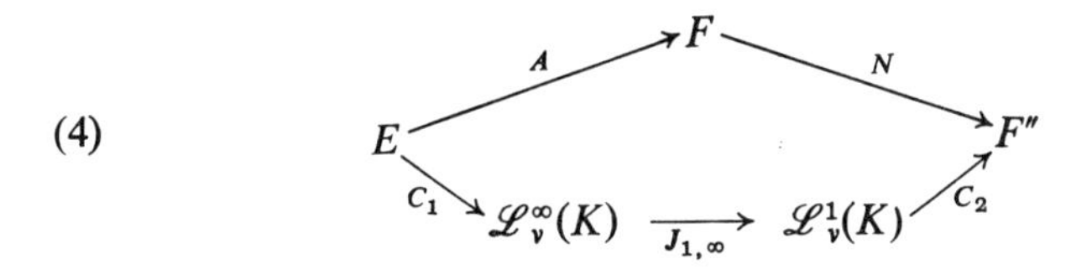

is commutative.

Proof. a) If NA has a factorization (4), then NA is integral by 4.(2) since $J_{1,\infty}$ is integral.

Since $\langle NAx, w\rangle$, $w \in F'''$, is the separately weakly continuous extension of $\langle Ax, v\rangle$, $v \in F'$, (1) implies that $\langle Ax, v\rangle$ is integral and therefore A also (and $\|A\|_I = \|NA\|_I$ by (1) and $\|NA\|_I \leqq 1$ by 4.(3)).

b) Conversely, let us assume that A is integral (and $\|A\|_I \leqq 1$ for normed spaces). Then the corresponding bilinear form $B(x, v) = \langle Ax, v\rangle$ on $E \times F'_b$ is integral and on the compact space $K = U^\circ \times V^{\circ\circ}$ (U, V absolutely convex neighbourhoods of $\mathfrak{o}$ resp. the unit balls in E and F) there exists a positive measure ν, $\nu(K) = 1$, such that by 4.(16)

$$(5) \qquad B(x, v) = \langle Ax, v\rangle = \int_K \langle u, x\rangle\langle z, v\rangle\, d\nu = J_\infty(A_1 x, A_2 v)$$

$$= \langle J_{1,\infty} A_1 x, A_2 v\rangle \quad \text{for all } x \in E,\ v \in F',$$

where $A_1 \in \mathfrak{L}(E, \mathscr{L}^\infty_\nu(K))$, $A_2 \in \mathfrak{L}(F'_b, \mathscr{L}^\infty_\nu(K))$ (and $\|A_1\| \leqq 1$, $\|A_2\| \leqq 1$).

Every continuous bilinear form $B(x, v)$ can be written as $\langle \tilde{B}x, v\rangle$, where $\tilde{B} \in \mathfrak{L}(E, F''_n)$. In our case obviously $\tilde{B} = NA$ follows from the first equation of (5), but (5) implies also $B(x, v) = \langle J_{1,\infty} A_1 x, A_2 v\rangle = \langle A'_2 J_{1,\infty} A_1 x, v\rangle$ for all x and v, where $A'_2 \in \mathfrak{L}(\mathscr{L}^\infty_\nu(K)', F''_n)$. If we write C_1 for A_1 and C_2 for the restriction of A'_2 to $\mathscr{L}^1_\nu(K)$, then $NA = C_2 J_{1,\infty} C_1$, which proves (3).

We remark that C_1 is defined as

$$C_1 x = f_x(u, z) = \langle u, x\rangle, \qquad C_1 \in \mathscr{L}^\infty_\nu(U^\circ \times V^{\circ\circ})$$

for every $x \in E$ and that $C_2 h(u, z) = \int_{U^\circ \times V^{\circ\circ}} h(u, z) z\, d\nu \in F''$ for every $h \in \mathscr{L}^1_\nu(U^\circ \times V^{\circ\circ})$.

We have the following corollary:

(6) a) *Let E, F be locally convex, $A \in \mathfrak{L}(E, F)$ integral. Then NA is weakly compact, where N is the canonical injection of F into F''.*

b) *If F is quasi-complete and $A \in \mathfrak{L}(E, F)$ integral, then A is weakly compact.*

Proof. The statement a) follows from the factorization (4) if $J_{1,\infty}$ is weakly compact. Let M be the closed unit ball of $\mathscr{L}^\infty_\nu(K) = \mathscr{L}^1_\nu(K)'$; therefore M is $\mathfrak{T}_s(\mathscr{L}^1_\nu(K))$-compact. We will show that $J_{1,\infty}$ is $\mathfrak{T}_s(\mathscr{L}^1_\nu)$–$\mathfrak{T}_s(\mathscr{L}^\infty_\nu)$-continuous. But then $J_{1,\infty}(M)$ is $\mathfrak{T}_s(\mathscr{L}^\infty_\nu)$-compact in $\mathscr{L}^1_\nu$ and this will be a).

A $\mathfrak{T}_s(\mathscr{L}^\infty_\nu)$-neighbourhood V of $\mathfrak{o}$ in $\mathscr{L}^1_\nu$ is of the form

$$\left\{h \in \mathscr{L}^1_\nu;\ \sup_{i=1,\ldots,n} \left|\int h f_i\, d\nu\right| < \varepsilon,\ f_i \in \mathscr{L}^\infty_\nu\right\}.$$

Let U be

$$\left\{f \in \mathscr{L}^{\infty}_{\nu};\ \sup_{i=1,\ldots,n} \left|\int f f_i \, d\nu\right| < \varepsilon\right\}$$

with the same $f_i \in \mathscr{L}^{\infty}_{\nu} \subset \mathscr{L}^{1}_{\nu}$; then $J_{1,\infty}(U) \subset V$ and this is the wanted continuity of $J_{1,\infty}$.

We assume now that F is quasi-complete and $A \in \mathfrak{L}(E, F)$ is integral. Let U be an absolutely convex neighbourhood of $\mathfrak{o}$ in E such that $NA(U)$ is relatively $\mathfrak{T}_s(F''')$-compact in F'' by a). The $\mathfrak{T}_s(F''')$-closure $\overline{NA(U)}$ is then $\mathfrak{T}_s(F''')$-compact in F''. But $NA(U) = A(U) \subset F$ and $\mathfrak{T}_s(F''') = \mathfrak{T}_s(F')$ on $A(U)$, so that $\overline{A(U)} \subset F$ since F is quasi-complete, and $\overline{A(U)}$ is $\mathfrak{T}_s(F')$-compact.

In the case of Hilbert spaces we saw in 2. that every integral mapping is nuclear, so that even a compact mapping need not in general be integral.

On the other hand, there exist integral mappings which are not compact. Let K be the interval $[0, 1]$ and ν the Lebesgue measure; then $J_{1,\infty}$ is not compact: Let f_n be the function which has the alternating values $+1, -1, +1, -1, \ldots$ on the intervals of length $1/2^n$ into which $[0, 1]$ is divided. We have $\|f_n\|_\infty = \|f_n\|_1 = 1$ and $\|f_n - f_m\|_1 = 1$ for $n \neq m$; hence $J_{1,\infty}(U)$ is not relatively compact in $L^1_\nu([0, 1])$, where U is the unit ball in $L^\infty_\nu([0, 1])$.

(7) a) *Let E, F be locally convex. If $A \in \mathfrak{L}(E, F)$ is integral, then $A' \in \mathfrak{L}(F'_b, E'_b)$ is integral.*

b) *If E is, moreover, quasi-barrelled, then A is integral if and only if A' is integral.*

c) *For* (B)*-spaces one has $\|A'\|_I = \|A\|_I$.*

We remark that b) includes the case that E is metrizable.

Proof. a) A is integral implies that $B_A(x, v) = \langle Ax, v\rangle$, $x \in E$, $v \in F'$, is integral on $E \times F'_b$ or that $\langle A'v, x\rangle$ is integral on $F'_b \times E$. Then by (1) $\langle A'v, z\rangle$, $z \in E''$, is integral on $E''_n \times F'_b$.

Now $\mathfrak{T}_n$ is weaker than the strong topology $\mathfrak{T}_b$ on E''; hence $\langle A'v, z\rangle$ is also integral on $F'_b \times E''_b$ and this implies that $A' \in \mathfrak{L}(F'_b, E'_b)$ is integral.

b) The converse is true if $\mathfrak{T}_n$ and $\mathfrak{T}_b$ coincide on E'', and by §23, 4.(4) this is the case for a quasi-barrelled E.

c) B_A and $B_{A'}$ have the same supremum on the corresponding unit balls.

Examples of integral bilinear forms. From § 44, 8. we recall the spaces $l^1_\mathrm{A} \tilde{\otimes}_\varepsilon E$ and $\lambda \tilde{\otimes}_\varepsilon E$ for complete locally convex spaces E and perfect sequence spaces λ. We showed that $l^1_\mathrm{A} \tilde{\otimes}_\varepsilon E$ is isomorphic (even norm isomorphic if E is a (B)-space) to the space $l^1_\mathrm{A}(E)$ of all summable sequences

$x = (x_\alpha)$, $\alpha \in \mathsf{A}$, $x_\alpha \in E$, and also that $\lambda \tilde{\otimes}_\varepsilon E$ is isomorphic to the space $\lambda(E)$ of all sequences $y = (y_n)$, $y_n \in E$, such that $(u_n y_n)$ is summable for every $\mathfrak{u} = (u_n) \in \lambda^\times$.

We determine $(l^1_{\mathsf{A}} \tilde{\otimes}_\varepsilon E)' = \mathfrak{J}(l^1_{\mathsf{A}} \times E)$ following PIETSCH. A subset M of E' is called prenuclear if there exists a neighbourhood U of $\mathfrak{o}$ in E and a positive Radon measure μ on U° such that $\sup_{v_0 \in M} |v_0 x_0| \leqq \int_{U^\circ} |u_0 x_0| \, d\mu$ for all $x_0 \in E$. A vector $v = (v_\alpha)$, $\alpha \in \mathsf{A}$, $v_\alpha \in E'$, is prenuclear if the set $\{v_\alpha; \alpha \in \mathsf{A}\}$ is prenuclear.

(8) *$l^1_{\mathsf{A}}(E)' = \mathfrak{J}(l^1_{\mathsf{A}} \times E)$ can be identified with the set of all prenuclear vectors $v = (v_\alpha)$, $v_\alpha \in E'$. The duality is given by*

$$vx = \sum_\alpha v_\alpha x_\alpha, \qquad x = (x_\alpha) \in l^1_{\mathsf{A}}(E).$$

Proof. a) Sufficiency. Let $v = (v_\alpha)$ be prenuclear. By assumption there exist U, μ such that $|v_\alpha x_0| \leqq \int_{U^\circ} |u_0 x_0| \, d\mu$. By using § 44, 8.(2), we obtain

$$\sum_\alpha |u_\alpha x_\alpha| \leqq \sum_\alpha \int_{U^\circ} |u_0 x_\alpha| \, d\mu = \int_{U^\circ} \sum_\alpha |u_0 x_\alpha| \, d\mu$$

$$\leqq \|\mu\| \sup_{u_0 \in U^\circ} \sum_\alpha |u_0 x_\alpha| = \|\mu\| \varepsilon_U(x) < \infty$$

and $vx = \sum_\alpha v_\alpha x_\alpha$ is continuous.

b) Necessity. Assume $v \in l^1_{\mathsf{A}}(E)'$. Then $|vx| \leqq \varepsilon_U(x) = \sup_{u_0 \in U^\circ} \sum_\alpha |u_0 x_\alpha|$ for some U in E and all $x = (x_\alpha) \in l^1_{\mathsf{A}}(E)$, $x_\alpha \in E$. Let e_α, $\alpha \in \mathsf{A}$, be the unit vectors in l^1_{A} and $x_0 \in E$. We define $v_\alpha x_0 = v(x_0 e_\alpha)$ and we have $v_\alpha \in E'$ since $|v_\alpha x_0| = |v(x_0 e_\alpha)| \leqq \sup_{u_0 \in U^\circ} |u_0 x_0|$. The summability of $x = (x_\alpha)$ implies $vx = \sum_\alpha v_\alpha x_\alpha$. It remains to show that $v = (v_\alpha)$ is prenuclear.

If K is the closed unit ball in l^∞_{A}, then it follows from $|vx| \leqq \varepsilon_U(x)$ and 4.(8) that there exists on $K \times U^\circ$ a Radon measure $\mu \geqq 0$, $\|\mu\| \leqq 1$ such that $vx = \int_{K \times U^\circ} \tilde{x} \, d\mu$, where $\tilde{x}$ is the continuous function on $K \times U^\circ$ corresponding to x, $\tilde{x} = (\xi_\alpha(u_0 x_\alpha))$, $(\xi_\alpha) \in l^\infty_{\mathsf{A}}$, $|\xi_\alpha| \leqq 1$ for $\alpha \in \mathsf{A}$, $u_0 \in U^\circ$. Hence $|v_\alpha x_0| = |v(x_0 e_\alpha)| = \left| \int_{K \times U^\circ} \xi_\alpha(u_0 x_0) \, d\mu \right| \leqq \int |u_0 x_0| \, d\mu = \mu(|u_0 x_0|)$.
Let $\tilde{\mu}$ be the restriction of the linear functional μ on $C(K \times U^\circ)$ to $C(U^\circ)$ defined by $\tilde{\mu}(f) = \mu(1 \times f)$ for $f \in C(U^\circ)$ and 1 the identity on K; then we obtain $|v_\alpha x_0| \leqq \int_{U^\circ} |u_0 x_0| \, d\tilde{\mu}$, i.e., the prenuclearity of $v = (v_\alpha)$.

Let λ be a perfect sequence space such that $\lambda^\times$ is the normal hull of vectors $\mathfrak{u} = (u_n)$, where all $u_n \neq 0$. By definition (§ 44, 8.) $\lambda(E)$ is the intersection of the spaces $\lambda_{\mathfrak{u}}(E)$ consisting of all vectors $x = (x_n)$, $x_n \in E$, such that $(u_n x_n)$ is summable in E for the chosen $\mathfrak{u}$. Obviously, $\lambda_{\mathfrak{u}}(E)$ is isomorphic to $l^1(E)$ by a diagonal transformation. Using (8), one sees easily that $\lambda(E)'$ consists of all vectors (v_n/u_n), where $\{v_n\}$ is a prenuclear set in E' and $\mathfrak{u} = (u_n)$ is some element of $\lambda^\times$, where all $u_n \neq 0$.

6. Nuclear and integral norms. Let E, F be (B)-spaces, A a nuclear mapping from E in F. In generalization of 1.(4) one has

(1) *Every nuclear A is integral and $\|A\|_I \leqq \|A\|_\nu$.*

Proof. Since A is nuclear, there exists for any $\delta > 0$ a representation $A = \sum_{n=1}^\infty u_n \otimes y_n$, $u_n \in E'$, $y_n \in F$, such that $\sum_{n=1}^\infty \|u_n\| \|y_n\| < (\|A\|_\nu + \delta)$. By 4.(1) we have

$$\|A\|_I = \sup_{\left\|\sum_{i=1}^m x_i \otimes v_i\right\|_\varepsilon \leqq 1} \left|\sum_i v_i(Ax_i)\right|.$$

We write $u_n = \|u_n\| u_n'$, $y_n = \|y_n\| y_n'$, and obtain

$$\left|\sum_{i=1}^m v_i(Ax_i)\right| = \left|\sum_{i=1}^m v_i \sum_{n=1}^\infty (u_n x_i) y_n\right|$$
$$\leqq \sum_{n=1}^\infty \|u_n\| \|y_n\| \left|\sum_i (v_i y_n')(u_n' x_i)\right| \leqq \sum_{n=1}^\infty \|u_n\| \|y_n\|$$

since $\left|\sum_{i=1}^m (v_i y_n')(u_n' x_i)\right| \leqq \left\|\sum_{i=1}^m v_i \otimes x_i\right\|_\varepsilon \leqq 1$. This implies $\|A\|_I < \|A\|_\nu + \delta$ for every δ.

In fact, in the most important cases one has even $\|A\|_\nu = \|A\|_I$. This is a consequence of the following characterization of the metric approximation property due to GROTHENDIECK [13]:

(2) *For a* (B)*-space E the following statements are equivalent:*
a) *E has the metric approximation property;*
b) *for any* (B)*-space F the canonical map of $E \tilde{\otimes}_\pi F$ into $\mathfrak{J}_I(E' \times F') = (E' \otimes_\varepsilon F')_b'$ is a norm isomorphism.*

Proof. a) implies b). We recall that $(E \otimes_\pi F)' = \mathscr{B}(E \times F)$ so that $\langle \mathscr{B}(E \times F), E \otimes_\pi F\rangle$ is a dual pair. It follows from 1.(5) that the closed unit ball V of $E' \otimes_\varepsilon F'$ is contained in the closed unit ball U of $\mathscr{B}_b(E \times F)$.

Using a) and § 43, 8.(1), one sees that for a given $\varepsilon > 0$, a given compact set $K \subset E$, and a given $A \in \mathfrak{L}_b(E, F') = \mathscr{B}_b(E \times F)$, $\|A\| \leqq 1$, there exists a $B \in \mathscr{F}(E, F') = E' \otimes_\varepsilon F'$, $\|B\| \leqq 1$, such that $\|(A - B)(K)\| \leqq \varepsilon$. Since $B \in V$, this implies that V is $\mathfrak{T}_s(E \otimes F)$-dense in U or $U = \overline{V} = V^{\circ\circ}$, where $\overline{V}$ denotes the $\mathfrak{T}_s(E \otimes F)$-closure of V in $\mathscr{B}(E \times F)$.

This implies $U^\circ = V^\circ$ in $E \otimes F$; hence the π-unit ball U° in $E \otimes F$ coincides with the $\mathfrak{T}_b(E' \otimes_\varepsilon F')$-unit ball V° in $E \otimes F$, which means that $E \otimes_\pi F$ is norm isomorphically embedded in $\mathfrak{J}_l(E' \times F')$. By completing $E \otimes_\pi F$ to $E \tilde{\otimes}_\pi F$ we obtain b).

b) implies a). The closed unit ball V_1 of $E' \otimes_\varepsilon E''$ is contained in the closed unit ball W of $\mathscr{B}_b(E \times E')$ by 1.(5). The assumption b) implies that the canonical mapping of $E \tilde{\otimes}_\pi E'$ into $\mathfrak{J}_l(E' \times E'') = (E' \otimes_\varepsilon E'')'_b$ is a norm isomorphism. Using polarity in $\langle E \otimes_\pi E', E' \otimes_\varepsilon E'' \rangle$ and in $\langle E \otimes_\pi E', \mathscr{B}(E \times E') \rangle$, we obtain as the closed unit ball in $E \otimes_\pi E'$ in the first case V_1° and in the second case W°. Hence one has $V_1^{\circ\circ} = \overline{V}_1 = W$, where $\overline{V}_1$ means the $\mathfrak{T}_s(E \otimes E')$-closure in $\mathscr{B}(E \times E')$.

Using 5.(1) b), we see that the closed unit ball V_0 of $E' \otimes_\varepsilon E$ is $\mathfrak{T}_s(\mathfrak{J}(E' \times E))$- and therefore also $\mathfrak{T}_s(E \otimes E')$-dense in V_1; hence V_0 is $\mathfrak{T}_s(E \otimes E')$-dense in W, which is by § 41, 3.(6) also the closed unit ball in $\mathfrak{L}_b(E, E'')$.

We have to show that V_0 is $\mathfrak{T}_c$-dense in $W_0 = W \cap \mathfrak{L}(E, E)$. So far we have proved that W_0 is the $\mathfrak{T}_s(E \otimes E')$-closure of V_0 in $\mathfrak{L}(E, E)$. We recall from § 39, 7.(2) that $\mathfrak{L}_s(E, E)' = E' \otimes E$; hence $\mathfrak{T}_s(E \otimes E')$ is the weak topology on $\mathfrak{L}_s(E, E)$, whereas $\mathfrak{T}_s$ is the simple topology. Hence $\overline{V}_0 = W_0$ for $\mathfrak{T}_s$ on $\mathfrak{L}(E, E)$. Since W_0 is equicontinuous in $\mathfrak{L}(E, E)$, $\mathfrak{T}_s$ and $\mathfrak{T}_c$ coincide on W_0 (§ 39, 4.(2)), which finally implies a).

As a corollary we state

(3) *For a* (B)*-space E the following statements are equivalent:*

a) *E' has the metric approximation property;*

b) *for every* (B)*-space F the canonical map I_1 of $E' \tilde{\otimes}_\pi F$ into $\mathfrak{J}_l(E \times F')$ is a norm isomorphism.*

Proof. We assume a). If $w = \sum_{n=1}^{N} u_n \otimes y_n \in E' \otimes_\pi F$, then $I_1 w \in \mathfrak{J}_l(E \times F')$ is defined for $x \otimes v \in E \otimes F'$ by $(I_1 w)(x \otimes v) = \sum_{n=1}^{N} (u_n x)(v y_n)$. The canonical injection I_2 of $\mathfrak{J}_l(E \times F')$ into $\mathfrak{J}_l(E'' \times F')$ is a norm isomorphism by 5.(1) a) and $I_2 I_1$ is a norm isomorphism by (2); hence I_1 is also a norm isomorphism.

From b) it follows that $I_2 I_1$ is a norm isomorphism of $E' \otimes_\pi F$ into $\mathfrak{J}_l(E'' \times F')$ and, using (2), we obtain a).

(4) *If E' has the metric approximation property, any integral $A \in \mathfrak{L}(E, F)$ is nuclear and $\|A\|_\nu = \|\dot{A}\|_\pi = \|A\|_I$.*

Proof. A can be identified with an element $\dot{A}$ of $E' \tilde{\otimes}_\pi F$ and $\|A\|_\nu = \|\dot{A}\|_\pi$. The corresponding bilinear form $\langle Ax, v\rangle$ is in $\mathfrak{J}(E \times F')$ and (3) b) implies $\|A\|_\nu = \|\dot{A}\|_\pi = \|A\|_I$.

In the same way as (2) implies (3), (3) implies

(5) *For a* (B)*-space E the following statements are equivalent:*

a) *E' has the metric approximation property;*

b) *for every* (B)*-space F the canonical map of $E' \tilde{\otimes}_\pi F'$ into $\mathfrak{J}_I(E \times F)$ is a norm isomorphism.*

The following example shows that in general $\mathfrak{N}(E, F)$ is a strict subspace of $\mathfrak{L}^I(E, F)$ even if the nuclear norm of the elements of $\mathfrak{N}(E, F)$ coincides with the integral norm.

Following the remarks preceding 5.(7), we see that the integral mapping $J_{1,\infty}$ of $L_\nu^\infty([0, 1])$ into $L_\nu^1([0, 1])$ is not nuclear; on the other hand, $(L_\nu^\infty([0, 1])')$ has the metric approximation property (see GROTHENDIECK [13], p. 185) and we have the situation described in (4).

7. When is every integral mapping nuclear? We saw in 6. that this is not always the case. Nevertheless, in the case of Hilbert spaces integral and nuclear mappings coincide; this is SCHATTEN's result 2.(4). So one looks for a generalization of SCHATTEN's theorem. The first decisive results were given in GROTHENDIECK's thesis [13]. So far these theorems have been proved only by using rather deep results on vector measures. During the last years some geometric properties of (B)-spaces have been found which are equivalent to the measure theoretic properties involved. All this material has been collected in the very recent book of DIESTEL and UHL [1'].

We will here indicate only a few of the first important results of GROTHENDIECK.

Let K be a compact space, F a (B)-space, μ a positive measure on K, $\mu(K) < \infty$. We introduced in § 41, 7. the space $L^1_{K,\mu}\{F\}$ of all absolutely μ-summable F-valued functions as the completion of the space $S\{F\}$ of all F-valued simple functions $s(t) = \sum_{i=1}^m \chi_i(t) y_i$, where $t \in K$, $y_i \in F$, and χ_i the characteristic functions of the μ-measurable sets K_i of a decomposition $K = \bigcup_{i=1}^m K_i$, $K_i \cap K_j = \varnothing$ for all $i \neq j$.

For $f \in L^1_{K,\mu}\{F\}$ there exists therefore a sequence of simple functions s_n such that $\|f - s_n\| = \pi(f - s_n) = \int_K \|f(t) - s_n(t)\| \, d\mu \to 0$.

For a simple function $s(t)$ the integral $\int_K s(t)\,d\mu$ is defined as $\sum_{i=1}^{m} \mu_i(K_i) y_i$ and for $f(t)$ by $\lim_n \int_K s_n(t)\,d\mu$ if s_n π-converges to f. This integral is called the Pettis-integral.

An F-valued function g on K is called μ-measurable if there exists a sequence s_n of simple functions such that $\lim_n \|g(t) - s_n(t)\| = 0$ μ-almost everywhere. We are now able to understand the meaning of the following theorem (for a proof see GROTHENDIECK [4'], p. 234):

(1) (DUNFORD–PETTIS–PHILLIPS). *Let K be compact, μ a positive measure on K, $\mu(K) < \infty$, E a* (B)*-space, T a weakly compact linear mapping from $L^1_{K,\mu}$ into E.*

Then there exists a μ-measurable E-valued function $g(t)$ on K such that

(2) $\|g(t)\| \leqq \|T\|$ *for all $t \in K$ and*

(3) $Tf = \int_K g(t)f(t)\,d\mu$ *for all $f \in L^1_{K,\mu}$.*

(1) will be needed in the proof of GROTHENDIECK's theorem:

(4) *Let E, F, G be* (B)*-spaces. If $A \in \mathfrak{L}(E, F)$ is integral, $B \in \mathfrak{L}(F, G)$ is weakly compact, then $BA \in \mathfrak{L}(E, G)$ is nuclear and*

(5) $$\|BA\|_\nu \leqq \|B\|\,\|A\|_I.$$

Proof. We assume that $\|A\|_I = 1$. Using the factorization 5.(3) for A, we obtain the following diagram:

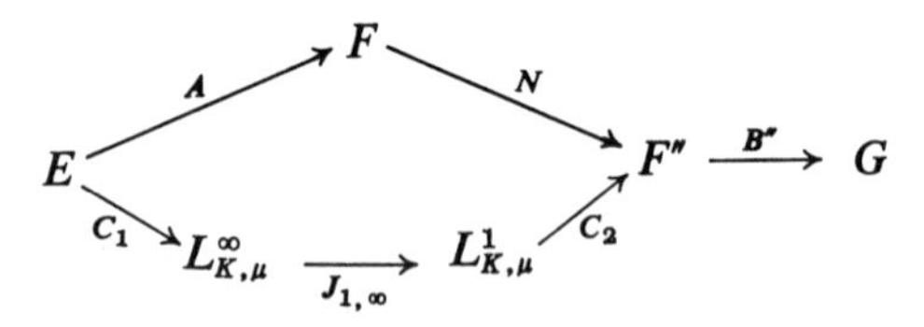

or $BA = B''NA = B''C_2J_{1,\infty}C_1$, where K is a compact space, μ a positive measure on K, $\mu(K) = 1$, and $\|C_1\| \leqq 1$, $\|C_2\| \leqq 1$. We remark that by assumption B maps the unit ball of F into a weakly compact subset of G; hence B'' is again weakly compact from F'' into G (§ 42, 2.(1)) and $B = B''N$.

Now $T = B''C_2$ is weakly compact from $L^1_{K,\mu}$ into G and has by (1) a representation $Tf = \int_K g(t)f(t)\,d\mu$, $f \in L^1_{K,\mu}$, $g \in L^1_{K,\mu}\{G\} = L^1_{K,\mu} \tilde{\otimes}_\pi G$. The last identity follows from § 41, 7.(8).

Since K is compact, $L^\infty_{K,\mu} = (L^1_{K,\mu})' \subset L^1_{K,\mu}$; hence $TJ_{1,\infty}$ is the restriction of T from $L^1_{K,\mu}$ to $L^\infty_{K,\mu}$. One has $TJ_{1,\infty}h = \int\limits_K gh\, d\mu$ for every $h \in L^\infty_{K,\mu}$ and $TJ_{1,\infty} \in L^1_{K,\mu} \tilde{\otimes}_\pi G \subset \mathfrak{L}(L^\infty_{K,\mu}, G)$. It follows that $TJ_{1,\infty}$ is nuclear since $L^1_{K,\mu}$ has the approximation property (§ 43, 7.(10)) and one has $\|TJ_{1,\infty}\|_\pi = \|TJ_{1,\infty}\|_\nu$ by a remark after § 42, 5.(6). Hence $BA = TJ_{1,\infty}C_1$ is also nuclear.

We investigate the norms. We recall from § 41, 7. the norm of g in $L^1_{K,\mu}\{G\}$ and find $\|TJ_{1,\infty}\|_\nu = \|g\|_\pi = \int \|g(t)\|\, d\mu \leqq \int \|T\|\, d\mu \leqq \|T\|$ by (2). Furthermore, one has $\|BA\|_\nu \leqq \|TJ_{1,\infty}\|_\nu \|C_1\| \leqq \|T\| \leqq \|B''\| = \|B\| \leqq \|B\|\,\|A\|_I$, which proves (5).

As a corollary to (4) we obtain a generalization of SCHATTEN's theorem:

(6) *Let E and F be* (B)*-spaces; let F, moreover, be reflexive. Then integral and nuclear mappings A of E in F coincide and one has $\|A\|_I = \|A\|_\nu$. Hence $\mathfrak{N}_\nu(E, F)$ is norm isomorphic to $\mathfrak{L}^I(E, F)$.*

Take for B in (5) the identity on F; then $\|A\|_\nu \leqq \|A\|_I$ by (5) for every integral mapping and 6.(1) implies $\|A\|_\nu = \|A\|_I$.

Another immediate corollary to (4) is

(7) *Let E, F, G be* (B)*-spaces. If $A \in \mathfrak{L}(E, F)$ and $B \in \mathfrak{L}(F, G)$ are integral, then $BA \in \mathfrak{L}(E, G)$ is nuclear.*

(6) is not the best result available. Fundamental for the understanding of the situation is a vector measure theoretical notion, the so-called RADON–NIKODYM property of (B)-spaces. We refer the reader again to the book of DIESTEL and UHL [1'], where this situation is explained in detail and with all the interesting ramifications in the different parts of Banach space theory.

In the second part of his thesis [13] GROTHENDIECK developed the theory of nuclear spaces which grew out in a natural way of his theory of tensor products and of nuclear mappings. This theory has been made into a theory in its own right with many deep results. Unfortunately, it seemed impossible to include an adequate presentation of this theory also in this volume. We refer the reader to the book of PIETSCH [10'] and a forthcoming book of MITIAGIN.

Bibliography

ADASCH, N.: [1′] *Über unstetige Abbildungen von lokalkonvexen Räumen.* Diplomarbeit, Frankfurt, 1968.

— [2′] Tonnelierte Räume und zwei Sätze von Banach. *Math. Ann.* **186**, 209–214 (1970).

— [3′] Eine Bemerkung über den Graphensatz. *Math. Ann.* **186**, 327–333 (1970).

— [4′] Der Graphensatz in topologischen Vektorräumen. *Math. Z.* **119**, 131–142 (1971).

— [5′] Vollständigkeit und der Graphensatz. *J. reine angew. Math.* **249**, 217–220 (1971).

— [6′] Über die Vollständigkeit von $L_\sigma(E, F)$. *Math. Ann.* **191**, 290–292 (1971).

—, and ERNST, B.: [1′] Lokaltopologische Vektorräume I. *Collectanea Math.* **25**, 255–274 (1975); II. *ibid.* **26**, 13–18 (1975).

ARONSZAJN, N., and SMITH, K. T.: [1′] Invariant subspaces of completely continuous operators. *Ann. Math.* **60**, 345–350 (1954).

BAKER, J. W.: [1′] On a generalized open-mapping theorem. *Math. Ann.* **172**, 217–221 (1967).

— [2′] Operators with closed range. *Math. Ann.* **174**, 278–284 (1967).

— [3′] Projection constants for $C(S)$ spaces with the separable projection property. *Proc. Amer. Math. Soc.* **41**, 201–204 (1973).

BATT, J.: [1′] Die Verallgemeinerungen des Darstellungssatzes von F. Riesz und ihre Anwendungen. *Jahresber. DMV* **74**, 147–181 (1973).

BIERSTEDT, K. D.: [1′] Gewichtete Räume vektorwertiger Funktionen und das injektive Tensorprodukt. *J. reine angew. Math.* **259**, 186–210 (1973).

—, and MEISE, R.: [1′] Bemerkungen über die Approximationseigenschaft lokalkonvexer Funktionenräume. *Math. Ann.* **209**, 99–107 (1974).

— [2′] Induktive Limites gewichteter Räume stetiger und lokalholomorpher Funktionen. *J. reine angew. Math.* **282**, 186–220 (1976).

BROWDER, F. E.: [1′] Functional analysis and partial differential equations I. *Math. Ann.* **138**, 55–79 (1959).

BUCHWALTER, H.: [1′] Produit topologique, produit tensoriel et c-replétion. *Bull. Soc. Math. France Mém.* **31–32**, 51–71 (1972).

CROSS, R. W.: [1′] Existence theorems for linear equations in an infinity of unknowns. *Quart. J. Math. Oxford* **14** (2), 113–119 (1963).

DAVIE, A. M.: [1′] The approximation problem for Banach spaces. *Bull. London Math. Soc.* **5**, 261–266 (1973).

DEAN, D. W.: [1′] The equation $L(E, X^{**}) = L(E, X)^{**}$ and the principle of local reflexivity. *Proc. Amer. Math. Soc.* **40**, 146–148 (1973).

De Wilde, M.: [1′] Sur le théorème du graphe fermé. *C.R. Acad. Sci. Paris* **265**, 376–379 (1967).
— [2′] Théorème du graphe fermé et espaces à réseaux absorbants. *Bull. Soc. Math. Roumaine* **11** (59), 2, 224–238 (1967).
— [3′] Réseaux dans les espaces linéaires à semi-normes. *Mém. Soc. Royale Sci. Liège* **18** (5), 2, 1–144 (1969).
— [4′] Opérateurs ouverts et sous-espaces complémentaires dans un espace ultrabornologique. *Bull. Soc. Royale Sci. Liège* **38**, 454–458 (1969).
— [5′] Ultrabornological spaces and the closed-graph theorem. *Bull. Soc. Royale Sci. Liège* **40**, 116–118 (1971).
— [6′] Quelques propriétés de permanence des espaces à réseau. *Bull. Soc. Royale Sci. Liège* **39**, 240–248 (1970).
— [7′] Vector topologies and linear maps on products of topological vector spaces. *Math. Ann.* **196**, 117–128 (1972).
— [8′] Finite codimensional subspaces of topological vector spaces and the closed graph theorem. *Arch. Math.* **23**, 180–181 (1972).
— [9′] *Closed Graph Theorems and Webbed Spaces.* Pitman, London, 1978.
Dieudonné, J.: [1′] On biorthogonal systems. *Michigan Math. J.* **2**, 7–20 (1953).
Diestel, J. and Uhl Jr., J. J.: [1′] *Vector Measures.* Math. Surveys No. 15, Amer. Math. Soc., Providence, Rhode Island, 1977.
Eberhardt, V.: [1′] Der Graphensatz von A. P. und W. Robertson für s-Räume. *Manuscr. Math.* **4**, 255–262 (1970).
— [2′] Durch Graphensätze definierte lokalkonvexe Räume. Dissertation München, 1972.
— [3′] Einige Vererbbarkeitseigenschaften von B- und B_r-vollständigen Räumen. *Math. Ann.* **215**, 1–11 (1975).
— [4′] Beispiele topologischer Vektorräume mit der Komplementärraumeigenschaft. *Arch. Math.* **26**, 627–636 (1975).
Edwards, R. E.: [1′] *Functional Analysis. Theory and Application.* Holt Rinehart and Winston, New York, 1965.
Eidelheit, M.: [1′] Zur Theorie der Systeme linearer Gleichungen. *Studia Math.* **6**, 139–148 (1936).
— [2′] Zur Theorie der Systeme linearer Gleichungen (II). *Studia Math.* **7**, 150–154 (1938).
Enflo, P.: [1′] A counterexample to the approximation problem in Banach spaces. *Acta Math.* **130**, 309–317 (1973).
Figiel, T., and Johnson, W. B.: [1′] The approximation property does not imply the bounded approximation property. *Proc. Amer. Math. Soc.* **41**, 197–200 (1973).
Fillmore, P. A., and Williams, J. P.: [1′] On operator ranges. *Advances in Math.* **7**, 254–281 (1971).
Garnir, H. G., De Wilde, M., and Schmets, J.: [1′] *Analyse fonctionelle.* Tome I. Birkhäuser Verlag, Basel, 1968.
Goldberg, S.: [1′] *Unbounded Linear Operators.* McGraw-Hill, New York, 1966.
Goodner, D.: [1′] Projections in normed linear spaces. *Trans. Amer. Math. Soc.* **69**, 89–108 (1950).
Grathwohl, M.: [1′] Ultrabornologische Räume und der Graphensatz. Dissertation Frankfurt, 1974.

GROTHENDIECK, A.: [1′] Résumé de la théorie métrique des produits tensoriels topologiques. *Bol. Soc. Mat. São Paulo* **8**, 1–79 (1956).

— [2′] Sur certaines classes de suites dans les espaces de Banach et le théorème de Dvoretzky–Rogers. *Bol. Soc. Mat. São Paulo* **8**, 80–110 (1956).

— [3′] La théorie de Fredholm. *Bull. Soc. Math. France* **84**, 319–384 (1965).

— [4′] *Topological Vector Spaces*. Gordon and Breach, New York, 1973 (English translation of Grothendieck [11]).

HASUMI, M.: [1′] The extension property of complex Banach spaces. *Tôhoku Math. J.* **10**, 135–142 (1958).

HENRIQUES, G.: [1′] Ein nicht *d*-separabler linearer Unterraum eines *d*-separablen tonnelierten Raumes. *Arch. Math.* **15**, 448–449 (1964).

HOGBE-NLEND, H.: [1′] Les espaces de Fréchet–Schwartz et la propriété d'approximation. *C.R. Acad. Sci. Paris* (*A*) **275**, 1073–1075 (1972).

— [2′] Techniques de bornologie en théorie des espaces vectoriels topologiques. *Lecture Notes* **331**, 84–162 (1973).

HOLLSTEIN, R.: [1′] σ-topologische Räume und projektive Tensorprodukte. *Collectanea Math.* **26**, 239–252 (1975).

— [2′] (DCF)-Räume und lokalkonvexe Tensorprodukte. *Arch. Math.* **29**, 524–531 (1977).

HOLUB, J. R.: [1′] Tensor product mappings I. *Math. Ann.* **188**, 1–12 (1970).

— [2′] Compactness in topological tensor products and operator spaces. *Proc. Amer. Math. Soc.* **36**, 398–406 (1972).

— [3′] Tensor product mappings II. *Proc. Amer. Math. Soc.* **42**, 437–441 (1974).

HORVÁTH, J.: [1′] *Topological Vector Spaces and Distributions*. Vol. I. Addison–Wesley, Reading, Massachusetts, 1966.

HUSAIN, T.: [1′] *The Open Mapping and Closed Graph Theorem in Topological Vector Spaces*. Friedr. Vieweg, ŭ. Sohn, Braŭnschweig, 1965.

— [2′] Two new classes of locally convex spaces. *Math. Ann.* **166**, 289–299 (1966).

ICHINOSE, T.: [1′] On the spectra of tensor products of linear operators in Banach spaces. *J. reine angew. Math.* **244**, 119–153 (1970).

— [2′] Operators on tensor products of Banach spaces. *Trans. Amer. Math. Soc.* **170**, 197–219 (1972).

JOHNSON, W. B.: [1′] On the existence of strongly series summable Markuschewich bases in Banach spaces. *Trans. Amer. Math. Soc.* **157**, 481–486 (1971).

— [2′] A complementary universal conjugate Banach space and its relation to the approximation problem. *Israel J. Math.* **13**, 301–310 (1972).

JÖRGENS, K.: [1′] *Lineare Integralgleichungen*. Teubner, Stuttgart, 1970.

KABALLO, W.: [1′] Liftingsätze für Vektorfunktionen und das ε-Tensorprodukt. Habilitationsschrift, to be published.

KALTON, N. J.: [1′] Some forms of the closed graph theorem. *Proc. Cambridge Phil. Soc.* **70**, 401–408 (1971).

— [2′] A barrelled space without a basis. *Proc. Amer. Math. Soc.* **26**, 465–466 (1970).

KATO, K.: [1′] Perturbation theory for nullity, deficiency and other quantities of linear operators. *J. d'Analyse Math.* **6**, 273–322 (1958).

KAUFMAN, R.: [1′] A type of extension of Banach spaces. *Acta Sci. Math. Szeged* **27**, 163–166 (1966).

KELLEY, J. L.: [1′] Banach spaces with the extension property. *Trans. Amer. Math. Soc.* **72**, 323–326 (1952).
— [2′] Hypercomplete linear topological spaces. *Michigan Math. J.* **5**, 235–246 (1958).
—, and NAMIOKA, I.: [1′] *Linear Topological Spaces.* Van Nostrand, New York, 1963.
KÖTHE, G.: [1′] Das Reziprokentheorem für zeilenabsolute Matrizen. *Monatshefte Math. Phys.* **47**, 224–233 (1939).
— [2′] Homomorphismen von (F)-Räumen. *Math. Z.* **84**, 219–221 (1964).
— [3′] General linear transformations of locally convex spaces. *Math. Ann.* **159**, 309–328 (1965).
— [4′] Über einen Satz von Sobczyk. *Anais Fac. Ciênc. Porto* **44**, 1–6 (1966).
— [5′] Hebbare lokalkonvexe Räume. *Math. Ann.* **165**, 181–195 (1966).
— [6′] Fortsetzung linearer Abbildungen lokalkonvexer Räume. *Jahresber. DMV* **68**, 193–204 (1966).
— [7′] Die Bildräume abgeschlossener Operatoren. *J. reine angew. Math.* **232**, 110–111 (1968).
— [8′] Zur Theorie der kompakten Operatoren in lokalkonvexen Räumen. *Portug. Math.* **13**, 97–104 (1954).
— [9′] Die Gleichungstheorie im Hilbertschen Raum. *Math. Z.* **41**, 153–162 (1936).
KRISHNAMURTHY, V.: [1′] On the state diagram of a linear operator and its adjoint. *Math. Ann.* **141**, 153–160 (1966).
—, and LOUSTAUNAU, J. O.: [1′] On the state diagram of a linear operator and its adjoint in locally convex spaces I. *Math. Ann.* **141**, 176–206 (1966).
LACEY, E., and WHITLEY, R. J.: [1′] Conditions under which all the bounded linear maps are compact. *Math. Ann.* **158**, 1–5 (1965).
LANDSBERG, M.: [1′] Über die Fixpunkte kompakter Abbildungen. *Math. Ann.* **154**, 427–431 (1964).
LINDENSTRAUSS, J.: [1′] Extension of compact operators. *Mem. Amer. Math. Soc.* **48** (1964).
—, and PEŁCZYNSKI, A.: [1′] Absolutely summing operators in $\mathscr{L}_p$ spaces and their applications. *Studia Math.* **29**, 275–326 (1968).
—, and ROSENTHAL, H. P.: [1′] The $\mathscr{L}_p$ spaces. *Israel J. Math.* **7**, 325–349 (1969).
—, and TZAFRIRI, L.: [1′] Classical Banach spaces. *Lecture Notes* **338** (1973).
— [2′] Classical Banach spaces I: Sequence spaces. *Ergebnisse Math.* **92** (1977).
LOMONOSOV, V. I.: [1′] Invariant subspaces for the family of operators which commute with a completely continuous operator. *Functional Analysis Appl.* **7**, 213–215 (1974).
LOTZ, H. T.: [1′] Lectures on topological tensor products, linear mappings and nuclear spaces. Notes, University of Illinois, 1971. Notes prepared by A. Peressini.
— [2′] Grothendieck ideals of operators on Banach spaces. Notes, University of Illinois, 1973. Notes prepared by A. Peressini and D. R. Sherbert.
LOUSTAUNAU, J. O.: [1′] On the state diagram of a linear operator and its adjoint in locally convex spaces II. *Math. Ann.* **176**, 121–128 (1968).
MCARTHUR, C. W.: [1′] Developments in Schauder basis theory. *Bull. Amer. Math. Soc.* **78**, 887–908 (1972).
MACINTOSH, A.: [1′] On the closed graph theorem. *Proc. Amer. Math. Soc.* **20**, 397–404 (1969).

MAHOWALD, M.: [1′] Barrelled spaces and the closed graph theorem. *J. London Math. Soc.* **36**, 108–110 (1961).

MARTI, J. T.: [1′] *Introduction to the theory of bases. Springer Tracts in Natural Philosophy* **18** (1969).

MARTINEAU, A.: [1′] Sur le théorème du graphe fermé. *C.R. Acad. Sci. Paris* **263**, 870–871 (1966).

— [2′] Sur des théorèmes de S. Banach et L. Schwartz concernant le graphe fermé. *Studia Math.* **30**, 43–54 (1968).

MOCHIZUKI, N.: [1′] On fully complete spaces. *Tôhoku Math. J.* **33**, 485–490 (1961).

NACHBIN, L.: [1′] Some problems in extending and lifting continuous linear transformations. *Proc. Symp. Linear Spaces, Jerusalem, 1961*, pp. 340–350.

NEUBAUER, G.: [1′] Zur Spektraltheorie in lokalkonvexen Algebren. *Math. Ann.* **142**, 131–154 (1961).

— [2′] Zur Spektraltheorie in lokalkonvexen Algebren II. *Math. Ann.* **143**, 251–263 (1961).

NEWNS, W. F.: [1′] On the representation of analytic functions by infinite series. *Phil. Trans. Royal Soc. London (A)* **245**, 429–468 (1953).

NIETHAMMER, W., and ZELLER, K.: [1′] Unendliche Gleichungssystem mit beliebiger rechter Seite. *Math. Z.* **96**, 1–6 (1967).

PEŁCZYNSKI, A.: [1′] Projections in certain Banach spaces. *Studia Math.* **19**, 209–228 (1960).

— [2′] On strictly singular and strictly cosingular operators I, II. *Bull. Acad. Sci. Math. Astr. Phys.* **13**, 31–41 (1965).

— [3′] On James's paper "Separable conjugate spaces". *Israel J. Math.* **9**, 279–284 (1971).

PERSSON, A.: [1′] A remark on the closed graph theorem in locally convex spaces. *Math. Scand.* **19**, 54–58 (1966).

PIETSCH, A.: [1′] Zur Theorie der topologischen Tensorprodukte. *Math. Nachr.* **25**, 19–31 (1963).

— [2′] *Nukleare Lokalkonvexe Räume*, 2nd ed. Akademie-Verlag, Berlin, 1969 (English: *Nuclear Locally Convex Spaces.* Springer-Verlag, Berlin and New York, 1972).

PITT, H. R.: [1′] A note on bilinear forms. *J. London Math. Soc.* **11**, 171–174 (1936).

POWELL, M.: [1′] On Kōmura's closed-graph theorem. *Trans. Amer. Math. Soc.* **211**, 391–426 (1975).

RAÍKOW, D. A.: [1′] Double closed-graph theorem for topological linear spaces. *Siber. Math. J.* **7**, 287–300 (1966).

RANDTKE, P. J.: [1′] Characterization of precompact maps, Schwartz spaces and nuclear spaces. *Trans. Amer. Math. Soc.* **165**, 87–101 (1972).

— [2′] A factorization theorem for compact operators. *Proc. Amer. Math. Soc.* **34**, 201–202 (1972).

— [3′] A structure theorem for Schwartz spaces. *Math. Ann.* **201**, 171–176 (1973).

ROBERTSON, A. P. and W. J.: [1′] On the closed graph theorem. *Proc. Glasgow Math. Ass.* **3**, 9–12 (1956).

— [2′] Topological vector spaces. *Cambridge Tracts* **53** (1964).

ROBERTSON, W.: [1′] On the closed graph theorem and spaces with webs. *Proc. London Math. Soc.* **24** (3), 692–738 (1972).

ROSENTHAL, H. P.: [1′] On quasi-complemented subspaces of Banach spaces with an appendix on compactness of operators from $L^p(\mu)$ to $L^r(\nu)$. *J. Functional Analysis* **4**, 176–214 (1969).
SCHAEFER, H.: [1′] *Topological Vector Spaces*. Macmillan, New York, 1966.
SCHATTEN, R.: [1′] A theory of cross-spaces. *Ann. of Math. Studies* **26** (1950).
— [2′] Norm ideals of completely continuous operators. *Ergebnisse Math.* **27** (1961).
SCHWARTZ, L.: [1′] Produits tensoriels topologiques d'espaces vectoriels topologiques. Espaces vectoriels topologiques nucléaires. Applications. Séminaire Schwartz 1953–1954. Faculté des Sciences de Paris.
— [2′] Sur le théorème du graphe fermé. *C.R. Acad. Paris (A)* **263**, 602–605 (1966).
— [3′] Théorie des distributions à valeurs vectorielles (I). *Ann. Inst. Fourier* **7**, 1–141 (1957).
SINGER, I.: [1′] *Bases in Banach Spaces I*. Springer-Verlag, Berlin and New York, 1970.
SLOWIKOWSKI, W.: [1′] On continuity of inverse operators. *Bull. Amer. Math. Soc.* **67**, 467–470 (1961).
— [2′] Quotient spaces and the open mapping theorem. *Bull. Amer. Math. Soc.* **67**, 498–500 (1961).
SULLEY, L. J.: [1′] On $B(\mathfrak{T})$ and $B_x(\mathfrak{T})$ locally convex spaces. *Proc. Cambridge Phil. Soc.* **68**, 95–97 (1970).
SZANKOWSKI, A.: [1′] Subspaces without approximation property. *Israel J. Math.* **30**, 123–129 (1978).
TAYLOR, A. E., and HALBERG, C. J.: [1′] General theorems about a bounded operator and its conjugate. *J. reine angew. Math.* **198**, 93–111 (1957).
TERZIOGLU, T.: [1′] Die diametrale Dimension von lokalkonvexen Räumen. *Collectanea Math.* **20**, 49–99 (1969).
— [2′] On Schwartz spaces. *Math. Ann.* **182**, 236–242 (1969).
— [3′] A characterization of compact linear mappings. *Arch. Math.* **22**, 76–78 (1971).
— [4′] On compact and infinite-nuclear mappings. *Bull. Soc. Math. Roumaine* **14**, 93–99 (1970).
— [5′] Remarks on compact and infinite-nuclear mappings. *Math. Balkanika* **2**, 251–255 (1972).
TOEPLITZ, O.: [1′] Die Jacobische Transformation der quadratischen Formen von unendlich vielen Veränderlichen. *Göttinger Nachr.* **1907**, 101–109.
TREVES, F.: [1′] *Topological Vector Spaces, Distributions and Kernels*. Academic Press, New York, 1967.
VALDIVIA UREÑA, M.: [1′] El teorema general de la grafica cerrada en los espacios vectoriales localmente convexos. *Rev. Real Acad. Cienc. Madrid* **52**, **3**, 545–551 (1968).
— [2′] El teorema general de la aplicación abierte en los espacios vectoriales topológicos localmente convexos. *Rev. Real Acad. Cienc. Madrid* **52**, **3**, 553–562 (1968).
— [3′] Sobre el teorema de la grafica cerrada. *Collectanea Math.* **27**, 51–72 (1971).
VEECH, W. A.: [1′] A short proof of Sobczyk's theorem. *Proc. Amer. Math. Soc.* **28**, 627–628 (1971).

WAELBROECK, L.: [1′] Le calcul symbolique dans les algèbres commutatives. *J. Math. Pures Appl.* **33**, 147–186 (1954).

— [2′] Note sur les algèbres du calcul symbolique. *J. Math. Pures Appl.* **37**, 41–44 (1958).

WILANSKY, A.: [1′] On a characterization of barrelled spaces. *Proc. Amer. Math. Soc.* **57**, 375 (1976).

WONG, Y. C.: [1′] Nuclear spaces, Schwartz spaces and tensor products. Notes, Department of Mathematics, The Chinese University of Hong Kong.

ZIPPIN, M.: [1′] The separable extension problem. *Israel J. Math.* **26**, 372–387 (1977).

Author and Subject Index

Grundlehren der mathematischen Wissenschaften

A Series of Comprehensive Studies in Mathematics

A Selection

114. Mac Lane: Homology
131. Hirzebruch: Topological Methods in Algebraic Geometry
144. Weil: Basic Number Theory
145. Butzer/Berens: Semi-Groups of Operators and Approximation
146. Treves: Locally Convex Spaces and Linear Partial Differential Equations
152. Hewitt/Ross: Abstract Harmonic Analysis. Vol. 2: Structure and Analysis for Compact Groups. Analysis on Locally Compact Abelian Groups
153. Federer: Geometric Measure Theory
154. Singer: Bases in Banach Spaces I
155. Müller: Foundations of the Mathematical Theory of Electromagnetic Waves
156. van der Waerden: Mathematical Statistics
157. Prohorov/Rozanov: Probability Theory. Basic Concepts. Limit Theorems. Random Processes
158. Constantinescu/Cornea: Potential Theory on Harmonic Spaces
159. Köthe: Topological Vector Spaces I
160. Agrest/Maksimov: Theory of Incomplete Cylindrical Functions and their Applications
161. Bhatia/Szegö: Stability of Dynamical Systems
162. Nevanlinna: Analytic Functions
163. Stoer/Witzgall: Convexity and Optimization in Finite Dimensions I
164. Sario/Nakai: Classification Theory of Riemann Surfaces
165. Mitrinovic/Vasic: Analytic Inequalities
166. Grothendieck/Dieudonné: Eléments de Géometrie Algébrique I
167. Chandrasekharan: Arithmetical Functions
168. Palamodov: Linear Differential Operators with Constant Coefficients
169. Rademacher: Topics in Analytic Number Theory
170. Lions: Optimal Control of Systems Governed by Partial Differential Equations
171. Singer: Best Approximation in Normed Linear Spaces by Elements of Linear Subspaces
172. Bühlmann: Mathematical Methods in Risk Theory
173. Maeda/Maeda: Theory of Symmetric Lattices
174. Stiefel/Scheifele: Linear and Regular Celestial Mechanic. Perturbed Two-body Motion—Numerical Methods—Canonical Theory
175. Larsen: An Introduction to the Theory of Multipliers
176. Grauert/Remmert: Analytische Stellenalgebren
177. Flügge: Practical Quantum Mechanics I
178. Flügge: Practical Quantum Mechanics II
179. Giraud: Cohomologie non abélienne
180. Landkof: Foundations of Modern Potential Theory
181. Lions/Magenes: Non-Homogeneous Boundary Value Problems and Applications I
182. Lions/Magenes: Non-Homogeneous Boundary Value Problems and Applications II
183. Lions/Magenes: Non-Homogeneous Boundary Value Problems and Applications III
184. Rosenblatt: Markov Processes. Structure and Asymptotic Behavior
185. Rubinowicz: Sommerfeldsche Polynommethode
186. Handbook for Automatic Computation. Vol. 2. Wilkinson/Reinsch: Linear Algebra

187. Siegel/Moser: Lectures on Celestial Mechanics
188. Warner: Harmonic Analysis on Semi-Simple Lie Groups I
189. Warner: Harmonic Analysis on Semi-Simple Lie Groups II
190. Faith: Algebra: Rings, Modules, and Categories I
191. Faith: Algebra II, Ring Theory
192. Mallcev: Algebraic Systems
193. Pólya/Szegö: Problems and Theorems in Analysis I
194. Igusa: Theta Functions
195. Berberian: Baer*-Rings
196. Athreya/Ney: Branching Processes
197. Benz: Vorlesungen über Geometric der Algebren
198. Gaal: Linear Analysis and Representation Theory
199. Nitsche: Vorlesungen über Minimalflächen
200. Dold: Lectures on Algebraic Topology
201. Beck: Continuous Flows in the Plane
202. Schmetterer: Introduction to Mathematical Statistics
203. Schoeneberg: Elliptic Modular Functions
204. Popov: Hyperstability of Control Systems
205. Nikollskii: Approximation of Functions of Several Variables and Imbedding Theorems
206. André: Homologie des Algèbres Commutatives
207. Donoghue: Monotone Matrix Functions and Analytic Continuation
208. Lacey: The Isometric Theory of Classical Banach Spaces
209. Ringel: Map Color Theorem
210. Gihman/Skorohod: The Theory of Stochastic Processes I
211. Comfort/Negrepontis: The Theory of Ultrafilters
212. Switzer: Algebraic Topology—Homotopy and Homology
213. Shafarevich: Basic Algebraic Geometry
214. van der Waerden: Group Theory and Quantum Mechanics
215. Schaefer: Banach Lattices and Positive Operators
216. Pólya/Szegö: Problems and Theorems in Analysis II
217. Stenström: Rings of Quotients
218. Gihman/Skorohod: The Theory of Stochastic Processes II
219. Duvaut/Lions: Inequalities in Mechanics and Physics
220. Kirillov: Elements of the Theory of Representations
221. Mumford: Algebraic Geometry I: Complex Projective Varieties
222. Lang: Introduction to Modular Forms
223. Bergh/Löfström: Interpolation Spaces. An Introduction
224. Gilbarg/Trudinger: Elliptic Partial Differential Equations of Second Order
225. Schütte: Proof Theory
226. Karoubi: K-Theory. An Introduction
227. Grauert/Remmert: Theorie der Steinschen Räume
228. Segal/Kunze: Integrals and Operators
229. Hasse: Number Theory
230. Klingenberg: Lectures on Closed Geodesics
231. Lang: Elliptic Curves: Diophantine Analysis
232. Gihman/Skorohod: The Theory of Stochastic Processes III
233. Stroock/Varadhan: Multi-dimensional Diffusion Processes
234. Aigner: Combinatorial Theory
235. Dynkin/Yushkevich: Markov Control Processes and Their Applications
236. Grauert/Remmert: Theory of Stein Spaces
237. Köthe: Topological Vector Spaces II
238. Graham/McGehee: Essays in Commutative Harmonic Analysis
239. Elliott: Probabilistic Number Theory I
240. Elliott: Probabilistic Number Theory II